COMETARY SCIENCE AFTER HALE–BOPP
VOLUME 2

Cometary Science after Hale–Bopp
Volume 2

Proceedings of IAU Colloquium 186
21–25 January 2002, Tenerife, Spain

Edited by

H. BOEHNHARDT
European Southern Observatory, Santiago, Chile

M. COMBI
University of Michigan, Ann Arbor, U.S.A.

M. R. KIDGER
Instituto de Astrofísica de Canarias, Tenerife, Spain

R. SCHULZ
European Space Agency, Noordwijk, The Netherlands

Reprinted from *Earth, Moon, and Planets*
Volume 90, Nos. 1–4, 2002

KLUWER ACADEMIC PUBLISHERS
DORDRECHT / BOSTON / LONDON

A C.I.P. Catalogue record for this book is available from the Library of Congress.

ISBN 1-4020-0978-X

Published by Kluwer Academic Publishers,
P.O. Box 17, 3300 AA Dordrecht, The Netherlands.

Sold and distributed in North, Central and South America
by Kluwer Academic Publishers,
101 Philip Drive, Norwell, MA 02061, U.S.A.

In all other countries, sold and distributed
by Kluwer Academic Publishers,
P.O. Box 17, 3300 AA Dordrecht, The Netherlands.

Printed on acid-free paper

Printed in the Netherlands

EARTH, MOON, AND PLANETS / *Vol. 90 Nos. 1–4 March–June 2002*

COMETARY SCIENCE AFTER HALE–BOPP

Proceedings of IAU Colloquium 186, 21–25 January, 2002, Tenerife, Spain

Edited by H. BOEHNHARDT, M. COMBI, M. R. KIDGER, and R. SCHULZ

* Contributed papers are indicated by (CP) and research notes by (RN).

First row crouching: 1. Hideo Kawakita, 2. Juan Rodriguez, 3. Salvador Sanchez, 4. Juan Rodriguez, 5. Nicolas Biver, 6. Jaume Llodra. **First row standing:** 7. Siybylle Peschke, 8. Mark Kidger, 9. Eva Bejarano, 10. Judith de Aaroz, 11. Dominique Bockelee-Morvan, 12. Julio Fernandez, 13. Paul Weissman, 14. Vera Rosenbush, 15. Klim Churyumov, 16. Tra-Mi Ho, 17. Beata Dziak, 18. Florence Henry, 19. Unknown, 20. Maria Teresa Capria, 21. Rita Schulz, 22. Edith Hadamick. **Second row standing:** 23. Bozewo Todorovic-Juchniewics, 24. Dina Prialnik, 25. Diane Wooden, 26. Sergiy Borysenko, 27. Fernanda Artigue, 28. Bernard Foing, 29. Ludmilla Kolokolova, 30. Anita Cochran, 31. Slawmira Szutowicz (?), 32. Hans Rickman, 33. Emmanuel Jehin, 34. Jacques Crovisier, 35. Jens Biele, 36. Jochen Kissel. **The Back:** 37. Unknown, 38. Bjoern Davidsson, 39. Hermann Boehnhardt, 40. Humberto Campins, 41. Norbert Koemle, 42. Mike A'Hearn, 43. Susan Lederer, 44. Ben Clark, 45. H.U. Keller, 46. Mike Combi, 47. Michel Festou, 48. Casey Lisse, 49. Joern Helbert, 50. Laurent Jorda, 51. Tanyu Bonev, 52. Ekkehard Kuehrt, 53. Unknown, 54. Stephen Lowry, 55. Dave Schleicher, 56. Jose Luis Ortiz, 57. Milos Tichy, 58. Laura Woodney, 59. Dan Boice, 60. Heike Rauer, 61. Jana Ticha, 62. Yan Fernandez, 63. Jun-Ichi Watanabe, 64. Daniela Lazzaro, 65. Gian Paolo Tozzi, 66. Imre Toth (?), 67. Pascale Ehrenfreund, 68. Leonid Shulman, 69. William Jackson, 70. Walter Harris, 71. Nalin Samarasinha, 72. Tony Farnham, 73. Piotr Dybczynski, 74. Ray Goldstein, 75. Teemu Makinen, 76. Jacek Leliwa-Kopystynski, 77. Michael Mueller, 78. Jack Brandt, 79. Mike Belton, 80. W. Jackson's son, 81. Umesh Joshi, 82. Gonzalo Tancredi, 83. Walter Huebner, 84. Steven Charnley, 85. Karoly Szego, 86. David Harker.

PREFACE

CONTRIBUTED PAPERS & RESEARCH NOTES (VOLUME 2 OF 2)

Two issues of "Earth, Moon, and Planets" are devoted to the proceedings of the International Astronomical Union IAU Colloquium No. 186 on "Cometary Science after Hale–Bopp", held at Puerto de la Cruz, Tenerife, Canary Islands during the week January 21–25, 2002. Following an earlier meeting, the First International Conference on Comet Hale–Bopp held at the same place in January 1998, this colloquium addressed the state-of-the-art knowledge in cometary science after the appearance of Comet C/1995 O1 Hale–Bopp and at the beginning of a new era of spacecraft explorations of comets.

In honor of his visionary scientific contributions to this field of research the conference was dedicated to Mayo Greenberg, Emeritus Professor for Laboratory Physics at the University of Leiden/The Netherlands (see the "In Memoriam and Dedication" written by Hans Rickman, Secretary General of the International Astronomical Union IAU, in the second volume of these proceedings).

The proceedings of the conference are published in two volumes: 15 invited papers are compiled in the first volume followed by the collection of contributed papers (39) and research notes (14) in this second volume. The latter papers comprise either condensed summaries and extensions of more comprehensive work published in several papers elsewhere or provide progress reports on on-going research with first preliminary results. The invited papers are given in the sequence of their presentation at the conference. The contributed papers and research notes are arranged according to the sessions to which they were assigned during the meeting, sorted in alphabetic order of the first authors. A short summary of the round table discussion held at the end of the conference completes the second volume.

The editors (undersigned) are grateful to many people, namely the referees of proceedings papers, the organizers of the meeting, the editor-in-chief of this journal, the Kluwer publishing office and – most important – the authors themselves who have submitted and revised in total 68 papers following a tight schedule that was aiming for a rapid publication of the volumes after the conference. Our special thanks are given to Andrea Lagarini of the European Southern Observatory Santiago de Chile, who served efficiently and proactively as secretary for the edition of these proceedings.

Hermann Boehnhardt
Mike Combi
Mark Kidger
Rita Schulz

J. MAYO GREENBERG (1922–2001)
IN MEMORIAM AND DEDICATION

It is a sad duty to write in memoriam of our good friend and colleague, J. Mayo Greenberg, but it is a task that I more than willingly undertake. Whenever I think of him, there is one association that inevitably springs to mind: That of a great scientist. Keeping in mind that this is not at all the whole story about a remarkable man of great culture and a warm heart, I nevertheless want to focus on that particular aspect. Not all of us, active researchers in astrophysics, have had the chance to get to know such persons, but I was fortunate enough to collaborate with and learn from Mayo, and I consider this to be one of the most important experiences of my career.

To me, Mayo impersonated the very idea of a great scientist: the deep knowledge, the never ending curiosity about how nature works, and the almost childish joy of asking new questions and putting pieces of the puzzle together. This was true all the time that I knew him, but especially at the very end, when, weakened by illness, he was as happy as ever to attend the honorary Symposium at Leiden, September 2001, which was organized in expectation of his 80th birthday. In my experience, Mayo never was indifferent to anything that truly had to do with comets, but at this particular time it was touching to part from him in the midst of discussing what to do in the future to elaborate on what we had already found.

In the world of cometary astronomers, Mayo was somewhat of a strange bird, who had come flying from the reaches of interstellar space – yes, knowing that comets and our Solar System were part of the same, universal story. I remember the time of the Halley exploration and the Heidelberg meeting in 1986, when a first summary of the results was made. The most impressive aspect was that, not only had Fred Whipple been proven right in that the solid nucleus did exist in the way he had predicted, but also, less obviously at first sight, Mayo Greenberg had been right when telling us what cometary material should be like – a porous mixture of grains consisting of silicates, organics and ices.

More recent bright comets, such as Hale–Bopp and Hyakutake, have tended to agree with this picture, which I believe can be termed a new paradigm of cometary science. Thus the issues of whence and how comets originated have come into much better focus, although many aspects, naturally, remain unsolved. But more importantly, we have learned that the study of comets, indeed, means looking back into the history of our Galaxy and is thus of higher significance than the mere explanation of how our system of planets and comets came into being. Mayo is the uncontested father of this paradigm, and he is the one who taught us that studying comets is even more interesting than most of us had realized!

I therefore think that the present meeting, summarizing most aspects of cometary research in the post-Hale–Bopp era and thus enjoying and profiting from the foundations laid by Mayo's work, is rightly to be dedicated to his bright memory. May we continue to be inspired and guided by his heritage in searching for the truth about these elusive and ephemeral objects, the comets!

HANS RICKMAN
Secretary General
International Astronomical Union IAU

THE 1995–2002 LONG-TERM MONITORING OF COMET C/1995 O1 (HALE–BOPP) AT RADIO WAVELENGTH

NICOLAS BIVER
ESA/ESTEC, Sci-Sb, Keplerlaan 1, 2200-AG Noordwijk, The Netherlands and LESIA, Observatoire de Paris, 5 pl. J. Janssen, F-92195 Meudon, France
E-mail: Nicolas.Biver@obspm.fr

DOMINIQUE BOCKELÉE-MORVAN, PIERRE COLOM, JACQUES CROVISIER, FLORENCE HENRY and EMMANUEL LELLOUCH
LESIA, Observatoire de Paris, 5 pl. J. Janssen, F-92195 Meudon, France

ANDERS WINNBERG and LARS E.B. JOHANSSON
Onsala Space Observatory, S-439 92 Onsala, Sweden

MARCUS GUNNARSSON and HANS RICKMAN
Uppsala Astronomiska Observatorium, Box 515, S-751 20 Uppsala, Sweden

FREDRIK RANTAKYRÖ
European Southern Observatory, Alonso de Cordova 3107, Vitacura, Santiago, Chile

JOHN K. DAVIES and WILLIAM R.F. DENT
Astronomy Technology Centre, Royal Observatory, Blackford Hill, Edinburgh, EH9 3HJ, UK

GABRIEL PAUBERT
IRAM, Avd. Divina Pastora, 7, 18012 Granada, Spain

RAPHAËL MORENO and JÖRN WINK†
IRAM, 300, rue de la Piscine, F-38406 Saint Martin d'Hères, France

DIDIER DESPOIS
Observatoire de Bordeaux, BP 89, F-33270 Floirac, France

DOMINIC J. BENFORD, MATT GARDNER, DARIUSZ C. LIS, DAVID MEHRINGER and THOMAS G. PHILLIPS
Caltech, MS 320-47, Pasadena , CA 91125, USA

HEIKE RAUER
DLR, Institut für Planetenerkundung, Rutherfordstrasse 2, 12489 Berlin-Adlershof, Germany

(Received 26 March 2002; Accepted 12 June 2002)

Abstract. The bright comet Hale–Bopp provided the first opportunity to follow the outgassing rates of a number of molecular species over a large range of heliocentric distances. We present the results of our observing campaign at radio wavelengths which began in August 1995 and ended in January 2002. The observations were carried out with the telescopes of Nançay, IRAM, JCMT, CSO and, since September 1997, SEST. The lines of nine molecules (OH, CO, HCN, CH_3OH, H_2CO, H_2S, CS, CH_3CN and HNC) were monitored. CS, H_2S, H_2CO, CH_3CN were detected up to r_h = 3–4 AU from the Sun, while HCN and CH_3OH were detected up to 6 AU. CO, which is the main driver of cometary activity at heliocentric distances larger than 3–4 AU, was last detected in August 2001, at r_h = 14 AU.

Earth, Moon and Planets **90:** 5–14, 2002.

The gas production rates obtained from this programme contain important information on the nature of cometary ices, their thermal properties and sublimation mechanisms. Line shapes allow to measure gas expansion velocities, which, at large heliocentric distances, might be directly connected to the temperature of the nucleus surface. Inferred expansion velocity of the gas varied as $r_h^{-0.4}$ within 7 AU from the Sun, but remained close to 0.4 km s^{-1} further away. The CO spectra obtained at large r_h are strongly blueshifted and indicative of an important day-to-night asymmetry in outgassing and expansion velocity. The kinetic temperature of the coma, estimated from the relative intensities of the CH_3OH and CO lines, increased with decreasing r_h, from about 10 K at 7 AU to 110 K around perihelion.

Keywords: C/1995 O1 (Hale–Bopp), comets, molecules, radio observations

1. Introduction

The exceptional comet C/1995 O1 (Hale–Bopp) provided us with an unprecedented opportunity to draw the link between distant activity of a comet and its outgassing close to the Sun. In the present paper, we summarize results of the long-term monitoring we performed from September 1995 to January 2002 at millimetre to submillimetre wavelengths. Nine molecules were followed and the heliocentric range covers $r_h = 0.9$ to 14 AU. This paper updates the previously published report of Biver et al. (1999a), which covers observations until January 1998. All details on the observational and analysis procedures will not be recalled here.

2. Observations

Comet Hale–Bopp has been observed on a regular basis since September 1995, with the IRAM radio telescopes (30-m single dish at Pico Veleta and the 15-m antennas of the Plateau-de-Bure interferometer used in single dish mode), and the JCMT and CSO submillimetre telescopes atop Mauna Kea, Hawaii. Since September 1997 and until January 2002, observations have been carried out at SEST in Chile, due to the negative declination of the comet. Half-power beam widths ranged from 10″ to 55″, depending on the antenna and observing frequency. This long-term monitoring concerns CO, CH_3OH, HCN, H_2S, H_2CO, CS, CH_3CN, and HNC. Their first detections were obtained between 6.7 AU for CO (Jewitt et al., 1996; Biver et al., 1996) and 2.4 AU for HNC (Bockelée-Morvan et al., 1996), as summarized in Biver et al. (1997). Since the report made by Biver et al. (1999a) concerning observations before January 1998, the most distant detections of HCN ($r_h = 6.4$ AU in Sept. 1998), CH_3OH ($r_h = 6.1$ AU in Aug. 1998) and CO ($r_h = 14$ AU in Aug. 2001) in any comet were achieved (Figures 1–2). Other species fell below the detection limit at distances of 4.2 AU or less (Biver et al., 1999a).

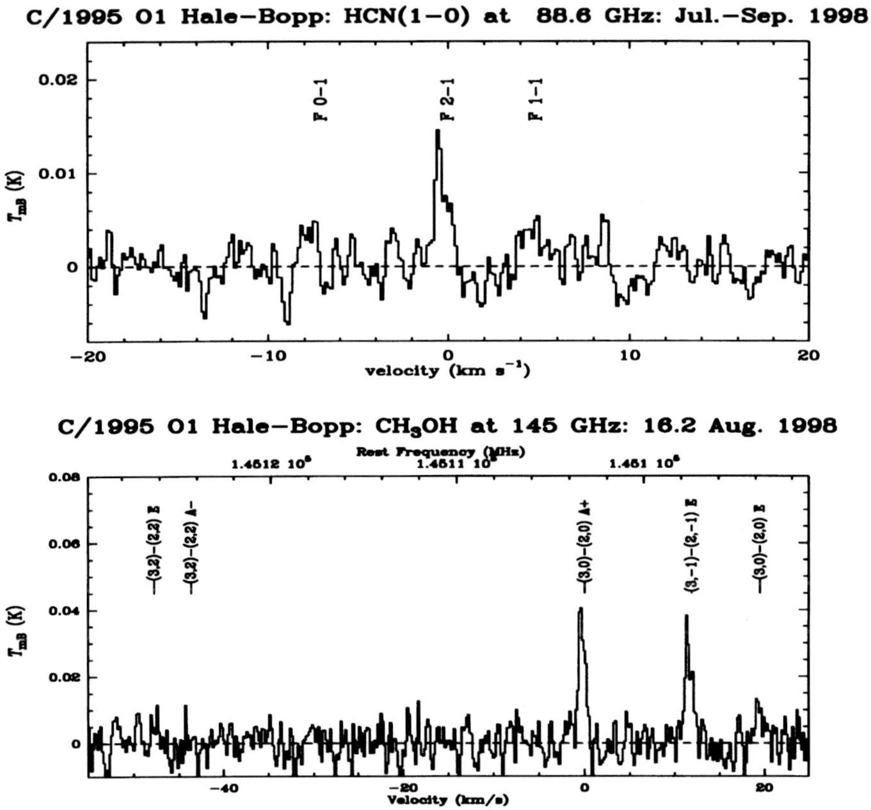

Figure 1. HCN $J(1\text{–}0)$ and CH_3OH 145 GHz spectra of comet C/1995 O1 (Hale–Bopp) observed at SEST. They correspond to the most distant detections of these molecules in any comet before. HCN was last detected in September 1998 at $r_h = 6.4$ AU and CH_3OH in August 1998 at 6.1 AU. The HCN $J(1\text{–}0)$ spectrum is the average of the last 3 runs when it was detected (July, August and September 1998).

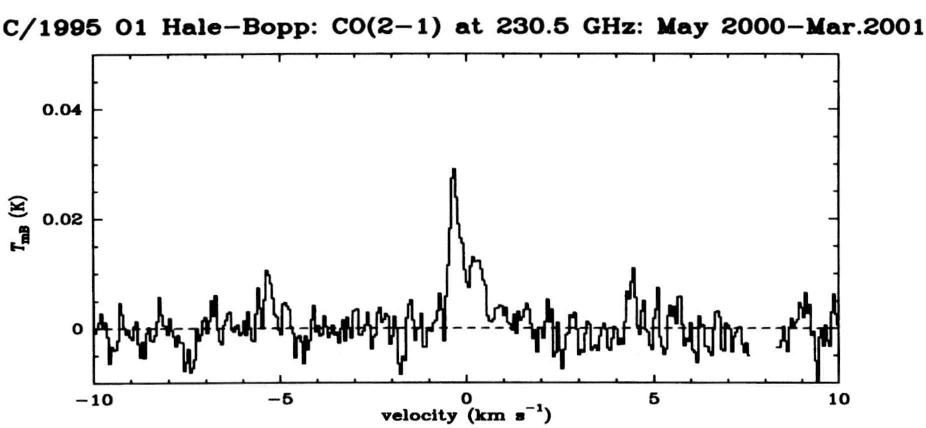

Figure 2. CO $J(2\text{–}1)$ line at 230 GHz in comet C/1995 O1 (Hale–Bopp) observed at SEST during the May 2000–March 2001 time interval: average of 6 periods with $11 < r_h < 13$ AU.

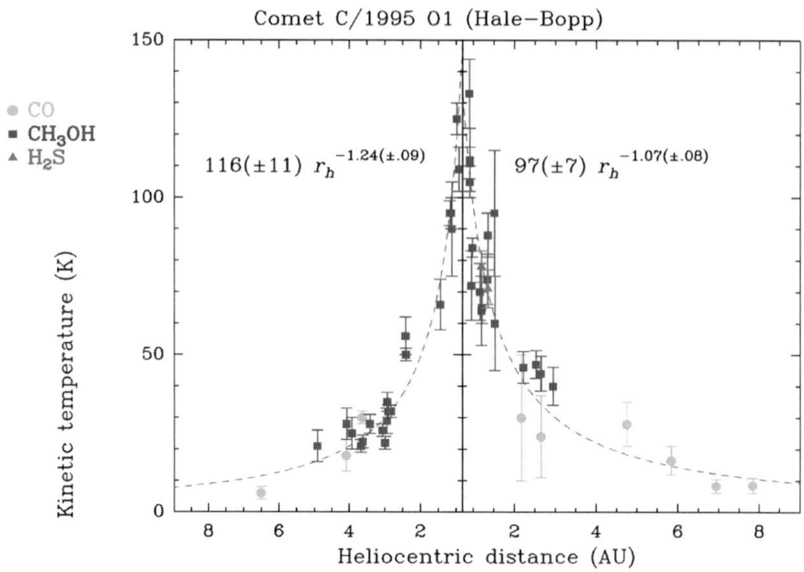

Figure 3. Evolution of the gas kinetic temperature in comet Hale–Bopp as measured from CO, H_2S and CH_3OH lines with fitted power laws (dashed lines). Left: Pre-perihelion data; right: Post-perihelion data.

3. Temperature and Expansion Velocity

The kinetic temperature of the coma and its expansion velocity are key parameters to derive accurate production rates. The simultaneous observations of several lines of a given species can help to constrain their values. CH_3OH (304, 307, 157 and 252 GHz) and CO (J = 1–0, 2–1, 3–2, 4–3) series of lines are of particular interest because their rotational temperature is very close to the kinetic temperature of the inner coma, unlike the CH_3OH (97 GHz and 145 GHz) series lines (Bockelée-Morvan et al., 1994). Figure 3 shows the evolution of the kinetic temperature from 6.5 AU to perihelion inbound and then to 7.9 AU outbound. Average of pre and post-perihelion data gives $T = 102\ (\pm 6)\ r_h^{-1.12(\pm 0.06)}$ K.

Beyond 7 AU post-perihelion, the CO $J(1–0)$ line was only marginally detected in the average of data taken at 7.9 ± 0.4 AU. Thereafter, this line was too weak to provide stringent constraints on the temperature of the coma. Taking a temperature of 5 K instead of the value of 10 K assumed for $r_h > 8$ AU, the CO production rates would be increased by about 60%. This is however going in the direction opposite to the outgassing pattern correction (cf. next Section).

The gas outflow velocity (v_{exp}) was estimated from the line shapes. As explained by Biver et al. (1999a), it was derived from the velocity at half-maximum of the lines, "half width at half maximum" (*HWHM*), on their blueshifted wing which corresponds to molecules moving preferentially from the day side of the nucleus. The evolution of this parameter, tracing the expansion velocity, is plotted in Figure 4, as a function of heliocentric distance. Average of pre- and post-perihelion data yields *HWHM*= 1.125 (±0.015) $r_h^{-0.42(\pm 0.01)}$ km s^{-1}. The expansion velocity value we actually used to derive production rates is reduced by 10%, to take into account thermal broadening. This leads to expansion velocities on the order of 1.05 km s^{-1} at perihelion, and 0.45 km s^{-1} at 6 AU, which is comparable to the value of 0.5 km s^{-1} measured in comet 29P/Schwassmann–Wachmann 1 at similar r_h (Crovisier et al., 1995).

From 7 to 14 AU after perihelion, the expansion velocity did not vary in a measurable way (Figure 4b). From the CO line shape we actually deduce a mean value of $v_{\rm exp} = 0.4$ km s^{-1} on the day side of the nucleus, that we used for the computation of the production rates. This velocity is intermediate between the initial CO sonic velocity expected near the surface, and the maximum velocity that the CO molecules could reach during adiabatic expansion, in case of a large fluid region (see Crifo et al., 1999). Indeed, assuming a slow-rotating body for the computation of the surface equilibrium temperature and a Bond albedo of 0.04, one computes an initial CO velocity between 0.23 and 0.19 km s^{-1} for r_h between 7 and 14 AU. Terminal velocities are 0.6 to 0.5 km s^{-1} for r_h = 7 to 14 AU. The expected decrease of the velocity in $r_h^{-0.25}$, which would reflect the decrease of the surface temperature as the comet recedes from the Sun, is not clearly seen in the data, but might be within the measurement uncertainties. The actual fit *HWHM* = 0.853 $r_h^{-0.25}$ shown in Figure 4b for CO data beyond 3 AU, is relatively poor with a correlation factor $\rho = 0.51$. Note that the asymmetrical shape of the CO line profile (Figure 2) suggests that the velocity of the CO molecules on the night side of the nucleus is smaller than that of the CO molecules on the day side. CO radio line profiles obtained in comet 29P/Schwassmann–Wachmann 1 show similar shapes (Crovisier et al., 1995; Gunnarsson et al., 2002). The sunward/anti-sunward asymmetry in the CO velocity was attributed to a day/night asymmetry in the nucleus temperature (Crifo et al., 1999). A detailed analysis of CO Hale–Bopp spectra is to be presented by Gunnarsson et al. in a forthcoming paper.

4. Production Rates

The models for converting line intensities into production rates use the previously determined temperatures and expansion velocities, with a spherically symmetric Haser density distribution. In addition, radiative decay and infrared or ultraviolet pumping by solar radiation are taken into account for all molecules (Biver et al., 1997). Photodissociative lifetimes were taken from Crovisier (1994). CS is assumed to be the photodissociation product of CS_2. H_2CO is assumed to be released from an extended source, as suggested by the comparison between observations made at centre and offset positions and from interferometric maps (Wink et al., 1999), and as observed in other comets (Colom et al., 1992; Meier et al., 1993; Biver et al., 1999b). Its parent equivalent lifetime is taken equal to 10^4 s at 1 AU.

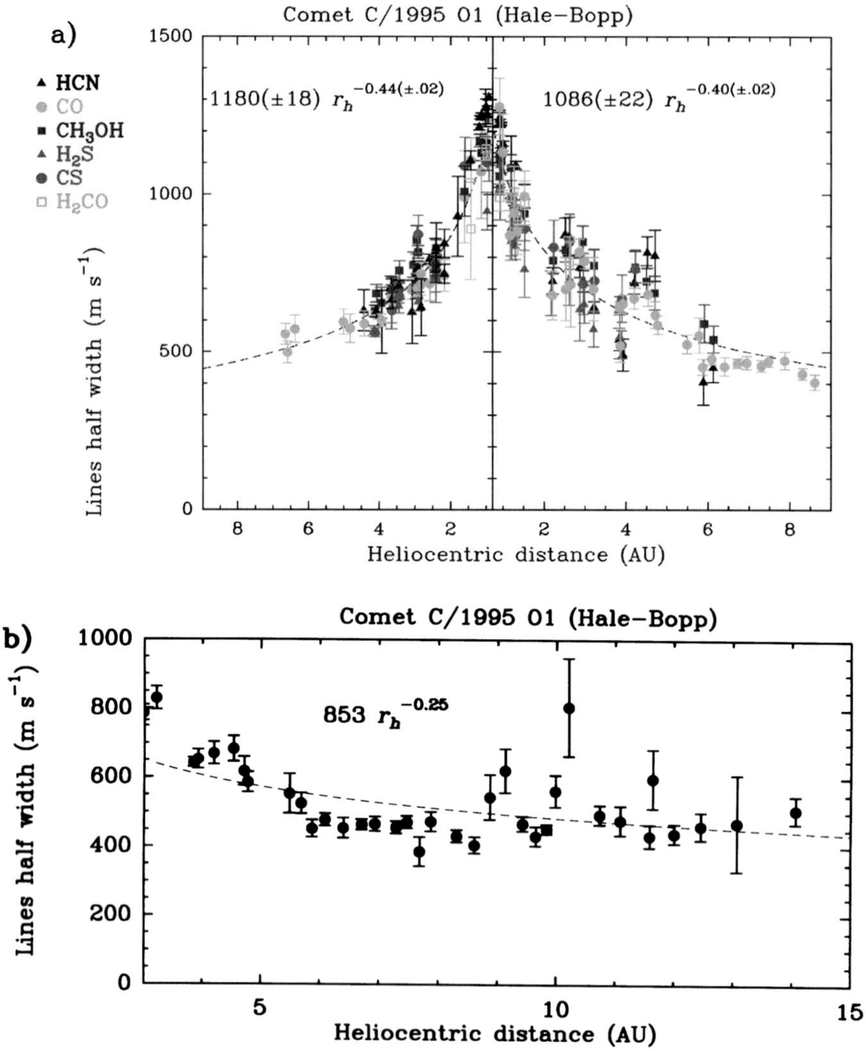

Figure 4. Upper plot: Velocity at half-maximum on the blue wing of the lines, which provides an estimate of the gas expansion velocity in comet Hale–Bopp. Fitted power laws are superimposed. Lower plot: Velocity at half-maximum on the blue wing of the CO lines, beyond 3 AU post-perihelion. A power law fit to these data is shown, but there are large deviations that suggest two distinct regimes, below and beyond 7 AU.

More accurate production rates would require the modelling of the outgassing pattern using the line shapes. The asymmetry of the lines observed beyond 3 AU could be interpreted as: (i) An outgassing in a restricted cone of about 120° at a constant expansion velocity (i.e., no outgassing around the antisolar point): This would increase the production rates by 10–15%; or (ii) an outgassing at a lower rate with lower velocity towards the anti-solar point, which is more likely (Crifo et al.,

1999). A two-component model with a larger expansion velocity and outgassing rate in a 45° cone along the comet-Sun direction, with respect to the remaining solid angle, would result in total outgassing rates reduced by up to 40%.

Figure 5 shows the gas production curves. Included are the OH production rates measured at Nançay from April 1996 to September 1997, as reported by Colom et al. (1999). Post-perihelion UV observations of OH (converted into water production rates) (Stern et al., 1999; Weaver et al., 1999), and infrared observations of water (Crovisier et al., 1999), are also plotted in Figure 5.

Power law fits to the production rates are shown in Figure 5. The pre-perihelion data have been split into 3 periods which show different trends for most molecules. Post-perihelion data do not show such distinct regimes.

- CO is the most volatile species and showed a moderate increase of its production rate at large heliocentric distances: slopes are $r_h^{-2.2}$ at $r_h > 3$ AU pre-perihelion, $r_h^{-1.9}$ at $r_h > 0.9$ AU pre-perihelion and $r_h^{-2.0}$ at $r_h > 0.9$ AU post-perihelion. Between 3 and 1.6 AU inbound, the CO production rate stalled or even decreased before exhibiting a steep increase: this stagnation is present in numerical simulations of comet Hale–Bopp's CO production (Enzian et al., 1998). Beyond 6 AU post-perihelion, the CO production rate decreased as r_h^{-2} in average, but exhibited some significant scatter. This heliocentric evolution is not yet well simulated by thermal models of comet Hale–Bopp nucleus (Capria et al., 2000).
- CH_3OH, HCN, CH_3CN, and H_2S molecules are less volatile and less abundant than CO. However, at large heliocentric distances, they are still overabundant with respect to water, when compared to their abundances near perihelion. There are not enough data points for H_2S and CH_3CN to distinguish distinct behaviours with respect to CH_3OH and HCN. Beyond 3 AU, all these species exhibit heliocentric variations similar to CO, although slightly steeper. Upper limits obtained for CH_3OH and HCN at 6.6 AU pre-perihelion and around 7 AU post-perihelion suggest that slopes may have been steeper beyond ≈ 6 AU than below. Post-perihelion CN production rates obtained by Rauer et al. (2002) show a similar change of slope at large r_h. This is the behaviour that one would expect for a compound with the volatility of pure HCN or CH_3OH ices ($T_{\mathrm{sublimation}} \approx 100$ K).

The production rates of CS (tracing CS_2), H_2CO and HNC displayed a much steeper evolution with heliocentric distance when compared to other species, though their volatility is similar.

- The CS/HCN ratio varied as $r_h^{-0.7}$, as observed in comet C/1996 B2 (Hyakutake) (Biver et al., 1999b) and other comets (Biver et al., 2002). The origin of this variation is unexplained, but may be linked to the presence of a distributed source of CS other than CS_2.
- The HNC/HCN ratios increased with decreasing r_h with a slope of -1.1, questioning the nuclear origin of this species. In the case of this highly pro-

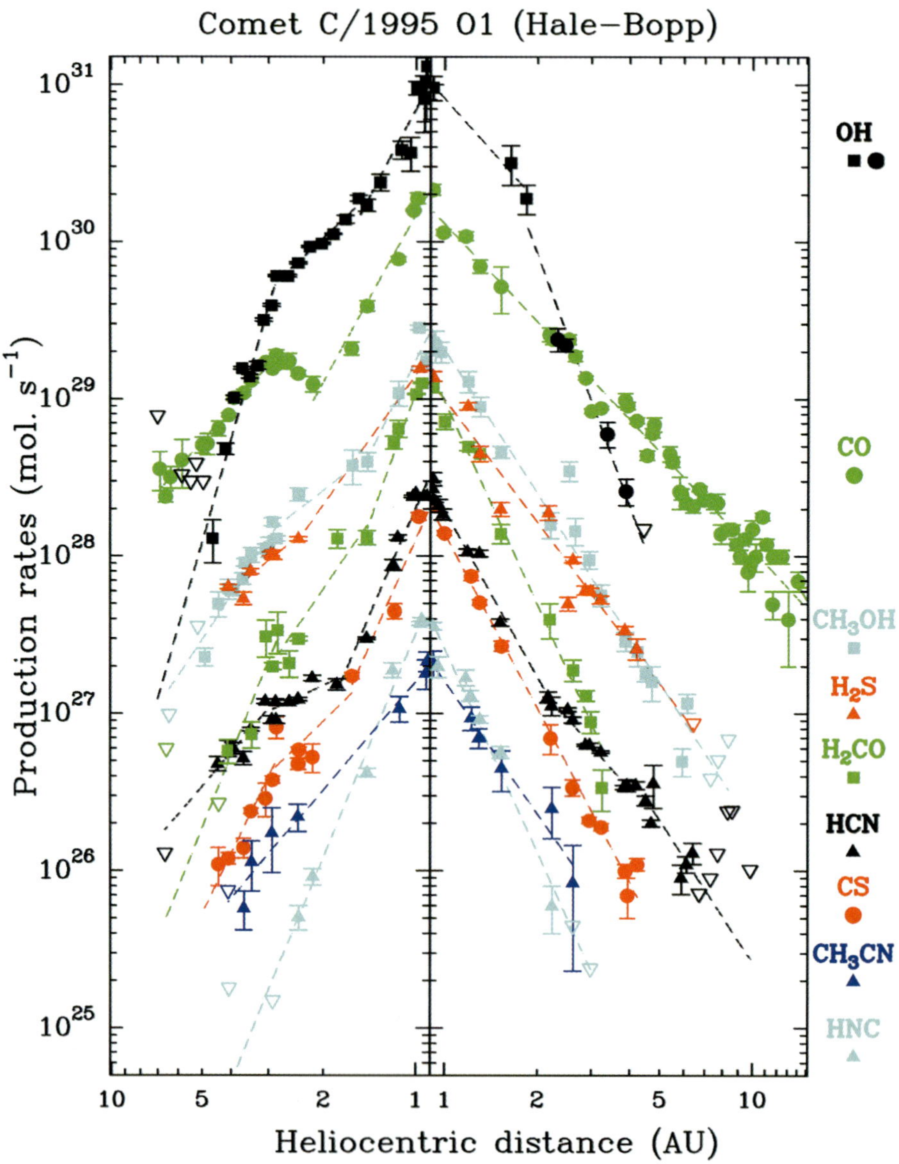

Figure 5. Evolution of production rate with heliocentric distance and fitted power laws (dashed lines). OH measurements are from Colom et al. (1999). Additional H_2O production rates measurements post-perihelion (filled circles and upper limit at 4.8 AU) are from Stern et al. (1999), Weaver et al. (1999) and Crovisier et al. (1999). Downward pointing triangles are 3-σ upper limits. Left: Pre-perihelion data; right: Post-perihelion data.

ductive comet, Irvine et al. (1999) and Rodgers and Charnley (1998) invoked the formation of HNC from chemical reactions in the coma involving HCN.

– H_2CO/HCN increased as $r_h^{-1.5}$ beyond 1.5 AU from the Sun. This suggests that the H_2CO source may be much less volatile than H_2CO itself. An alternative explanation is that H_2CO comes from the thermal degradation of polymers on grains (Cottin et al., 2001). The increased efficiency of this process close to the Sun, together with the increasing grain production, at least qualitatively agrees with the observed evolution of the H_2CO production rate with r_h.

Acknowledgements

We are grateful to the IRAM, JCMT, CSO and SEST staff for scheduling and assistance during the observations of these long-term projects. SEST is operated jointly by the Swedish National Facility for Radio Astronomy, and by the European Southern Observatory. IRAM is an international institute cofunded by the CNRS, the Max-Planck-Gesellschaft and the Instituto Geogràfico Nacional, Spain. The JCMT is operated on behalf of the Particle Physics and Astronomy Research Council of the United Kingdom, the Netherlands Organization for Scientific Research, and the National Research Council of Canada. The CSO is supported by National Science Foundation grant AST 99-80846. The Nançay Radio Observatory is operated by the Unité Scientifique de Nançay of the Observatoire de Paris, associated with the CNRS, and also gratefully acknowledges the financial support of the Conseil Régional of the Région Centre in France.

References

Biver, N., Bockelée-Morvan, D., Colom, P. et al.: 1997, 'Evolution of the Outgassing of Comet Hale–Bopp (C/1995 O1) from Radio Observations', *Science* **275**, 1915–1918.

Biver, N., Bockelée-Morvan, D., Colom, P. et al.: 1999a, 'Long-Term Evolution of the Outgassing of Comet Hale–Bopp from Radio Observations', *Earth Moon Planets* **78**, 5–11.

Biver, N., Bockelée-Morvan, D., Crovisier, J. et al.: 1999b, 'Spectroscopic Monitoring of Comet C/1996 B2 (Hyakutake) with the JCMT and IRAM Radio Telescopes', *Astron. J.* **118**, 1850–1872.

Biver, N., Bockelée-Morvan, D., Crovisier, J. et al.: 2002, 'Chemical Composition Diversity among 24 Comets Observed at Radio Wavelengths', these proceedings.

Biver, N., Rauer, H., Despois, D. et al.: 1996, 'Substantial Outgassing of CO from Comet Hale–Bopp at Large Heliocentric Distance', *Nature* **380**, 137–139.

Bockelée-Morvan, D., Crovisier, J., Colom, P., and Despois, D.: 1994, 'The Rotational Lines of Methanol in Comets Austin 1990 V and Levy 1990 XX', *Astron. Astrophys.* **287**, 647–665.

Bockelée-Morvan D., Moreno, R., Biver, N. et al.: 1996, 'Comet C/1995 O1 (Hale–Bopp)', *IAUC* 6511.

Capria, M. T., Coradini, A., De Sanctis, M. C., and Orosei, R.: 2000, 'CO Emission Mechanisms in C/1995 O1 (Hale–Bopp)', *Astron. Astrophys.* **357**, 359–366.

Colom, P., Crovisier, J., Bockelée-Morvan, D., Despois, D., and Paubert, G.: 1992, 'Formaldehyde in Comets I. Microwave Observations of P/Brorsen–Metcalf (1989 X), Austin (1990 V) and Levy (1990 XX)', *Astron. Astrophys.* **264**, 270–281.

Colom, P., Gérard, E., Crovisier, J. et al.: 1999, 'Observations of the OH radical in Comet Hale–Bopp with the Nançay Radio Telescope', *Earth Moon Planets* **78**, 37–43.

Cottin, H., Gazeau, M. C., Benilan, Y., and Raulin, F.: 2001, 'Polyoxymethylene as Parent Molecule for the Formaldehyde Extended Source in Comet Halley', *Astrophys. J.* **556**, 417–420.

Crifo, J.-F., Rodionov, A. V., and Bockelée-Morvan, D.: 1999, 'The Dependence of the Circumnuclear Coma Structure on the Properties of the Nucleus. III. First Modeling of a CO-Dominated Coma, with Application to Comets 46P/Wirtanen and 29P/Schwassmann–Wachmann 1', *Icarus* **138**, 85–106.

Crovisier, J.: 1994, 'Photodestruction Rates for Cometary Parent Molecules', *J. Geophys. Res.* **99–E2**, 3777–3781.

Crovisier, J., Biver, N., Bockelée-Morvan, D. et al.: 1995, 'Carbon Monoxide Outgassing from Comet P/Schwassmann–Wachmann 1', *Icarus* **115**, 213–216.

Crovisier, J., Leech, K., Bockelée-Morvan, D. et al.: 1999, 'The Spectrum of Comet Hale–Bopp as seen by ISO', *ESA-SP* **427**, 137–140.

Enzian, A., Cabot, H., and Klinger, J.: 1998, 'Simulation of the Water and Carbon Monoxide Production Rates of Comet Hale–Bopp Using a Quasi 3D Nucleus Model', *Planet. Space Sci.* **46**, 851–858.

Gunnarsson, M., Rickman, H., Festou, M. C., Winnberg, A., and Tancredi, G.: 2002, 'An Extended CO Source around Comet 29P/Schwassmann–Wachmann 1', *Icarus*, in press.

Irvine, W. M., Schloerb, F. P., Dickens, J. E. et al.: 1999, 'The HNC/HCN Ratio in Comets', *Earth Moon Planets* **78**, 29–35.

Jewitt, D., Senay M., and Matthews H.: 1996, 'Observations of Carbon Monoxide in Comet Hale–Bopp', *Science* **271**, 1110–1113.

Meier, R., Eberhardt, P., Krankowsky, D., and Hodges, R. R.: 1993, 'The Extended Formaldehyde Source in Comet Halley', *Astron. Astrophys.* **277**, 677–690.

Rauer, H., Helbert, J., Arpigny, C. et al.: 2002, 'Long-Term Optical Spectrophotometric Monitoring of Comet C/1995 O1 (Hale–Bopp)', *Astron. Astrophys.*, in press.

Rodgers, S. D. and Charnley, S. B.: 1998, 'HNC and HCN in Comets', *Astrophys. J. Lett.* **501**, L227–L230.

Stern, S. A., Colwell, W. B., Festou, M. C. et al.: 1999, 'Comet Hale–Bopp (C/1995 O1) near 2.3 AU Post-Perihelion: Southwest Ultraviolet Imaging System Measurements of the H_2O and Dust Productions', *Astron. J.* **118**, 1120–1125.

Weaver, H. A., Feldman, P. D., A'Hearn, M. F., Arpigny, C., Brandt, J. C., and Stern, S. A.: 1999, 'Post-perihelion HST Observations of Comet Hale–Bopp (C/1995 O1)', *Icarus* **141**, 1–12.

Wink J., Bockelée-Morvan D., Despois, D. et al.: 1999, 'Evidences for Extended Sources and Temporal Modulations in Molecular Observations of C/1995 O1 (Hale–Bopp) at IRAM Interferometer', *Earth Moon Planets* **78**, 63.

LARGE-SCALE STRUCTURES IN COMET HALE–BOPP

Latitudinal Variations and Monster Disconnection Event

J. C. BRANDT
Institute for Astrophysics, Department of Physics & Astronomy, University of New Mexico, Albuquerque, NM, USA
E-mail: jbrandt@as.unm.edu

M. SNOW
Laboratory for Atmospheric and Space Physics, University of Colorado, Boulder, CO, USA
E-mail: snow@lasp.colorado.edu

Y. YI
Department of Astronomy & Space Science, Chungnam National University, Daejon 305-764, South Korea
E-mail: euyiyu@hanbat.chungnam.ac.kr

S. M. LARSON
Lunar & Planetary Laboratory, University of Arizona, USA
E-mail: slarson@jupiter.lpl.Arizona.edu

H. MIKUZ
Crni Vrh Observatory, Slovenia
E-mail: herman.mikuz@uni-lj.si

C. C. PETERSEN
Loch Ness Productions, Groton, MA, USA
E-mail: carolyn@lochness.com

W. LILLER
Instituto Isaac Newton, Chile
E-mail: wliller@compuserve.com

(Received 2 March 2002; Accepted 16 June 2002)

Abstract. The plasma tails of comets clearly show the demarcation of the solar wind into distinct equatorial and polar regions (Brandt and Snow (2000), *Icarus* **148**, 52–64). The boundary is determined by the maximum extent in latitude of the heliospheric current sheet (HCS). The observational record contains many well-observed equatorial comets, but observations of comets in the polar region are relatively rare.

In addition to its size and brightness, comet Hale–Bopp had an orbital inclination of 89.4° and was well observed for months in the polar region. We document the comet's large-scale appearance throughout the apparition, including the polar region and its transition into the equatorial region. The bright dust tail hampered observations of the plasma tail, particularly near the head, but images taken with a CO^+ filter show a very large disconnection event (DE) on May 7 and May 8, 1997. The time of disconnection is estimated at approximately May 4.0. This DE is associated with a crossing of the HCS. The model calculations of the HCS indicate that other crossings might have occurred in late April, but given the uncertainty in the calculation, the comet might have missed the HCS. Sparse observational coverage and the bright dust tail prevent further investigation of the potential earlier

Earth, Moon and Planets **90:** 15–33, 2002.

HCS crossings. The plasma tail shows anomalous orientations at the highest latitudes and possible explanations are discussed.

Introduction

Comets in the inner solar system are enveloped in the solar wind. Radiation from the sun sublimates the ices in the comet nucleus to produce coma gases, and some of this gas is ionized. The interaction of cometary plasma with the magnetized solar wind produces the plasma (Type I) tail and other plasma features. The cometary ions, such as H_2O^+ and CO^+, are trapped onto the magnetic field lines. This fact allows changes in the magnetic field structure and orientation to be observed from earth.

The *Ulysses* spacecraft (launched in 1990) has produced measurements that showed the heliosphere at approximately 1 AU heliocentric distance to consist of two distinct regions (Phillips et al., 1995; McComas et al., 1998; Smith et al., 2001). In the *equatorial region*, average properties are a radial speed W_r of 450 km s^{-1} and a density $N_e = N_p = 9$ cm^{-3}. Variations in these quantities can be quite large. The equatorial region contains the heliospheric current sheet (HCS). In the *polar region*, average properties are a radial solar-wind speed of 750 km s^{-1} and a density of 3 cm^{-3}. Variations in these quantities are small. The HCS does not extend into the polar region. The boundary between the equatorial and polar regions is approximately determined by the maximum latitudinal extent of the HCS and changes with phase of the solar cycle. This description is based on measurements obtained away from solar maximum, and refinements or restatement may be necessary for times near solar maximum.

Because of the interaction between the solar wind and the cometary plasma, comets are natural, *in situ* solar-wind probes. The plasmas in comets should show the following behavior, the Ulysses Comet Watch (UCW) paradigm:

1. Appearance. Comet tails in the equatorial region are exposed to a highly variable solar wind and should appear disturbed. Tails in the polar region are exposed to a steady solar wind and should appear relatively undisturbed.
2. Disconnection Events (DEs). On the view of DEs occurring when a comet crosses the HCS, they should occur only in the equatorial region and not in the polar region.
3. Orientation of plasma tail. The orientation of the plasma tail is determined by the solar-wind speed by dynamical aberration. The general pattern should be a low-speed regime (equatorial, W_r = 450 km s^{-1}) with high scatter versus a high-speed regime (polar, W_r = 750 km s^{-1}) with lower scatter.

The UCW paradigm was described in detail by Brandt and Snow (2000) and tested against comet observations for essentially the entire 20th century. They also found that observations of comets in the polar region prior to 1995 were rare.

For example, the Belton and Brandt (1966) catalogue of comet-tail orientations covering the time interval 1889 to 1963 had 700 measurements useful for plasma studies, and only about 58 of these were polar. Comets Perrine (1898b), Borrelly (1903c), Brooks (1911c), and Mrkos (1957d) were the only comets observed in the polar region. Even with the appearance of additional comets that sampled the polar region – de Vico (C/122P) in 1995 and Hyakutake (C/1996 B2) – observations of comets in the polar region remained rare. Historically, most observations were of comets in the equatorial region.

Comet Hale–Bopp (C/1995 O1) with its exceptional size and brightness also had an orbital inclination of 89.4°. The combination of these circumstances provided an extraordinary opportunity to observe a comet in the polar region, in transition from the polar to the equatorial region, and in the equatorial region. One of the purposes of this paper is to document the comet's large-scale structure throughout the apparition. Earlier reports were given by Kinoshita et al. (1997), Fischer et al. (1997), Jones (1997), and Rauer (1997).

The other purpose of this paper is to discuss the observations in the context of the latitudinal structure of the solar wind. The large-scale appearance in the images is entirely consistent with the paradigm described above. Although the bright dust tail hampered observations of the plasma tail, particularly near the head, a very large DE was observed at large distances from the head on May 7 and 8, 1997. The large distances from the head for this DE required the development of a new kinematical equation to describe the motion of the DE. The time of disconnection is estimated as May 4.0 with an uncertainly of about 1 day. The DE occurred well within the equatorial solar-wind region and is associated with a crossing of the HCS. The orientation of the plasma tail is consistent with the paradigm except at the highest latitudes.

1. Large-Scale Structure

The dominant structures in the comet during late 1996 were reasonably stable dust features. These are shown in Figures 1 and 2. Plasma structures were not generally visible in the comet during late 1996, although Jockers et al. (1996) report a transient plasma feature beginning on August 17, 1996.

Plasma structure was certainly visible by January 30, 1997, and plasma turn-on probably occurred in mid-January. We present the overall large-scale structure of the comet in two figures of sequences. The sequence of color images in Figure 3 shows the comet from February 19, 1997 to April 24, 1997. The dust and plasma tails are shown dramatically. Figure 4 shows the comet from March 10, 1997 to May 4, 1997. The overall appearance and changes are well documented by these figures.

By early June 1997, the plasma tail was not visible and the dust tail had faded considerably. The main feature was a neck-line structure (Kimura and Liu, 1977;

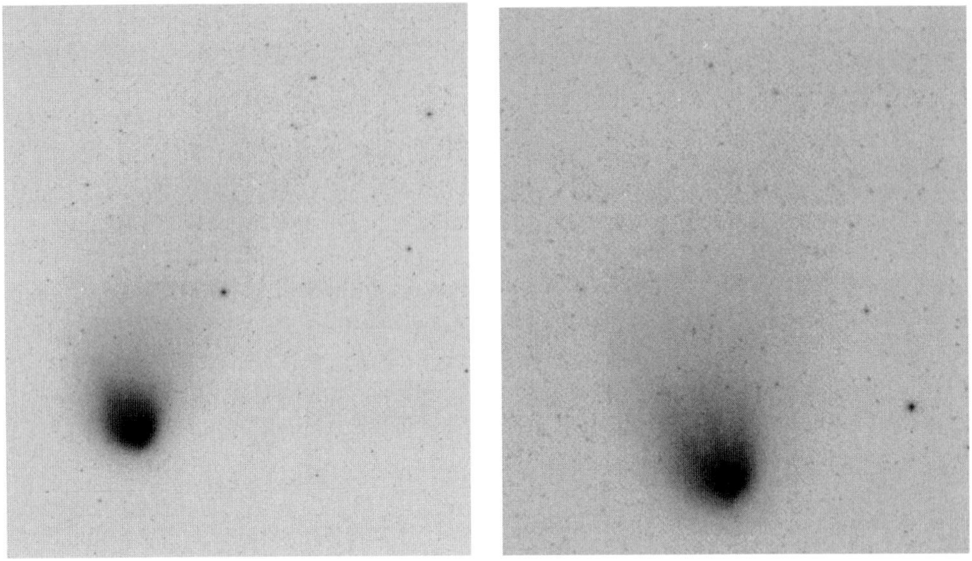

a. September 27, 1996 b. November 1, 1997

Figure 1. The appearance of comet Hale–Bopp in late 1996. (a) Comet Hale–Bopp on September 27, 1996. The exposure was 3 minutes beginning at 02:19:30 UT and was during a total lunar eclipse. The width of the image shown is approximately 9.7 arc min. (b) Comet Hale–Bopp on November 1, 1997. The exposure was 5 minutes beginning at 00:03:12 UT. The width of the image shown is approximately 7.3 arc min. These ESO Schmidt images were obtained by G. and O. Pizarro using Kodak 4415 emulsion and a GG 385 filter for the Liller–Brandt program. Only dust structures are seen.

Pansecchi et al., 1987; Fulle, 1987; Cremonese and Fulle, 1989) as shown in Figure 5, taken on June 6, 1997. Essentially the same structure was recorded on images taken on June 7 and June 8.

2. Latitudinal Changes and Disconnection Event

2.1. Appearance

The summary sequences in Figures 3 and 4 show the appearance of the comet while at high latitudes in the polar solar-wind region and the contrast with the appearance in the equatorial solar-wind region. During late February and early March 1997, the comet was high in the polar region and the comet had a beautiful and ethereal appearance. The plasma tail (blue in the color plates) had some filamentary fine structure but little of the gross structure (knots, kinks, etc.) common in equatorial plasma tails. By late April 1997, the comet showed the general kind of fine structure associated with the equatorial region. The contrast is illustrated in Figure 6 by

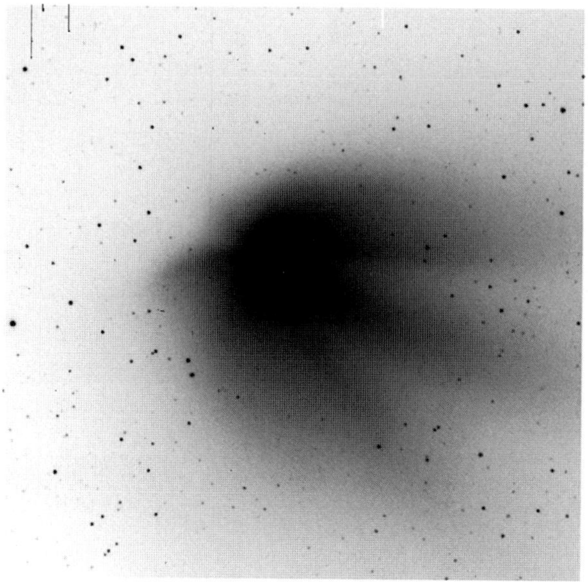

Figure 2. Comet Hale–Bopp imaged on November 12, 1996. The 10 second exposure started at 05:04:36 UT. (O. R. Hainaut, K. J. Meech, J. Bauer, and C.C. Petersen, using the 2.2-meter University of Hawaii telescope at Mauna Kea.)

images taken as part of the Ulysses Comet Watch by E. Kolmhofer and H. Raab near Linz, Austria on March 2.15 and May 2.83, 1997. The contrast is dramatic.

Note that comet Hale–Bopp may be unusual in its equatorial structure too. The structure as seen in the images shows the plasma tail as several prominent streamers. Bear in mind that the structure near the nucleus is obscured by the bright dust tail, except in the CO^+ images (Figure 7). These streamers are clearly shown in the CO^+ images.

The boundary between the polar and equatorial solar-wind regions was crossed by comet Hale–Bopp in late April 1997. We have examined the available images and fix the date at April 26, corresponding to latitudes of approximately 9°N (solar) or 12°N (ecliptic). We estimate the uncertainty in the date at about one day. The boundary date is consistent with the calculated position of the heliospheric current sheet (Figure 10) and the change of radial solar-wind speed evident in the orientations of the plasma tail (Figure 11).

2.2. Disconnection Event

A monster disconnection event (DE) was recorded in comet Hale–Bopp as shown in Figure 7. The DE was observed by author SML on May 7 and May 8, 1997 using a filter to isolate the light of CO^+. These are the only images to unambiguously show the DE. Although the UCW obtained other images in the relevant time period that might have shown the DE, visibility was hampered for two reasons: (1) The

a. February 19, 1997. B. Yen.

b. March 1, 1997. P. Candy.

c. March 20, 1997.
A. Dimai.& D. Ghirardo

d. April 2, 1997. H. Mikuz.

e. April 15, 1997. H. Mikuz.

f. April 24, 1997. H. Mikuz.

Figure 3. The overall appearance of comet Hale–Bopp in the polar solar-wind region from February 19, 1997 to April 24, 1997 as recorded by the Ulysses Comet Watch observers. See Appendix A for details.

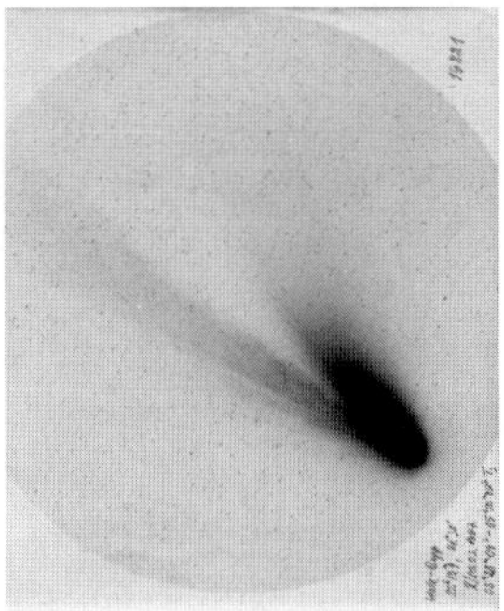

a. March 10, 1997. A. Alksnis.

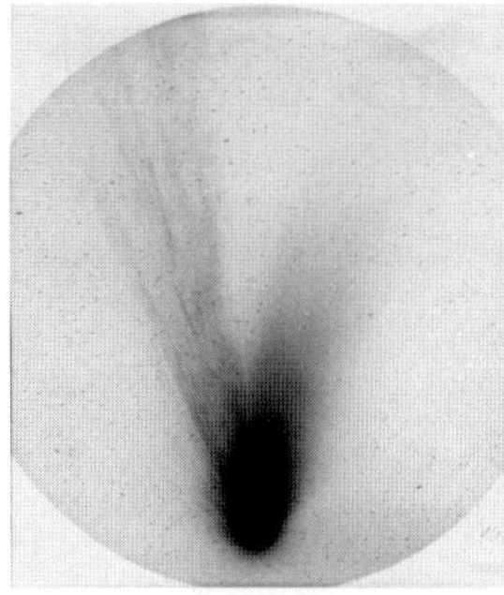

b. March 30, 1997. A. Alksnis.

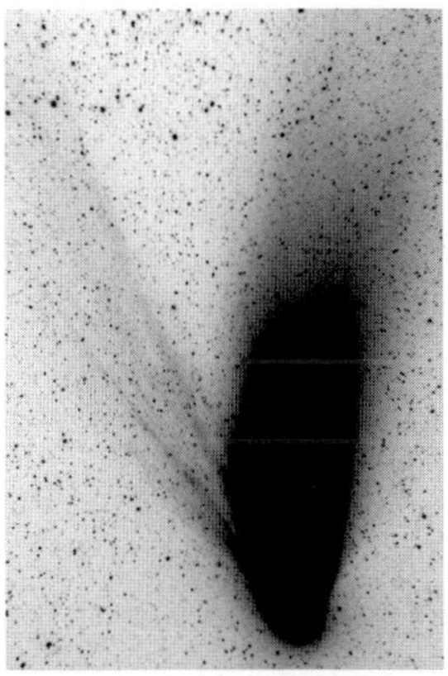

c. April 7, 1997. H. Mikuz.

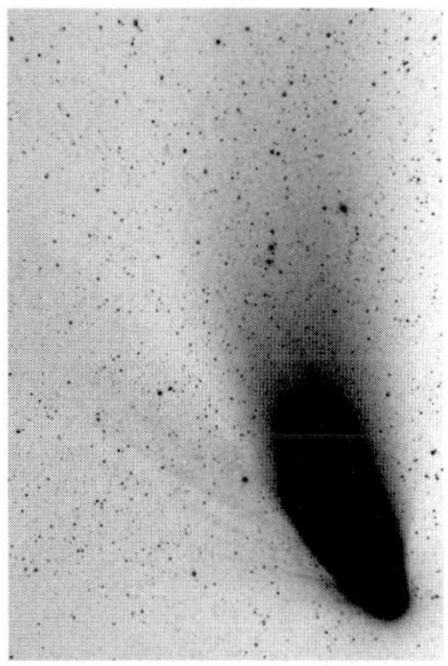

d. April 24, 1997. H. Mikuz.

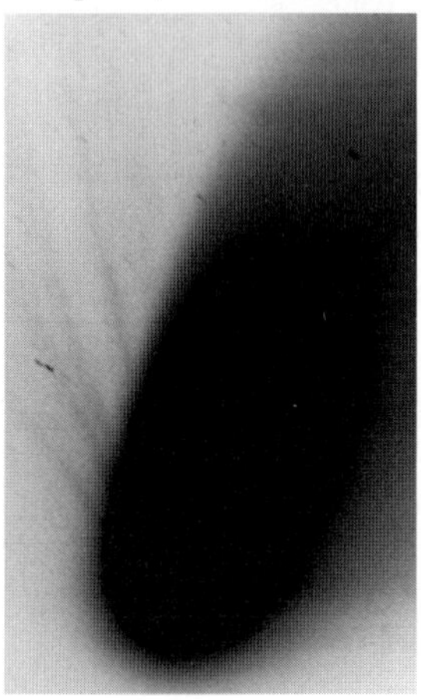

e. May 1, 1997. D. Hanon.

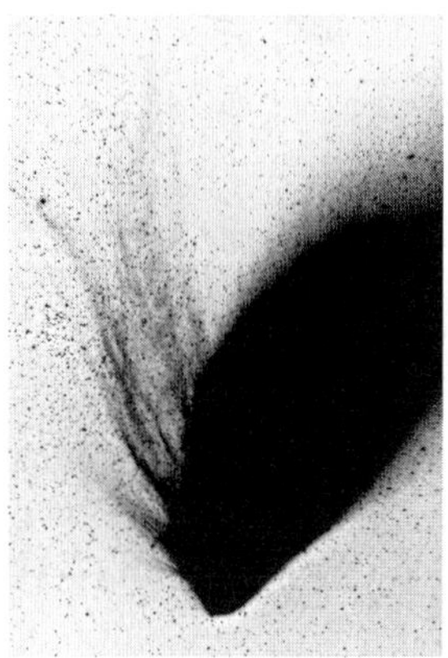

f. May 4, 1997. L. Tan.

Figure 4. The overall appearance of comet Hale–Bopp in the polar solar-wind region from March 10, 1997 to April 24, 1997 and in the equatorial solar-wind region on May 1 and May 4, 1997 as recorded by the Ulysses Comet Watch observers. See Appendix A for details.

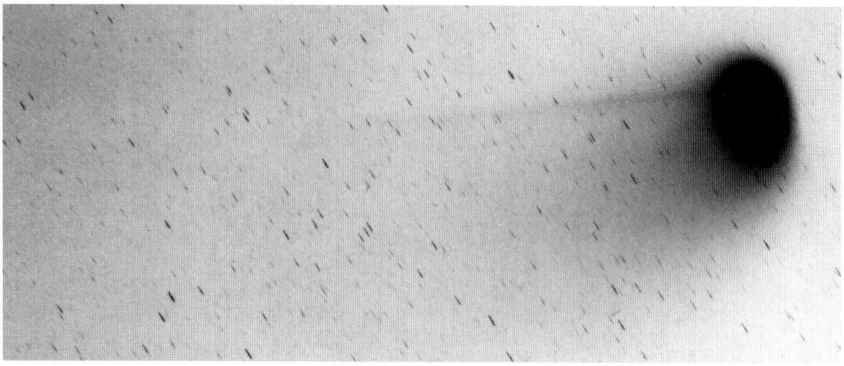

Figure 5. The neck-line structure in comet Hale–Bopp on June 6, 1997. The 5 minute exposure began at 23:03:00 UT. The width of the image shown is approximately 30 arc min. This ESO Schmidt image was obtained by G. Pizarro using Kodak 4415 emulsion and a GG 385 filter for the Liller–Brandt program.

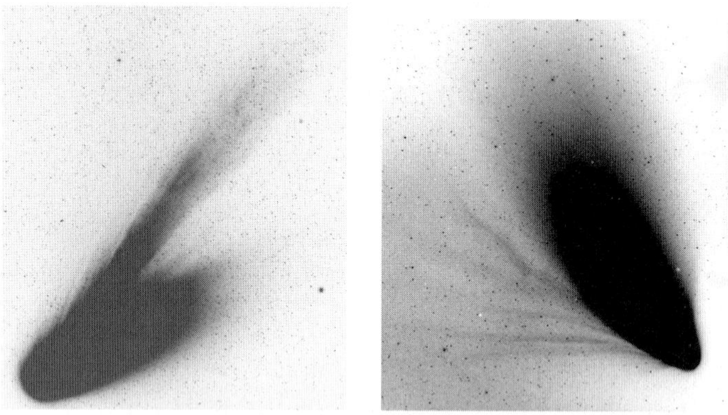

a. March 2.15 b. May 2.83

Figure 6. A comparison of the appearance of comet Hale–Bopp in the polar and equatorial regions of the solar wind. Left. Comet Hale–Bopp in the polar region, ecliptic latitude +87° (solar latitude +84°) on March 2.15, 1997. The dust tail (b) is smooth and the plasma tail (a) is relatively undisturbed. Image taken using a Schmidt camera and Kodalith film. Right. Comet Hale–Bopp in the equatorial region, ecliptic latitude +5° (solar latitude +1°) on May 2.83, 1997. The dust tail (b) is smooth and the plasma tail (a) is disturbed. Image taken using a Schmidt camera and Tech pan 6415 film with a W02B filter. The contrast between the two images is dramatic. Both images obtained as part of Ulysses Comet Watch activities by E. Kolmhofer and H. Raab, near Linz, Austria.

near-nuclear region was obscured by the dust tail in wide-band observations, and (2) observing conditions in late April and early May 1997 were difficult because the comet was close to the sun in the sky. The comet's apparent solar elongation was 35° on April 23 and decreased to 30° on May 6. The elongation remained below 30° for the remainder of the time period of interest.

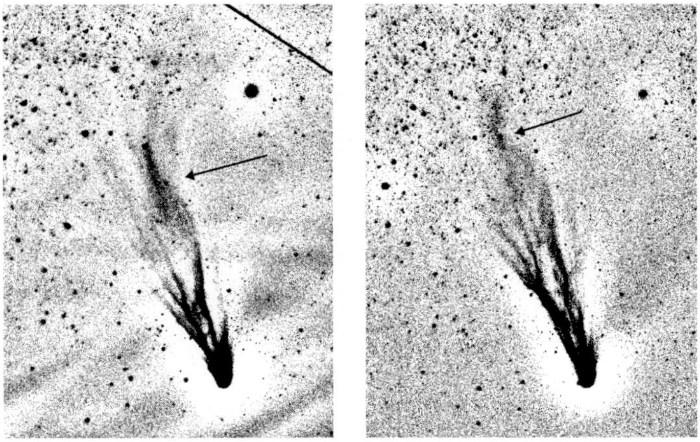

May 7, 1997 May 8, 1997

Figure 7. The monster DE in comet Hale–Bopp as imaged with a CO^+ filter. Arrows indicate the DE. The images were taken with the standard NASA-funded Hale–Bopp CO^+ filter. This filter minimized the contamination from continuum and emission. A simple, large kernel spatial filter was used to suppress the twilight sky brightness gradient. There was no additional processing; specifically, a dust image was not subtracted. The field of view for these images is approximately 9.7° x 12.9°. (S. M. Larson, Lunar and Planetary Laboratory, University of Arizona.)

The DE may be visible on an image obtained on May 4.88 through an H_2O^+ filter (Figure 8). The plasma tail protruding from the dust tail on this image does not resemble the plasma tail in early May as seen in Figure 4 and 6. The May 4 image in Figure 4 was not taken with a filter, but the plasma tail seems quite disturbed. The DE may be visible as the right-hand part of the plasma tail. The thick plasma feature in Figures 4 and 8 resembles the DE as shown in Figure 7. While this conclusion is not definitive, the feature shown in Figures 4 and 8 is probably the DE.

On May 7, 1997, the DE was almost one-half AU ($\approx$ 70 million km) from the nucleus and was approximately 30 million km long. On May 8, the DE was approximately 115 million km from the nucleus. The difference in positions gives a radial speed for the DE of about 520 km s^{-1}. This was approximately the solar-wind speed at the comet. Specifically, the solar wind corresponding to May 7 and May 8 at the comet was seen at WIND spacecraft on May 16 and 17. The measured speed was approximately 475 km s^{-1}. We adopt 500 km s^{-1}as the solar-wind speed in the discussion below. These figures assume that the DE is located along the prolonged radius vector (PRV). This should be a good assumption near the head, but may not be as valid far from the head.

Because the DE was observed only at unusually large distances from the head, the usual technique of assuming a constant acceleration (Niedner, 1981; Brandt et al., 1999, Snow et al., 2002: Comet Hyakutake (C/1996 B2): Spectacular Dis-

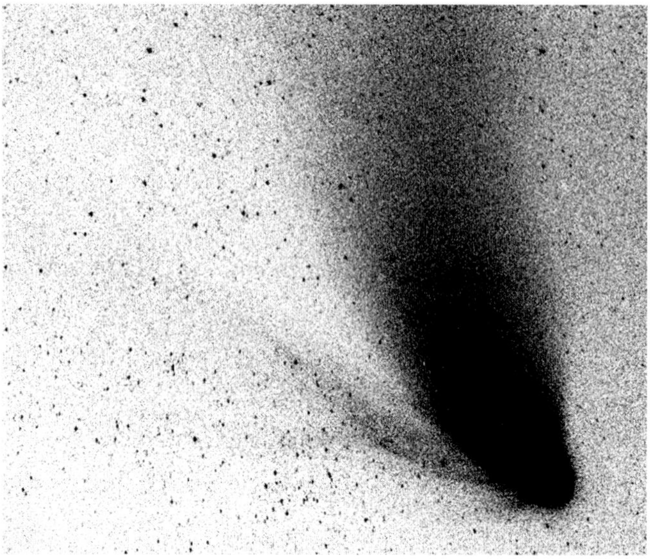

Figure 8. Comet Hale–Bopp as imaged on May 4.88, 1997 through an H_2O^+ filter. See text for discussion. (Geraint Jones, Imperial College, London.)

connection Event and the Latitudinal Structure of the Solar Wind, in preparation) is not applicable. A new kinematical representation is needed. The observation that the DE was moving at approximately the solar-wind speed prompts us adopt an approach that assumes that the speed of the DE asymptotically approaches the solar-wind speed, W_r, or simply that the acceleration is proportional to the difference between the speeds of the solar wind, W, and the DE, V. Thus, the acceleration is

$$a = \frac{dV}{dt} = k(W - V) \tag{1}$$

Integrating Equation (1) yields the speed:

$$V(t) = W - (W - V_0)e^{-kt} \tag{2}$$

where

$$k = \frac{a_0}{W - V_0} \tag{3}$$

Here a_0 is the initial acceleration and V_0 is the initial velocity. Integrating Equation (3) gives the downstream distance:

$$x(t) = Wt + \frac{(W - V_0)^2}{a_0}(e^{-kt} - 1) \tag{4}$$

Equation (4) has been used to obtain a rough fit to the observations a shown in Figure 9a, and Equation (2) has been used to calculate the recession velocity as a function of time as shown in Figure 9b. The adopted parameters are $V_0 = 100$ km s^{-1}, $W = 500$ km s^{-1}, and $a_0 = 2 \times 10^{-3}$ km s^{-2}. Because the fit is only to two points, it must be regarded as only approximate.

Nevertheless, the fit can be used to estimate the time of disconnection at approximately May 4.0, 1997. If the adopted solar-wind speed of 500 km s^{-1} is used as the speed for the DE since its origin, the latest it could have possibly occurred is approximately May 5.5, 1997. This fact and the uncertainty in the kinematics leads us to adopt May 4.0 $\pm$ 1.0 as the time of disconnection. This date is also consistent with the images in Figures 4 and 8.

The adopted parameters compare with typical values determined by Niedner (1981) in his study of 72 DEs. The average initial velocity $\langle V_0 \rangle$ in Niedner's sample was 44 km s^{-1}. The comet Hale–Bopp value of 100 km s^{-1} is higher than the mean, but Niedner (1981; Figure 3) found several values greater than 100 km s^{-1}. The value of $\langle a_0 \rangle$ in Niedner's sample was 0.5×10^{-3} km s^{-2}. This is a factor of 4 less than the comet Hale–Bopp value. The differences are probably not profound, particularly when the vastly differing circumstances are considered. Niedner's sample contained mostly observations $\leq$ 10 million km from the nucleus whereas the comet Hale–Bopp observations were taken at distances $\geq$ 70 million km.

The Carrington-Niedner diagram for the time around May 4.0 is shown in Figure 10; the comet (solid straight line with diamonds), spacecraft at earth (solid straight line with asterisks), and the heliospheric current sheet (solid and dashed curves) are plotted as a function of location of the Parker spiral on the source surface in Carrington longitude (assuming a radial solar-wind speed of 450 km s^{-1}) versus solar latitude. On either computation of the HCS (see below), comet Hale–Bopp was quite close to it around May 4, 1997, and it was well within the equatorial solar-wind region.

Spacecraft measurements can verify the location of the HCS. It was measured by the WIND spacecraft on about May 11.25, 1997. This is close to the time expected (Figure 10) on the basis of the "classical" calculation (solid curve) and also close but about 2° southward of the expected location for the preferred "radial" calculation (dashed curve). If the radial curve is adopted, it perhaps should be about 2° southward. The HCS curves used here are from the Wilcox Solar Observatory: *http://quake.stanford.edu/~wso/wso.html.*

None of this discussion changes the close association of this DE with the HCS. But, it does bear on the question of HCS crossings in the April 24 to May 1, 1997 time period. If there were crossings, DEs should have been produced on the physical picture adopted in this paper. This question cannot be answered definitively. If the dashed HCS curve is moved southward by 2°, the comet could have missed the HCS. If the solid HCS curve is valid, the comet should have crossed the HCS. No additional DEs were recorded. This could be due to poor observing conditions (see above) and the obscuration of the near-nuclear part of the plasma tail by the bright

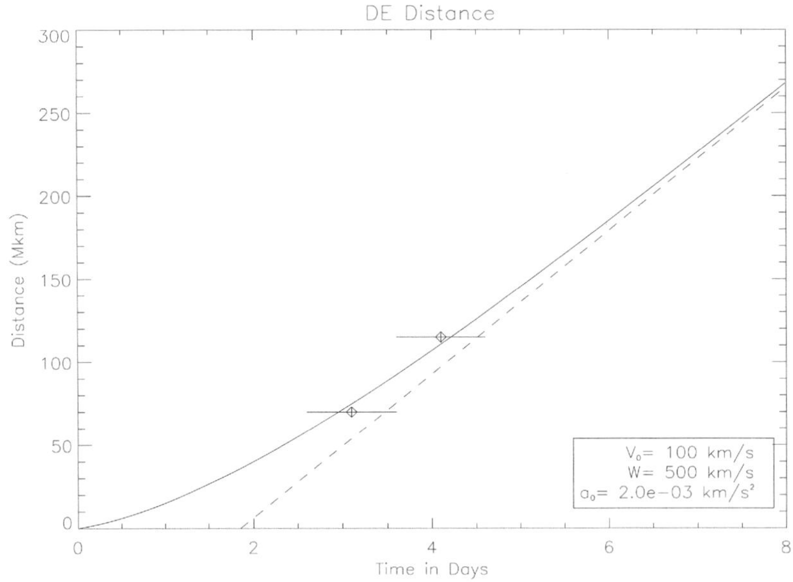

a. Distance of Disconnection Event

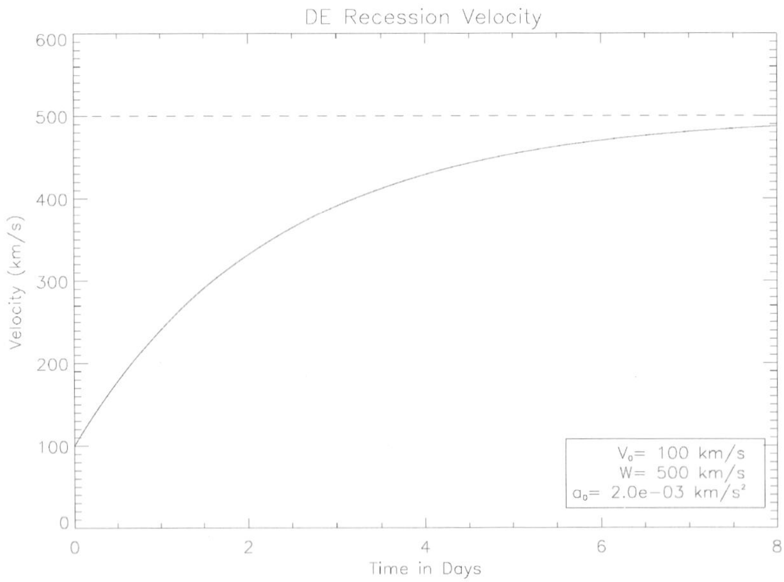

b. Velocity of DE

Figure 9. Kinematics of the DE in comet Hale–Bopp based on the observations and equations described in the text. Above. Distance of the DE from the head vs. time. The dashed line gives the result for a constant velocity of 500 km s^{-1}. Below. Velocity of the DE vs. time.

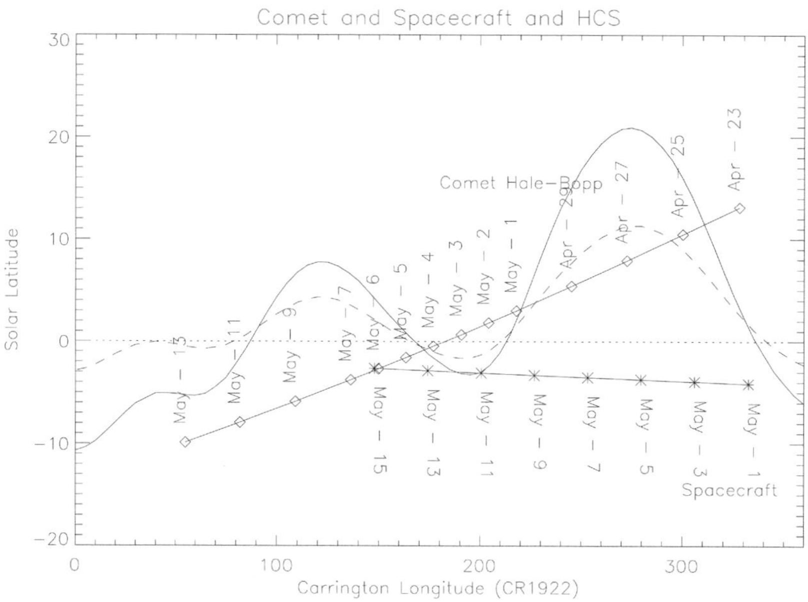

Figure 10. Locations of the comet, spacecraft (Earth), and the heliospheric current sheet in solar latitude and in solar longitude (Carrington) of the appropriate Parker spiral on the coronal source surface. See text for discussion.

dust tail. Unfortunately, insufficient information is available to further address the issue raised by possible, additional HCS crossings.

However, the comet probably came close to the HCS on or near April 27. Images taken with a CO^+ filter by C. Hergenrother and S. M. Larson (Lunar and Planetary Laboratory, University of Arizona) showed an increased level of plasma disturbance compared with images taken on April 28 and 29. The disturbances could result from structure associated with the HCS (e.g., Brandt and Snow, 2000).

The comet and the WIND spacecraft were separated by less than 3° *in latitude* on May 4.0. Thus, the measured conditions at WIND approximately 9 days later (see Figure 10) could be indicative of solar-wind radial speed and density at the comet. Prior to May 6.0 at the comet, the speed is close to 300 km s^{-1} and the density is generally below 20 cm^{-3}. Close to May 6.0 at the comet , the radial speed rapidly increases to 400 km s^{-1} and is close to 500 km s^{-1} by May 7.0. The density increases to about 35 cm^{-3} for a short time after May 6.0. These increases occurred about 2 days too late to have been responsible for the DE. Thus, the suggestion by Wegmann (1995, 2000) that solar-wind disturbances (rather than crossings of the HCS) produce DEs is not supported by the evidence for the major DE in comet Hale–Bopp.

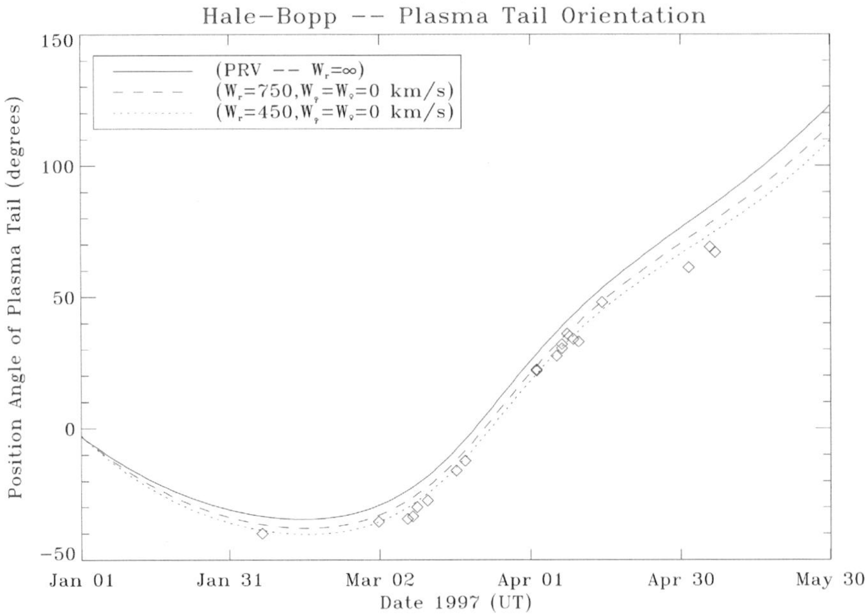

Figure 11. Measured position angles for the plasma tail and calculated values for assumed solar-wind speeds as a function of date. See text for discussion.

2.3. PLASMA TAIL ORIENTATION

The orientation of the plasma tail is determined by the solar-wind velocity (Biermann, 1951; Brandt et al., 1972) via dynamical aberration. The direction of the plasma tail is given by **W** − **V**, where **W** is the solar-wind velocity and **V** is the comet's velocity. For an assumed **W**, the expected position angle, Θ, can be calculated. For a large value of W_r, Θ should lie close to the angle for the prolonged radius vector (PRV). For lower values of W_r, Θ moves away from the PRV (see Brandt et al. 1997).

Figure 11 shows the calculated curves and measured position angles for comet Hale–Bopp as a function of date. For the dates after April 2, the data follow the trend expected for the paradigm with higher speeds (and low scatter) in the polar region and lower speeds (and higher scatter) in the equatorial region. The measured position angles are consistent with the polar-equatorial transition date of April 26, 1997.

However, the data for the highest latitudes (the ecliptic latitude was 89.3° on March 4) show an anomalous orientation, one characteristic of the equatorial region. This phenomenon has occurred before for one point in comet Perrine (1898b) when the comet was at ecliptic latitude of approximately 80° (Belton and Brandt, 1966; Brandt and Snow, 2000). There are several possible explanations. (1) If non-radial motions were present in the solar wind, these could produce departures from the expected position angles for zero non-radial speeds, ($W_\phi = W_\theta = 0$).

While such flows are expected close to the sun (Suess and Smith, 1996; Smith et al., 2001), large values are not expected beyond about 10 $R_{\odot}$; the systematic values needed ($\approx$ 25 km s^{-1} or more) are not believed to be a general property of the solar wind at high latitudes (B. Goldstein 2002 and E. Smith 2002, private communications). First, well away from the sun, non-radial motions decrease as $1/r$. Hence, they must be very small by the time the solar wind reaches comet Hale–Bopp or the *Ulysses* spacecraft. Second, the systematic non-radial velocities that appear in the *Ulysses* data files (*http://nssdc.gsfc.nasa.gov/space*), for example, monthly averages of about 20 km s^{-1} in July/August 1995 (when *Ulysses* was at its highest latitudes) are not considered reliable. Error sources such as misalignments with respect to the sun's rotation axis could produce biases. The relative deflections are regarded as reliable. (2) A temporal event could be involved. The departures from expected values occur over a considerable time period (again, several weeks) and we find no evidence to support this idea. (3) This is a very large comet with a wide plasma tail. Could measurement error produce this effect? The wide plasma tail is present at all latitudes and is quite wide at lower latitudes. This explanation is unlikely. (4) The orientation of the solar-wind magnetic field might produce a deflection of the plasma tail. Most models use a magnetic field oriented at 45° to the flow direction (Gombosi et al., 1996, 1997) or at 90° to the flow direction (Yi et al., 1996). Perhaps a magnetic field slightly inclined to the flow direction, the case at high latitudes, could produce a deflection. This explanation can probably be ruled out. M. Combi (2002, private communication) reports that the calculations by the University of Michigan modeling group have never shown evidence for a plasma tail departing from the relative direction of the solar-wind flow when the mass loading is spherically symmetric, even for the case of a small angle between the magnetic field and the flow direction. (5) However, models run for comet Borrelly (Hansen et al., 2001; M. Combi 2002, private communication) show that an asymmetrical source can shift the plasma tail well away from the expected direction. If there were a significant asymmetry in the gas production of comet Hale–Bopp at high latitudes, this could provide a possible explanation for the anomalous plasma tail orientations. Verification would require a gas distribution for the comet at high latitudes derived from observations and model runs using this distribution. Again, the asymmetrical source would have to persist for weeks.

While a possible explanation for the high latitude behavior is available, it needs to be verified. For now, this phenomenon is unexplained.

3. Summary and Conclusion

Determining the detailed properties of comets in the equatorial and polar solar-wind regions and calibrating the plasma tails of comets as solar-wind probes requires continual monitoring of the large-scale structure for many comets, additional studies to establish the solar-wind conditions responsible for plasma phenomena in comets, advances in modeling the comet/solar-wind interaction, and in situ measurements of fields and plasmas at comets. Advances are being made in all areas. Recently, comets Hyakutake (C/1996 B2) and Hale–Bopp (C/1995 O1) were bright comets that probed both the equatorial and polar solar-wind regions and were well studied. The opportunities continue with C/2000 WM1 (which may assist is delineating the situation near solar maximum) and further opportunities will come in the future. Correlation studies (Delva et al., 1991; Yi et al., 1994; Voelzke and Matsuura, 1998; Brandt et al., 1999, Voelzke: Disconnection events in comets, these proceedings) continue and are helping constrain the theoretical efforts. Model codes (Gombosi et al., 1996, 1997; Yi et al., 1996; Bagdonat and Motschmann, 2002) are providing advances in details and physical understanding. Finally, *in situ* measurements by the Deep Space 1 spacecraft at Comet 19P/Borrelly (Nordholt et al., 2002; Young et al., 2001) provide critical space truth.

Comet Hale–Bopp, with its orbital inclination of 89.4°, spent months at high latitudes in the polar region. Figures 3 and 4 document the comet's large-scale structure and its changes. The comet crossed from the polar solar-wind region to the equatorial solar-wind region close to April 26, 1997. A monster DE was recorded on May 7 and May 8, 1997. The time of disconnection was estimated as May 4.0 and the DE was closely associated with a crossing of the HCS. Other DEs were possible in principle (in terms of HCS crossings), but they were not required and no other DEs were observed. Properties of the solar wind measured by WIND provide no support for the view that solar-wind disturbances could be responsible for DEs.

The properties of comet Hale–Bopp followed the Ulysses Comet Watch paradigm in terms of appearance, location of DEs, and plasma tail orientation with one significant exception: The orientation of the plasma tail at high latitudes. Several possible explanations are discussed, but the only plausible (but as yet unproven) one is asymmetrical gas distribution in the coma. This view receives some impetus from measurements of comet Borrelly (Nordholt et al., 2002) and the models used to explain them (Hansen et al., 2001).

Thus, comet Hale–Bopp, the most widely studied comet since comet Halley, has contributed greatly to our understanding of comets. It is certainly the best documented and studied comet in the polar solar-wind region in history. It also re-emphasizes the individuality of comets.

Acknowledgements

We thank the Ulysses Comet Watch observers for their expert observational effort and generous cooperation. We thank Drs. Edward Smith and Bruce Goldstein, Jet Propulsion Laboratory, for conversations and communications concerning possible non-radial solar-wind motions at polar latitudes. We thank Dr. Michael Combi, University of Michigan for conversations and communications concerning model calculations and possible mechanisms for deflecting plasma tails. We thank Mr. Jon Giorgini, Jet Propulsion Laboratory, for assistance with the Horizons Ephemeris Generator. We thank the data provider, K. Oglivie, NASA-Goddard Space Flight Center, and CDA/Web for the WIND plasma data. We thank the data provider, R. Lepping, NASA-Goddard Space Flight Center, and CDA/Web for the WIND magnetic field data. We thank the data provider P.H. Scherrer at the Wilcox Solar Observatory. We thank Drs. Tony Farnham, University of Texas at Austin; Geraint Jones, Imperial College, London; and Dave Schleicher, Lowell Observatory for providing their images of Comet Hale–Bopp for our examination.

This research was partially supported by JPL Grant 959349 to John C. Brandt as an Interdisciplinary Scientist on the Ulysses project.

Appendix A

TABLE A.I

Images of comet Hale–Bopp used in this investigation

Date	Observer	Solar latitude	Ecliptic latitude	R (AU)	Δ (AU)
Sept. 27.10, 1996	G. & O. Pizzaro	...	18°	2.9	3.0
Nov. 1.00, 1996	G. & O. Pizzaro	...	25°	2.5	3.1
Nov. 12.21, 1996	O. R. Hainaut, K. J. Meech, J. Bauer, & C.C. Petersen	...	27°	2.4	3.0
Feb. 19.54, 1997	B. Yen	78°	75°	1.2	1.7
Mar. 1.15, 1997	P. Candy	84°	86°	1.1	1.5
Mar. 2.15, 1997	E. Kolmhofer & H. Raab	84°	87°	1.1	1.5
Mar. 10.14, 1997	A. Alksnis	78°	83°	1.0	1.4
Mar. 16.14, 1997	A. Dimai & D. Ghirardo	69°	74°	1.0	1.3
Mar. 30.79, 1997	A. Alksnis	48°	53°	0.9	1.3
Apr. 2.80, 1997	H. Mikuz	43°	48°	0.9	1.4
Apr. 7.80, 1997	H. Mikuz	35°	39°	0.9	1.4
Apr. 15.83, 1997	H. Mikuz	23°	28°	0.9	1.5
Apr. 24.81, 1997	H. Mikuz	11°	15°	1.0	1.7
Apr. 24.82, 1997	H. Mikuz	11°	15°	1.0	1.7
May 1.14, 1997	D. Hanon	3°	6°	1.1	1.8
May 2.93, 1997	E. Kolmhofer & H. Raab	1°	4°	1.1	1.8
May 4.16, 1997	L. Tan	−1°	2°	1.1	1.8
May 4.88, 1997	G. Jones	−1°	2°	1.1	1.8
May 7.13, 1997	S. Larson	−4°	0°	1.1	1.9
May 8.15, 1997	S. Larson	−5°	−1°	1.1	1.9
Jun. 6.96, 1997	G. Pizzaro	...	−26°	1.5	2.4

References

Bagdonat, T. and Motschmann, U.: 2002, 'From a Weak to a Strong Comet – 3D Global Hybrid Simulation Studies', these proceedings.

Belton, M. J. S. and Brandt, J. C.: 1966, 'Interplanetary Gas. XII. A Catalog of Comet-Tail Orientations', *ApJS* **13**, 125–332.

Biermann, L.: 1951, 'Kometenschweife und Solar Korpuskularstrahlung', *Z. Astrophys.* **29**, 279–286.

Brandt, J. C. and Snow, M.: 2000, 'Heliospheric Latitude Variations of Properties of Cometary Plasma Tails: A Test of the Ulyesses Comet Watch Paradigm', *Icarus* **148**, 52–64.

Brandt, J. C., Caputo, F. M., Hoeksema, J. T., Niedner, Jr. M. B., Yi, Y., and Snow, M.: 1999, 'Disconnection Events (DEs) in Halley's Comet 1985–1986: The Correlation with Crossings of the Heliospheric Current Sheet (HCS)', *Icarus* **137**, 69–83.

Brandt, J. C., Roosen, R. G., and Harrington, R. S.: 1972, 'Interplanetary Gas. XVII. An Astrometric Determination of Solar-Wind Velocities from Orientations of Ionic Comet Tails', *ApJ* **177**, 277–284.

Brandt, J. C., Yi, Y., Petersen, C. C., and Snow, M.: 1997, 'Comet de Vico (122P) and Latitude Variations of Plasma Phenomena', *Planet. Space Sci.* **45**, 813–819.

Cremonese, G. and Fulle, M.: 1989, 'Photometrical Analysis of the Neck-Line Structure of Comet Halley', *Icarus* **80**, 267–279.

Delva, M., Schwingenshuh, K., Niedner, Jr. M. B., and Gringauz, K. I.: 1991, 'Comet Halley Remote Plasma Tail Observations and *In Situ* Solar Wind Properties: VEGA-1/2 IMF Plasma Observations from 1 December 1995 to 1 May 1996', *Planet. Space Sci.* **39**, 697–708.

Fischer, C. T., Haerendel, G., and Bogdanov, A. T.: 1997, 'Dynamics of Structures in the Plasma Tail of Comet Hale–Bopp', *Earth Moon Planets* **77**, 279.

Fulle, M.: 1987, 'A Possible Neck-Line Structure in the Dust Tail of Comet Halley', *Astron. Astrophys.* **181**, L13–L14.

Gombosi, T. I., DeZeeuw, D. L., Häberli, R. M., and Powell, K. G.: 1996, 'Three-Dimensional Multiscale MHD Model of Cometary Plasma Environments', *J. Geophys. Res.* **101**, 15,233–15,253.

Gombosi, T. I., Hansen, K. C., DeZeeuw, D. L., Combi, M. R., and Powell, K. G.: 1997, 'MHD Simulation of Comets: The Plasma Environment of Comet Hale–Bopp', *Earth Moon Planets* **79**, 179–207.

Hansen, K. C., Combi, M. R., Crary, F. J., De Zeeuw, D. L., Gombosi, T. I., and Young, D. T.: 2001, 'Global MHD Simulations of Comet Borrelly's Plasma Environment: Effects of a Strong Neutral Jet', *BAAS* **33**, 1075.

Jockers, K., Credner, T., Karpov, N., and Sergeev, A.: 1996, 'Comet C/1995 O1 (Hale–Bopp)', *IAUC* 6468.

Jones, G.: 1997, 'The Interaction of C/1995 O1 Hale–Bopp with the Solar Wind as Recorded in CoCam Images: A Progress Report', *Earth Moon Planets* **77**, 281.

Kimura, H. and Liu, C. P.: 1977, 'On the Structure of Cometary Dust Tails', *Chinese Astr.* **1**, 235–264.

Kinoshita, D., Watanabe, J., Fukushima, H., Sekiguchi, T., Yamamoto, N., and Abe, S.: 1997, 'Wide Field Imaging of Ion Tail of Comet C/Hale–Bopp', *Earth Moon Planets* **77**, 265–269.

McComas, D. J., Bame, S. J., Barraclough, B. L., Feldman, W. C., Funsten, H. O., Gosling, J. T., Riley, P., Skoug, R., Balogh, A., Forsyth, R., Goldstein, B. E., and Neugebauer, M.: 1998, 'Ulysses Return to the Slow Solar Wind', *Geophys. Res. Lett.* **25**, 1–4.

Niedner, Jr. M. B.: 1981, 'Interplanetary Gas. XXVII. A Catalogue of Disconnection Events in Cometary Plasma Tails', *ApJS* **46**, 141–157.

Nordholt, J. E., Reisenfeld, D. B., Wiens, R. C., Crary, F., Delapp, D. M., Elphic, R. C., Funsten, H. O., Gary, S. P., Hanley, J. J., Lawrence, D. J., McComas, D. J., Shappirio, M., Steinberg, J. T., Wang, J., and Young, D. T.: 2002, 'Deep Space 1 Encounter with Comet 19P/Borrelly: Ion Composition Measurements by the PEPE Mass Spectrometer', *Geophys. Res. Lett.*, in press.

Pansecchi, L., Fulle, M., and Sedmak, G.: 1987, 'The Nature of Two Anomalous Structures Observed in the Dust Tail of Comet Bennett 1970 III: A Possible Neck-Line Structure', *Astron. Astrophys.* **176**, 358–366.

Phillips, J. L., Bame, S. J., Goldstein, F. W. C., B. E., Gosling, J. T., Hammond, C. M., McComas, D. J., Neugebauer, M., Scime, E. E., and Suess, S. T.: 1995, 'Ulysses Solar Wind Plasma Observations at High Southerly Latitudes', *Science* **268**, 1030–1033.

Rauer, H.: 1997, 'Ion Composition and Solar Wind Interaction: Observations of Comet C/1995 O1 (Hale–Bopp)', *Earth Moon Planets* **79**, 161–178.

Smith, E. J., Balogh, A., Forsyth, R. J., and McComas, D. J.: 2001, 'Ulysses in the South Polar Cap at Solar Maximum: Heliospheric Magnetic Field', *Geophys. Res. Lett.* **28**, 4159–4162.

Suess, S. T. and Smith, E. J.: 1996, 'Latitudinal Dependence of the Radial IMF Component: Coronal Imprint', *Geophys. Res. Lett.* **23**, 3267–3270.

Voelzke, M. R.: 2002, 'Disconnection Events in Comets', these proceedings.

Voelzke, M. R. and Matsuura, O. T.: 1998, 'Morphological Analysis of the Plasma Structures of Comet P/Halley', *Planet. Space Sci.* **46**, 835–841.

Wegmann, R.: 1995, 'MHD Model Calculations for the Effects of Interplanetary Shocks on the Plasma Tail of a Comet', *A & A* **294**, 601–614.

Wegmann, R.: 2000, 'The Effect of Some Solar Wind Disturbances on the Plasma Tail of a Comet: Models and Observations', *A & A* **358**, 759–775.

Yi, Y., Brandt, J. C., Randall, C. E., and Snow, M.: 1994, 'The Disconnection Event of Comet Halley on 1986 March 16.0', *AJ* **107**, 1591–1596.

Yi, Y., Walker, R. J., Ogino, T., and Brandt, J. C.: 1996, 'Global Magnetohydrodynamic Simulation of a Comet Crossing the Heliospheric Current Sheet', *J. Geophys. Res.* **101**, 27,585–27,601.

Young, D. T., Crary, F. J., Nordholt, J. E., Hanley, J. J., Burch, J. L., McComas, D. J., Goldstein, R., Boice, D., Wiens, R. C., Lawrence, D. R., Eviatar, A., Sauer, K., and Meier, R.: 2001, 'Solar Wind Interaction with the Coma of Comet 19P/Borrelly', *BAAS* **33**, 1087.

4. The relative change of the solar distance $r(t)$ is negligible during one nucleus rotation period P_{rot}, i.e.:

$$\frac{r(t + P_{rot}) - r(t)}{r(t)} \ll 1. \tag{2}$$

5. Thermal conductivity within the nucleus is neglected (Desvoivres et al., 2000). Our calculations thus concern the maximum sublimation flux, since the whole influx of energy is used for sublimation and none for heating the sub-surface layers of the nucleus. Comparing with the results of Möhlmann (2002) who calculated several mass and energy fluxes involved in the physical processes in the surface layer of the cometary nucleus, we estimate that our assumption does not introduce errors larger than about 30%.

1.2. Model Equations

The local sublimation flux $Z = Z[T(z, \varphi)]$ depends only on the local, time-dependent temperature T. Here φ is the cometographic latitude. The zenith distance z is given by the formula

$$\cos z = \cos\theta \cos\varphi \cos\delta + \sin\varphi \sin\delta, \tag{3}$$

where θ is the hour angle and δ is the declination of the Sun.

Z is given by the temperature T of the nucleus surface and by the temperature-dependent water-vapor saturation pressure $p(T)$:

$$Z(T) = p(T)\sqrt{\frac{m_{H_2O}}{2\pi kT}} \quad [kg\ m^{-2}\ s^{-1}]. \tag{4}$$

Here: $m_{H_2O} = 2.988 \cdot 10^{-26}$ kg denote the molecular mass of H_2O, and $k = 1,38 \cdot 10^{-23}$ J K^{-1} is the Boltzmann constant. Following Fanale and Salvail (1984) the water-vapor saturation pressure is

$$p(T[in K]) = 3.56 \cdot 10^{12} \exp\frac{-6141.667}{T} \quad [Pa]. \tag{5}$$

The energy balance equation for the nucleus surface is

$$S\frac{1}{r^2}(1 - A)\cos z = \epsilon\sigma T^4 + H(T)Z(T) \quad [J\ m^{-2}\ s^{-1}]. \tag{6}$$

Here, A is the albedo, ϵ is the emissivity, $S = 1360$ J m^{-2} s^{-1} is the solar constant, and $\sigma = 5.67 \cdot 10^{-8}$ W m^{-2} K^{-4} is the Stefan–Boltzmann constant. The latent heat of sublimation H depends linearly on the surface temperature T (Delsemme and Miller, 1971):

$$H(T) = 2.888 \cdot 10^6 - 1116T \quad [J\ kg^{-1}]. \tag{7}$$

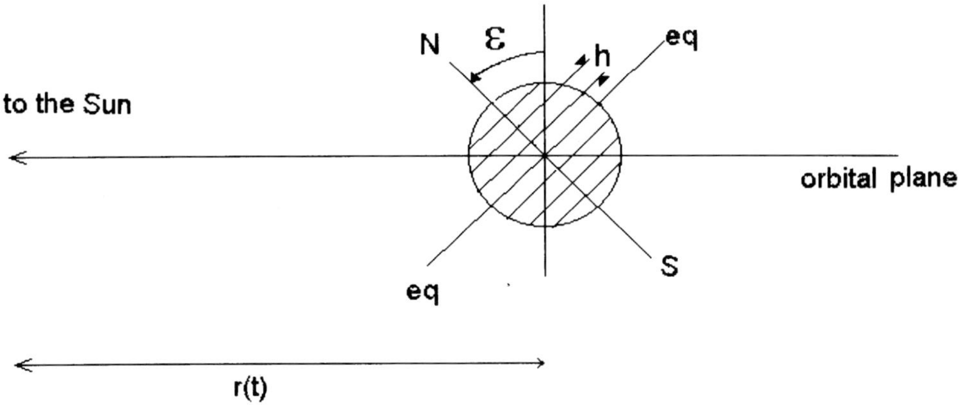

Figure 1. Spin axis orientation and division of the nucleus surface into the latitudinal strips of equal surface area; thus, the "thickness" of all strips is the same.

The sublimation flux Z is calculated versus time t elapsed from perihelion passage. Z is a function of the zenith distance z. The declination δ in Equation (3) depends on the inclination of the rotation axis ε to the normal of the orbital plane (see Figure 1). According to formula (4) the calculation of the local temperature T is crucial for estimating the sublimation flux Z. In order to calculate T, we divided the nucleus surface in N parallel strips of equal surface area, corresponding to different intervals of $\triangle\varphi$. The surface area of such a strip is given by

$$B = 2\pi Rh. \qquad (8)$$

Here, R is the radius of the sphere and h is the axial "thickness" of the strip. Hence the strips of equal surface area should have equal "thickness" h (see Figure 1). The solar energy flux is assumed to be φ-independent within each strip. We take its value at mid-h.

The value of the sublimation flux Z can be found numerically by solving the system of Equations (4)–(7), for Z, p, T, and H. This system is supplemented by Equation (3) and by the equations of motion of the nucleus and of its rotation. They contain the angles z, δ, and θ as well as the solar distance r. All of them are time-dependent. To obtain the time-dependent mass loss due to sublimation, $Z_t(\varphi)$ kg m^{-2}, we have to perform the integration of Z over time t from the perihelion passage ($t = 0$, $r = q$) to the position $(t, r(t))$ in the comet orbit: the mass loss $Z_P(\varphi)$ over the whole orbital period P, we integrate the sublimation flux $Z(\varphi)$ until $t = P/2$ and multiply the result by 2 (since in our model the sublimation is symmetrical with respect to perihelion passage).

The decrease of the nucleus radius, $\triangle R(\varphi)$, follows from the sublimation loss of mass, $Z_P(\varphi)$ through formula:

$$\triangle R(\varphi) = \frac{Z_P(\varphi)}{\rho} \cdot \frac{1}{1 - C} \quad [m]. \qquad (9)$$

TABLE I
Model parameters of comets C/1995 O1 Hale–Bopp and 46P/Wirtanen

Parameter	Symbol	Value	Reference
C/1995 O1 Hale–Bopp			
Eccentricity	e	0.995124	Marsden and Williams, 1997
Semi-major axis	a	187.4820 AU	Marsden and Williams, 1997
Orbital period	P	2500 yrs	Assumed
Rotational period	P_{rot}	10 h	Assumed
Radius	R	20 km	Weaver and Lamy, 1999
Albedo	A	0.03	Assumed from Prialnik, 1997
Emissivity	ϵ	1	Kührt and Keller, 1994
Density	ρ	500 kg m^{-3}	Kührt and Keller, 1994
Composition factor	C	0.5	Assumed
Inclination of spin axis	ε	4 different values assumed	
46P/Wirtanen			
Eccentricity	e	0.656770	Marsden and Williams, 1997
Semi-major axis	a	3.0993 AU	Marsden and Williams, 1997
Orbital period	P	5.46 yrs	Marsden and Williams, 1997
Rotational period	P_{rot}	6 h	Assumed
Radius	R	600 m	Lamy et al., 1998
A, ϵ, ρ, C and ε as for 1995 O1 Hale–Bopp			

ρ denotes the uniform density of the nucleus. For the nucleus, made of water ice and dust, the density is given by:

$$(1 - \psi)/\rho = (1 - C)/\rho_{ice} + C/\rho_{dust}. \qquad (10)$$

Here, ψ is the porosity (the fraction of the voids in the unit volume), ρ_{ice} is the bulk density of non-porous ice, ρ_{dust} is the bulk density of the dust grains, and C is defined by Equation (1). The denominator (1 - C) appearing in Equation (9) suggests that $\triangle R(\varphi)$ goes to infinity when C approaches to 1. However, this formal interpretation of Equation (9) has no physical meaning. One can imagine that for cometary nuclei C is equal to 0.5 or so, see Table I. Therefore the volume ratio *ice* : *dust* is $\approx 3:1$. Assuming that the 'typical' porosity of the nucleus is $\psi \geq 0.5$, we get the volume ratio *voids* : *ice* : *dust* = 4 (or more): 3 : 1. The sublimating water ice drags away the dust grains and therefore the surface of the nucleus recedes. However, this phenomenon cannot be effective for C close to 1, since in that case the volume ratio *ice* : *dust* $\ll 1$ and the gas drag forces are not sufficient to blow away the dust.

For a given (calculate) $\triangle R(\varphi)$ the total moment of inertia I is:

$$I = I_0 \left[1 - \sum_n \left(\frac{\Delta I}{I_0} \right)_n \right] \quad [kg\ m^2] \text{ where } I_0 = \frac{2}{5} M R^2, \tag{11}$$

where n is the number of the surface strip and $(\frac{\Delta I}{I_0})_n$ corresponds to the fractional change of the dimensionless moment of inertia calculated for the n-th strip.

2. Applications of the Model

We have applied our model to two comets: C/1995 O1 Hale–Bopp and 46P/Wirtanen. Their orbital and physical data are summarized in Table I.

It is evident that the long-period comet Hale–Bopp sublimates efficiently only during a small fraction of its orbital period. On the other hand, the sublimation of the nucleus of 46P/Wirtanen may occur during a large part of its orbit. One may expect that the mass-loss of 46P/Wirtanen will influence considerably the moment of inertia of its nucleus, and therefore the rotation properties, on time scale of several orbital periods.

3. Results

In all particular modeling runs we assumed that the winter solstice on the northern hemisphere happened when the comet was passing perihelion. The inclination angle ε of the rotation axis is a parameter. The number of the strips $N = 40$. Figure 2 presents the partial mass-loss due to sublimation Z_t (in kg m^{-2}) versus solar distance r (in AU). When the rotation axis is inclined to the orbital plane, the sublimation pattern is evidently asymmetric. Curves are labeled by the cometographic latitude φ of the middle of the n-th strip. For clarity the results are shown only for selected strips, $n = \pm1, \pm2, \pm5, \pm8, \pm11, \pm14, \pm17$ and ±20. The strips on the southern hemisphere have negative numbers. Note that the numbers $|n|$ increase as we go from the equator to the pole. Therefore, $n = \pm1$ for the near-equatorial strip and $n = \pm20$ for the near-polar strip.

In Figure 3 the mass losses over a complete orbital revolution are presented. The figure compares the sublimation losses for the nuclei of both comets Hale–Bopp and Wirtanen. The values of the radius decrease $\triangle R(\varphi)$ allowed us to calculate the changes of the moment of inertia of the nucleus. We note that sublimation mass loss can change the moment of inertia of the nucleus of comet Wirtanen much more than that of comet Hale–Bopp.

Figure 4 illustrates the result for the fractional dimensionless moment of inertia over the whole nucleus. We believe that due to simplifying assumptions of our model the results give upper limits for the sublimation flux and therefore also upper

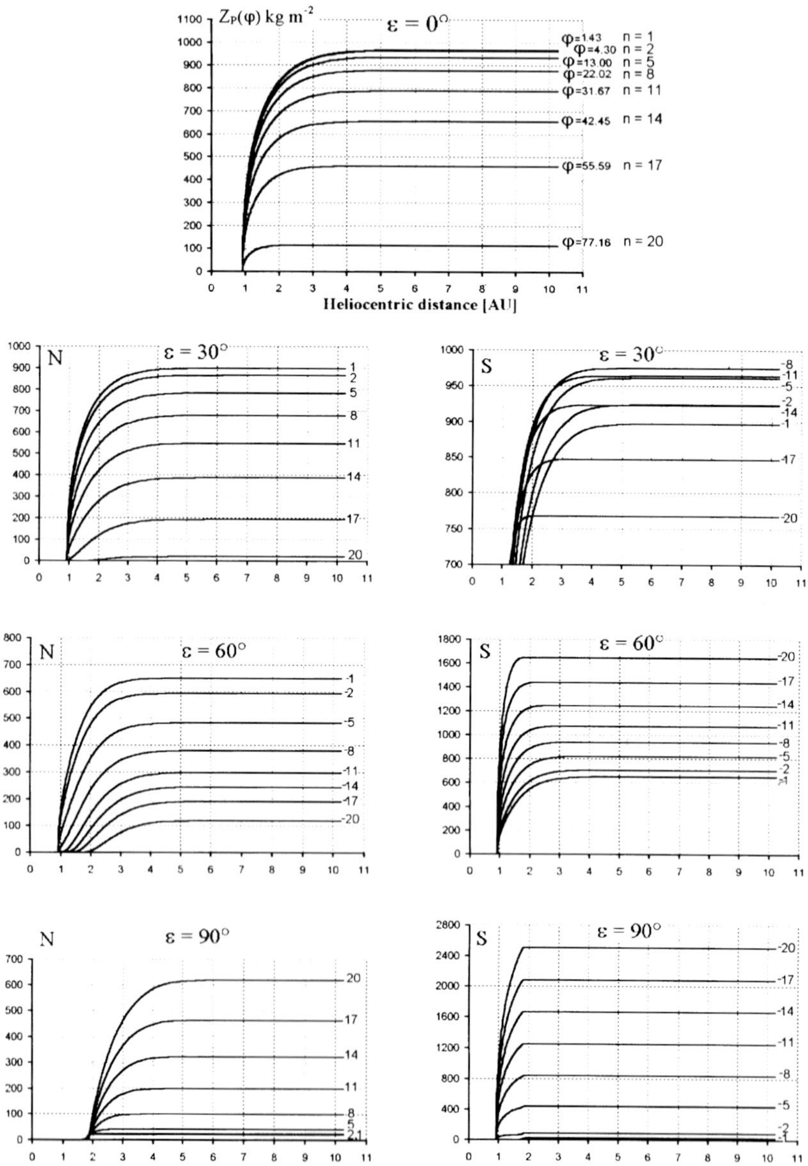

Figure 2. Hale–Bopp nucleus: The partial mass loss due to sublimation Z_t (in kg m^{-2}, vertical axis) versus solar distance r (in AU, horizontal axis). Cometographic latitude φ (i.e., the strip number n) is given as line label. Upper panel is for rotation axis of nucleus perpendicular to the orbital plane, $\varepsilon = 0°$. Lower panels are for $\varepsilon = 30°$, $60°$, and $90°$. The left column is for northern hemisphere, the right column is for southern hemisphere. Winter solstice on the northern hemisphere is assumed to be when the comet passes perihelion. When the rotation axis is inclined, the sublimation pattern is evidently asymmetric.

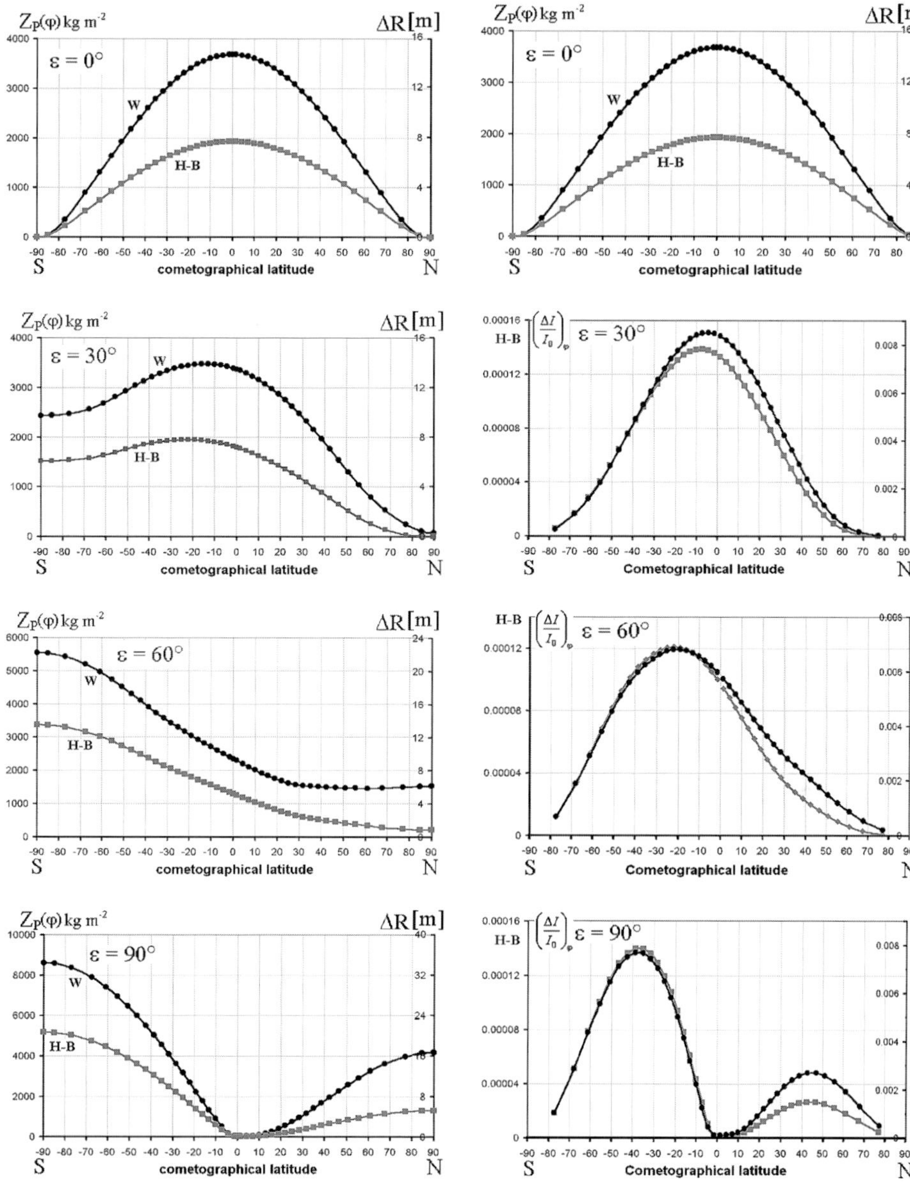

Figure 3. Modelling results for comets Hale–Bopp (line with squares) and Wirtanen (line with circles). The horizontal pairs correspond to the same inclination ε of the nucleus rotation axis. Left panels: Mass-loss due to sublimation Z_P (in kg m^{-2}, left-hand-side vertical axis) and decrease of radius $\triangle R$ (in meters, right-hand-side vertical axis) versus cometographic latitude. The symbols Z_P and $\triangle R$ correspond to the whole orbital period P. Right panels: The changes of the fractional dimensionless moment of inertia $(\frac{\triangle I}{I_0})_n$ of the n-th strip. The scale on the left-hand-side vertical axis is valid for Hale–Bopp, that on the right-hand-side for Wirtanen.

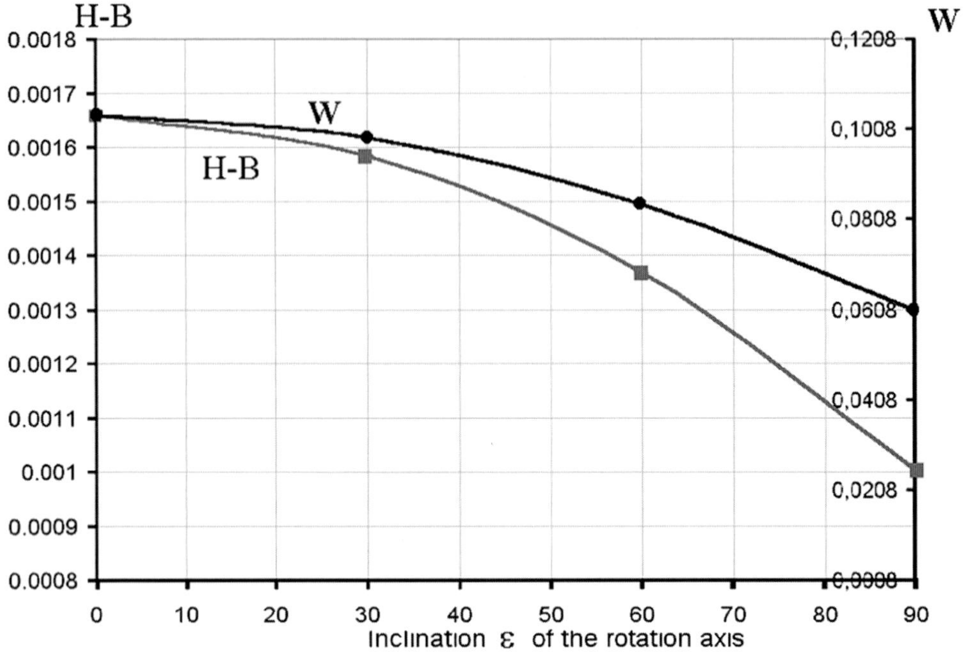

Figure 4. Sublimation induced changes of the total (dimensionless) moment of inertia $\sum(\Delta I/I_0)_n$ versus inclination ε of the rotation axis. Line with squares for Hale–Bopp (H–B, left-hand vertical scale), line with circles for Wirtanen (W, right-hand vertical scale).

limit for the changes of the moment of inertia. However, an important finding is that the sublimation of a small nucleus (e.g., Wirtanen) leads to significant changes of its moment of inertia. Therefore, the rotation parameters (orientation of the rotation axis and period) can be affected as well.

4. Conclusions

The integration of the sublimation rate $Z(\varphi)$ kg m^{-2} s^{-1} over the whole surface of a nucleus leads to the total production rate $Q(H_2O)$ kg s^{-1} or mol s^{-1} of water escaping from the nucleus. Our results of $Q(H_2O)$ for Hale–Bopp are in good agreement with the observations (see Table II). Thus, our simple model seems to be justified.

The calculations of $Z_P(\varphi)$ allow us to obtain the local, φ-dependent, value of the decrease of nucleus radius $\Delta R(\varphi)$. We assumed a nucleus density $\rho = 500$ kg m^{-3} and a nucleus composition with $C = 0.5$. From ΔR the moment of inertia can be calculated (see the left panels of Figure 3). Therefore, our results can be considered as a first step toward a more advanced analysis of the rotational behavior of cometary nuclei.

TABLE II

Sublimation rate $Q(H_2O)$ in molecules per second from the whole nucleus of Hale–Bopp

r	$Q(H_2O)$, observed		$Q(H_2O)$, this work
[AU]	mol s^{-1}	Reference	mol s^{-1}
4.8	1.5×10^{28}	Weaver et al., 1996	2.5×10^{28}
4.5	4.1×10^{28}	Weaver et al., 1996	5.2×10^{28}
2.88	4.7×10^{29}	Bockelée-Morvan et al., 1996	8.8×10^{29}
2.41	1.1×10^{30}	Bockelée-Morvan et al., 1996	1.7×10^{30}
1.37	2×10^{30}	Flammer et al., 1997	8.5×10^{30}
1.07	3.16×10^{30}	Flammer et al., 1997	1.5×10^{31}
0.914	2×10^{31}	Enzian et al., 1998; estimate	2.2×10^{31}

A comparison of the calculated sublimation pattern, $Z_P(\varphi)$ and $\triangle R(\varphi)$, for the nuclei of comets Hale–Bopp and Wirtanen is presented in Figure 3. However, we note that the assumption of uniform nucleus structure may be much more realistic for comet C/1995 O1 Hale–Bopp than for 46PWirtanen. The crust of the nucleus of comet Hale–Bopp most probably differs considerably from the crust of 46P/Wirtanen.

Acknowledgement

This work was supported by the Polish Committee of Scientific Research (through KBN grant 2.P03D.002.18).

References

Bockelée-Morvan, D. et al.: 1996, *IAUC* 6511.
Delsemme, A. H. and Miller, D. C.: 1971, *Planet. Space Sci.* **19**, 1229–1257.
Desvoivres, E. et al.: 2000, *Icarus* **144**, 172–181.
Enzian, A., Cabot, H., and Klinger, J.: 1998, *Planet. Space Sci.* **46**, 851–858.
Fanale, F. P. and Salvail, J. R.: 1984, *Icarus* **60**, 476–511.
Kűhrt, E. and Keller, H. U.: 1994, *Icarus* **109**, 121–132.
Flammer, K. R., Mendis, D. A., and Houpis H. L. F.: 1998, *Astroph. J.* **494**, 822–827.
Lamy, P. L., Toth, I., Jorda, L., Weaver, H. A., and A'Hearn, M.: 1998, *Astron. Astrophys.* **335**, L25–L29.
Marsden, B. G. and Williams, G. V.: 1997, *Catalog of Cometary Orbits*, edn. 12.
Möhlmann, D.: 2002, *Adv. Space Res.* **29**, 691–704.

Prialnik, D.: 1997, *Astrophys. J.* **478**, 107–110.
Weaver H. A. and Lamy, P. L.: 1999, *Earth Moon Planets* **79**, 17–33.
Weaver, H. A., Feldman, P. D., A'Hearn, M. F., Arpigny, C., Brandt, J. C., and Randall, C. E.: 1996, *IAUC* 6376.

WIDE FIELD IMAGING AND THE VELOCITY STRUCTURE IN THE COMA OF HALE–BOPP

WALTER M. HARRIS, JEFFREY P. MORGENTHALER, FRANK SCHERB and CHRISTOPHER ANDERSON
University of Wisconsin-Madison, 1150 University Ave. Madison, WI 53706, USA
E-mail: wharris@sal.wisc.edu

RONALD J. OLIVERSEN
Goddard Spaceflight Center, Code 681, Greenbelt, MD 20771, USA

(Received 18 March 2002; Accepted 3 July 2002)

Abstract. The comae of very active comets have a substantially more complex coma than their weaker cousins. The primary cause of this is photolytic heating and collisions that occur over an ever-larger volume of the coma as Q_{H_2O} increases. Collisions with the photochemical daughters of water in this region modify the radial distributions and outflow velocity of each species, excite and quench metastable emissions, and introduce velocity gradients from photolytic heating. Comet Hale–Bopp was the first comet for which the collisional coma was both spatially resolvable and comparable in extent to the scale lengths of major coma species. In the case of this object, the classical assumptions that make it possible to invert radial emission line profiles, brightnesses, and lineshapes to production rate and velocity either do not hold or require adjustment to work. Here we describe how a large collision zone modifies the coma, how it affects the classical methods for obtaining production rate and velocity, and discuss how wide field imaging may be combined with modified versions of simple models to address the complications and extract some structural information.

1. Introduction

Comet Hale–Bopp was among the most active comets ever observed, with a peak water production rate of $Q_{H_2O} = 10^{31}$ s^{-1} near perihelion (Colom et al., 1999; Combi et al., 2000; Dello Russo et al., 2000; Morgenthaler et al., 2001; Harris et al., 2002), that was several times greater than the highest production rate observed from 1P/Halley in 1986 (e.g., Craven and Frank, 1986; McFadden et al., 1986; Feldman et al., 1986). Such a high production rate has both positive and negative implications for determining the characteristics of the coma. In particular, high water production rates correlate with an increased ability to detect minor coma species, map detailed small-scale structures in the inner coma at high signal to noise, and to measure distributions of major species out to greater nuclear distances. On the other hand, high gas production was known before Hale–Bopp to correlate with increasing coma outflow velocities (Bockelée-Morvan et al., 1990), due presumably to photolytic heating over an increasingly extended, but not spatially resolved, region near the nucleus. Hale–Bopp represents an extreme case

Earth, Moon and Planets **90:** 45–56, 2002.

where the collisional region becomes large enough to be observed directly, and, more significantly, begins to approach the photochemical scale lengths of major coma species including, H_2O, OH, CO_2, and HCN. When this occurs, the velocity distribution of the coma ceases to be determinable with the standard assumptions of parent-daughter energy/velocity breakdown, vectorial flow, and a quasi iso-velocity ballistic coma that are used to invert distributions and aperture averaged line profiles directly to gas production and dynamics.

This paper discusses the background simplifying assumptions used to derive Q_{gas} for weaker comets and how the extreme nature of Hale–Bopp's coma affects them. We then examine how wide field data on active comets, which provide crucial information on the outer coma distribution and the velocity distribution of key species, can be modeled either with modified spherical expansion simulations or advanced codes that account for collisional effects.

2. Inversion Techniques and Assumptions

For the typical low activity comet, gas production rates and velocities can be determined reliably through a combination of spatial brightness mapping and simple models.

Production Rate. A species brightness distribution can be obtained using either variable-width aperture photometry (Morgenthaler et al., 2001; Schultz et al., 1993) or ring-sum extraction of the radial brightness away from the nucleus (e.g., Harris et al., 2002; Festou, 1981). If the emissions are prompt (metastable), then the column densities follow directly from the brightness (Schultz et al., 1993), while resonance line column densities are determined from the heliocentric velocity corrected fluorescence efficiencies (e.g., Schleicher and A'Hearn, 1988). These column densities are then inverted back to a species production rate using radial (Haser, 1957) or vectorialized (Festou, 1981) spherical expansion models that rely on known values of photochemical lifetime and parent-daughter branching ratios. Despite the simplicity of these models, they consistently produce results that compare favorably to those from more detailed simulations.

Outflow Velocity. The *average* velocity of a particular species can be directly measured with very high spectral resolution observations of the field-averaged emission line width (Bockelée-Morvan et al., 1990; Biver et al., 1999; Colom et al., 1999; Morgenthaler et al., 2001). Acceleration in the flow can be detected by comparing profiles obtained over fields of view with different radii. It is also possible to obtain the mean outflow velocity by fitting the radial brightness distribution with spherical expansion models. However, expansion models will provide unique results only when either the entire profile is measured, when the velocity parameter space is bounded by other measurements (Harris et al., 2002), or if the outflow velocity is pegged to an outside measurement or relationship such as the $V = 0.87\ R_H^{-0.5}$ (where R_H is heliocentric distance) formula of Budzien et al. (1994).

For the above techniques to provide unique, physically descriptive results that compare well with those of far more detailed photochemical simulations (Combi et al., 1993), several assumptions must hold. These include:

- The flow is completely ballistic.
- The velocity of each inner coma parent species is equal to the bulk water outflow.
- Daughter species velocity distributions are the convolution of the water velocity with the spherically symmetric vectorialized velocity that comes from excess energy of photodissociation.
- The mean velocity of a species is approximately uniform throughout the coma.
- Metastable emissions are neither quenched nor stimulated to a significant degree beyond the immediate vicinity of the nucleus.

3. The Role of Collisions

For most comets, there is a region known as the *collision sphere* near the nucleus, inside which collisions dominate and neutrals and ions of different energies are mixed together into a single thermalized flow. Provided that this is not spatially resolvable from the nucleus and is small compared with the photochemical scale lengths of the major coma species, no velocity gradients are introduced and the daughter species velocity distributions are not affected. In this unresolved case, the collision sphere acts only to increase the bulk coma outflow velocity along with Q_{H_2O} (Bockelée-Morvan et al., 1990; Cochran and Schleicher, 1993). However, as gas production increases, the size of the collision sphere does as well. This increase can be described by a simple relationship

$$R_C = \frac{\sigma\, Q_{H_2O}}{4\pi V}, \tag{1}$$

where σ is the ballistic cross section and V is the outflow velocity (Whipple and Huebner, 1976). This equation is a simplification, because, while collisional mixing occurs throughout the sphere, it does not account for differences between the inner 10% of the sphere that is thermalized (Combi et al., 1999) and the outermost regions where the flow is nearly ballistic and collisions are an occasional occurrence.

For even highly active comets, such as P/Halley, the collision sphere is very small in extent compared with relevant distance scales in the coma. With the above equation, $R_C = 10^4$ km for the highest measured activity from Halley ($Q_{H_2O} = 1.5 \times 10^{30}\ s^{-1}$) (Feldman and Brune, 1976), a value confirmed with *in situ* measurements and models (Hodges, 1990). However, circumstances were very different for comet Hale–Bopp, where there is general agreement that Q_{H_2O} was $= 1.5 \times 10^{31}\ s^{-1}$ at perihelion (e.g., Harris et al., 2002) and the collision sphere radius was therefore $R_C = 10^5$ km. For the geocentric distance of Hale–Bopp at perihelion (1.4 AU), such a collision sphere would have an easily resolvable angular diameter of roughly 3 archmin.

4. Consequences for the Active Comet Coma

The implications of $R_C > 10^5$ km for the structure of a coma are significant. Such a collision sphere is large enough to be both spatially resolvable and comparable to the photochemical lifetimes of major species. This is particularly important for water, which has a lifetime of $\tau_{H_2O} = 8.6 \times 10^4$ s at solar minimum and 1 AU (Huebner et al., 1992), and is the most abundant coma constituent. Even allowing for significant acceleration to >1 km/s across this region, R_C is comparable to the water scale length ($S_{H_2O} = \tau_{H_2O} V_{H_2O}$), and thus much of it is dissociated in an at least partially collisional environment. Furthermore the OH lifetime is also short ($\tau_{OH} = 1.3 \times 10^5$ s (Huebner et al., 1992), and much of it is *also* dissociated in the collisional region. The end result is that H_2O ends up thermalized with the products of its own photochemistry, including OH, H, O, H_2, and H_2O^+.

These circumstances generally invalidate the necessary assumptions described above for the determination of Q and V, especially as they relate to water and its daughter products. The outflow cannot be approximated as ballistic, nor is it uniform with increasing radial distance. The amount of energy being re-thermalized from water daughter products increases across the sphere, resulting in a steady increase in coma outflow velocity with cometocentric distance (e.g., Combi et al., 1999). Because the daughters are thermalized back into the flow, it is also not valid to assume that their velocity distribution is either vectorialized or directly related to the excess energy of dissociation. Indeed, aperture averaged line profile measurements will contain elements of the vectorial and thermalized velocity distributions. Given that the thermalized velocity changes with distance, both the velocity line shape *and* the radial brightness distribution inside the aperture must be known to correctly interpret the data, particularly in the inner comawhere the gradients are largest (Biver et al., 1999).

Collisions have extreme consequences for metastable emissions as well. For very strongly quenched features, such as 18 cm line of OH (Schloerb, 1988; Schloerb et al., 1999), observations sample only the outer regions of coma, effectively biasing their measurement to regions beyond the collision zone and any acceleration. In the case of Hale–Bopp, the radio OH quenching radius was larger than 5 $\times$ 10^5 km, which was well past the edge of the collision sphere. The use of other diagnostic prompt emissions such as OI (Morgenthaler et al., 2001; Schultz, et al., 1993) and CI (Oliversen et al., 2002), to determine their parent production rates is also complicated by collisional quenching and to some extent by both chemical

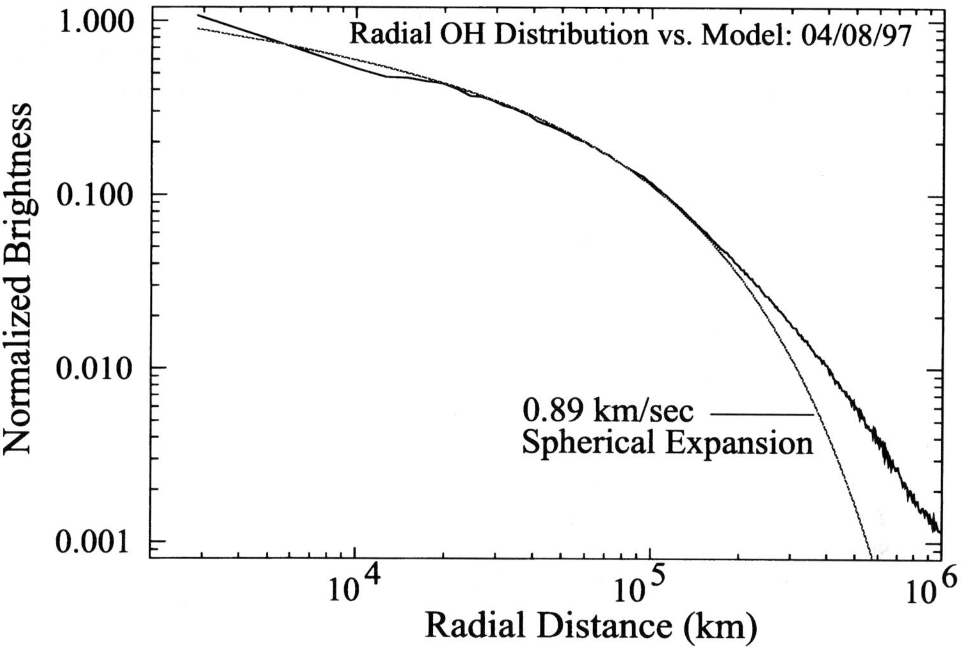

Figure 1. The azimuthally averaged radial distribution of ground state NUV OH emission from comet Hale–Bopp is compared with a two-component spherical expansion simulation for a typical outflow velocity of 1 km/s. Note the failure of the model to produce the observed emission in the outer coma.

reactions (Komitov, 1989) and excitation from collisions with neutral and charged particles inside the sphere (e.g., Bhardwaj, 1999).

5. Observational Evidence for Collisions in Hale–Bopp

Evidence for strong collisional effects in the coma of Hale–Bopp is pervasive in the data that has been presented to date. Clear signatures of this are found in the radial distribution of neutral emissions, the brightness of diagnostic metastable emission lines, the measured acceleration of ions, and in the emission line widths of key diagnostic species. In particular:

Radial Distributions. The radial extent of the coma in all primary species is consistent with a significantly higher outflow velocity than previous comets. In particular, OH cannot be fit with a 1 km/s outflow (Figure 1), but requires V_{OH} between 2.5 and 3.5 km/s to reach the radial distances where it is observed (Harris et al., 2002). Similar wide field measurements of C (λ1657) (Harris et al., 1999) and metastable $O(^1D)$ (Morgenthaler et al., 2001) show comparable radial extensions consistent with substantial acceleration in the coma.

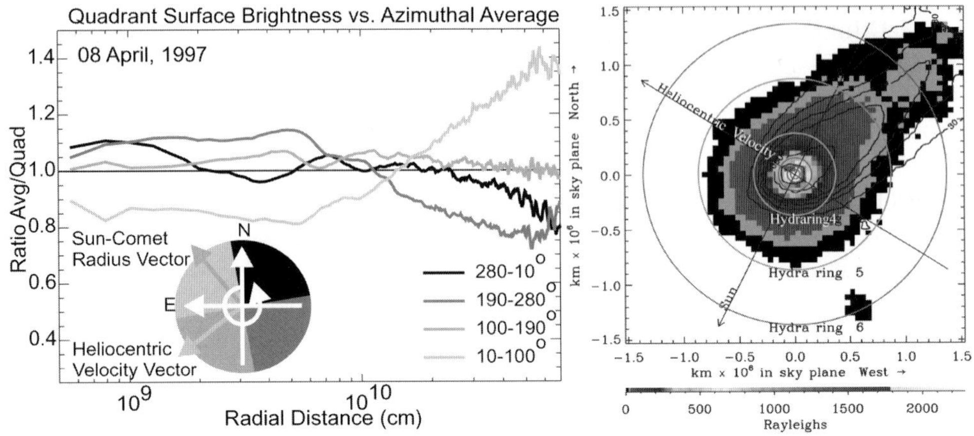

Figure 2. A continuum subtracted image of O(^{1}D) emission from Hale–Bopp and normalized radial quadrant ring summation of the OH distribution both show evidence of extended emissions in a region between the orbit trailing and anti-sunward directions. In neither case can the emission be tied to a secondary gas production source or dust contamination, but instead appear to be the result of a vectored acceleration in this area of the coma.

Metastable Emissions. Because the de-excitation of atoms and molecules in metastable states occurs very close to the location where they are formed, they effectively trace location of photochemistry in the coma. The radial distribution of metastable O(^{1}D), which is produced primarily from H_2O and OH dissociation was measured in Hale–Bopp. In addition, C(^{1}D), which is produced from CO dissociation was observed with a course aperture summation technique at points on and off the nucleus (Oliversen et al., 2002). Both species support the presence of a substantial collisional coma. The inner coma spatial distribution of O(^{1}D) can only be fit by models that include substantial quenching of the emission inside the collision sphere (Morgenthaler et al., 2001), while the C(^{1}D) emission was sharply *peaked* on the nucleus despite a substantially far longer metastable lifetime (4077 s) than O(^{1}D) (150 s). The aperture summation technique used for C(^{1}D) contains no radial information inside the aperture, and so it was not possible to determine how the C(^{1}D) emission was distributed in the inner coma. However, Oliversen et al. (2002) note that several collisional mechanisms could stimulate the emission at rate equal to or faster than quenching (Bardhwaj, 1999), an idea first proposed for C (λ1931) emission from comet West (Feldman, 1978).

Ion Acceleration. Observations of H_2O^+ velocity gradients in the inner coma suggest acceleration less than half that measured from Halley and other comets (Anderson, 1999). Anderson (1999) suggests that mass loading of solar wind by comet ions could account for the retardation, however, Harris et al. (2002) note that ion-neutral interactions in the collision sphere might produce a similar effect, and could account for an asymmetry seen in the anti-sunward coma radial distributions of OH and O(^{1}D) (Figure 2).

Velocity Profiles. Velocity resolved observations of radio emission lines were obtained from several species in the coma of Hale–Bopp, including OH, HCN, Hα, and others (Biver et al., 1999; Colom et al., 1999). All of these show signatures of acceleration. Coma averaged measurements of the average HCN emission line width inside 10^4 km (Biver et al., 1999) translate to velocities (V = 1.4 km/s) that are >50% greater than weaker comets, while the outer coma measurements of ($R > 5 \times 10^5$ km) OH (Colom et al., 1999; Schloerb et al., 1999) give velocities (V = 2.2 km/s) that are nearly twice as large. These results are both consistent with the presence of acceleration over much of the coma. High spectral resolution measurements of the Hα emission line from Hale–Bopp, which has components with excess velocities of 8 km/s from OH dissociation (Keller, 1976) and 25 km/s from H_2O (Combi et al., 2000), show a line shape that is narrower than expected based on the known rates of H_2O and OH photochemistry (Morgenthaler et al., 2002b). Such a narrowing is also consistent with thermalization of very fast H atoms with slower neutrals in the collisional coma.

6. Wide Field Observations

Wide field observations offer a means to determine the gas production rate and velocity structure from the coma brightness distribution and modified versions of spherical expansion models. The principal advantage of a very wide field is that the entire coma is included in the observation. This removes ambiguity about the total fraction of the emission being beyond the sampled region, the amount of which depends on the unknown outflow velocity. When the entire coma is sampled, then simple field summed photometry can be inverted directly to production via:

$$Q_x = \frac{10^6 I_x \Omega \Delta^2}{\tau_x g_x}, \tag{2}$$

where Ω is the solid angle of the FOV, Δ is the geocentric distance, τ_x is the lifetime of species x, and I_x is the brighness across the field of view, and g_x is emission line fluorescence efficiency. For metastable O(^{1}D) emission, which emits only once, Equation (2) becomes

$$Q_{\mathrm{O}(^1\mathrm{D})} = \frac{4}{3} 4\pi \Omega \Delta^2 I_{\lambda 6300} \tag{3}$$

(Schultz et al., 1993). In each case the production rate depends on factors unrelated to the velocity structure and spatial distribution of the coma.

Examples (Table I) of wide-field aperture summation production rate inversion in Hale–Bopp include C (λ1657) imaging (Harris et al., 1997), O(^{1}D) imaging

TABLE I

Hale–Bopp gas production rates from wide field aperture summation

Species	Date	Q_x^{30} ($\times 10^{30}$ s^{-1})
OH(NUV)[a]	04/08/97	$Q_{\rm OH}^{30}$ 9.3 ($\pm$ 1.4)
OH(radio)[b]	04/02–08/97	$Q_{\rm OH}^{30}$ 9.4 ($\pm$ 1.5)
C(λ1657)[c]	04/08/97	$Q_{\rm C}^{30}$ 2.6 ($\pm$ 1.2)
H(λ1215)[d]	04/07/97	$Q_{\rm H}^{30}$ 9.5 ($\pm$ 0.9)

[a] Harris et al., 2002.
[b] Colom et al., 1999.
[c] Harris et al., 1997.
[d] Combi et al., 2000.
Adapted from their published $Q_{\rm H_2O}$.

and integral field spectroscopy (Morgenthaler et al., 2001), H Ly-α with SOHO (Combi et al., 2000), and OH in the NUV (Harris et al., 2002) and radio (corrected for quenching) (Colom et al., 1999). The $Q_{\rm O(^1D)}$ can be used to determine the H_2O production rate by adding a term to the equation that includes the branching ratios ($B_{\rm H_2O}$ and $B_{\rm OH}$) from its parents. Except for radio OH measurements, these observations were all 2-dimensional images with velocity information in the radial extent of the emission. Once the production rate is determined from the full coma inversion, it serves as a boundary condition for using expansion models to determine the outflow velocity.

Modeling the Velocity Distribution. When the imaging data contains a full radial brightness distribution, the velocity structure can be constrained using modified versions of simple, spherical expansion models that account for collisional acceleration. The outer portion of the profile shape, which depends most strongly on the outflow velocity, combines with the boundary imposed by the production rate determination using the summation method to converge the results of model simulations toward a single result. In the wide field images obtained as part of the University of Wisconsin–Goddard Space Flight Center Hale–Bopp observing campaign (Morgenthaler et al., 2002a), Harris et al. (2002) used two variations of the two-component Haser model to determine the amount and extent of the acceleration, along with the velocities at the boundaries of the OH and O(^{1}D) radial profiles (Harris et al., 2002; Morgenthaler et al., 2001).

In one approach the radial profiles may be treated using a classic two-component Haser model (Krishna Swamy, 1997), where the acceleration is treated as a difference between fixed OH and H_2O velocities. The results of this model will converge tightly to a small range of values shown in Figure 3, however they imply a greater production rate ($Q_{\rm H_2O\text{-}model} = 1.4 \times 10^{31}$ s^{-1} vs. $Q_{\rm H_2O\text{-}aperture} = 1.1$

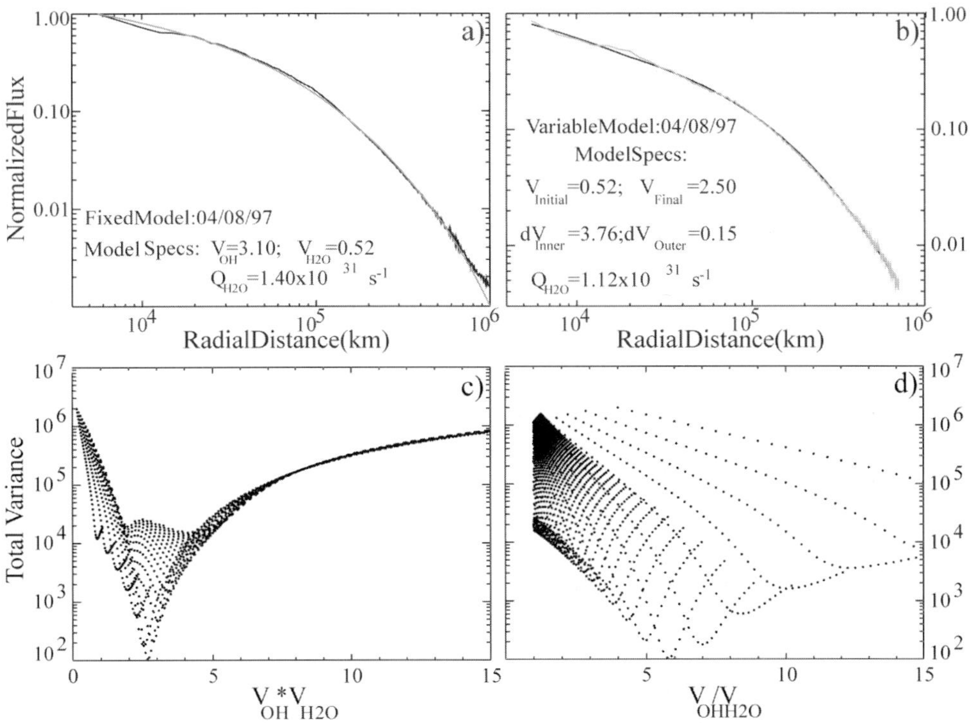

Figure 3. The fit to the observed radial distribution of OH for each model is shown in a and b, while c and d show a graphical representation of the convergence to a small range of values in the fixed case. Both models fit the profile, the accelerating model produces velocities and production rate values more consistent with direct measurements of these quantities (Table II). Below the profile matches the quality of the point-by-point variance of the comparison between various model runs and the data are shown as a function of the velocity product c) and ratio d). Note the sharp convergence to a velocity product of 2.5 that indicates acceleration from the inner to outer coma.

$\times\ 10^{31}\ s^{-1}$) and both a smaller initial velocity (0.5 km/s vs. 1.4 km/s) and higher final velocity (3.35 km/s vs. 2.2 km/s) than were observed in the radio (Biver et al., 1999; Colom et al., 1999).

An improvement over the fixed velocity model results is obtained if the 2 component equations are modified to permit linear accelerations with different gradients inside and outside the region where collisions fully dominate, all while keeping $V_{OH} = V_{H_2O}$. The addition of the collision sphere diameter as a third parameter reduces the convergence of the results compared to the fixed case. However, the quality of the fit is better overall and both the production rates and velocities are more consistent with other estimates (Table II, Figure 3). Confidence in the veracity of this method is bolstered by the fact that the closest match to the radial profile also inverts to the observed production rate, and by the fact that models reproduce the aperture averaged velocities obtained in radio measurements (Colom et al., 1999; Biver et al., 1999).

TABLE II

Characteristics of water production in an accelerating coma from spherical expansion models are compared with summation measurements of brightness and velocity. The model results in the inner and outer coma are isolated to regions comparable to those sampled in radio observations

Source	$V_{\mathrm{Inner\ Coma}}$	$V_{\mathrm{Outer\ Coma}}$	$Q^{31}_{H_2O}$ ($\times 10^{31}$ s^{-1})
OH summation[a]	–	–	1.1 ($\pm$ 0.16)
OH radio[b]	–	2.2 km/s	–
HCN radio[c]	1.2 km/s	–	–
Fixed model[a,d]	0.5 km/s	3.5 km/s	1.4
Accel. model[a]	1.4 km/s	2.3 km/s	0.93

[a] Harris et al., 2002.
[b] Colom et al., 1999.
[c] Biver et al., 1999.
[d] Morgenthaler et al., 2001.

7. Summary and Conclusions

By virtue of its very high gas production rate, comet Hale–Bopp provided a unique opportunity to observe the effects of collisional heating on the velocity, thermal, and chemical characteristics of the coma. While these effects are easily identifiable in this case, the important points to be taken are that the collision sphere should be accounted for in *any* active comet and that the manner in which the collision sphere affects the coma depends on a synergy of processes (e.g., gas production rate, gas species photochemical lifetime, opacity to solar UV radiation, heliocentric velocity and distance). As a result each comet must be evaluated on the basis of its individual circumstances at the time observation, a significant departure from the one-size-fits-all approach that works for less active bodies. There is limited observational evidence from a steady increase of the coma averaged outflow velocity (Bockelée-Morvan et al., 1990; Cochran and Schleicher, 1993) that the point where collisions begin to take hold is for $Q_{H_2O} \gtrsim 10^{29}$ s^{-1}. From that turn up point to the Hale–Bopp production rate, the size of the sphere will expand until it encompasses the entire water scale length, at which point the coma will be largely thermalized. In between, the observed velocity will be a mix of a thermalized component and the vectorialized ballistic flow seen in weak objects. For extreme (as yet unobserved) cases with $Q_{H_2O} \gtrsim 10^{31}$ s^{-1}, opacity effects may prevent significant photochemistry in the deep collision sphere, which may actually reduce the effect compared to Hale–Bopp.

Acknowledgements

The authors wish to acknowledge the efforts of many observers who participated in UW-GSFC observing program that produced the data set that forms the basis of this work. These team members include E. Mierkiewicz, M. Vincent, C. Woodward, N. Doane, and F. Roesler. This research was supported under NASA grant NAGW-7952 to the University of Wisconsin-Madison.

References

Anderson, C. M.: 1999, *Earth Moon Planets* **78**, 99–104.
Bhardwaj, A. 1999, *J. Geophys. Res.* **104**, 1929–1942.
Biver, N., et al.: 1999, *Earth Moon Planets* **78**, 5–11.
Bockelée-Morvan, D., Crovisier, J., and Gerard, E. 1990, *Astron. Astrophys.* **238**, 382–400.
Budzien, S. A., Festou, M. C., and Feldman, P. D.: 1994, *Icarus* **107**, 164–188.
Cochran, A. L. and Schleicher, D. G.: 1993, *Icarus* **105**, 235–253.
Colom, P., Gerard, E., Crovisier, J., Bockelée-Morvan, D., Biver, N., and Rauer, H.: 1999, *Earth Moon Planets* **78**, 37–43.
Combi, M. R., Bos, B. J., and Smyth, W. H.: 1993, *Ap. J.* **408**, 668–677.
Combi, M. R., Kabin, K., DeZeeuw, D. L., Gombosi, T. I., and Powell, K. G.: 1999, *Earth Moon Planets* **79**, 275–306.
Combi, M. R., Reinard, A. A., Bertaux, J.-L., Quemerais, E., and Mäkinen, T.: 2000, *Icarus* **144**, 191–202.
Craven, J. D. and Frank, L. A.: 1987, *Astron. Astrophys.* **187**, 351–356.
Dello Russo, N., Mumma, M. J., DiSanti, M. A., Magee-Sauer, K., Novak, R., and Rettig, T. W.: 2000, *Icarus* **143**, 324–337.
Feldman, P. D.: 1978, *Astron. Astrophys.* **70**, 547–553.
Feldman, P. D. and Brune, W. H.: 1976, *Ap. J.* **209**, L45–L48.
Feldman, P. D. et al.: 1987, *Astron. Astrophys.* **187**, 325–332.
Festou, M. C.: 1981, *Astron. Astrophys.* **95**, 69–79.
Harris, W. M., Nordsieck, K. H., Scherb, F., and Mierkiewicz, E. J.: 1999, *Earth Moon Planets* **78**, 161–167.
Harris, W. M., Scherb, F., Mierkiewicz, E. J., Oliversen, R. J., and Morgenthaler, J. P.: 2002, *Ap. J.* **578**, 996.
Haser, L.: 1957, *Bull. Soc. Roy. Sci. Liege* **43**, 740–750.
Hodges, R. R.: 1990, *Icarus* **83**, 410–433.
Huebner, W. F., Keady, J. J., and Lyon, S. P.: 1992, *Astrophys. Space Sci.* **195**, 7–125.
Keller, H.: 1976, *Space Sci. Rev.* **18**, 641–684.
Komitov, B.: 1989, *Adv. Space Res.* **9**, 177–179.
Krishna Swamy, K. S.: 1997, *Physics of Comets*, World Scientific, Singapore, 310 pp.
McFadden, L. A., A'Hearn, M. F. Edsall, D. M., Feldman, P. D., Roettger, E. E., and Butterworth, P. S.: 1987, *Astron. Astrophys.* **187**, 333–338.
Morgenthaler, J. P., Harris, W. M., Scherb, F., Anderson, C. M., Oliversen, R. J., Doane, N. E., Combi, M. R., Marconi, M. L., and Smyth, W. H.: 2001, *Ap. J.* **563**, 451–461.
Morgenthaler, J. P., Harris, W. M., Roesler, F. L., Scherb, F., Anderson, C. M., Doane, N. E., and Oliversen, R. J.: 2002a, these proceedings.
Morgenthaler, J. P., Harris, W. M., Scherb, F., Doane, N. E., and Oliversen, R. J.: 2002b, these proceedings.

Oliversen, R. J., Doane, N. E., Scherb, F., Harris, W. M., and Morgenthaler, J. P.: 2002a, *Ap. J.* **581**, in press.
Schleicher, D. G. and A'Hearn, M. F.: 1988, *Ap. J.* **331**, 1058–1077.
Schloerb, F. P.: 1988, *Ap. J.* **332**, 524–530.
Schloerb, F. P., Devries, C. H., Lovell, A. J., Irvine, W. M., Senay, M., and Wootten, H. A.: 1999, *Earth Moon Planets* **78**, 45–51.
Schultz, D., Li, G. S. H., Scherb, F., and Roesler, F. L.: 1993, *Icarus* **101**, 95–107.
Whipple, F. L. and Huebner, W. F.: 1976, *Annu. Rev. Astron. Astrophys.* **14**, 143–172.

OBSERVATIONS OF ROTATING JETS OF CARBON MONOXIDE IN COMET HALE–BOPP WITH THE IRAM INTERFEROMETER

FLORENCE HENRY*, DOMINIQUE BOCKELÉE-MORVAN and JACQUES CROVISIER

Observatoire de Paris, 5 place Jules Janssen, 92195 Meudon, France

JÖRN WINK†

IRAM, 300 Rue de la Piscine, 38406 Saint Martin d'Hères, France

(Received 14 February 2002; Accepted 25 May 2002)

Abstract. CO was observed on March 11, 1997 in comet Hale–Bopp with the IRAM Plateau de Bure interferometer. The maps show evidence for asymmetrical patterns, due to the Existence of CO jets. Analysis of the spectra and their velocity shifts shows that there is a spiral CO jet rotating in a plane almost perpendicular to the sky plane. This is the first time that rotating jets are observed for parent molecules. We have developed a 3-D model simulating rotating spiral jets of CO gas. We present here the comparison between the observations and our model.

Keywords: Comet C/1995 O1 (Hale–Bopp), comets, interferometry, molecules, radio lines

Comet Hale–Bopp was observed at the IRAM Plateau de Bure Interferometer (PdBI) between March 6 and March 22, 1997, while it was near its perihelion. The CO $J(2-1)$ and $J(1-0)$ lines were observed on March 11, from 4 to 15 h UT. During the observations, the five 15-m antennas were in a compact configuration, providing baselines** from 20 to 150 m. Observations were performed either in the interferometric mode (for each couple of antennas, the signal received by one antenna is correlated with the other one) or in the single-dish mode, also called the *on-off* mode (for each antenna, the signal received is correlated with itself). $60'' \times 60''$ maps of the CO $J(2-1)$ line were obtained with $2'' \times 1.4''$ spatial resolution (~1500 km in the coma) and 0.11 km s^{-1} spectral resolution.

Figure 1a presents all the individual $J(2-1)$ spectra obtained with the on-off mode of the interferometer. The velocity shift (the first order moment of the spectrum) is moving from positive to negative velocities. This is indicative of a jet which direction is moving from anti-Earthward to Earthward. The time evolution of the line velocity shift is plotted in Figure 1b. This curve is well fitted by a sinusoid with a period equal to the nucleus rotation period ($P = 11.35$ h from Jorda et al., 1999) and centred on $v_0 = -0.05$ km s^{-1}. The spectrum velocity shift represents the projection of the jet mean velocity on the line of sight. As the jet rotates with the nucleus, the velocity shift follows a sinusoid which amplitude and mean value

* Author for correspondence. E-mail: florence.henry@obspm.fr

† Deceased.

** A baseline is the vector between 2 antennas.

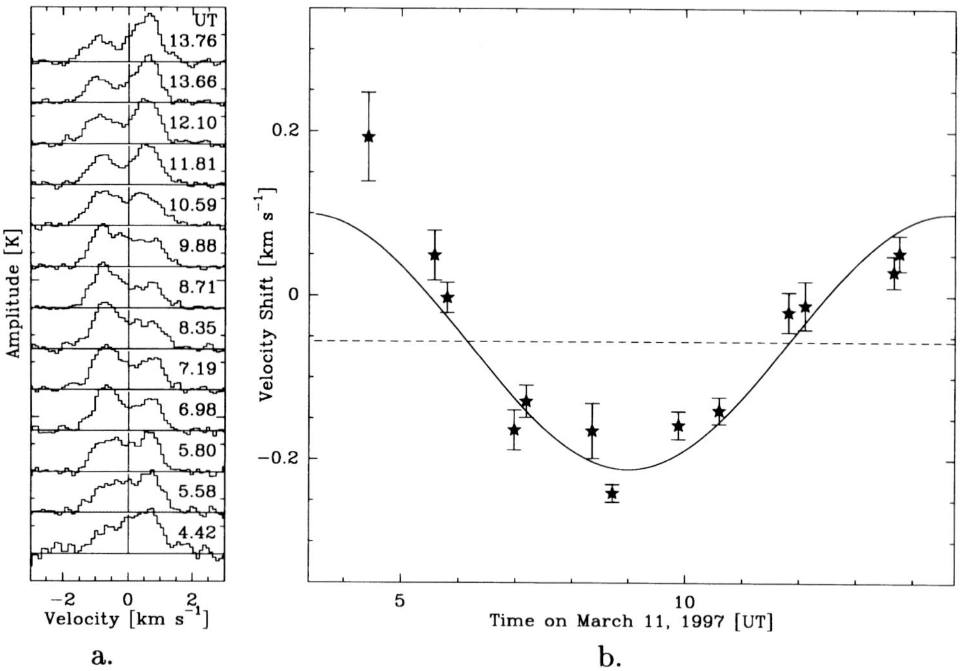

Figure 1. (a) CO *J(2-1)* on-off spectra on March 11, 1997. The integration time is 5 min for each spectrum. The UT time is given in hour on the right side. One graduation is 1 K. These spectra have been occasionally taken between the runs made in interferometric mode. (b) Time evolution of the velocity shifts of the individual spectra. A sinusoid of period 11.35 h and centred on $v = -0.05$ km s^{-1} (dashed line) is drawn.

v_0 depend both on the rotation axis coordinates and on the jet latitude, in addition to the jet strength. Since the comet rotation axis was close to the sky plane (Jorda et al., 1999), the small negative value $v_0 = -0.05$ km s^{-1} that is observed is indicative of a northern jet located slightly above the equator.

The average CO $J(2-1)$ map, computed from the whole interferometric data set (Figure 2), presents anisotropic features that cannot be due to the beam shape. We split this data set into several subsets of 1 h each, and we obtained the maps presented in Figure 3a. The direct interpretation of these maps is hampered by highly elliptical instrumental beams. Nevertheless, an interesting feature is to be noted. We have computed the mean photometric centre from the peak intensity on the map derived from the whole data set (last frame in Figure 2). Similarly, we computed the individual centres for each map of Figure 3a. These centres follow a counterclockwise trajectory around the mean photometric centre (Figure 3b). Given the finite angular resolution of the observations, any anisotropic feature will offset the brightness distribution maximum toward its direction. Then, the displacement of the photometric centre reflects the evolution of the CO coma morphology, due to the rotation, as seen on the sky plane. In the case of a single rotating jet, this

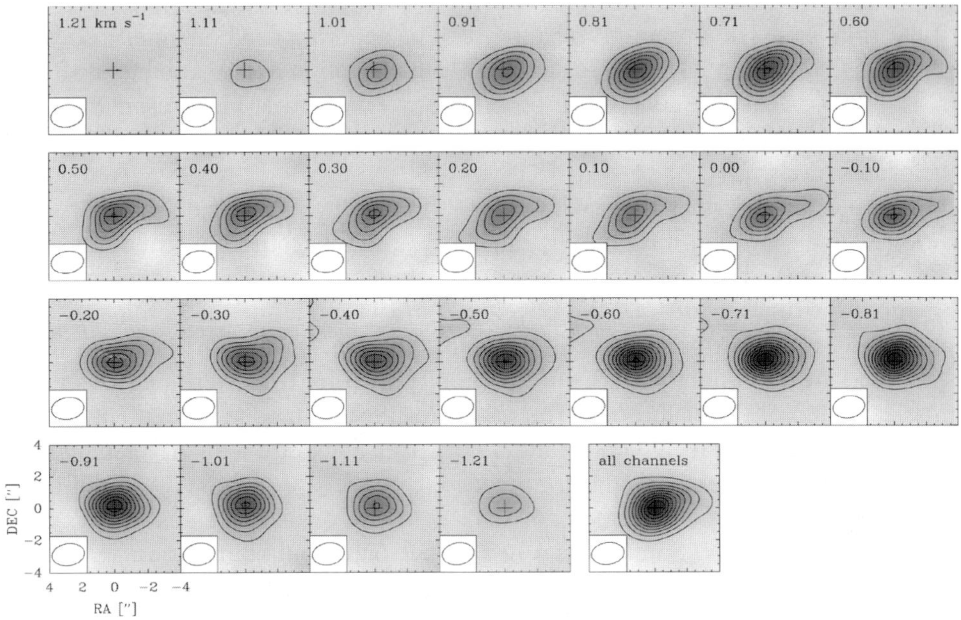

Figure 2. CO *J(2-1)* spectral maps. Each map corresponds to a velocity (in the nucleus frame). The beam shape is shown in the bottom left corner. The last frame is the result of co-adding all the individual spectral maps.

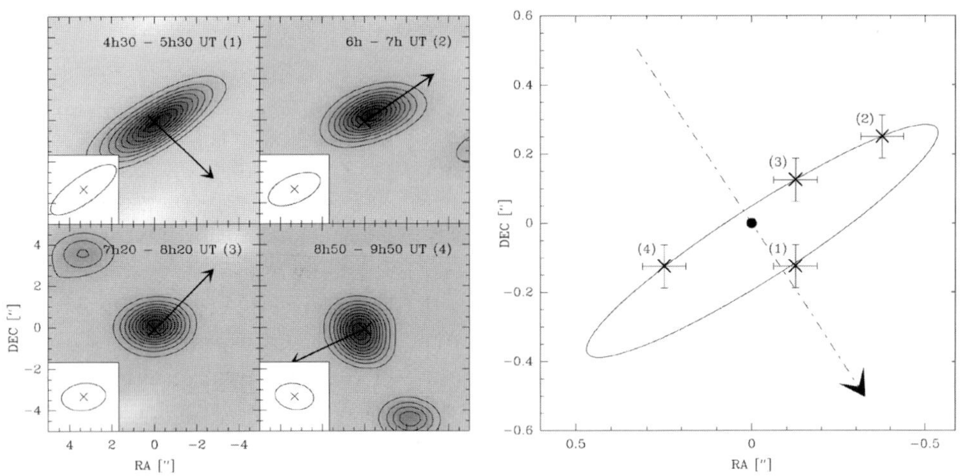

Figure 3. (a) Individual maps for data subsets of 1 h. Crosses locate the mean photometric centre (MPC) determined from the whole data set (last frame in Figure 2). For each map, the arrow represents the direction of the individual photometric centre. The beam shape is shown in the bottom left corner. (b) Time evolution of the photometric centres. Crosses represent the photometric centres, the black disk locates the MPC, the dash-dotted arrow is the nucleus spin vector, and the ellipse drawn is the projection of a parallel of latitude 6° North of a sphere centred on the MPC, with aspect angle = 80° and position angle = 213°.

displacement should follow an ellipse which short axis has the position angle *P.A.* of the spin axis. The observed trajectory can be superimposed on an ellipse which short axis has *P.A.* $=213°$, corresponding to the rotation axis parameters of Jorda et al. (1999). This excludes those proposed by Farnham et al. (presented at IAU Colloquium 186 in Tenerife, January 2002) providing *P.A.* $= 322°$ at the time of the observations.

In order to go further in the analysis of the data, we studied directly the visibilities, which represent the Fourier transform of the brightness distribution on the sky. The observed visibilities show strong deviations from those expected for an isotropic distribution of CO. We developed a 3-D model of the coma consisting of an isotropic contribution plus a conical spiralling jet. Our goal was to understand and fit the velocity shift curves (both on-off and interferometric) and the time evolution of the visibilities. The model results are in good agreement with the observations for a jet located at a latitude of $15° \pm 5°$ North, with an opening angle of $30° \pm 10°$. Furthermore, the jet outgassed $40 \pm 10\%$ of the CO in the coma. These results are in good agreement with the wide dust jet located at $10° \pm 10°$ North proposed by N. Samarasinha (2000; personal communication), in order to explain the coma morphology observed in the visible near the comet perihelion. The real structure of the jet may be more complicated than assumed, especially within the first thousand kilometres from the nucleus. The question of its physical origin remains open. Here, we observed that nearly half of the CO was released in a small opening angle. It is then hard to explain how a highly volatile molecule can expand in such a way from a porous nucleus, as expected for comet Hale–Bopp. More details about the model and its results will be published in Astronomy & Astrophysics.

References

Jorda, L., Rembor, K., Lecacheux, J., Colom, P., Colas, F., Frappa, E., and Lara, L. M.: 1999, *Earth Moon Planets* **77**, 167–180.

Samarasinha, N.: 2000, *ApJ* **529**, L107–L110.

FROM HALE–BOPP'S ACTIVITY TO PROPERTIES OF ITS NUCLEUS

E. KÜHRT
DLR, Institute of Planetary Research and Space Sensor Technology, Rutherfordstr. 2, 12489 Berlin, Germany

(Received 4 April 2002; Accepted 15 June 2002)

Abstract. Because of the outstanding brightness of comet Hale–Bopp reliable production rates of several molecules could be measured over a wide range of heliocentric distances. The H_2O-activity of a comet is closely connected with the energy balance and the composition of its surface. It is often assumed that more volatile molecules originate from subsurface regions. The CO observations provided an important data base for a better understanding of this matter. A thermal model has been applied to calculate gas production rates of Hale–Bopp. The strong obliquity of the spin axis causes pronounced seasonal effects and, therefore, an asymmetry of water activity with respect to perihelion. By comparing measured and calculated rates properties of the nucleus have been derived.

Keywords: Cometary nucleus, CO-production, Hale–Bopp's activity, thermal model, water production

1. Introduction

The exceptional data base for Hale–Bopp's production rates of several volatiles over a wide range of heliocentric distances provides an powerful tool for the analysis of cometary activity and physical properties (for a review, see Bockelée-Morvan and Rickman, 1999). Water molecules probably come directly from the surface. The production rates of this species offer the simplest way to obtain information about the nucleus. The analysis of extremely volatile species like CO is more complex. It is not completely understood where they are released, but observations seem to prove that CO is released both from the nucleus and from extended sources (DiSanti et al., 2001). In the models of nucleus activity it is usually assumed that the nucleus is highly porous and that the sublimation fronts of the strongly volatile ices are in a depth of several meters below the surface. However, such models can't describe the observations adequately or need complex assumptions. A thermal model is applied to calculate production rates. The influence of heat conduction and of the nucleus' rotational properties is discussed. The results are compared with experimental data published by different groups and conclusions on the nucleus properties are given.

Earth, Moon and Planets **90:** 61–65, 2002.

TABLE I
Model parameters

Known from measurements:	
Spin period	11.35 h (Jorda et al., 1998)
Obliquity of the spin axis	84° (Jorda et al., 1998)
Unknown but not very critical:	
Bond albedo	0.04
Emissivity	0.9
Porosity	0.5
Pore size	1 mm
Unknown and critical:	
Thermal conductivity	0.001 ... 1 W K^{-1} m^{-1}

2. Data Base

Weaver et al. (1997, 1999) analysed IUE and HST spectra and deduced the water production from OH-data. Crovisier et al. (1999) detected H_2O-molecules directly with the ISO-PHT-S at 2.7 μm. Colom et al. (1999) obtained spectra at 18 cm from ground-based radio telescopes and detected OH. Unfortunately, the data reduction is not straight forward but needs additional model assumptions. CO-data were derived by Biver et al. (1999) from radio observations. DiSanti et al. (2001) found from IR-spectroscopy near 4.7 μm that CO comes both directly from the nucleus and within 2 AU from an additional distributed source.

3. Results

The thermal model applied for the computations is described in Kührt et al. (1994). The used parameters are given in Table I. From a physical point of view the outgassing of a comet is the result of allocation of the available energy to competing processes as sublimation, re-radiation, and heat transport. These processes depend strongly nonlinear and in rather different way on the temperature and, therefore, on physical and rotational properties of the nucleus. Whereas the re-radiation means energy loss for sublimation the heat transport only redistributes the energy in time. This complex matter determines the gas production. Figure 1 shows measured and calculated H_2O-production rates. Active areas statistically distributed over the nucleus are assumed. A low heat conductivity of $K = 0.01$ W/Km fits the observational data but higher values do not. The best fit in Figure 1 was achieved with an active surface area of about 2000 km^2. This area corresponds to a sphere

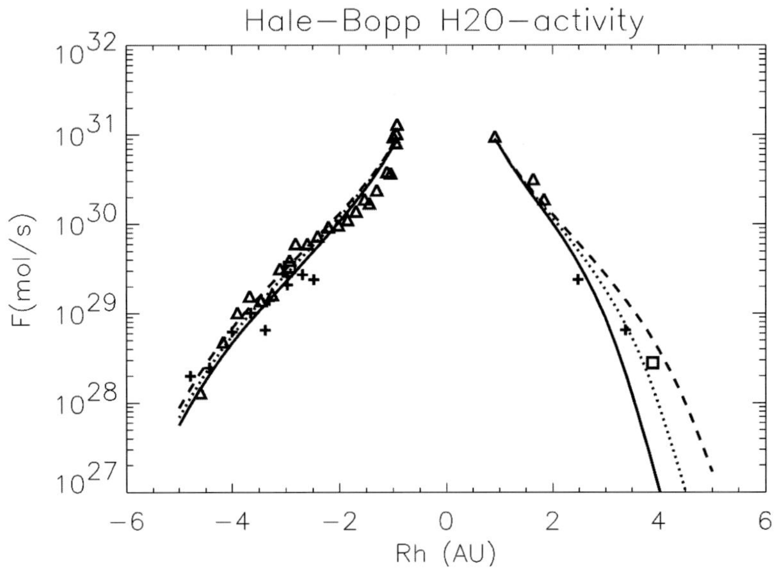

Figure 1. H_2O production of Hale–Bopp. Solid line: $K = 0.1$ W/Km; pointed line: $K = 0.01$ W/Km; dashed line: $K = 0.001$ W/Km. +, data by Weaver et al.; △, data by Colom et al.; □, data by Croivisier et al.

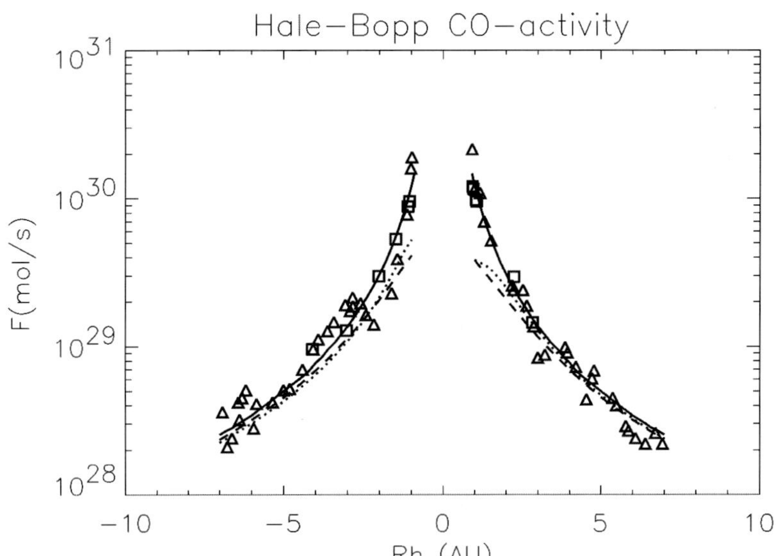

Figure 2. CO production of Hale–Bopp. Solid line: Represents sublimation from the surface (independent on K) and from shallow depths(<1 skin depth) with $K > 0.1$ W/Km; pointed line: Sublimation from 10 skin depths (1 m) with $K = 1$ W/Km; dashed line: Sublimation from 1 skin depth with $K = 0.01$ W/Km. △, data by Biver et al.; □, CO from the nucleus by DiSanti et al.

of pure ice with a radius of 13 km and gives a lower limit for the size of Hale–Bopp. Computations with zero obliquity and/or a high thermal conductivity fail to describe the data. This may be the reason why Enzian (1999) and Capria et al. (2000) did not obtain agreement between the modelled and measured H_2O-activities of Hale–Bopp without postulating an extended source and/or amorphous ice in the nucleus. Figure 2 compares the calculated CO-curves with the measured production rates. Because of the low sublimation temperature (<40 K) and the negligible re-radiation the heat conductivity only weakly influences the results for sublimation from the surface. As a consequence the curves have a shape that is different from that in Figure 1. It can be seen that CO-sources on the surface with an active area of 40 km^2 well describe the observations. A good agreement to the measurements is also found if one assumes that the CO-ice sublimates from moderate depths ($\sim$1 skin depth) and that the total heat conductivity is at least 0.1 W/Km. Both scenarios generate questions to the origin and structure of a cometary nucleus. However, large depths of the CO sublimation front – as found in other models – fail to describe the Hale–Bopp data. Possibly, the microporosity is not as high as often assumed resulting in a limited range of the majority of pores into the depth. Consequently, the CO sublimation front could not move too far away from the water sublimation front on the surface.

4. Conclusions

- At large heliocentric distances the water production on the surface of a comet is higher the more pronounced the temperature differences over the surface are. Consequently, low thermal conductivity, strong seasons, and slow rotation enhance the total activity at heliocentric distances greater than 2 AU. Generally, there is a tendency for higher water production rates after perihelion as a consequence of the thermal inertia of the cometary material. However, overlapping seasonal effects can reverse this trend.
- A thermal model with a low heat conductivity of about 0.01 W K^{-1} m^{-1} fits the experimental H_2O-data over a wide range of heliocentric distances. A lower limit of 13 km for the nuclear radius of comet Hale–Bopp has been derived from comparison of model results with measurements.
- Even at large heliocentric distances the production of very volatile molecules as CO on the surface or from shallow depths is mainly driven by the energy input (heliocentric distance) and scarcely by the properties of the nucleus.
- The CO-production of Hale–Bopp is consistent with the release from a CO source of 40 km^2 at the surface or at shallow depths ($\sim$1 skin depth) with a heat conductivity of 0.1 … 1 W/Km. One reason that CO may be stored near the surface could be that the majority of micropores in a comet has a limited range into the depth.

5. Future Work

The thermal model will be developed to higher complexity (more components, improved heat transport algorithms) and applied to other observed molecules. By using the percolation theory and distinguishing between closed and open pores the influence of the porosity on the water ice/CO ice layering will be investigated in more detail.

References

Biver, N. et al.: 1999, 'Outgassing of Comet Hale–Bopp from Radio Observations', *Earth Moon Planets* **78**, 5–11.

Bockelée-Morvan, D. and Rickman, H.: 1999, 'C/1995 O1 (Hale–Bopp) Gas Production Curves and their Interpretation', *Earth Moon Planets* **79**, 55–77.

Capria, M. T., Coradini, A., De Sanctis, M. C., and Orosei, R.: 2000, 'CO Emission Mechanisms in C/1995 O1 (Hale–Bopp)', *Astron. Astrophys.* **357**, 359–366.

Colom, P., Gérard, E., Crovisier, J., Bockelée-Morvan, D., Biver, N., and Rauer, H.: 1999, 'Observations of the OH Radicals in Comet C/1995 O1 (Halle–Bopp) with the Nancy Radio Telescope', *Earth Moon Planets* **78**, 37–43.

Crovisier, J., Leech, K., Bockelée-Morvan, D., Lellouch, E, Brooke, T. Y., Hanner, M. S., Altieri, B., Keller, H. U., and Lim, T.: 1999, 'The Spectrum of Comet Hale–Bopp as seen by ISO', *ESA SP-427* 137–140.

DiSanti, M. A., Mumma, M. J., Dello Russo, N., and Magee-Sauer, K.: 2001, 'Carbon Monoxid Production and Excitation in Comet C/1995 O1 (Hale–Bopp): Isolation of Native and Distributed CO Sources', *Icarus* **153**, 361–390.

Enzian, A.: 1999, 'On the Prediction of CO Outgassing from Comets Hale–Bopp and Wirtanen', *Space Sci. Rev.* **90**, 131–139.

Jorda, L., Rembor, K., Lecacheux, J., Colom, P., Colas, F., Frappa, E., and Lara, L. M.: 1998, 'The Rotational Parameters of Hale–Bopp (C/1995 O1) from Observations of the Dust Jets at Pic du Midi Observatory', *Earth Moon Planets* **77**, 167–180.

Kührt, E. and Keller, H. U.: 1994, 'The Formation of Cometary Surface Crusts', *Icarus* **109**, 121–132.

Weaver, H. A., Feldman, P. D., A'Hearn, M. F., Arpigny, C., Brandt, J. C., Festou, M. C., Haken, M., McPhate, J. B., Stern, S. A., and Tozzi, G. P.: 1997, 'The Activity and Size of the Nucleus of Comet Hale–Bopp (C/1995 O1)', *Science* **275**, 1900–1904.

Weaver, H. A., Feldman, P. D., A'Hearn, M. F., Arpigny, C., Brandt, J. C., and Stern, S. A.: 1999; 'Post-Perihelion HST Observations of Comet Hale–Bopp', *Icarus* **141**, 1–12.

THE SHADOW OF COMET HALE–BOPP IN LYMAN-ALPHA

An Absolute Measurement of H Production Rate with SOHO/SWAN[*]

ROSINE LALLEMENT and JEAN-LOUP BERTAUX
Service d' Aéronomie du CNRS, BP 3, F-91371 Verrieres-le-Buisson, France
E-mail: lallement@aerov.jussieu.fr

KARÖLY SZEGÖ and SZILVIA NEMETH
KFKI Research Institute for Particles and Nuclear Physics, PO Box 49, H-1525 Budapest, Hungary

(Received 22 March 2002; Accepted 12 July 2002)

Abstract. For a few months around perihelion, the central part of the Hale–Bopp hydrogen cloud has been optically thick to the solar Lyα radiation, and has significantly reduced the solar flux available for the resonance glow of interstellar hydrogen beyond the comet. This shadowing effect on the interstellar gas is the first ever observed comet shadow. It is modeled and compared with SWAN observations. Shadow modelling will help to constrain the comet water production and radiative transfer effects in the interstellar ionisation cavity.

Keywords: Comets, interstellar medium, solar system

1. Introduction

We present the first ever observation of the shadow of a comet on the background of interstellar Hydrogen cloud flowing through the Solar System. We also provide a simple model for the shadowing effect. The source of light is the Sun and its strong Lα emission; the mask (the object which casts the shadow) is the huge cloud of H atoms produced from H_2O and possibly other molecules from photodissociation in the coma of Comet C/1995 O1 Hale–Bopp. The cloud of interplanetary H atoms acts as the "screen" on which the shadow is projected. These H atoms are of interstellar origin and pass continuously through the Solar System where they are ionised mainly by charge exchange with solar wind protons. Near the Sun this ionisation mechanism creates a cavity with a lower density of detectable H atoms. Thus, the study of the shadow will provide information both on the mask (the cometary H cloud) and the "screen" (the interplanetary H atoms).

This shadowing effect was detected by the SWAN instrument on board the SOHO spacecraft. The main objective of SWAN is to determine the latitude distribution of the solar wind and its evolution with the solar cycle. This is done by recording – typically three times per week – the Lα emission of the interplanetary

[*] SOHO is a co-operative mission of ESA and NASA. SWAN was financed in France by CNES with support from CNRS and in Finland by TEKES and the FMI.

Earth, Moon and Planets **90:** 67–76, 2002.

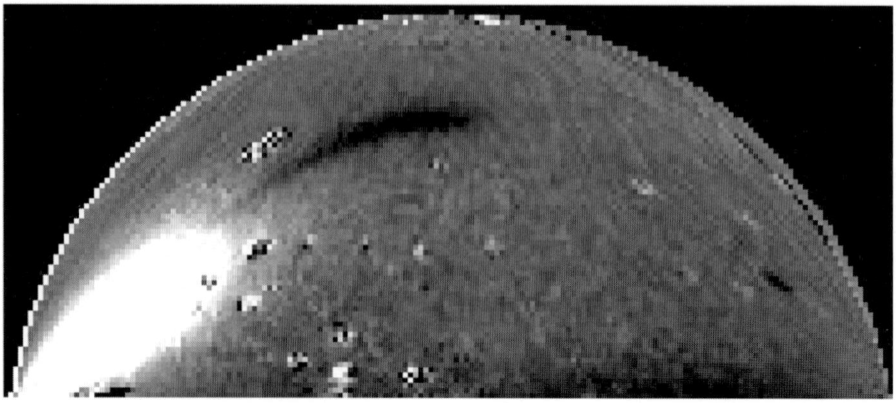

Figure 1. Hale–Bopp H cloud (bright patch) and its shadow (dark feature): Hemispherical map made from SWAN Lyα data. For each line of sight the ratio of the intensities recorded in 1997 and 1996 for the same Earth longitude (April 17) is shown.

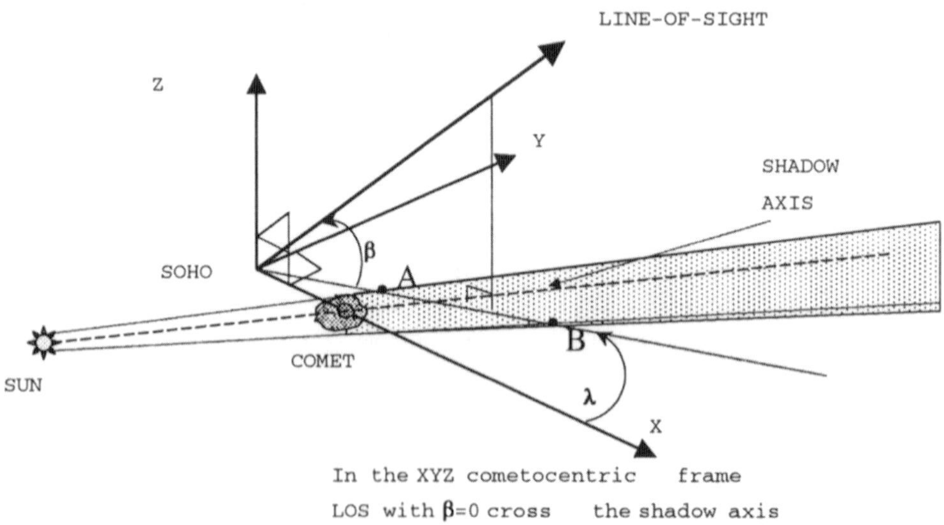

Figure 2. The geometry of the shadow observations

H atoms over the whole sky. A description of the instrument and first results can be found in Bertaux et al. (1997).

In addition, the Lα emission of many comets was observed by SWAN and water production rates of these comets could be derived (i.e., Mäkinen et al., 2001). Comet C/1995 O1 Hale–Bopp produced a gigantic cloud of H atoms that extended over more than 10^8 km and was detectable through its cometary Lα emission over more than 40 ° in the sky. The geometry of the shadow observations is illustrated on Figure 2. The cometary H cloud scatters the Lα solar photons, which are thus no longer available to illuminate the interplanetary H atoms further downstream, creating a shadow cone of infinite extension and with focus point in the Sun. The size of the mask (the cometary H cloud) is much larger than the Sun, therefore penumbra effects (like for solar and lunar eclipses) are negligible.

2. A Simple Quantitative Model of the Shadow

As seen from SWAN/SOHO (Figure 2), only the line of sight (LOS) which intersects the cone shadow will show a deficit ΔIp of interplanetary Lα emission Ip. Let S be the position of SOHO, A and B the points of intersections of the LOS with the shadow cone. The emission rate Ip can be written as integral of the Lα emissivity ϵ_0 (number of photons cm^{-3} s^{-1}) along the LOS:

$$\mathrm{Ip} = \int_0^\infty \epsilon_0 dl = \int_0^A \epsilon_0 dl + \int_A^B \epsilon_0 dl + \int_B^\infty \epsilon_0 dl. \tag{1}$$

Ip denotes the intensity in the case of no shadow, and Ips the one for the case when a shadow is present. We define the shadow intensities deficit ΔIp as the difference between Ip and Ips:

$$\Delta\mathrm{Ip} = \mathrm{Ip} - \mathrm{Ips} = \int_A^B \epsilon_0 dl. \tag{2}$$

Here we have assumed that the interplanetary hydrogen is optically thin, with only direct scattering of solar photons ϵ_0 contributing to the observed emission rate.

For a particular LOS that intersects the shadow cone, the length of segment AB, thus the deficit ΔIp, will depend on the size of the comet mask which is directly related to the H production rate Q_H of the comet. Therefore, measurements of this deficit and its geometrical extent on the sky provide a new method to determine Q_H, completely independent of the usual method of comparing the Lα cometary emission to a model for the H distribution produced from H_2O photo-dissociation. In principle, other sources of H can also contribute to the shadow.

More precisely, the angular extent of the geometrical shadow (defined by ΔIp $\neq 0$) across the Sun-comet line depends only on Q_H (i.e., the size of the mask), while the magnitude of the deficit ΔIp depends on the size of the mask AND the exact value of ϵ_0:

$$\Delta\mathrm{Ip} = \int_A^B \epsilon_0 dl \approx AB\epsilon_0, \tag{3}$$

where the length AB is only a function of the LOS geometry and Q_H, while ϵ_0, the primary excitation of interplanetary H atoms, is a direct function of the density n of these atoms:

$$\epsilon_0 = \frac{g_0}{r^2} n, \tag{4}$$

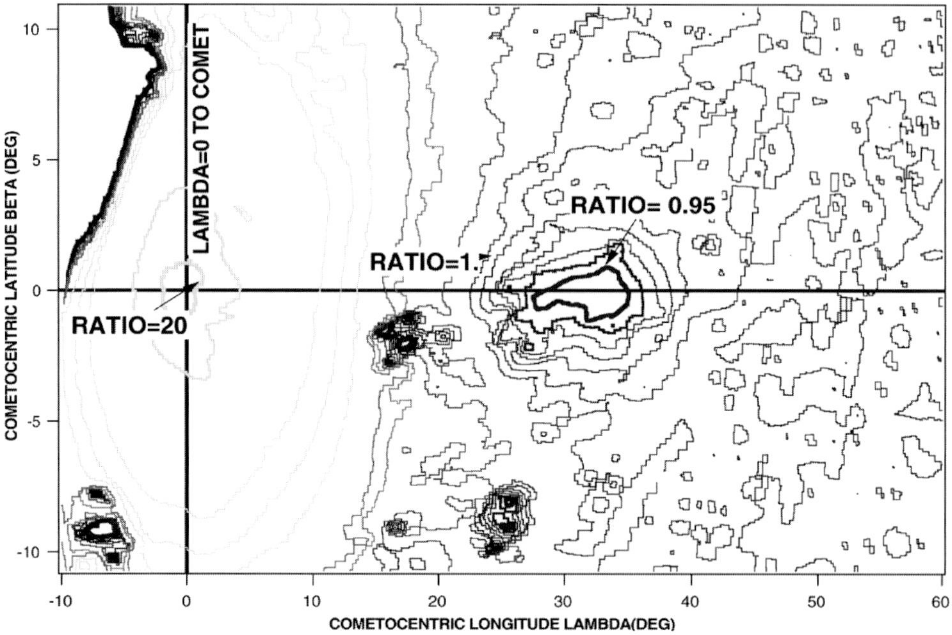

Figure 3. Iso-ratio contours in the comet-shadow region in the reference frame of Figure 2.

where g_0 is the excitation rate of H atoms at 1AU ($g_0 = 1.8 \times 10^{-3}$ s^{-1} for a solar flux of 3.3 10^{11} photons cm^{-2} s^{-1}Å^{-1}) and r is the distance to the Sun (in AU) for the point in the shadow. Equation (3) shows that, once Q_H and therefore the length AB are determined, the measured deficit ΔIp allows to determine ϵ_0 = ΔIp/AB. This is a unique way to get a local measurement of ϵ_0 and n, since otherwise the Lα sky emissions are integrated along LOS extending to infinity. This is very important, since the 3-D distribution of H in the solar system depends on some factors which are not very well known: (1) The H density "at infinity" n_∞ (before significant ionisation, say at $\approx$ 50 AU), (2) the magnitude of ionisation by solar wind and solar EUV. Therefore, the quantitative study of the comet shadow will not only serve to provide an independent determination of Q_H, but it will also provide new and important data on the distribution of interplanetary H in the solar system.

For Comet Hale–Bopp the shadow was detected over several weeks, covering various geometrical configurations, and probing n at various places in the solar system.

One complication arises from the fact that, in reality, the single scattering approximation in the interplanetary H is not fully valid, because the density is of the order of 0.1 to 0.2 H atom cm^{-3}, yielding an optical thickness of 0.1 to 0.2 per AU outside of the ionised region. In our simple model, this effect is not taken into account.

3. The Data

On each sky map recorded before and after perihelion the huge cloud of H atoms of the comet is a conspicuous bright feature which makes it difficult to reveal the shadow. However, when compared to a map of the sky recorded one year earlier, subtle differences show up. Indeed, the map of Figure 1 has been obtained by dividing the Lα intensity map recorded by SWAN on 17 April 1997 by the intensity recorded in April 1996.

The one-year interval is taken because SOHO was at the same position in the solar system as one year before, i.e., for the same viewing geometry for the pattern of interplanetary Lyα emission. In the intensity ratio map the shadow of the comet is visible as a dark area that is elongated into the anti-sunward direction as seen from the comet. The Lyα emission of the cometary coma appears as a bright elliptic feature.

It is more convenient to plot these ratio data in a new coordinate system, defined by its origin O at SOHO, axis OX directed towards the comet's nucleus, and the plane OX-OY containing the Sun (see Figure 2). In principle, the OX-OY plane should be the plane of symmetry between cometary coma and shadow, and indeed, this is what is found in the contour ratio map plotted in Figure 3. High ratios are found for the comet itself at left. Outside the coma emission and its "shadow", the initial ratio is different from 1. This is caused by a combination of several small factors: (1) The Lα solar flux F_s has increased from 1996 to 1997, (2) the interplanetary density distribution has changed due to a change in the ionisation pattern, (3) the calibration factor of the SWAN instrument might have been different from what was assumed to convert the counting rates in Rayleighs units, and – possibly more important – (4) the strong Lα emission of the comet is an additional source illuminating also the interplanetary H atoms, thus contributing to their emissivity and to the intensity displayed in the map of 17 April 1997. A detailed calculation of this contribution Ic is deferred to a later study, but it probably does not exceed about 20 % of the Sun-induced interplanetary emission Ip. In addition, there is no shadow on this extraneous emission. In order to measure the actual depth of the shadow, all intensity ratios have been normalized to 1 outside of the shadow region.

4. Modelling the Comet Absorption of the Solar Flux and the Effect on the Interplanetary Emission

4.1. The Comet Model

The H atoms in the coma have a certain velocity distribution, and therefore they absorb the solar Lα radiation very selectively, i.e., at a single rest frame wavelength $\lambda_0 = 1215.66$ Å. The wavelength integrated absorption cross-section is $\sigma_\lambda = 0.544 \times 10^{-14}$ cm^{2}Å. The wavelength scale may be also determined as a velocity scale with the appropriate Doppler shift conversion, and the velocity integrated cross-section becomes:

$$\sigma_v = 1.3410^{-12}\ \mathrm{cm}^2(\mathrm{km/s}). \quad (5)$$

Therefore, the optical thickness $\tau(V_x)$ of the comet H cloud at velocity V_x is: $\tau(V_x) = \sigma_v dN/dVx$ with dN/dV_x is the integrated line density of H atoms per cm^2 per unit velocity (1 km/s). The integration in the comet cloud is along the line from the Sun to the point where the shadow deficit must be calculated.

The attenuation of the solar flux is defined by the transmission of the cloud: $T(V_x) = exp(-\tau(V_x))$. We are using a simplified comet model, the Haser model which assumes isotropic and homogeneous outflow of the gas species through the coma. Therefore, the coma is spherically symmetric, and the quantity dN/dV_x is determined only by $Q(H)$, by the impact parameter p of the line (distance to the nucleus center) and by the velocity distribution of H atoms. In the following it is assumed that H atoms have two radial velocities, 20 and 8 km/s, corresponding to the photo-dissociation of H_2O and OH. Let v be one of these velocities. Along a line of integration, for this particular population of H atoms, there is only one point (if any) where the projected velocity is V_x. It can be shown with some simple maths that:

$$dN = n(r)\frac{p}{v}\frac{1}{(1-(\frac{v_x}{v})^2)^{\frac{3}{2}}}dV_x \quad (6)$$

where the local density n(r) is computed according to the Haser's formula, found for instance in Bertaux et al (1998) for the analysis of the emission in Comet Hyakutake Lα. The transmission of each population, and the total transmission (product of both transmissions) is computed as a function of $Q(H_2O)$ and the impact parameter. The 20 km/s population has a broader absorption, but shallower than the 8 km/s population, which has an optical thickness larger than unity even at 10^7 km from the nucleus. This whole absorption pattern has then to be displaced by the Doppler shift of the comet nucleus, i.e., 7.5 k ms^{-1} radial heliocentric velocity at the time when our data were taken.

4.2. The Model of Interplanetary Hydrogen and Lα Emission

The distribution of H atoms in the solar system $n(rs, \theta)$ is of the so-called hot model type, with formulas that are established in Lallement et al (1985). It is a function of the radial distance to the sun, rs, and the angle θ with the wind axis. The H distribution is axi-symmetric around the wind vector. The following parameters were taken: temperature T of interstellar H at infinity, i.e., T = 11000 K as derived from an analysis of SWAN maps (Costa et al., 1999); direction of incoming wind (upwind direction seen from the Sun): λ; β = 252; 7 deg.; density n_∞ and lifetime

T_D of H against ionisation are free parameters, typically around 0. 13 cm^{-3} and 1.2×10^6 s, respectively.

The computation of the emissivity at a point given by (rs, θ), takes into account the actual velocity distribution of H atoms at this point, and the solar profile, modified by the cometary absorption up to the point (rs, θ) as explained above. The solar line shape in the self reversal region is fit to a parabola according to the last SOHO/SUMER profile measurements (Lemaire, private communication). The model emission rate (intensity) seen from SOHO is obtained by integrating the emissivity along the line of sight. In addition, the self-absorption by interplanetary H atoms along the line of sight is taken into account. Except for this last attenuation factor, the intensity is therefore calculated with a pure optically thin approximation. It was found by experience that the total intensity calculated with a full radiative transfer model (Quemerais, 2000) models this self-absorption attenuation in a better way.

5. Comparison of Model to Data

The model intensity was computed for both, the case with (I_{m1}) and without the comet (I_{m0}). The absolute deficit $I_{m0} - I_{m1}$, the relative deficit $\delta m = (I_{m0} - I_{m1})/I_{m0}$ and the simple deficit ratio I_{m1}/I_{m0} were derived.

Figure 4 shows the measured intensity ratio map along a cut perpendicular to the shadow axis, at an angle of 32 degrees behind the comet. The intersection of the LOS with the Sun-comet line is at a distance of $\approx$ 5 AU from the comet and $\approx$ 6 AU from the Sun. The shadow is clearly seen as a depression on an otherwise smoothly varying background that is approximated by a linear fit. The map has then been divided by this linear function; this allows to determine the relative shadow deficit δ as fraction of the sky intensity.

The model was run for two different values of the production rate of the comet: $Q = 0.9 \times 10^{31}$ H_2O mol.s^{-1} (the quantity estimated from the Lα emission of the comet recorded by SWAN, Combi et al., 2000), and twice this value. Indeed, the model with the higher production rate shows a larger deficit (smaller minimum ratio or larger relative absorption), and also a wider angular size. Clearly, the results favor the value of Combi et al.

In Figure 5 we compare the data with the model, now along a longitudinal cut at a fixed latitude of 0 ° (along the projected shadow axis). The upper panel shows the solar distance of LOS intersection with the shadow axis in AU, extending to infinity at a longitude of 38 °. The lower panel shows both the data and the model for the same value of $Q = 0.9 \times 10^{31}$ mol.s^{-1}, but this time for two different values of the ionisation lifetime, i.e., for $T_D = 1 \times 10^6$ and 2×10^6 s, thus showing the influence of the interplanetary H density.

The data is clearly in contradiction with the models, showing a smaller maximum absorption than the model predicts, and the maximum occurs when the LOS

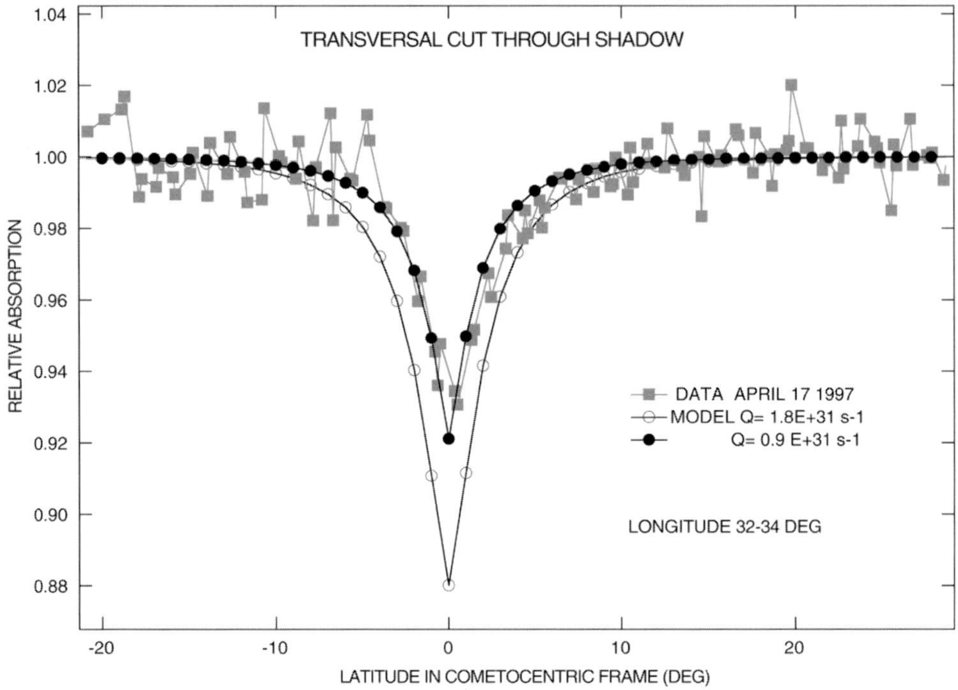

Figure 4. Transversal cut through the shadow: Data and model.

crosses the shadow about 3 degrees closer than predicted. Varying the water production (not shown here), or the ionisation modifies somewhat the pattern but does not help in reproducing the location of the maximum absorption. Clearly some physical effect is not taken into account in our modeling (see below).

6. Conclusions

There is an important discrepancy between the model and the data when the LOS intersects the shadow axis at more than 6–10 AU, where the observed deficit is smaller than the modelled one. We interpret this as a deficiency of the model which assumes an optically thin interplanetary H medium. The atoms in the shadow of the comet are not illuminated by the Sun, but they are still illuminated by all the other atoms of the interplanetary space which are not in the shadow, after second and higher orders scattering. This effect is not taken into account in our model. It increases with the H atom density at infinity, i.e., the larger n_∞, the lower the contribution of first scattering to the total emission rate, the only one obscured by the comet. Potentially, one can derive the true value of n_∞ from an analysis of the comet shadow. Clearly, one has to use a more sophisticated radiative transfer model (Quémerais, 2000) that takes into account all orders of scattering. This, however, was beyond the scope of this first simple model for the comet shadow,

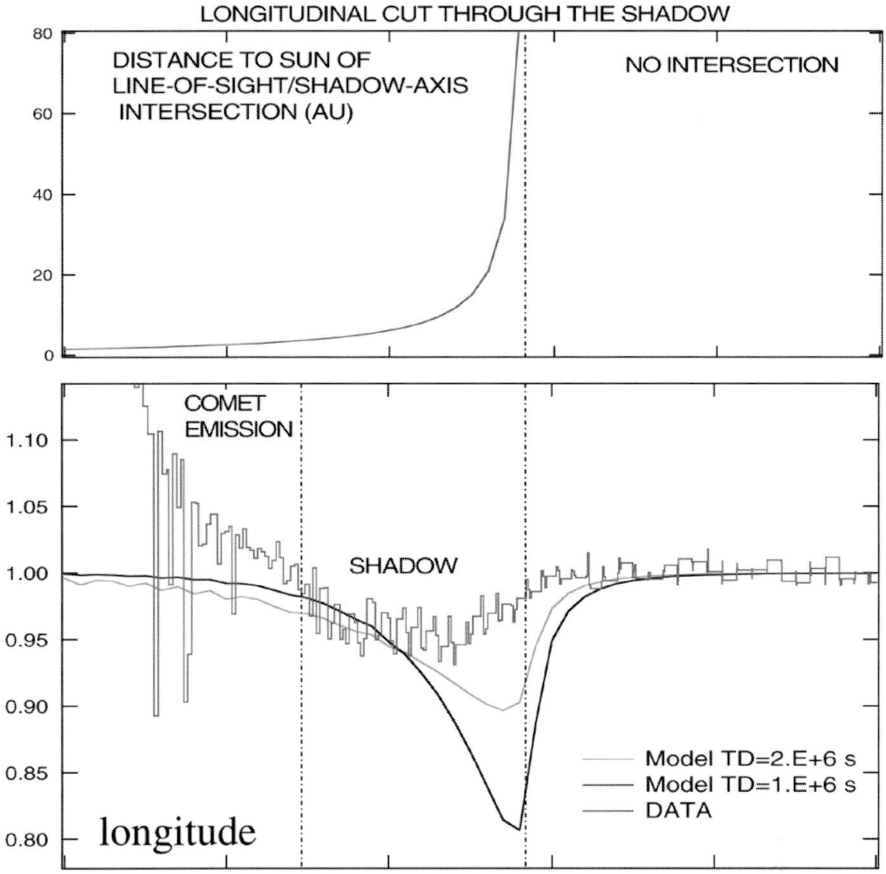

Figure 5. Longitudinal cut through the shadow (bottom) and distance to Sun of the LOS intersection with the shadow (top).

a phenomenon observed with SWAN for Comet Hale–Bopp thanks to its gigantic production rate. The data of the absolute and relative flux deficit spanning two months in time and various positions in the solar system, will be instrumental to disentangle the 3 unknowns that we hope to derive from this exercise: The production rate Q, the density of interstellar H at infinity n_∞, and the ionisation lifetime T_D under the action of the solar wind and solar EUVE radiation.

References

Bertaux, J. L., Costa, J., Quémerais, E., Lallement, R., Berthé, M., Kyrölä, E., Schmidt,W., Summanen,T., Mäkinen, T., and Goukenleuque, C.: 1998, *Planet. Space Sci.* **46**, 555.

Bertaux, J. L., Quémerais, E., Lallement, R., Kyrölä, E., Schmidt, W., Summanen, Goutail J. P., Berthé, M., Costa J., and Holzer T.: 1997, *Solar Phys.* **175**, 737.

Combi, M. R., Reinard, A. A., Bertaux, J.-L., Quemerais, E., and Mäkinen, T.: 2000, *Icarus* **144**, 191.

Costa, J., Lallement, R., Quémerais, E., Bertaux, J.-L., Kyrölä, E., and Schmidt, W.: 1999, *Astron. Astrophys.* **349**, 660.

Lallement, R., Bertaux, J. L., and Dalaudier, F.: 1985, *Astron. Astrophys.* **150**, 21.

Mäkinen, J. T. T., Bertaux, J.-L., Pulkkinen, T. I., Schmidt, W., Kyrölä, E., Summanen, T., Quémerais, E., and Lallement, R.: 2001, *Astron. Astrophys.* **368**, 292.

Quémerais, E.: 2000, *Astron. Astrophys.* **358**, 353.

THE GAS PRODUCTION RATE AND COMA STRUCTURE OF COMET C/1995 O1 (HALE–BOPP)

JEFFREY P. MORGENTHALER* and WALTER M. HARRIS
University of Wisconsin–Madison, Space Astronomy Laboratory, 1150 University Ave., Madison, WI 53706, USA; currently at NASA/Goddard Space Flight Center, Code 681, Greenbelt, MD 20771, USA

FREDERICK L. ROESLER and FRANK SCHERB
University of Wisconsin–Madison, Department of Physics, 1150 University Ave., Madison, WI 53706, USA

CHRISTOPHER M. ANDERSON and NATHANIEL E. DOANE
University of Wisconsin–Madison, Department of Astronomy, 1150 University Ave., Madison, WI 53706, USA

RONALD J. OLIVERSEN
NASA/Goddard Space Flight Center, Code 681, Greenbelt, MD 20771, USA

(Received 1 March 2002; Accepted 26 June 2002)

Abstract. The University of Wisconsin–Madison and NASA–Goddard conducted a comprehensive multi-wavelength observing campaign of coma emissions from comet Hale–Bopp, including OH 3080 Å, [O I] 6300 Å, H_2O^+ 6158 Å, H Balmer-α 6563 Å, NH_2 6330 Å, [C I] 9850 Å CN 3879 Å, C_2 5141 Å, C_3 4062 Å, C I 1657 Å, and the UV and optical continua. In this work, we concentrate on the results of the H_2O daughter studies. Our wide-field OH 3080 Å measured flux agrees with other, similar observations and the expected value calculated from published water production rates using standard H_2O and OH photochemistry. However, the total [O I] 6300 Å flux determined spectroscopically over a similar field-of-view was a factor of $3-4$ higher than expected. Narrow-band [O I] images show this excess came from beyond the H_2O scale length, suggesting either a previously unknown source of [O I] or an error in the standard $OH + \nu \rightarrow O(^1D) + H$ branching ratio. The Hale–Bopp OH and [O I] distributions, both of which were imaged to cometocentric distances $> 1 \times 10^6$ km, were more spatially extended than those of comet Halley (after correcting for brightness differences), suggesting a higher bulk outflow velocity. Evidence of the driving mechanism for this outflow is found in the Hα line profile, which was narrower than in comet Halley (though likely because of opacity effects, not as narrow as predicted by Monte-Carlo models). This is consistent with greater collisional coupling between the suprathermal H photodissociation products and Hale–Bopp's dense coma. Presumably because of mass loading of the solar wind by ions and ions by the neutrals, the measured acceleration of H_2O^+ down the ion tail was much smaller than in comet Halley. Tailward extensions in the azimuthal distributions of OH 3080 Å, [O I], and [C I], as well as a Doppler asymmetry in the [O I] line profile, suggest ion-neutral coupling. While the tailward extension in the OH can be explained by increased neutral acceleration, the [O I] 6300 Å and [C I] 9850 Å emissions show 13% and > 200% excesses in this direction (respectively), suggesting a non-negligible contribution from dissociative recombination of CO^+ and/or electron collisional excitation. Thus, models including the effects of photo- and collisional chemistry are necessary for the full interpretation of these data.

Keywords: Carbon monoxide, C/1995 O1 (Hale–Bopp), comets, dissociative recombination, electron collisional excitation, hydrogen Balmer$-\alpha$ coma, hydroxyl radical photodissociation, metastable oxygen coma, water photochemistry, water production rate

* Author for correspondence. E-mail: jpmorgen@alum.mit.edu

1. Introduction

One of the major goals of cometary studies is to find the relative abundances of materials in the nucleus in order to answer fundamental questions of the origin and evolution of comets and the Solar System. Until the detailed bulk properties of several comets are probed by space craft visits, we must rely on remote sensing studies of the comet nucleus surface, coma, and dust to determine relative abundances. The scheme for using remote sensing studies to derive cometary abundance ratios goes something like this:

- Count all the photons produced in a few key coma emission lines.
- Understand how these photons relate to parent populations.
- Understand how outgassing rates relate to intrinsic nuclear abundance ratios.
- Derive abundance ratios.

Spectroscopy and narrow-band imaging are the tools used for isolating coma lines. Because most telescopes have narrow fields of view (FOVs) compared to the typical cometary coma, full interpretation of the data depends on models of the comet atmosphere and coma emissions. The smaller the telescope FOV, the more tenuous the connection between the measured flux and the intrinsic production rate of the target species (e.g., Cochran and Schleicher, 1993).

For three decades, the University of Wisconsin–Madison has specialized in using Fabry–Pérot (FP) spectrometers for sensitive studies of faint, diffuse emissions from the Earth's atmosphere, objects in the solar system, including comets and the Io plasma torus, and the Galaxy (e.g., Nossal et al., 2001; Magee-Sauer et al., 1988; Woodward et al., 1994; Reynolds, 2002). FPs have the advantage of providing very high spectral resolution and sensitivity with a very wide FOV (Roesler et al., 1995). FPs can be used as tuneable narrow-band filters for imaging applications (e.g., Figure 2) or for wide-field spectroscopy (e.g., Morgenthaler et al., 2002). Thus, they are an excellent tool for measuring coma emissions. FP data can be used to constrain the models needed to relate coma emissions to parent populations and the models needed for the interpretation of data collected with more conventional observation techniques.

In most cometary atmospheres, interactions with the solar wind and radiation field are the dominant mechanisms for shaping the coma: collisions between atmospheric constituents in typical comets only occur within a few $\times 10^3$ km of the nucleus (Whipple and Huebner, 1976). Comet C/1995 O1 (Hale–Bopp) was not typical. With a water production rate, $Q(H_2O) > 1 \times 10^{31}$ molecules s^{-1}, at perihelion, Hale–Bopp's collision sphere extended to $r \sim 10^5$ km (Combi et al., 1997; Harris et al., 2002), which is comparable to the photodissociation scale length of H_2O at 1 AU. Thus, rather than simple Haser (1957) or vectorial (Festou, 1981) models, in most cases, sophisticated hydrodynamic or Monte-Carlo models (e.g., Crifo and Rodionov, 1999; Combi and Smyth, 1988) are needed to fully interpret Hale–Bopp's coma emissions.

Our collaborative team from the University of Wisconsin–Madison and NASA–Goddard conducted a comprehensive ground-based study of Hale–Bopp during its perihelion passage with the goal of mapping the intensity distributions and spectral line shapes of several species, notably H_2O daughters and $C(^1D)$. In this work, we briefly summarize all of the observations and results to date.

2. Observations

Nine instruments on eight telescopes were used in the Wisconsin/GSFC observing campaign. Table I shows which telescopes each instrument was coupled to. Several of the telescopes were located on Kitt Peak, including the 0.6 m siderostat and 150 mm FP imaging spectrometer that comprise the Wisconsin Hα Mapper (WHAM; Tufte, 1997), the McMath-Pierce (MMP) solar, Wisconsin Indiana, Yale, NRAO (WIYN), and Burrell Schmidt telescopes. The Halfwave Polarimeter (HPOL), an imaging spectropolarimeter (see, e.g., Meyer et al., 2002), was located at the Pine Bluff Observatory (PBO) in Pine Bluff, Wisconsin and the ADOPT adaptive optics imager was coupled to the 100″ Hooker telescope at Mt. Wilson. The Wide-field Imaging Survey Polarimeter (WISP) is a UV sounding rocket payload which was flown from White Sands Missile Range on 1997 April 8. As indicated in Table II, WISP recorded spectropolarimetric images of Hale–Bopp in C I 1657 Å and the UV continuum at 2700 Å. The configuration of this instrument and the results of these observations are presented by Harris et al. (1999). Images of Hale–Bopp were recorded by the Burrell Schmidt and Hooker 100″ telescopes using the narrow-band filters indicated in Tables II–III on the dates indicated in Figure 1. The Burrell had a 1° FOV and imaged the full spatial extent of the species observed; the ADOPT system had a 22″ FOV and captured the detailed structure of the inner coma. Both the Densepak and Hydra MOSs were used to observe Hale–Bopp over a 300 Å bandpass centered on 6250 Å. The Densepak MOS is a hexagonal array of ninety-one 3″ fibers on 4″ centers. The Hydra MOS has ninety-six fibers, each with a 3″ FOV which can be arranged with very few constraints within a 1°diameter FOV. For these observations, Hydra was configured with the fibers in concentric rings centered on the nucleus, as indicated in Figurc 2.

We operated the three FPs in one of two modes: imaging or spectroscopic. In the first case, the sky is imaged onto a CCD camera through the FP. An iris can be used to adjust the bandpass of the image, from the free spectral range of the instrument down to the intrinsic resolving power. The central wavelength of the image can be adjusted to a precision much better than the intrinsic resolving power of the instrument. In spectral mode, the FP ring pattern is imaged directly onto the CCD, capturing the entire free spectral range of the instrument in one exposure. Although changing between imaging and spectral mode is a simple matter of inserting or removing lenses, we recorded data with the 150 mm FP on the MMP only in image

TABLE I
Wisconsin/GSFC key to instruments and telescopes

Instrument	Telescope
WISP (UV imaging polarimeter)	0.3 m sounding rocket 36.157 UL
Burrell Schmidt (wide-field imager)	0.6 m Schmidt, Kitt Peak
WIYN MOS (Hydra and Densepak MOS)	3.5 m WIYN, Kitt Peak
WHAM FP (imager/spectrometer)	0.6 m Wisconsin Hα Mapper, Kitt Peak
MMP 150 mm FP (imager)	0.9 m McMath-Pierce West Auxiliary
MMP 50 mm FP (spectrometer)	2.0 m McMath-Pierce Main, North Port
PBO HPOL (imaging spectropolarimeter)	0.9 m Pine Bluff Observatory
ADOPT (adaptive optics imager)	2.5 m (100″) Hooker, Mt. Wilson

TABLE II
Wisconsin/GSFC key to observations

Instrument	Emission source	λ (Å)	R[a]
WISP	C I	1657 ± 50	
WISP	UV Continuum	2700 ± 35	
Burrell Schmidt	CN, C_2, C_3,		
	OH, Cont. (blue, green)	Table III	
Hooker 100″ ADOPT	CN, C_2, H_2O^+,		
	Cont. (blue), R-band	Table III	
WIYN MOS	H_2O^+, NH_2, [O I], C_2	6100–6400	15,000
WHAM 150 mm FP	[O I]	6300	25,000[b]
WHAM 150 mm FP	H_2O^+	6170	25,000[b]
MMP 150 mm FP	H_2O^+	6158	15,000
MMP 50 mm FP	[O I]	6300	60,000
MMP 50 mm FP	Hα	6563	60,000
MMP 50 mm FP	[C I]	9850	40,000
PBO HPOL	Continuum	3200–10,500	500

[a]Resolving power ($\lambda/\Delta\lambda$).
[b]Imaging and spectral capabilities.

mode and the 50 mm FP only in spectral mode. WHAM data was recorded in both modes.

TABLE III
Filter specifications

Filter	λ^1
OH	3090 ± 31
CN	3870 ± 31
C_3	4062 ± 31
Blue Cont.	4450 ± 33
C_2	5141 ± 59
Green cont.	5260 ± 28
R-band	6450 ± 700
H_2O^+	7020 ± 85

[1]Center ± bandpass (Å).

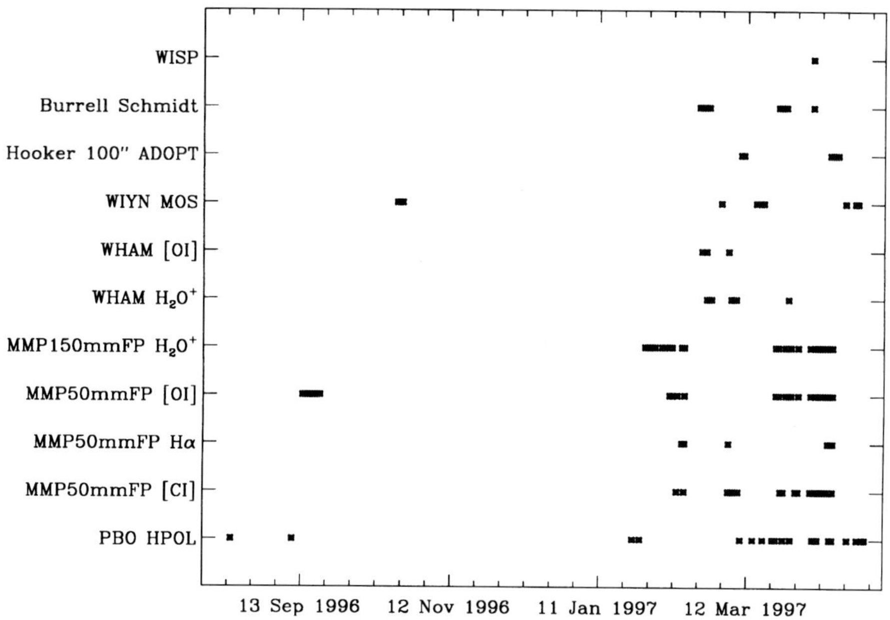

Figure 1. Summary of Wisconsin/NASA–GSFC comet Hale–Bopp observations.

3. Data Reduction

The Burrell Schmidt OH images, WHAM [O I] spectra and images, the WISP UV images, and most of the WIYN MOS spectra have been fully and optimally reduced. The results of the WIYN MOS NH_2 analyses are reported elsewhere (Glinski et al., 2001). The Hα, [C I], and [O I] spectra from the 50 mm FP at the McMath-Pierce telescope are fully reduced, though refinements in these reductions discussed by Morgenthaler et al. (these proceedings) may lead to more stable surface brightness values and, in the case of Hα, detailed line shapes. The HPOL data have been reduced by pipeline software, which merges the entire 1′ FOV. Extracting the spatially varying components of the dust polarization as a function of wavelength within that 1′ FOV requires software that has yet to be developed. The narrow-band filter images taken at Mt. Wilson have been preliminarily reduced and show structure similar to that seen by other observers, but we are unsure of the effect of the diffuse nature of Hale–Bopp's inner coma on the ADOPT system and the resulting image quality. None of the WHAM H_2O^+ data and only one night of the McMath-Pierce 150 mm FP H_2O^+ data have been reduced. Similarly, except for OH, the narrow-band filter images taken with the Burrell Schmidt telescope remain unreduced.

4. Results

4.1. OH AND O(^1D)

The Burrell Schmidt OH images and the WHAM [O I] spectra were taken over a sufficiently large FOV (1°, or 1×10^6 km for Hale–Bopp at perihelion) that essentially all the emission from these species was detected. Using the OH 3080 Å g-factor, Harris et al. (these proceedings) find OH production rates, Q(OH), consistent with other results. Using standard H_2O and OH photochemistry, Harris et al. convert these Q(OH) values into Q(H_2O) values that also agree with other results (e.g., Combi et al., 2000). Using the Combi et al. Q(H_2O) values and standard H_2O and OH photochemistry (Huebner et al., 1992; Van Dishoeck and Dalgarno, 1984), [O I] spectra from four instruments on three telescopes taken over the span of two months show evidence of 3 – 4 times more [O I] emission than expected (Morgenthaler et al., 2001). Morgenthaler et al. (2001) present radial profiles of the WHAM [O I] image shown in Figure 2 from regions excluding the tailward extension (discussed in section 4.5) to demonstrate that this excess came from beyond the H_2O scale length. This suggests either a previously unknown source of O(^1D) or an error in the $OH + \nu \rightarrow O(^1D) + H$ branching ratio of Van Dishoeck and Dalgarno (1984). Using the experimentally determined OH photodissociation cross section at Lyα of Nee and Lee (1984) as a guide, Morgenthaler et al. propose a modification to the OH cross section that predicts the observed [O I] without undue modification to the total OH photodissociation lifetime. Because of differing excess energies of photodissociation, modification of the branching ratio implies a change in the Hα line shape which may be detectable in our Hα data (Section 4.2).

4.2. Hα

As described by Morgenthaler et al. (2002), we measure Hale–Bopp's Hα line width to be 11 – 13 km s^{-1} (FWHM) for a $\sim 2.7 \times 10^5$ km FOV centered $\sim 3.3 \times 10^5$ km sunward of the nucleus. This is significantly broader than the intrinsic line width of 4.5 km s^{-1} predicted by Monte-Carlo techniques (Combi, 2002, private communication). Opacity in the Hα line is likely the reason for this discrepancy. As shown by Smyth et al. (1993), the Hα line profiles predicted by these Monte-Carlo simulations are quite complicated; H atoms are produced in photodissociation events with a variety of excess energies and are thermalized with varying efficiencies, depending on the local coma density. The 5 km s^{-1} resolution of our Hale–Bopp Hα measurements is sufficient to reveal some of this predicted structure and may be useful in testing the hypothesis posed by Morgenthaler et al. (2001), that the van Dishoeck and Dalgarno (1984) OH + $\nu \rightarrow$ O(^1D) + H branching ratio may be too low (Section 4.1).

4.3. [C I] 9850 Å

Because of Hale–Bopp's high production rate and resulting brightness, many emission lines previously undetected or marginally detected in comets were observed and studied in detail. The first ground-based detection of cometary [C I] emission was reported by Münch et al. to the International Halley Watch hotline (Tozzi et al., 1998). Oliversen et al. (2002) were able to obtain quantitative [C I] emission measurements on 11 nights, including measurements off the nucleus in the sunward and anti-sunward directions. Oliversen et al. find a factor two more [C I] emission tailward compared to sunward for $\sim 2.7 \times 10^5$ km diameter FOVs centered at cometocentric radii of $\sim 3.4 \times 10^5$ km. The emission observed in the tailward direction is roughly consistent with a photodissociation model but since the distribution of other species show tailward extensions rather than sunward deficits, Oliversen et al. propose that the spatial distribution of [C I] is primarily determined by effects other than photodissociation. For instance, dissociative recombination of CO^+ was cited as the major source of [C I] 1931 Å emission in comet C/1975 V1-A (West; Feldman, 1978). As discussed in Section 4.5, electron collisional excitation of C is also a possible mechanism. As a result of these effects, even with the relatively large FOV (4′.1) of the 50 mm FP, it is not possible to simply invert the [C I] flux values to obtain an estimate of Q(CO), as done for [O I] and Q(H_2O). Rather, a sophisticated global coma model is needed to describe the complex photo- and collisional chemistry of the neutrals and ions as a function of position in the coma.

4.4. H_2O^+ VELOCITY DISTRIBUTION

Preliminary analysis of the MOS data (Anderson, 1997) shows that the acceleration of H_2O^+ down the tail of Hale–Bopp was 16–19 cm s^{-2} compared to 30–300 cm s^{-2} in Halley (Scherb et al., 1990). This suggests that mass loading of the solar wind and/or collisions between the ions and neutrals play a more important role in Hale–Bopp than Halley. Using velocity resolved H_2O^+ images recorded with WHAM and the 150 mm FP at the McMath-Pierce telescope, we expect to map the coma ion velocity distribution on up to 20 nights between 1997 Jan 30 and Apr 16. These maps will constrain global models of the neutral/ion interactions and may provide a clue to the cause of the asymmetries seen in our OH, [O I] 6300 Å, and [C I] 9850 Å data (Section 4.5).

4.5. Tailward Asymmetry

The OH 3080 Å, [O I] 6300 Å, and [C I] 9850 Å data show extensions in the tailward direction. Harris et al. (2002a, b) show that in OH, this extension lies somewhere between the ion and dust tails. Harris et al. argue that the total amount of OH in this azimuthal slice of the coma is the same as in the other directions, but it is accelerated more than elsewhere, resulting in a lower surface brightness close to the nucleus and a higher surface brightness further away. Harris et al. suggest that collisions between ions and neutrals in Hale–Bopp's dense coma are responsible for both the low H_2O^+ acceleration (Section 4.4; Anderson, 1997) and the increased acceleration of OH in the anti-sunward direction.

Figures 2 and 3 show a similar result for the [O I] data. Furthermore the high resolution [O I] spectra recorded by the 50 mm FP over its 2′ radius FOV (centered on the nucleus) show red wings (Morgenthaler et al., 2001, Figure 4), which, given the viewing geometry, is consistent with $O(^1D)$ atoms flowing down the ion tail. This provides partial support for the ion collisional acceleration mechanism proposed by Harris et al. However, unlike the OH emission, the [O I] emission shows a 13% excess in the tailward quadrant. Because there is no excess OH emission, we can assume that there is no additional source of H_2O or OH in this region, and that the extra $O(^1D)$ is coming from some other source. The likely sources are: electron collisional excitation of O and dissociative recombination of oxygen-bearing species.

Support for electron collisional excitation of [O I] 6300 Å is provided by the detection of [O I] 1356 Å emission at cometocentric distances of 1.6×10^4 km (McPhate et al., 1999), since this transition is not excited by dissociation. Although collisional excitation cross sections are highly energy dependent, it is reasonable to assume a broad enough electron energy distribution in the coma so that collisional excitation of [O I] 6300 Å (and [C I] 9850 Å) is also implied. Support for dissociative recombination of CO^+ comes from the strong tailward asymmetry seen in our [C I] 9850 Å data (§4.3). This effect would also contribute to the tailward excess in the [O I] 6300 Å emission.

5. Conclusions

Particle densities in a cometary coma and dust tail span the range from near terrestrial atmospheric to interplanetary and are unperturbed by the gravitation of the nucleus. Thus, the coma and dust tail provide excellent laboratories for study of gas collisional chemistry, solar photochemistry, the solar wind, metastable state emissions, dust polarization, etc. For proper characterization of these effects, it is

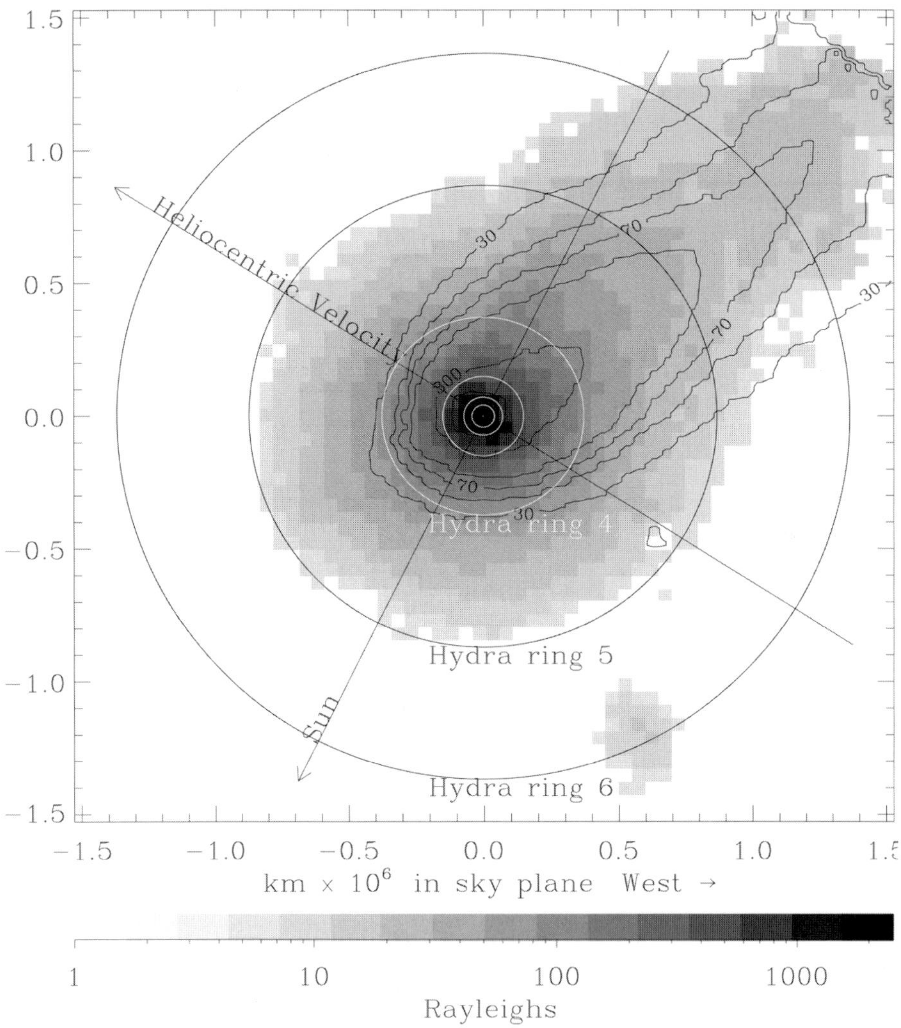

Figure 2. Hale–Bopp March 5 image with [O I] 6300 Å emission shown in gray scale, dust in contours, and circles showing positions of the Hydra annuli plotted in Figure 3. The edge of the 1° WHAM FOV can be seen in the dust contours in the upper right hand corner of the image. The angular radii of the Hydra rings are: $0'.67$, $1'.15$, $2'.4$, $6'$, $14'$, and $22'$.

essential to observe diffuse coma emissions from the major cometary volatiles and their daughter products to distances beyond the photodissociation scale lengths of the emitting species. The high sensitivity, wide-field, narrow-band imaging and spectroscopic capabilities of Fabry-Pérot spectrometers used by the Wisconsin/GSFC team, complemented by conventional multi-object spectroscopy and narrow-band filter imaging provided the necessary set of data for studying detailed coma physics in Hale–Bopp. With a more complete understanding of cometary

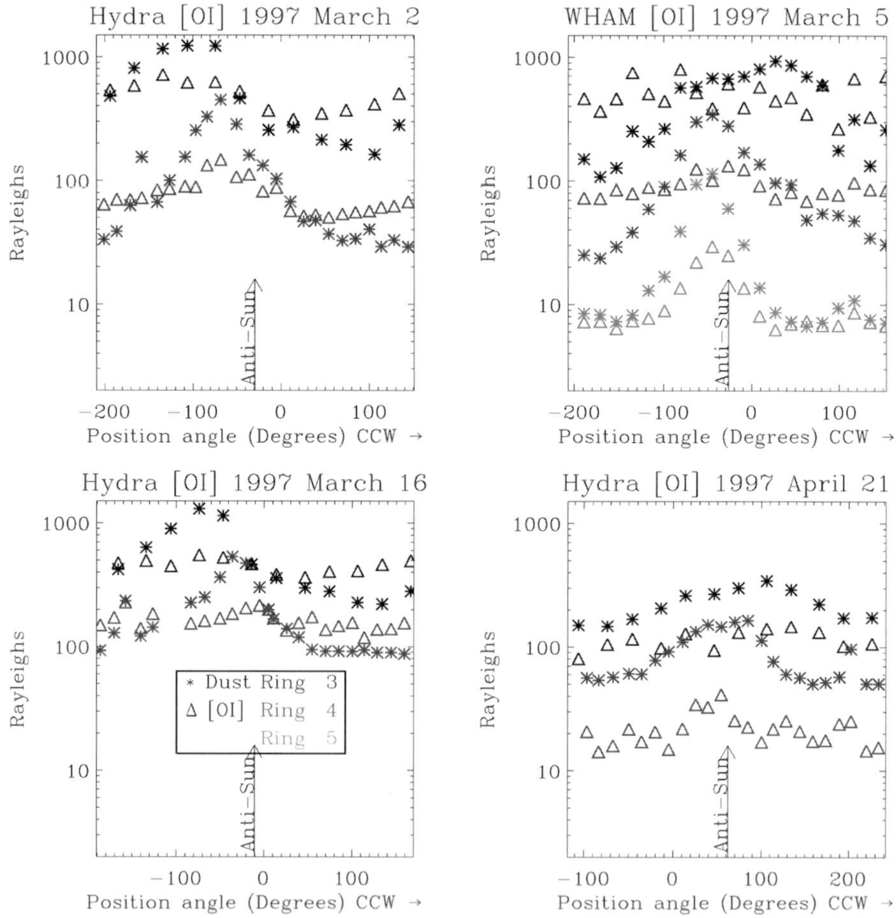

Figure 3. Comparison of the WHAM and Hydra [O I] data. The triangles show the azimuthal distribution of the [O I] surface brightness, the asterisks show the dust surface brightness. The Hydra surface brightness values are plotted directly, with the greyscale and legend indicating which ring in Figure 2 the data correspond to. The values for the points in the top right plot were derived from the WHAM image by dividing it into 5-pixel wide rings (1 pixel = 0.′8) centered on the Hydra rings and finding the average pixel value in 20 azimuthal bins.

comæ, the fundamental questions of cometary composition and origin can be more accurately studied with remote sensing techniques.

References

Anderson, C. M.: 1999, *Earth Moon Planets* **78**, 99.

Cochran, A. L. and Schleicher, D. G.: 1993, *Icarus* **105**, 235–253.

Combi, M. R. and Smyth, W. H.: 1988, *ApJ* **327**, 1026–1043.

Combi, M. R., Kabin, K., De Zeeuw, D. L., Gombosi, T. I., and Powell, K. G.: 1999, *Earth Moon Planets* **79**, 275–306.

Combi, M. R., Reinard, A. A., Bertaux, J., Quemerais, E., and Mäkinen, T.: 2000, *Icarus* **144**, 191–202.

Crifo, J. F. and Rodionov, A. V.: 1999, *Planet. Space Sci.* **47**, 797–826.

Feldman, P. D.: 1978, *A&A* **70**, 547–553.

Festou, M. C.: 1981, *A&A* **95**, 69–79.

Glinski, R. J., Post, E. A., and Anderson, C. M.: 2001, *ApJ* **550**, 1131–1139.

Harris, W. M., Scherb, F., Mierkiewicz, E., Oliversen, R., and Morgenthaler, J. P.: 2002a, *ApJ* **578**, 996–1008.

Harris, W. M., Morgenthaler, J. P., Scherb, F., Anderson, C., and Oliversen, R.: 2002b, these proceedings.

Harris, W. M., Nordsieck, K. H., Scherb, F., and Mierkiewicz, E. J.: 1999, *Earth Moon Planets* **78**, 161–167.

Haser, L.: 1957, *Bull. Soc. Roy. Sci. Liège* **43**, 740.

Huebner, W. F., Keady, J. J., and Lyon, S. P.: 1992, *Astrophys. Space Sci.* **195**, 1–294.

Magee-Sauer, K., Roesler, F. L., Scherb, F., Harlander, J., and Oliversen, R. J.: 1988, *Icarus* **76**, 89–99.

McPhate, J. B., Feldman, P. D., McCandliss, S. R., and Burgh, E. B.: 1999, *ApJ* **521**, 920–927.

Meyer, J. M., Nordsieck, K. H., and Hoffman, J. L.: 2002, *AJ* **123**, 1639–1646.

Morgenthaler, J. P., Harris, W. M., Scherb, F., Anderson, C. M., Oliversen, R. J., Doane, N. E., Combi, M. R., Marconi, M. L., and Smyth, W. H.: 2001, *ApJ* **563**, 451–461.

Morgenthaler, J. P., Harris, W. M., Scherb, F., Doane, N. E., and Oliversen, R. J.: 2002, these proceedings.

Nee, J. B. and Lee, L. C.: 1984, *J. Chem. Phys.* **81**, 31–36.

Nossal, S., Roesler, F. L., Bishop, J., Reynolds, R. J., Haffner, M., Tufte, S., Percival, J., and Mierkiewicz, E. J.: 2001, *J. Geophys. Res.* **106**, 5605–5616.

Reynolds, R. J.: 2002, *Sci. Amer.* **286**, 32–41.

Roesler, F. L., Reynolds, R. J., and Scherb, F.: 1995, in *Tridimensional Optical Spectroscopic Methods in Astrophysics*, ASP, pp. 95–106.

Scherb, F., Magee-Sauer, K., Roesler, F. L., and Harlander, J.: 1990, *Icarus* **86**, 172–188.

Smyth, W. H., Marconi, M. L., Scherb, F., and Roesler, F.: 1993, *ApJ* **413**, 756–763.

Tozzi, G. P., Feldman, P. D., and Festou, M. C.: 1998, *A&A* **330**, 753–763.

Tufte, S. L.: 1997, 'The WHAM Spectrometer, Design, Performance Characteristics and First Results', Ph.D. Thesis, University of Wisconsin–Madison.

Van Dishoeck, E. F. and Dalgarno, A.: 1984, *Icarus* **59**, 305–313.

Whipple, F. L. and Huebner, W. F.: 1976, *ARA&A* **14**, 143–172.

Woodward, R. C. J., Scherb, F., and Roesler, F. L.: 1994, *Icarus* **111**, 45–64.

VELOCITY-RESOLVED OBSERVATIONS OF Hα EMISSION FROM COMET C/1995 O1 (HALE–BOPP)

JEFFREY P. MORGENTHALER* and WALTER M. HARRIS
University of Wisconsin–Madison, Space Astronomy Laboratory, 1150 University Ave., Madison, WI 53706, USA; Currently at NASA/Goddard Space Flight Center, Code 681, Greenbelt, MD 20771, USA

FRANK SCHERB
University of Wisconsin–Madison, Department of Physics, 1150 University Ave., Madison, WI 53706, USA

NATHANIEL E. DOANE
University of Wisconsin–Madison, Department of Astronomy, 1150 University Ave., Madison, WI 53706, USA

RONALD J. OLIVERSEN
NASA/Goddard Space Flight Center, Code 681, Greenbelt, MD 20771, USA

(Received 1 March 2002; Accepted 26 June 2002)

Abstract. We present hydrogen Balmer-α spectra of comet C/1995 O1 (Hale–Bopp) recorded on 5 nights from 1997 February 1 to April 19 by a high-resolution ($\Delta v = 5$ km s^{-1}) Fabry–Pérot spectrometer for a $4'.1$ ($\sim 2.7 \times 10^5$ km) FOV centered $5'$ sunward of the nucleus. The Hα line profile is an important diagnostic of photolytic heating in cometary atmospheres. Extraction of the spectra from the Fabry–Pérot ring images was complicated by obscuration of the telescope FOV due to Hale–Bopp's low elevation, but the measured H-α line widths of 11–13 km s^{-1} (FWHM) are insensitive to the spectral extraction technique. The line widths are consistent with estimates derived from a successful model of Hale–Bopp's hydrogen Lyman-α coma assuming the inner coma is opaque to Hα. We discuss methods for improving the spectral extraction technique and deriving a precise instrument profile which will allow the detailed shape of the line profile to constrain coma models.

Keywords: C/1995 O1 (Hale–Bopp), comets, Fabry–Pérot spectrometers, hydrogen Balmer-α coma, line widths, photolytic heating

1. Introduction

When a comet approaches within a few AU of the sun, sublimation of icy volatiles, particularly H_2O, CO, and CO_2 create an atmosphere which is free to expand into interplanetary space. At the extreme base of the atmosphere, the excess energy of sublimation and collisions between the sublimation products determine the velocity distribution of the expanding atmosphere. Where the atmosphere becomes transparent to solar UV radiation, cometary molecular species begin to photodissociate.

* Author for correspondence. E-mail: jpmorgen@alum.mit.edu

At this point, the atmosphere is still dense enough for frequent ballistic collisions to occur, so the excess energy of photodissociation becomes the driver of bulk outflow acceleration. The dominant cometary volatile, H_2O, and its daughter, OH, have excess energies of photodissociation ranging from a few to several tens of eV (Huebner et al., 1992). Conservation of momentum necessitates that this energy is transferred to the hydrogen atom. Thus, it is ballistic collisions with these hot hydrogen atoms that determines the majority of cometary atmosphere outflow velocity structure.

Cometary outflow velocity can be determined in several ways, for example: *in situ* measurements, extraction of Doppler information from emission line profiles, the Greenstein effect, and measurement of radial surface brightness variation of a particular coma line. This last technique requires accurate knowledge of the lifetime against photodestruction of the emitting species and, if the field of view (FOV) of the observation is not large enough, previous knowledge of the production rate of the comet in that species (Cochran and Schleicher, 1993).

Extraction of Doppler information from emission line profiles is the most common method for determining comet atmosphere bulk outflow velocities (e.g., Bockelée-Morvan et al., 1990). Since typical outflow velocities are ~ 1 km s^{-1}, the high sensitivity and spectral resolving power in the radio make it the natural spectral range for these measurements. In cometary comæ, only molecular transitions are available in the radio, so direct observation of the hot hydrogen atoms that drive the outflow must be done at shorter wavelengths.

Excited by solar Lyman-β radiation, hydrogen Balmer-α (Hα) is the brightest line available to ground-based observers for spectroscopic determination of the cometary hydrogen velocity distribution. A collaboration from the University of Wisconsin–Madison and NASA Goddard Space Flight Center recorded high-resolution Hα spectra on five nights during the perihelion passage of comet C/1995 O1 (Hale–Bopp). The measured line widths of 11–13 km s^{-1} (FWHM), are significantly larger than the intrinsic velocity dispersion of 4.5 km s^{-1} predicted by a successful model of the Lyα coma (Combi et al., 2000). Radiative transfer calculations based on this model suggest that the inner regions of the comet atmosphere are opaque to Hα, resulting in a loss of photons from the line center and a corresponding increase in the line width (M. R. Combi, 2002; private communication). Preliminary calculations suggest a good match between the measured and modeled line widths, but further work on both fronts is necessary in order extract all the information these data have to offer on the driving mechanism of cometary atmosphere outflow.

2. Observations

The Hα observations presented here were part of a comprehensive observing campaign of comet Hale–Bopp conducted by the University of Wisconsin–Madison and NASA Goddard Space Flight Center (Morgenthaler et al., 2001, 2002; Glinski et al., 2001; Anderson, 1999; Harris et al., 2002a, b). A 50 mm dual-etalon pressure-tuned Fabry–Pérot (FP) spectrometer coupled to a CCD detector (Roesler, 1974; Coakley and Roesler, 1994) located at the north port of the main McMath–Pierce Solar telescope on Kitt Peak recorded high-resolution spectra in [C I] 9850 Å, [O I] 6300 Å, and Hα. The instrument configuration and the [CI] results are presented by Oliversen et al. (2002). The instrument field-of-view (FOV) is derived by Morgenthaler et al. (2001, Figure 5) and has an effective diameter of $4'.1$.

Hα spectra of the comet were recorded on 7 nights, 5 of which were usable (Table I). The intended observing procedure was to record one or two spectra with the telescope FOV centered on the nucleus and the rest centered $5'$ sunward. As expected, the nucleus-centered spectra were dominated by the H_2O^+ line that is coincident with Hα. The sunward observations showed no evidence of this emission. On two nights, sunward observations could not be taken. Because of the limited FOV of the telescope guider, we estimate that the FOV drifted $30'' \sim 60''$ during the sunward observations (Oliversen et al., 2002).

As this pressure-tuned dual-etalon FP was used as part of a multiwavelength campaign, the gas pressure in the chambers was changed by $\sim$0.5 atmospheres every night. This induced a slight bowing of the pressure chambers in which the individual etalons were mounted. The two glass plates of each etalon remained highly parallel (a necessary condition to proper FP functioning), but the orientation of the etalons with respect to each other would change slightly, which induces second-order effects. The chambers (since replaced) were originally part of a pressure-scanned system in which the FP bullseye was imaged onto a photomultiplier. This configuration was not very sensitive to misalignment because the individual bullseyes cover a large range in angle space. By imaging the FP ring pattern onto a CCD, many spectral elements can be sampled simultaneously, greatly increasing instrument sensitivity and observing efficiency (Coakley and Roesler, 1994). However the FP rings become narrower with increasing radius (decreasing wavelength), which translates into an increase in sensitivity to the relative alignment of the etalons. The effect of misalignment is to decrease the intensity of the primary transmission peak of the system since the output of the first etalon is no longer efficiently coupled to the input of the second. This in turn increases the width of the instrument response and decreases the contrast between the primary transmission peak and the higher order peaks (ghosts). The last effect complicates the fitting of spectral lines whose relative spacing happen to fall near intervals of the higher-resolution etalon free-spectral-range ($\sim$50 km s^{-1} for these observations).

We have developed a physically realistic model of the dual-etalon FP system that accounts for all these effects. An interface between this model and the `curvefit` routine of IDL is being written that will allow us to use our extensive set of post-campaign calibration data to determine the invariant parameters of the FP system, such as the etalon spacings and reflectivities. Next we will fit the model to the nightly calibration data to determine a reasonable starting point for the parameters likely to vary through the night, such as the relative orientation of the etalons, their relative wavelength tune, and the FP ring pattern center. For each comet FP ring image, we will repeat the planned spectral extraction and fitting procedures detailed in Section 3 for a grid of these parameters. Using the geocoronal Hα line present in each cometary spectrum and accurate measurements and predictions of its line shape and intensity (Nossal et al., 1997, 1998; Bishop et al., 2001), we will find the best-fit values for these parameters. This technique will also be applied to our [O I] 6300 Å and [C I] 9850 Å data, likely resulting in a significant decrease in the scatter seen in the intensity values of these data (Morgenthaler et al., 2001; Oliversen et al., 2002).

3. Reduction

We have completed preliminary reduction of the Hale–Bopp Hα data set using the software developed for the [C I] analysis of Oliversen et al. (2002). Figure 1 shows a sample FP ring image of Hale–Bopp Hα emission processed by standard bias subtraction, flat-fielding and cosmic-ray removal algorithms. The various telescope obscuration effects described by Oliversen et al. are indicated. The noticeable left-right gradient in the image is not understood at this time: it is seen in most of the Hale–Bopp Hα data but not in the [O I] or [C I] comet images or any of the calibration data. Since the Hα observations were generally taken at less favorable viewing geometries than the [C I] and [O I] observations, this may be an exaggerated effect of white light scattering from the telescope superstructure: scattering of atmospheric emission lines into the telescope beam was noted in the other observations. In an attempt to avoid these effects, we selected small areas of each image, such as that indicated in Figure 1, from which to extract the FP spectra (e.g., Figure 2). The vertical lines at each point of Figure 2 are the error bars derived from the number of CCD electrons in each spectral bin. We experimented with extracting spectra from different portions of the ring image and found that the Hα linewidth was insensitive to the region used but the continuum level and shape, as well as the area under the Hα line were quite sensitive to the selected region.

In order to make the grid search of FP parameters discussed in Section 2 practical, we are designing a program that will automatically extract the best spectrum from each image assuming a particular orientation of the etalons. The first step in the program will be to extract spectra from small pie-slices arranged azimuthally about the assumed FP ring pattern center. The small statistical error bars in Figure 2 indicate that many azimuthal bins can to be used. Since each spectral bin should be invariant as a function of angle or, if there is etalon misalignment, vary in a prescribed way, data beyond the edges of the usable image, and any severe problems from gradients, can be easily removed. The remaining data will then be

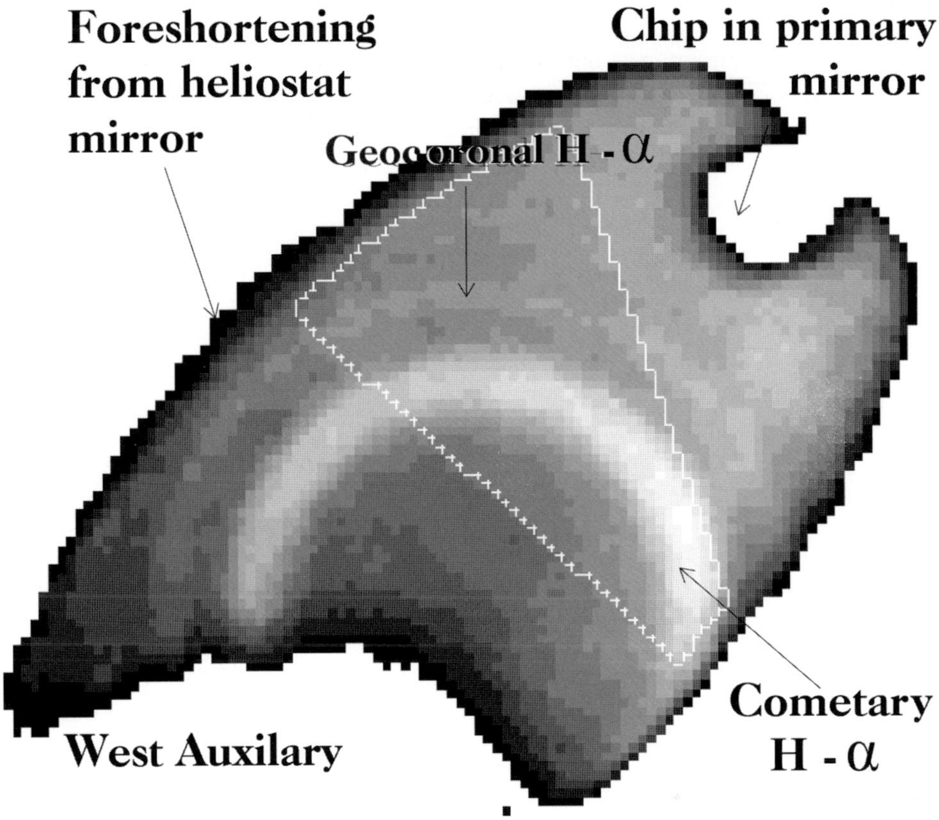

Figure 1. Fabry–Pérot (FP) ring image of Hale–Bopp in Hα recorded UT 1997 April 16 03:18. The primary (#2) mirror was imaged directly onto the CCD detector and was conjugate with the FP ring pattern. The #2 mirror was not illuminated uniformly, hence the incomplete rings. The gradient across the image is likely from scattered light and affects the continuum shape but not the comet or geocoronal line widths. The outline shows the section of the image used to make the spectrum in Figure 2.

fitted by the assumed azimuthal variation of the ring, resulting in a best-fit value and uncertainty for each spectral bin.

We plan to use the model of the FP instrument response discussed in Section 2 for our future spectral analysis. For the purposes of this work we approximated the instrument response with a Voigt profile fit to the Ce 6300.21 Å line of our post-campaign calibration data. The line had to be fit in order to remove the ghost of the Ce 6295.58 Å line. The resulting instrument profile, shown in Figure 3, has a width of 4.2 km s^{-1} (FWHM). We checked this result by fitting the airglow [OI] line recorded on 1997 April 14 (Morgenthaler et al., 2001, Figure 4, 4.7 km s^{-1}) and the Th-Ar 6564.44 Å calibration line recorded on 1997 March 6 (4.8 km s^{-1}). Although the airglow line may just be resolved by this instrument, variation in the orientation of the etalons is the most likely cause of the variations in these widths.

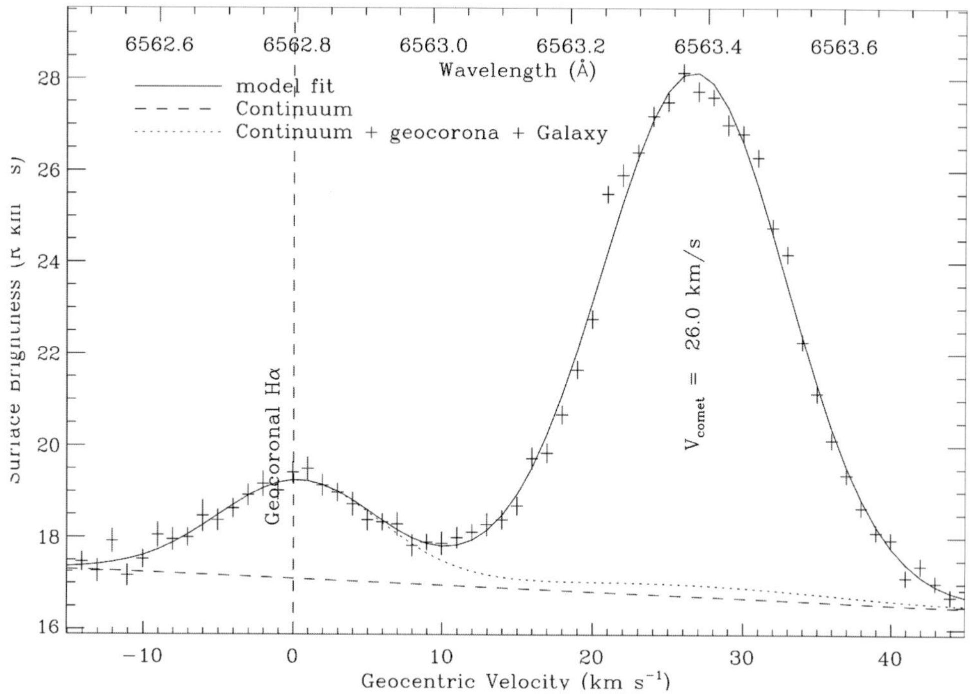

Figure 2. Fabry–Pérot (FP) spectrum, with statistical error bars, of the portion of the ring image shown in Figure 2. The spectrum was fitted using the instrument response shown in Figure 3 and a model composed of two variable Gaussians, a sloping continuum and two fixed Gaussians representing the Galaxy.

We used the instrument response shown in Figure 3 to fit Voigt profiles to the geocoronal and cometary emissions (solid line in Figure 2). Galactic emission determined from the Wisconsin Hα Mapper (WHAM) survey data (Reynolds, 1997, http://www.astro.wisc.edu/wham) and shown as the dotted line in Figure 2, was included as a fixed part of the continuum in the fits. The Gaussian line widths from these fits are given in Table I. The Lorentzian components of the comet fits always converged to small values; for the geocorona, they often had to be set to zero.

Preliminary results from the detailed FP response modeling discussed in Section 2 suggest that the difficulty we had constraining the geocoronal Hα line widths was due to the proximity of the this line to a ghost of the cometary Hα line. Using fits to the nightly hydrogen-deuterium calibration spectra, we see evidence for an increase of the instrument profile width to 8 km s^{-1} in February with a corresponding decrease in the instrument sensitivity. This corresponds to a change of 0.5 km s^{-1} in the cometary Hα line width. We expect to be able to model and completely remove these effects in the final analysis.

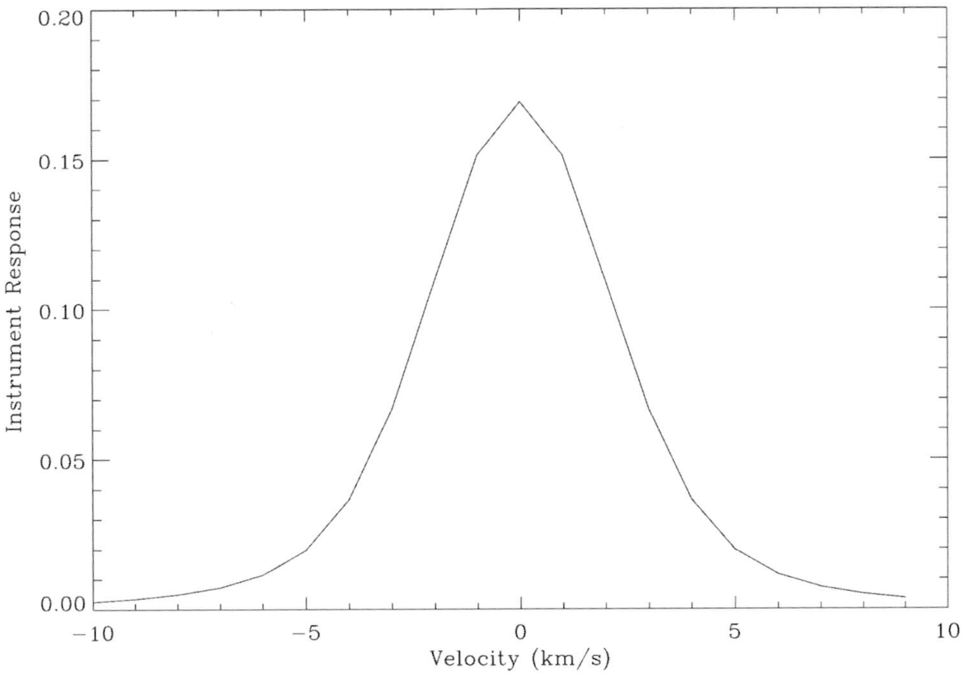

Figure 3. Estimate of the spectral response of the 50 mm FP spectrometer at Hα.

TABLE I

Preliminary Hα results

UT (1997)	R[a]	$\dot{R}$[b]	Δ[c]	$\dot{\Delta}$[d]	FOV[e]	dv_g[f]	dv_c[g]
Feb 15 12:03	1.2017	−18.75	1.7209	−33.09	3.01	5.0	14.4
Mar 5 12:28	1.0291	−13.85	1.4273	−21.89	2.57	8.1	13.6
Apr 14 02:31	0.9431	7.58	1.4970	24.95	2.70	6^8	11.4
Apr 14 02:37	0.9431	7.58	1.4971	24.95	2.70	6.7	11.0
Apr 14 02:48	0.9431	7.58	1.4972	24.96	2.70	6^8	10.5
Apr 15 03:39	0.9478	8.14	1.5122	25.57	2.72	9.8	13.9
Apr 16 02:48	0.9525	8.64	1.5264	26.10	2.75	6^8	11.9
Apr 16 03:05	0.9525	8.65	1.5266	26.11	2.75	10.2	13.0
Apr 16 03:18	0.9526	8.65	1.5267	26.12	2.75	10.5	13.0

[a] Heliocentric distance in AU.
[b] Heliocentric velocity in km s^{-1}.
[c] Geocentric distance in AU.
[d] Geocentric velocity in km s^{-1}.
[e] Field-of-view diameter in units of 10^5 km.
[f] Geocoronal Hα line width (FWHM) in km s^{-1}.
[g] Cometary Hα line width (FWHM) in km s^{-1}.
[h] Not well constrained.

4. Results

In spite of the planned improvements in the Hα spectral extraction and instrument profile generation, we do not expect much change in the line width results, summarized in the "dv_c" column of Table I, since we found these values to vary by no more than 0.5 km s^{-1} due to any of the effects discussed above. Thus, Hale–Bopp's measured Hα line width is 11–13 km s^{-1} (FWHM) for a $\sim$2.7 $\times$ 10^5 km FOV centered $\sim$3.3 $\times$ 10^5 km sunward of the nucleus.

The Hα line profile of comet Halley, measured with a similar instrument, was successfully modeled using a sophisticated Monte-Carlo code that traces the evolution of individual particles in the comet atmosphere (Combi and Smyth, 1988a; Smyth et al., 1993). This code, which has also successfully matched observations of the Lyα coma in comets 1P/1982 U1 (Halley; Combi and Smyth, 1988b), C/1996 B2 (Hyakutake; Richter et al., 2000), and Hale–Bopp (Combi et al., 2000) predicts an intrinsic hydrogen velocity dispersion of 4.5 km s^{-1} for our FOV (M. R. Combi, 2002; private communication). Scaling the radiative transfer calculations of Richter et al. (2000) to the conditions in Hale–Bopp at the position of our FOV suggests that the top half of the Hα line is lost due to the optical depth effects (Combi, 2002, private communication). This results in an effective line width of 10–12 km s^{-1}, which is consistent with our measurements.

We intend to continue our analysis of the Hale–Bopp Hα data using the techniques discussed in Section 3 to reduce the magnitude of the relative systematic errors between data points to the level of the statistical error bars shown in Figure 2 or better. With the techniques discussed in Section 2, we expect to derive the instrument profile to similar accuracy. The resulting line profiles will provide a powerful constraint to detailed models of cometary atmosphere dynamics and coma emissions.

Acknowledgements

We would like to thank F. Roesler, E. Mierkiewicz, and M. Combi for valuable discussions. This work has been supported under NSF grant AST-9615625, NASA grant NAG 5-7952, and NASA contract NASW-97020.

References

Anderson, C. M.: 1999, *Earth Moon Planets* **78**, 99.
Bishop, J., Harlander, J., Nossal, S., and Roesler, F. L.: 2001, *J. Atmos. Sol.-Terr. Phys.* **63**, 341–353.
Bockelée-Morvan, D., Crovisier, J., and Gerard, E.: 1990, *A&A* **238**, 382–400.
Coakley, M. M. and Roesler, F. L.: 1994, *Proc. SPIE* **2266**, 122–132.
Cochran, A. L. and Schleicher, D. G.: 1993, *Icarus* **105**, 235–253.
Combi, M. R. and Smyth, W. H.: 1988a, *ApJ* **327**, 1026–1043.

Combi, M. R. and Smyth, W. H.: 1988b, *ApJ* **327**, 1044–1059.
Combi, M. R., Reinard, A. A., Bertaux, J., Quemerais, E., and Mäkinen, T.: 2000, *Icarus* **144**, 191–202.
Glinski, R. J., Post, E. A., and Anderson, C. M.: 2001, *ApJ* **550**, 1131–1139.
Harris, W. M., Morgenthaler, J. P., Scherb, F., Anderson, C., and Oliversen, R.: 2002a, these proceedings.
Harris, W. M., Scherb, F., Mierkiewicz, E. J., Oliversen, R. J., and Morgenthaler, J. P.: 2002b, *ApJ* **578**, 996–1008.
Huebner, W. F., Keady, J. J., and Lyon, S. P.: 1992, *Astrophys. Space Sci.* **195**, 1–294.
Morgenthaler, J. P., Harris, W. M., Scherb, F., Anderson, C. M., Oliversen, R. J., Doane, N. E., Combi, M. R., Marconi, M. L., and Smyth, W. H.: 2001, *ApJ* **563**, 451–461.
Morgenthaler, J. P., Harris, W. M., Roesler, F. L., Scherb, F., Anderson, C. M., Doane, N. E., and Oliversen, R. J.: 2002, these proceedings.
Nossal, S., Roesler, F. L., and Coakley, M. M.: 1998, *J. Geophys. Res.* **103**, 381–390.
Nossal, S., Roesler, F. L., Coakley, M. M., and Reynolds, R.: 1997, *Geophys. Res.* **102**, 14541–14554.
Oliversen, R. J., Doane, N. E., Scherb, F., Harris, W. N., and Morgenthaler, J. P.: 2002, *ApJ*, in press.
Reynolds, R. J.: 1997, *Science* **277**, 1446–1447.
Richter, K., Combi, M. R., Keller, H. U., and Meier, R. R.: 2000, *ApJ* **531**, 599–611.
Roesler, F. L.: 1974, in N. P. Carleton (ed.), *Methods of Experimental Physics*, Vol. 12, part A, Academic Press, New York, pp. 531–569.
Smyth, W. H., Marconi, M. L., Scherb, F., and Roesler, F.: 1993, *ApJ* **413**, 756–763.

AN EASY-TO-USE MODEL FOR THE OPTICAL THICKNESS AND AMBIENT ILLUMINATION WITHIN COMETARY DUST COMAE

MICHAEL MÜLLER⋆, SIMON F. GREEN⋆⋆ and NEIL MCBRIDE⋆⋆⋆
Planetary and Space Sciences Research Institute, The Open University, Walton Hall, Milton Keynes, MK7 6AA, UK

(Received 6 March 2002; Accepted 21 May 2002)

Abstract. We define a procedure which allows estimation of the optical thickness of a cometary dust coma and the ambient illumination of the nucleus for any given comet, if estimates of the nucleus radius and the dust activity ($Af\rho$) are available. The calculation is performed for a singly scattering coma with a $\cos(\vartheta)$ distribution of dust over its day side. We find that the ambient illumination is of the same order as the incident sunlight if the optical thickness is of order one. The optical thickness increases, all else equal, linearly with the nucleus radius. Therefore the effect of the presence of the coma may be neglected for small ($\approx$ 1 km diameter) comets, but is important for comets such as 1P/Halley and Hale–Bopp.

Keywords: Comets, dust, optical thickness, radiative transfer

1. Introduction

The optical thickness of cometary dust comae is of great importance for the understanding of observations of cometary nuclei and their activity. The optical thickness reduces the visibility of cometary nuclei (Hellmich and Keller, 1981) and can influence the comet activity (Hellmich, 1981). Despite its importance relatively little is known about the optical thickness of cometary dust comae.

The optical thickness at the comet surface is also an important parameter for thermal models of the nucleus (e.g., Weissman and Kieffer, 1981; Fanale and Salvail, 1984; Enzian et al., 1998). However, most models are only applicable if the optical thickness of the dust coma is negligible (e.g., Fanale and Salvail, 1984; Enzian et al., 1998). An exception is the work by Weissman and Kieffer (1981) who consider a multiply scattering radially symmetric dust coma. However, Herman and Salo (1987) pointed out that Weissman and Kieffer (1981) made a mistake in the calculation of the radiative transfer in the coma. Therefore, it seems that the lack of an easy-to-use radiative transfer model prevented the energy input by the dust coma being properly incorporated into these models.

Several models for the radiative transfer in cometary comae exist (Marconi and Mendis, 1984; Hellmich, 1981; Herman and Salo, 1987; Salo, 1988). study the

⋆ Author for correspondence. E-mail: m.mueller@open.ac.uk
⋆⋆ E-mail: s.f.green@open.ac.uk
⋆⋆⋆ E-mail: n.m.mcbride@open.ac.uk

Earth, Moon and Planets **90:** 99–108, 2002.

special case of 1P/Halley during the fly-by of the Giotto and Vega spacecraft. Their results are dependent on assumed dust-to-gas ratios, dust size distributions and gas activities. Application of these models to other comets would require a re-run of these models with a modified set of parameters. More general are the works by Herman and Salo (1987) and Salo (1988), who present tabulated values as a function of the coma optical thickness. These results could readily be applied to any comet, if the optical thickness of the coma were known.

In the present work we introduce an expression which allows the estimation of the optical thickness of a dust coma from an $Af\rho$ value and a nucleus radius. The $Af\rho$ value is a measure of the dust activity of a comet and was introduced by A'Hearn et al. (1984). $Af\rho$ values and estimates of nucleus sizes are available for various comets (A'Hearn et al., 1995 and Fernández et al., 1999). Therefore it is now possible to introduce an easy-to-use model for the energy influx on the nucleus surface due to the dust coma. The purpose of this work is twofold. Firstly, we want to point out that the optical thickness increases with the nucleus size and that for most small ($\approx$ 1 km diameter) comets it is justified to neglect the ambient illumination. Secondly, we introduce a model for the ambient illumination of a cometary nucleus, which can easily be used as input in thermal models. The model is based on the assumption of single scattering and uses the dust particle properties given by Divine (1981).

2. Dust Environment Model

Landgraf et al. (1999) introduced a model for the dust environment in the vicinity of a spherical nucleus. The model is restricted to the innermost part of the dust coma where the influence of radiation pressure is negligible and the gas drag dominates the motion of the dust particles. Under the assumption of a Sun driven activity on the nucleus surface, the number density of dust in the coma is calculated physically consistently. The number density calculated with this model can be approximated well by

$$n_d(\mathbf{r}) = \frac{C}{G} \max\left[\frac{\cos(\vartheta)}{r^2}, 0\right], \tag{1}$$

where r is the distance from the nucleus centre and ϑ is the angle of the vector $\mathbf{r}$ with respect to the Sun direction $\mathbf{e}_{\text{sun}}$. G is the geometric particle cross section and C is a constant which is specified below.

Landgraf et al. (1999) considered various dust classes which represent dust particles of different sizes and masses. However, as far as the model of the radiation environment around the nucleus is concerned, it is sufficient to consider one dust class only that has the *ensemble* properties of particles of all sizes and masses. The dust particle properties that are important in the present work are the particle

geometric albedo $p = 0.04$, the phase function $j(\alpha)$ (see Müller, 1999) and the particle temperature:

$$T_d(r_h) = T_0 r_h^{-\gamma} \tag{2}$$

with $T_0 = 310$ K and $\gamma = 0.58$ (Divine, 1981). All these parameters were derived from ground based observations under the assumption that the same particles that dominate the scattered light also dominate the thermally emitted radiation (Divine, 1981).

The temperature of particles measured from ground based observations is dominated by particles in the optically thin part of the coma, where the incident intensity is $I = f_{\text{sun}} r_h^{-2}$. For the purpose of Section 4, where we consider a position dependent optical thickness of the coma, we re-write the dust temperature as a function of the incident intensity I:

$$\tilde{T}_d(I) = T_0 \left(\frac{I}{f_{\text{sun}}} \right)^{\gamma/2} . \tag{3}$$

An essential feature of the model is that the overall dust activity of the comet is adjusted such that an observed dust activity is matched. This can be achieved by setting the constant C in Equation (1) to

$$C = \frac{A(0)f\rho}{8p}, \tag{4}$$

where $A(0)f\rho$ is the observed dust activity if the comet were observed at a phase angle of zero degrees. The $Af\rho$ value is dependent on the phase angle during the observation. This is because of the non-isotropic phase function of the individual particles and a geometric projection effect on the image plane. For the present simplified coma model the $Af\rho$ value at zero phase angle can be calculated from an $Af\rho$ value measured at phase angle α by $A(0)f\rho = A(\alpha)f\rho / j_{geo}(\alpha)$, where j_{geo} is the 'geometric phase function' introduced in Müller (1999).

3. Optical Thickness of the Dust Coma

The optical thickness between an observer at position $\mathbf{r}_{\text{obs}}$ and a point at distance Δ from the observer, in the direction of the line of sight $\mathbf{e}_{\text{los}}$ is given by

$$\tau_{los,i}(\mathbf{r}_{\text{obs}}, \mathbf{e}_{\text{los}}, \Delta) = Q_{ext,i} G \int_0^{\Delta} n_d(\mathbf{r}_{\text{los}}(\Delta))\, d\Delta, \tag{5}$$

where $\mathbf{r}_{\text{los}}(\Delta) = \mathbf{r}_{\text{obs}} + \Delta \mathbf{e}_{\text{los}}$. The index i indicates either thermal (th) or scattered (sc) components. In the present study we set $Q_{\text{ext,sc}} = 1$ for scattered visual radiation and $Q_{\text{ext,th}} = 2$ for thermally emitted radiation. For the present simplified coma model the optical thickness of a point $\mathbf{r}$ with respect to the Sun is given by

$$\tau_{sun}(\mathbf{r}) = \tau_{los,sc}(\mathbf{r}, \mathbf{e}_{\text{sun}}, \infty) = Q_{ext,sc} \frac{C}{r^*}, \tag{6}$$

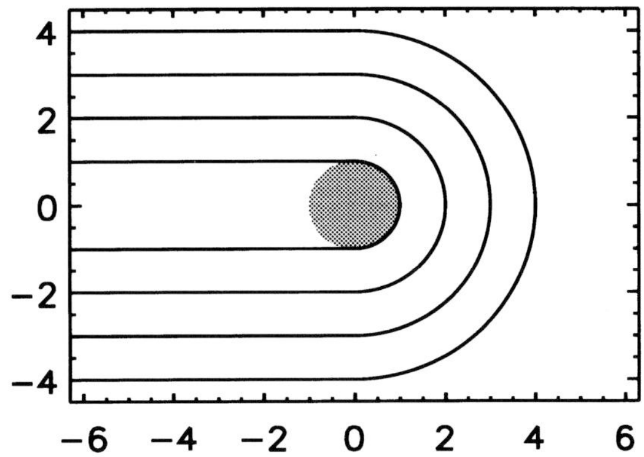

Figure 1. Lines of constant optical thickness. The grey circle indicates the nucleus. The scale is in nucleus radii. The Sun is to the right.

where r^* is the distance from the nucleus centre if the point is over the nucleus day side. If the point is over the nucleus night side r^* is the distance from the symmetry axis. Lines of constant optical thickness are shown in Figure 1.

Over the comet day side the optical thickness is constant at any given distance r from the nucleus centre and the optical thickness at the nucleus surface ($r - r_n$) is given by:

$$\tau_n = Q_{ext,sc}\frac{C}{r_n} = \frac{Q_{ext,sc}A(0)f\rho}{8pr_n}. \tag{7}$$

Assuming that the dust particle properties (i.e., scattering properties and extinction efficiency) do not change with heliocentric distance, Equation (7) states that the optical thickness on the comet surfaces varies with the heliocentric distance like the $Af\rho$ parameter. Furthermore, for different comets at the same heliocentric distance it can be seen that the optical thickness at the comet surface increases roughly with the size of the nucleus. This can easily be understood by making the zero-order-approximations that the dust velocity does not change with the size of the nucleus and that every nucleus produces (on average) the same amount of dust per unit surface area. With these assumptions the $Af\rho$ parameter increases with the square of the nucleus radius. As the optical thickness is proportional to the ratio $Af\rho/r_n$ the optical thickness increases linearly with the nucleus radius.

The optical thicknesses for the comets 46P/Wirtanen, 1P/Halley and Hale–Bopp (C/1995 O1) estimated using Equation (7) are given in Table I. The $Af\rho$ values for 46P/Wirtanen and Hale–Bopp were taken from Farnham and Schleicher (1997) and Churyumov et al. (1997), respectively. For 1P/Halley the post-perihelion value given by Schleicher et al. (1998) was used and an effective nucleus radius of 5 km was considered. Assuming a geometric albedo of 0.04, the nucleus radius of 700 m for 46P/Wirtanen is consistent with the magnitude given by Fernández et al. (1999). Weaver and Lamy (1997) estimate that the probable range for the radius

TABLE I
Optical thickness of the dust coma of three comets close to 1 AU

Comet	r_h [AU]	α [deg]	$A(\alpha) f\rho$ [cm]	$A(0) f\rho$ [cm]	r_n [km]	τ_n
46P/Wirtanen	1.07	40	1.2×10^2	2.6×10^2	0.7	0.02
1P/Halley	0.97	65	2.14×10^4	6.2×10^4	5	0.8
Hale–Bopp	0.91	48	10^6	2.4×10^6	35	4.3

of the nucleus of Hale–Bopp is 20–35 km. Using the value of 35 km, we find the significant value of $\tau_n \approx 4$ for the optical thickness of Hale–Bopp's dust coma near perihelion. If the nucleus of Hale–Bopp has a radius of only 20 km then we find the even higher value of $\tau_n = 7.5$. For 1P/Halley, we calculate the optical thickness to be 0.8. The lower optical thickness is mostly due 1P/Halley's smaller size. For the much smaller comet 46P/Wirtanen we derive an optical thickness of $\tau_n = 0.02$. However, if the optical thickness is scaled linearly with the nucleus radius the difference in optical thickness cannot be explained by the much smaller nucleus radius alone. The ratio of the $Af\rho$ value and the hydroxide production rate is an indicator for how 'dusty' a comet is. For Hale–Bopp and 1P/Halley close to 1 AU this ratio is $\log(Af\rho/Q_{\mathrm{OH}}) \approx -25$ (Schleicher et al., 1997, 1998), where $Af\rho$ is measured in cm and Q_{OH} is in molecules s^{-1}. For 46P/Wirtanen this ratio is only ≈ -25.9 (A'Hearn et al., 1995). Therefore our small dataset of three comets confirms the anticipated trend that the optical thickness increases with the nucleus size, but also shows differences due to the diversity in the dust production of different comets.

4. Ambient Illumination of the Comet Surface

The optical thickness of the coma reduces the amount of sunlight that reaches the nucleus directly. However, because the absorbed and scattered light can reach the nucleus indirectly, it is not clear, *a priori*, whether the presence of the dust coma reduces or increases the total amount of radiation that reaches the nucleus. In this section we calculate the power incident on the nucleus under the assumption of single scattering.

The radiation received from a line of sight can be written

$$J_{\mathrm{los},i} = \int_0^\infty e^{-\tau_{\mathrm{los},i}(\Delta)} G n_d(\mathbf{r}_{\mathrm{los}}(\Delta)) D_{\mathrm{los},i}(\Delta)\, d\Delta . \tag{8}$$

The function $D_{\mathrm{los},i}(\Delta)$ specifies the amount of radiation scattered (i = sc) or thermally emitted (i = th) by a particle towards the respective point on the nucleus surface. These functions are given by

$$D_{\mathrm{los,sc}}(\Delta) = \frac{pj(\alpha_{\mathrm{los}})}{\pi} \frac{f_{\mathrm{sun}}}{r_h^2} e^{-\tau_{\mathrm{sun}}(\mathbf{r}_{\mathrm{los}}(\Delta))}, \tag{9a}$$

$$D_{\mathrm{los,th}}(\Delta) = \frac{\sigma}{\pi} \tilde{T}_d^4 \left(\frac{f_{\mathrm{sun}}}{r_h^2} e^{-\tau_{\mathrm{sun}}(\mathbf{r}_{\mathrm{los}}(\Delta))} \right) = \frac{\sigma T_d^4(r_h)}{\pi} e^{-2\gamma \tau_{\mathrm{sun}}(\mathbf{r}_{\mathrm{los}}(\Delta))}. \tag{9b}$$

In the previous equation the Sun is the only light source. Hence the present work is restricted to single scattering. However, we take into account that a dust particle receives *attenuated* sunlight due to the optical thickness of the coma.

If θ denotes the angle of the line of sight with the local surface normal and ϕ is a corresponding azimuthal angle the intensity received at a point ϑ at the nucleus surface is given by

$$I_{\mathrm{amb}}(\vartheta) = \int_0^{2\pi} \int_0^1 (J_{\mathrm{los,th}} + J_{\mathrm{los,sc}}) \cos(\theta)\, d\cos(\theta)\, d\phi. \tag{10}$$

The total power received due to ambient illumination P_{amb} can easily be found by integrating $I_{\mathrm{amb}}(\vartheta)$ over the nucleus surface. Figure 2 shows the power due to the ambient illumination scaled to the power the nucleus would receive if the dust coma were not present (i.e., $P_{\mathrm{amb}}/(\pi r_n^2 f_{\mathrm{sun}}/r_h^2)$). The calculation of the ambient illumination was performed for a coma at 1 AU heliocentric distance and a nucleus of 40 km radius. From equations (8)–(10) it can be seen that the results shown in Figure 2 do not depend on the nucleus radius. Therefore the results are applicable to a nucleus radius of any size. The ambient illumination is dominated by the thermal radiation i.e., it scales approximatively with the exponent -4γ with heliocentric distance. γ is the exponent which describes the variation of the dust temperature with heliocentric distance (see Equation (2)). As this exponent is approximately $\gamma \approx 0.5$ the ratio of the ambient illumination and incident sun light, as shown in Figure 2, is only moderately dependent on the heliocentric distance. Therefore the results can be applied as a first approximation to a range of heliocentric distances. For optical thicknesses $\tau_n \ll 1$ the power due to ambient illumination increases linearly with the optical thickness. For an optical thickness of $\tau_n \approx 1$ the ambient illumination reaches a maximum and even decreases for large optical thicknesses. This is because the light incident on dust particles close to the nucleus is significantly decreased for $\tau_n > 1$. For an optical thickness $\tau_n \approx 3$ the nucleus receives as much radiation from the dust coma as from direct sunlight. The total intensity received by the nucleus is always smaller than the power that would reach the nucleus if the dust coma was not present.

The distribution of the incident ambient illumination on the nucleus day side can, to first order, be approximated by a cosine $I_{amb}(\vartheta) \approx I_{amb}(0)\cos(\vartheta)$. As the

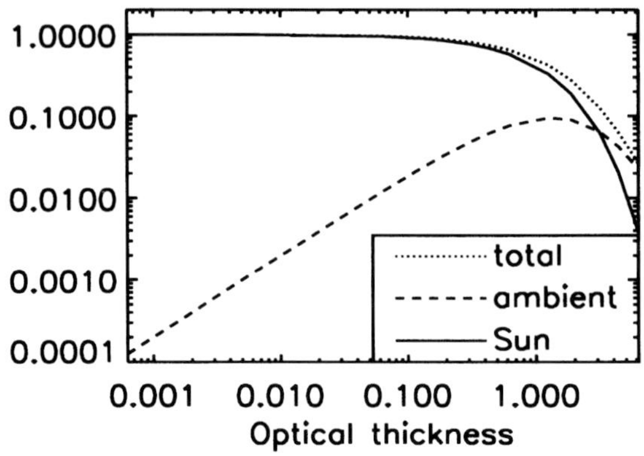

Figure 2. Ambient radiation that reaches the nucleus scaled to the power of incident sunlight that would reach the nucleus surface without the presence of the dust coma.

incident solar intensity decreases like a cosine as well, it makes sense to define an *effective* optical thickness which describes both direct sunlight and the gain due to the presence of the dust coma:

$$\tau_{\rm eff}(\tau_n) = -\ln\left(e^{-\tau_n} + \frac{P_{\rm amb}}{\pi r_n^2 f_{\rm sun}/r_h^2}\right). \tag{11}$$

On the basis of the effective optical thickness τ_{eff} the intensity incident on a point on the nucleus day side at an angle ϑ from the subsolar point can then easily be calculated by

$$I_{tot}(\vartheta) = \frac{f_{\rm sun}}{r_h^2}\cos(\vartheta)e^{-\tau_n} + I_{amb} \approx \frac{f_{\rm sun}}{r_h^2}\cos(\vartheta)e^{-\tau_{eff}}. \tag{12}$$

For various optical thicknesses τ_n, values for τ_{eff} are given in Table II. Also given in Table II are the corresponding values of the $A(0)f\rho/r_n$-ratios.

5. Summary

The optical thickness on the nucleus surface is proportional to the ratio of the $Af\rho$ value and the nucleus radius. At a given heliocentric distance, $Af\rho$ values of different comets increase approximatively with the square of the nucleus radius. Therefore optical thickness increases linearly with the nucleus radius.

For a model with a $\cos(\vartheta)$-distribution of dust over the nucleus day side, we find that the optical thickness with respect to the Sun is constant on the comet day side. For comets 46P/Wirtanen, 1P/Halley and Hale–Bopp near 1 AU, the optical thicknesses are 0.02, 0.8 and 4.3, respectively. The difference in the values is mainly explained by the different sizes of the nuclei.

TABLE II

Effective optical thicknesses. For explanation see text

$\frac{A(0)f\rho}{r_n}$	τ_n	τ_{eff}	$\frac{A(0)f\rho}{r_n}$	τ_n	τ_{eff}
$\tau_{eff,th}$					
0.001	0.006	0.004	0.100	0.625	0.431
0.010	0.063	0.043	0.200	1.250	0.860
0.020	0.125	0.086	0.300	1.875	1.278
0.030	0.188	0.130	0.500	3.125	2.072
0.050	0.313	0.216	0.700	4.375	2.776
0.070	0.438	0.303	1.000	6.250	3.640

We calculated the ambient illumination of the nucleus surface under the assumption of single scattering in the coma. The ambient illumination is comparable with direct sunlight incident on the nucleus surface, for an optical thickness $\tau_n \approx 1$. The calculations take into account that the dust particle close to the nucleus only receive attenuatted Sun light. With the $\cos(\vartheta)$-model for the dust number density, we find that the ambient illumination on the comet day side also varies as $\cos(\vartheta)$. With the additional simplification of the $\cos(\vartheta)$-model that the optical thickness is constant on the comet surface, we are able to define an easy-to-use model for the estimation of the ambient illumination of the nucleus surface. This estimation can be performed in the following four steps: (i) Get an estimate for the nucleus radius and the dust activity $A(\alpha)f\rho$ for the respective comet at the respective heliocentric distance (see e.g., Fernández et al., 1999; A'Hearn et al., 1995). (ii) Scale the $Af\rho$ value to zero degrees phase angle ($A(0)f\rho = A(\alpha)f\rho/j_{\rm geo}(\alpha)$; values for $j_{\rm geo}$ can be found in Müller, 1999). (iii) Read the corresponding effective optical thickness $\tau_{\rm eff}$ from Table II. (iv) Calculate the ambient illumination using Equation (12). This procedure may be applied to comets of any size and to a large range of heliocentric distances.

The calculations were performed assuming the optical and thermal properties given by Divine (1981) for 1P/Halley. It is worth mentioning how the results depend on these parameters. The optical thickness is inversely proportional to the dust particle albedo (see Equation (7)) and can therefore easily be scaled to values different from $p = 0.04$. For comparison with other values given in the literature it is worth noting that the albedo $A(0)$ which is used in the definition of the $Af\rho$-value (A'Hearn et al., 1984) is related to the geometric albedo by $A(0) = 4p$. If all parameters apart from the albedo remain unchanged then the ambient illumination is only a function of the coma optical thickness τ_n. Hence for a geometric albedo different from $p = 0.04$ the effective optical thickness $\tau_{\rm eff}$ can still be read from Table II by using the scaled value of the optical thickness τ_n. As the ambient il-

lumination is dominated by the thermal radiation, the ambient illumination scales linearly with the radiation emitted by the dust (σT_d^4). If the emission spectrum of the dust is not well represented by the spectrum of a blackbody (e.g., in the case of superheat), a dust temperature should be used which represents the total amount of energy emitted over the entire spectrum. As superheat does not change the total amount of radiation emitted (it only changes the distribution over the spectrum), the results should not be affected significantly. Also, the size distribution of the dust particles only effects the results if the effective optical properties of the ensemble is affected.

Acknowledgements

This work was funded by the UK Particle Physics and Astronomy Research Council.

References

A'Hearn, M. F., Millis, R. L., Schleicher, D. G., Osip, D. J., and Birch, P. V.: 1995, 'The Ensemble Properties of Comets: Results from Narrowband Photometry of 85 Comets, 1976–1992', *Icarus* **118**, 223–270.

A'Hearn, M. F., Schleicher, D. G., Feldman, P. D., Millis, R. L., and Thompson, D. T.: 1984, 'Comet Bowell 1980b', *Astron. J.* **89**, 579–591.

Churyumov, K. I., Kleshchonok, V. V., and Kravtsov, F. I.: 1997, 'Narrowband Electrophotometry of Comet Hale–Bopp (C/1995 O1)', *Earth Moon Planets* **78**, 155–160.

Divine, N.: 1981, 'A Simple Radiation Model of Cometary Dust of P/Halley', *ESA SP* **174**, 25–30.

Enzian, A., Cabot, H., and Klinger, J.: 1998, 'Simulation of the Water and Carbon Monoxide Production Rates of Comet Hale–Bopp Using a Quasi 3-D Nucleus Model', *Planet. Space Sci.* **46**, 851–858.

Fanale, F. P. and Salvail, J. R.: 1984, 'In Idealized Short-Period Comet Model: Surface Insolation H_2O Flux, Dust Flux and Mantle Evolution', *Icarus* **60**, 476–511.

Farnham, T. and Schleicher, D. G.: 1997, 'Comet 46P/Wirtanen', *IAUC* 6595.

Fernández, J. A., Trancredi, G., Rickman, H., and Licandro, J.: 1999, 'The Populations, Magnitudes, and Sizes of Jupiter Family Comets', *Astron. Astrophys.* **352**, 327–340.

Hellmich, R.: 1981, 'The Influence of the Radiation Transfer in Cometary Dust Halos on the Production Rates of Gas and Dust', *Astron. Astrophys.* **93**, 341–346.

Hellmich, R. and Keller, H. U.: 1981, 'On the Visibility of Nuclei of Dusty Comets', *Icarus* **47**, 325–332.

Herman, G. and Salo, H.: 1987, 'Light Scattering in Cometary Dust Comae', *Earth Moon Planets* **39**, 51–74.

Landgraf, M., Müller, M., and Grün, E.: 1999, 'Prediction of the *In-Situ* Dust Measurements of the Stardust Mission to Comet 81P/Wild 2', *Planet. Space Sci.* **47**, 1029–1050.

Marconi, M. L. and Mendis, D. A.: 1984, 'The Effects of the Diffuse Radiation Fields Due to Multiple Scattering and Thermal Reradiation by Dust on the Dynamics and Thermodynamics of a Dusty Cometary Atmosphere', *Astrophys. J.* **287**, 445–454.

Müller, M: 1999, 'A Model of the Inner Coma of Comets with Applications to the Comets P/Wirtanen nad P/Wild 2', Ph.D. Thesis, Ruprecht-Karls-Universität, Heidelberg, Germany.

Salo, H.: 1988, 'Monte Carlo Modeling of the Net Effects of Coma Scattering and Thermal Reradiation on the Energy Input to Cometary Nucleus', *Icarus* **76**, 253–269.

Schleicher, D. G., Lederer, S. M., Millis, R. L., and Farnham, T. L.: 1997, 'Photometric Behavior of Comet Hale–Bopp (C/1995 O1) Before Perihelion', *Science* **275**, 1913–1915.

Schleicher, D. G., Millis, R. L., and Birch, P. V.: 1998, Narrowband Photometry of Comet P/Halley: Variation with Heliocentric Distance, Season, and Solar Phase Angle', *Icarus* **132**, 397–417.

Weaver, H. A. and Lamy, P. A.: 1997, Estimating the Size of Hale–Bopp's Nucleus, *Earth Moon Planets* **79**, 17–33.

Weissman, P. R. and Kieffer, H. H.: 1981, 'Thermal Modeling of Cometary Nuclei', *Icarus* **47**, 302–311.

HOW FAST THE NUCLEUS OF COMET HALE–BOPP WAS ROTATING?

L. M. SHULMAN
Main Astronomical Observatory of the National Academy Science of Ukraine, 27 Zabolotny st, 03680, Kyiv, Ukraine
E-mail: shulman@mao.kiev.ua

M. KIDGER, M. SERRAT-RICART and R. TORRES-CHICO
Instituto Astrofisico de Canarias, La Laguna, Tenerife, Spain, E-mail: mrk@iac.es

(Received 25 February 2002; Accepted 22 July 2002)

Abstract. At the First International Conference on Comet Hale–Bopp (Canaries, 1998) many astronomers agreed with the value of the period of axial rotation of its nucleus, which was obtained by observers by a comparison of the two sequential patterns of the dust arcs. It was noted that the pattern had been repeated each ~12 hours. The value 11.35^{h} was taken as the period of axial rotation. However, alternative periods of 4 and 8 days have also been suggested. The possible reasons for this discrepancy are discussed. It is shown that a short period could be obtained by a misidentification of the same arc on two sequential images and, as a result, an overestimate of the radial velocity of dust in the arcs. To obtain the half day period one must take the projected velocity of the arc ~ 1.6 km/s, which exceeds the H_2O gas expansion velocity. If one takes a lower expansion velocity a longer period would be obtained.

Keywords: Comet Hale–Bopp, cometary nucleus axial rotation, jet activity of distant comets

1. Introduction

Comet Hale–Bopp was observed at the Teide Observatory while still beyond 6 AU. Kidger et al. (1995) and Kidger (1995a, b, c) observed three spiral jets between 1995 August and October. All the phenomena were also observed by many other astronomers. It is clear that a comet at such a large heliocentric distance cannot have a coma created by the sublimation of water ice. According to radio astronomy data by Biver et al. (1997) Comet Hale–Bopp had a CO dominated coma beyond 5 AU. However, water was ten times more abundant in this comet than CO, so water was ejected from the nucleus as a jet of icy crystals dragged by the CO outflow.

The observed velocity of expansion of CO of $V_{\mathrm{exp}} \approx 500$ m/s published by Biver et al. (1997) cannot be explained by simple sublimation. The equilibrium temperature of CO at 7 AU is ≈ 34.5 K and the velocity of sublimating CO molecules should be $\sqrt{2kT/(\pi m_{co})} = 82$ m/s. If the CO gas be ejected by an ideal Laval's nozzle then the theoretical limit of the velocity should be reached: $V_{\mathrm{lim}} = \sqrt{2c_pT} = 273$ m/s. This value corresponds to the thermocapacity $c_p = 3.5k/m_{\mathrm{CO}}$, where k is the Boltzmann's constant. One can see that the observed value exceeds its theoretical limit by a factor 2 or more. This discrepancy

can be removed if we suppose that CO was evaporated not under the equilibrium temperature of CO but under the equilibrium temperature of water ice ($\sim$ 130 K at 7 AU). Details of the physical picture of the jet generation are discussed in the previous publication by Shulman et al. (1999).

Before the First International conference on Comet Hale–Bopp, and just after the first HST observations published by Weaver, Sekanina found a much longer period $\approx 8^{\mathrm{d}}$ than the 11.34 hour period that has generally been accepted. Shulman obtained another value of the period from the observations of Kidger et al., namely $\approx 4^{\mathrm{d}}$. Sekanina (1996) gave a model of the jets interpreting them as a plain dust-and-gas atmosphere formed by sublimation of CO due to insolation. Unfortunately, no details of the model are presented in the cited paper so it is impossible to evaluate the results, and the velocity anomally is a good reason to revise the problem. We thus try to solve the inverse problem: to derive the circumstances of the jet formation from the observed shape of the spiral arms.

2. How Was the Long Period Obtained?

The long spin period was derived from a long series of observations taken at Teide Observatory between 1995 August 10th and October 31st. The August spiral jet was selected for a detail investigation because six consequent positions of this single jet were observed and measured. It is important to take into consideration that only this jet is traced from the nucleus to its disappearance in the sky background. The spiral structure is neither a synchrone of the classic mechanical theory by Bessel-Bredikhin, nor a plain dust tail which, roughly speaking, is the trajectory of a dust particle in the field of the solar radiation pressure. At 7 AU all the effects connected to the orbital movement of a comet are negligible on a time scale of a few months.

The geometry of the event is shown in the Figure 1. Let X and Y (Y-axis directed along the central meridian of the nucleus) be the co-ordinates of a point of the jet in the image plane. The circumstances of the dust ejection are defined by the angles:
ϕ – the nucleocentric latitude of the active region;
ϕ_o – the nucleocentric latitude of the observer;
θ_0 – the position angle of the central meridian of the comet "disk";
$\lambda(t) = \omega(t - t_0)$ – the longitude of the active region in the moment t,
t_0 – is the moment when the active region passes the central meridian.

If the active region was ejecting dust with a velocity $v(\tau)$ at the moment of ejection τ, then at that moment t an observer on Earth sees the curve whose equation may be written as follows:

$$X(t, \tau) = (t - \tau)v(\tau) \cos\phi \sin(\omega(\tau - t_0)), \qquad (1)$$

$$Y(t, \tau) = (t - \tau)v(\tau) [\sin\phi \cos\phi_o - \cos\phi \sin\phi_o \cos(\omega(\tau - t_0))] . \qquad (2)$$

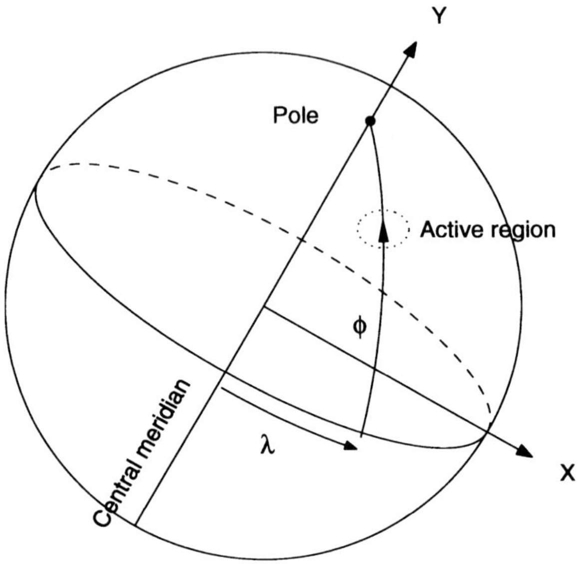

Figure 1. Geometry of the ejection.

This parametric curve can be described in polar coordinates in the image plain by the equations:

$$X = \rho \cos\theta, \; Y = \rho \sin\theta. \tag{3}$$

where $\theta = \theta_{\rm obs} - \theta_0 + \pi/2$ is the polar angle of a certain point in the jet, $\theta_{\rm obs}$ is the observed position angle of the same point, and ρ is its projected distance in the image plane. The relationship

$$\rho(t,\tau) = \sqrt{x^2 + y^2} = V_{\rm pr}(\tau)(t-\tau) = (t-\tau)v(\tau)\times$$
$$\Big[(\cos\phi + \cos\phi_o)^2 - \frac{1}{2}\sin 2\phi \sin 2\phi_o \cos(\omega(\tau - t_0))$$
$$+\cos^2\phi \cos^2\phi_o \sin^2(\omega(\tau - t_0))\Big]^{1/2} \tag{4}$$

is defined, where $V_{\rm pr}$ is the projected velocity and

$$\tan\theta(\tau) = y/x = \frac{\tan\phi\cos\phi_o}{\sin(\omega(\tau - t_0))} - \sin\phi_o \cot(\omega(\tau - t_0)). \tag{5}$$

Here x and y are the observed rectangular coordinates which are measured, respectively, to the East and to the North. The ejected dust moved from the nucleus with a velocity that does not depend on time but depends on the polar angle in the image plane. Using Equation (4) for a set of the polar angles one can find the projected

TABLE I
The circumstances of the jet generation

Parameter of the nucleus and source of dust	Quantity	Value
The active region passed the central meridian of the nucleus at	t_0	Oct 24.46
Positional angle of the central meridian	θ_0	130°
Nucleocentric latitude of the source	ϕ	68°
Nucleocentric latitude of the observer	ϕ_o	55°
The period of rotation	P	$3^{\mathrm{d}}.96$

velocity V_{pr} as well as the moment of ejection τ for each polar angle. These quantities can be calculated directly from the measured coordinates of points in the jet. One can estimate the axial rotation period $P = 2\pi/\omega$ because the nucleocentric angles covered by the jet correspond to approximately half a period, i.e., to the difference between the moments of ejection of the first and last points in the jet. It is $\sim 4^d$. It is essential that this value of the period is obtained from pure kinematics and is not model dependent.

After calculation of $V_{\mathrm{pr}}(\theta)$ and $\tau(\theta)$, we use Equation (5) and obtain the complete set of parameters: $\omega, \phi, \phi_o, t_0$. Two of these parameters, namely ϕ and ϕ_o, control the shape of the jet. A set of model calculations for different ϕ and ϕ_o allows us to choose the best values of these angles to fit the observed shape of the jet. There is a small uncertainty in this procedure but it cannot change the period of rotation. Other parameters control the orientation of the jet in the image plane. Finally, the values given in the Table I are obtained. One can calculate the spatial velocity of the ejected dust as a function of time. The result is plotted in Figure 2. The maximum value of the velocity $\sim$ 60 m/s agrees with the radio astronomy data (Biver et al., 1997) which gives the CO velocity of $\sim$ 500 m/s. The CO could thus have been the propellent that dragged the water ice grains to form the jet. Thus the four-day period could be in accordance both with the direct observed kinematics of the spiral jet and with the data on the movement of the gas which accelerated the grains.

The model image fits the observations well as we see in Figure 3. A physical mechanism of jet generation, which is able to produce a jet of icy crystals and to match the CO velocity, is considered by Shulman et al. (1999).

3. How to Get a Short Period?

The accepted period (~0.5 days) was obtained when the comet approached the Martian orbit. At that time the comet produced many jets and arcs. It was often possible to observe several arcs simultaneously. Farnham et al. (1998) found a

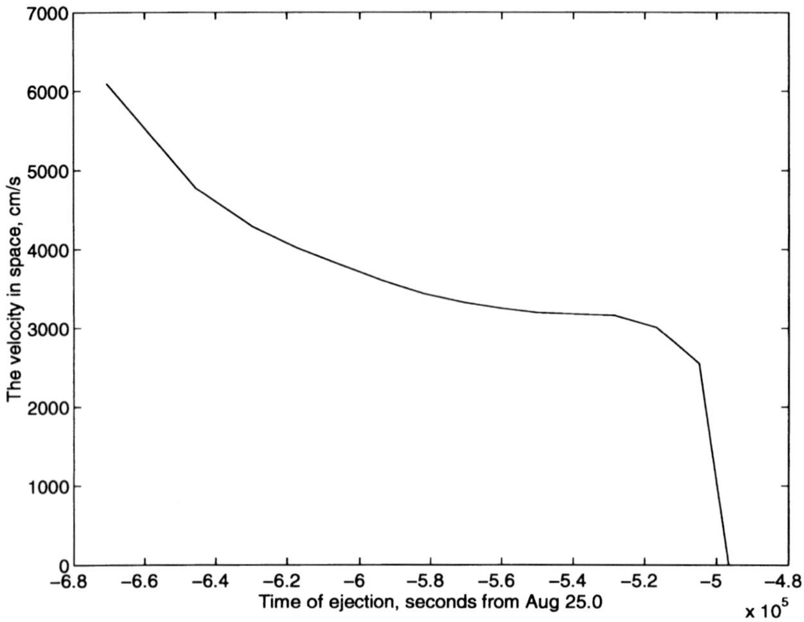

Figure 2. The absolute value of the velocity of dust in the August jet.

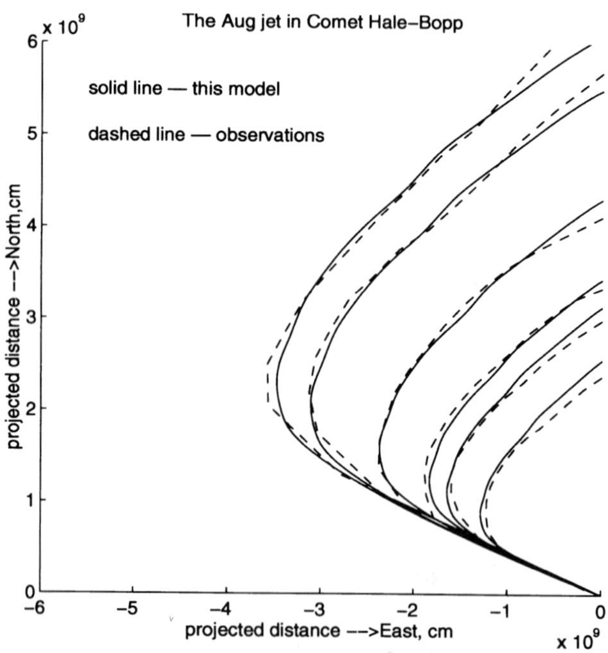

Figure 3. Comparison of this model with the observations.

period of 11.35 ± 0.05 hours from the movement of dust arcs from March 25 through April 1, 1997. The same value was been found by Licandro et al. (1998) from a 16 days series of narrow band near infrared images obtained during April 1997. If one assumes that both periods are correct one mechanism for explaining the change is the spinning-up of the nucleus by the angular momentum of ejected matter. More plausible would be the existence of more than one active regions on the surface of the nucleus (N_{ar}) each of which produced dust arcs.

If this were the case the rotation period found from observation of the arcs would be a sub-multiple of the true period: $P_{obs} = P_{real}/N_{arcs}$, so there may be a difference between the periods derived in 1995 and 1997, i.e., $P_{obs}(1997) = (P_{real}(1995) - \Delta P)/N_{ar}$. The discrepancy disappears if we take $N_{arcs}(1997) = 8$ and $\Delta P = P_{real}(1995) - P_{real}(1997) = 4.24$ hours so the angular velocity in 1997 exceeded its 1995 value by 4.5%. This small difference may be an artifact of data processing. Given the mass of the nucleus it is not plausible to spin up a nucleus with the observed gas jets. In the case of multiple active regions they would not have a specific appearance. Generally they would not be approximately equidistant in nucleocentric longitude.

The authors cited above (Farnham et al., 1998; Licandro et al., 1998) noted that the pattern of jets repeated approximately each 12 hours. However, as the comet's coma is not rotating as a solid body the periodicity of the pattern may not necessarily be an effect of spin of the nucleus. The effect of axial rotation on the coma appearance exists, but is rather complex and depends on many circumstances, thus coma patterns may not always be due directly to the spin of the nucleus.

A third possible reason for obtaining a period of rotation that is many times shorter than the real period is the possibility of misidentification of different arcs in two images separated by approximately one day, as two positions of the same arc is shown in Figure 4.

The real velocity of expansion is given by the relationship

$$V_{real} = \frac{R_2 - R_1}{N P_{real}}, \tag{6}$$

where $R_{1,2}$ are the nucleocentric distances of the arcs, N is the number of revolutions between the arcs, P_{real} is the true rotation period. In the case of a misidentification one would get (see Figure 5) a false estimate of the velocity.

$$V_{false} = \frac{R_2 - R_1}{\Delta t}, \tag{7}$$

where $\Delta t \approx 86400$ s is the approximately one day interval between two images. If the difference between the nucleocentric distances of two opposite edges of the same arc is

$$\Delta R = V P \frac{\psi}{2\pi}, \tag{8}$$

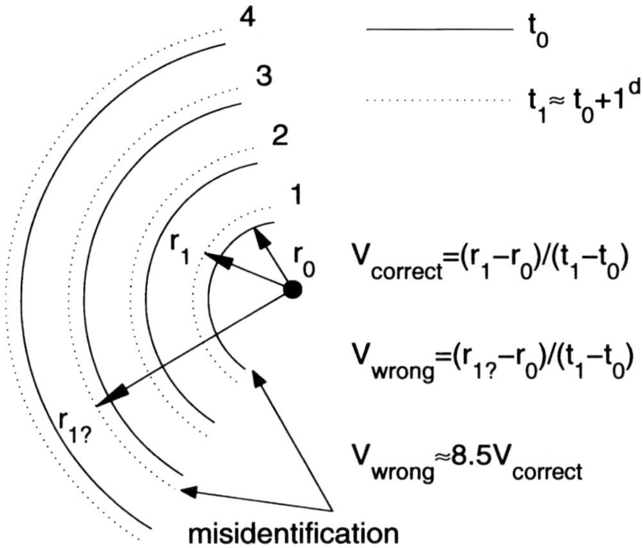

Figure 4. Misidentification of the same arc in two consecutive images.

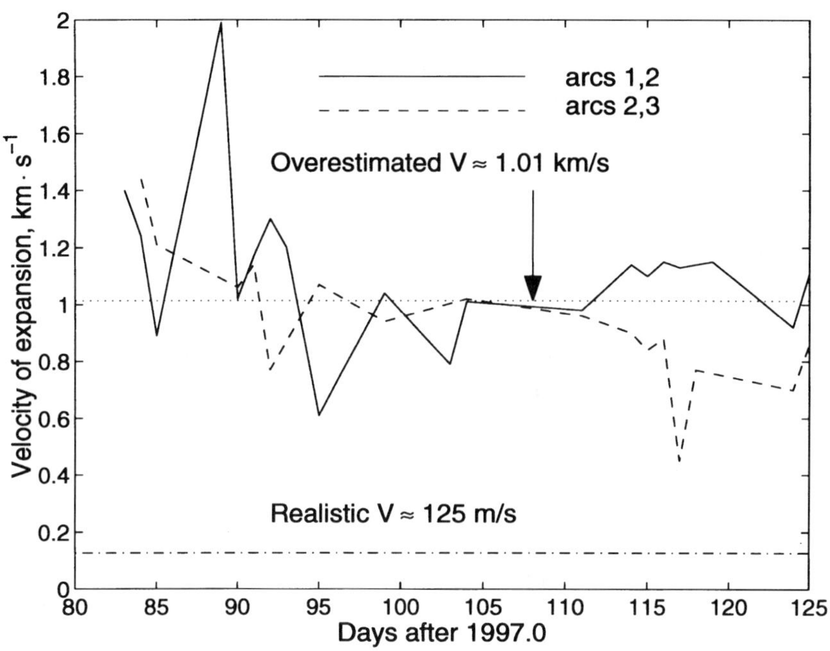

Figure 5. Overestimation of the velocity of expansion according to Yevtushevsky et al. (2001).

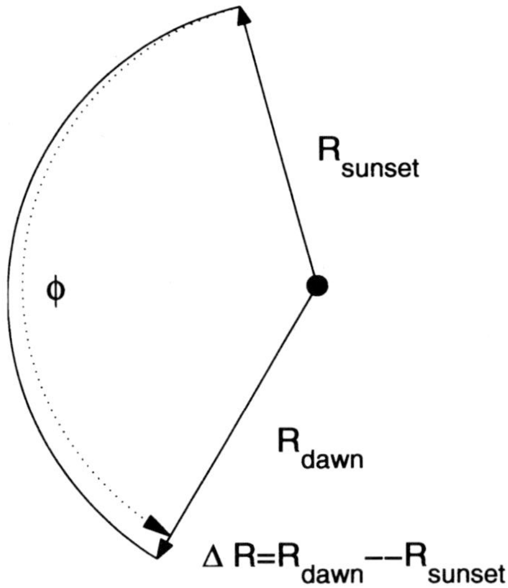

Figure 6. Determination of the period.

where ψ is the angle size of the arc then one could get a false period using the relationship (7)

$$P_{\text{false}} = \frac{\Delta R \cdot 2\pi \, \Delta t}{\psi (R_2 - R_1)} = \frac{\Delta t}{N}. \tag{9}$$

So misidentification can give a false period which may be an integer fraction of a day. This does not depend on the conditions of projection because both the distances and velocity are projected and so the function of the angle is self-eliminated. Note that this approach cannot give the true period as any true period would fit the relationships for (6) and (8). One thus has to obtain the velocity in another way, namely from the kinematics of an individual arc. It is important to remember that the physical nature of the arcs may have been different close to perihelion to those observed when the comet was at the heliocentric distance about 7 AU. When the comet crossed the Jovian orbit water vapour became the main constituent of its coma. So instead of icy grains propelled by CO, close to perihelion dust was dragged outward by water vapour. If we take the half day period of rotation we would expect the arc to expand with a velocity of 1–2 km/sec. This value exceeds by an order of magnitude the possible velocity of dust grains accelerated by gas outflow.

The discrepancy between our suggested 4-day period and Sekanina's 8-day period could be explained by the circumstances of the observations. Sekanina (1996) used Hubble Space Telescope data of images of the inner circum-nuclear region where a spiral jet could be misidentified over two revolutions.

Acknowledgements

This work was partially supported by State Foundation for Basic Researches of the Ministry Education and Science of Ukraine. Many thanks are due to the Instituto de Astrofísica de Canarias for arranging the visit of L. Shulman to IAC to carry out this study.

References

Biver, N., Bockelèe-Morvan, D., Colom, P., Crovisier, J., Davies, J. K., Dent, W. R. F., Despois, D., Gérard, E., Lellouch, E., Rauer, H., Moreno, R., and Paubert, G.: 1997, *Science* **275**, 1915–1919.

Farnham et al.: 1998, *First International Conference on Comet Hale–Bopp* (Tenerife, Feb. 2—5, 1998). Abstracts, p. 8.

Kidger, M.: 1995a, *IAUC* 6240.

Kidger, M.: 1995b, *IAUC* 6244.

Kidger, M.: 1995c, *Astronomer* **32**, 155.

Kidger, M., Serra-Ricart, M., and Caas, R.: 1995, *IAUC* 6220.

Licandro, J. et al.: 1998, *Astrophys. J. Lett.* **501**, 221–225.

Sekanina, Z.: 1996, *Astron. Astrophys.* **314**, 957–965.

Shulman, L.M., Kidger, M., Serra-Ricart, M., and Torres-Chico, R.: 1999, in K. I. Churyumov and V. G. Kruchinenko (eds.), *Proceedings of the International Conference "Physics and Dynamics of Solar System Small Bodies"*, Kyiv, pp. 40–47.

Yevtushevsky, O. M., Kravtsov, F. I., and Churymov, K. I.: 2001, in K. I. Churyumov (ed.), *Proceedings of the International Astronomical Conference "The 4th Vsekhsvyatsky Readings. Modern Problems of Physics and Dynamics of the Solar System"*, Kyiv, pp. 113–122 (in Ukrainian).

A STUDY OF NON-GRAVITATIONAL EFFECTS OF COMET C/1995 O1 HALE–BOPP

SŁAWOMIRA SZUTOWICZ[1,*], MAŁGORZATA KRÓLIKOWSKA[1] and GRZEGORZ SITARSKI[1,2]

[1] *Space Research Centre of Polish Academy of Sciences, Bartycka 18A, 00–716, Warsaw, Poland;*
[2] *Institute of Theoretical Physics, University in Bialystok, Lipowa 41, 15-424 Białystok, Poland*

(Received 28 February 2002; Accepted 16 July 2002)

Abstract. The role of non-gravitational forces in the evolution of orbital motion of C/1995 O1 (Hale–Bopp) has been investigated. In orbital calculations the observational material covering the period from April 1993 up to August 2001 was used. To model the non-gravitational acceleration, observed and theoretical profiles of the H_2O production rates were employed. A set of forced precession models of a rotating cometary nucleus consistent with the observed spin axis orientation was fitted to positional observations. The non-gravitational models allowed us to constrain the mass and radius of the comet. The orbital evolution of Comet Hale–Bopp was investigated over ±400 k y using two sets of randomly varied orbital elements well representing all positional observations in the pure gravitational case, as well as in the non-gravitational case. The calculations showed that the comet's motion is predictable only over an interval of a few orbital periods. The statistical conclusions change significantly when non-gravitational effects are included in the analysis.

Keywords: Comet Hale–Bopp, long-period comet, non-gravitational effects

1. Introduction

C/1995 O1 (Hale–Bopp) has provided an unprecedented opportunity to follow the evolution of activity and orbital motion of a long-period comet over a very large range of heliocentric distances. The comet was discovered on 23 July 1995 at 7 AU from the Sun in a state of major activity. The observed high water production rates indicate that non-gravitational activity should influence the orbital motion of the comet.

2. Heliocentric Dependence of Production Rates

As comets approach the Sun, the solar radiation heats the surface layers of their nuclei and induces sublimation of molecular ices. At large heliocentric distances the activity is driven by sublimation of carbon monoxide, whereas sublimation of water ice from the surface layer becomes efficient at about 3 AU from the Sun. To

* E-mail: slawka@cbk.waw.pl

Earth, Moon and Planets **90:** 119–130, 2002.

analyze the profile of activity of Comet Hale–Bopp, 104 measurements of the water production rates made between 7 April 1996 and 29 December 1997 were used. Several groups observed the comet over a very broad spectral range from ultraviolet to infrared. Figure 1 shows the variation of observed water outgassing rates versus heliocentric distance. Different symbols denote observations as they were given by the authors with the exception of the measurements of the OH radical, which were converted to the water production rate ($Q_{H_2O} = Q_{OH}/0.856$). The curves and lines represent theoretical sublimation rates and profiles of the water release rate fitted to the observations.

The general form of the water sublimation rate as a function of the heliocentric distance, r(AU), can be written as:

$$Q_{H_2O}(r) = Q_m \cdot g^*(r).$$

The values of the constant Q_m(mol./s) for the profiles studied are given in Table I. For the dimensionless function $g^*(r)$ two types of expressions are proposed:

$$g^*(r) = \alpha \left(\frac{r}{r_o}\right)^{-m} \left[1 + \left(\frac{r}{r_o}\right)^n\right]^{-k} \quad \text{or} \quad g^*(r) = h(r) = r^{-n}.$$

The power law r^{-n} was used to fit the pre- and post- perihelion observations separately by least-squares to $Q_m \cdot h(r)$ (see Table I). The former expression $g^*(r)$ stands for one of the functions: $g(r)$, $g'(r)$ or $f(r)$ which are representative in the following three cases:

- Water sublimation rate for an isothermal cometary nucleus. Then $g^*(r) = g(r)$, where the function $g(r)$ is commonly used in orbital calculations for non-gravitational effects.
- Outgassing restricted to subsolar point only. Then $g^*(r) = g'(r)$ and the appropriate coefficients for the function $g'(r)$ are taken from
- Water sublimation rate as observed for Comet Hale–Bopp. In this case $g^*(r) = f(r)$ and the production rates are fitted by least-squares to the expression $Q_m \cdot f(r)$.

The values of exponents: n, m, k, the scale distance r_o and the coefficient α for all mentioned functions are listed in Table I.

3. Non-Gravitational Models

In orbital studies the widely used form of the non-gravitational acceleration is (Marsden et al., 1973):

$$a_i = A_i \cdot g^*(r), \quad i = 1, 2, 3,$$

where a_i represent the radial, transverse and normal components, respectively. Hereafter the expression $g^*(r)$ will signify one of the functions $g(r)$, $f(r)$ or $h(r)$.

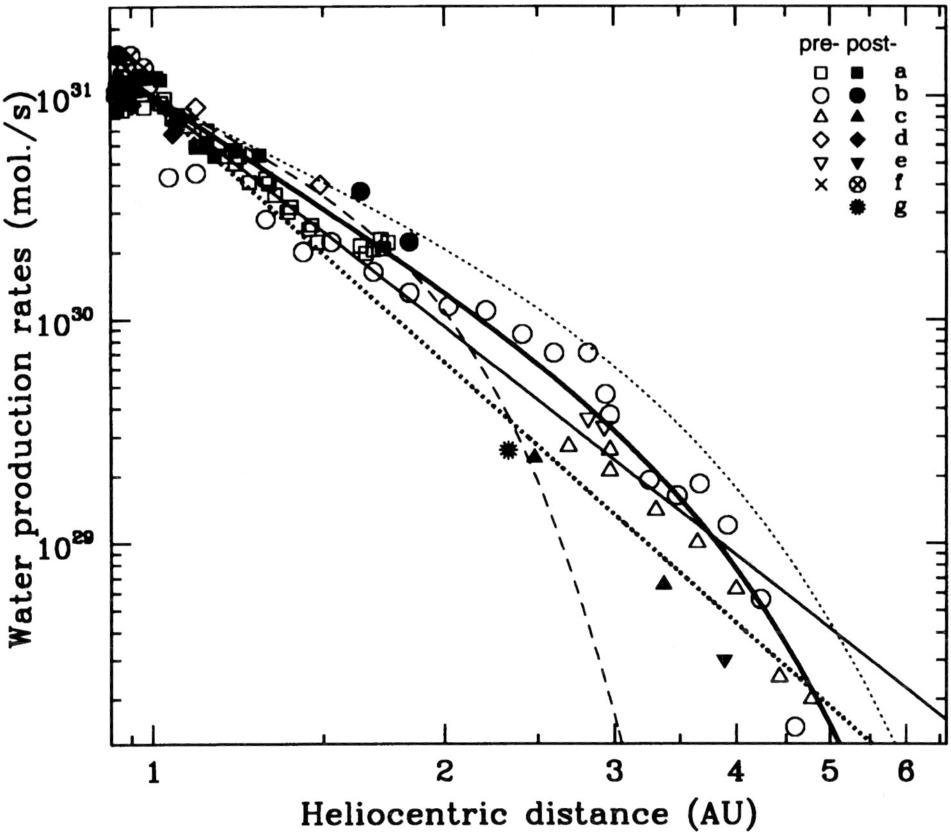

Figure 1. Heliocentric variation in the production rates of water. The open and filled symbols represent measurements performed before and after perihelion passage, respectively. (a) SWAN observation of Lyman-α (Combi et al., 2000). (b) Observation of OH at Nancay (Colom et al., 1997; Biver et al., 1997). (c) IUE and Hubble Space Telescope (Weaver et al., 1999). (d) IR observations (Dello Russo et al., 2000). (e) Infrared Space Observatory (Crovisier et al., 1999). (f) OH observations in UV, SOLSTICE (Woods et al., 2000). (g) Southwest Ultraviolet Imaging System (Stern et al., 1999). The dotted curve shows the theoretical sublimation rate at subsolar point ($Q_m \cdot g'(r)$), and the dashed one is the surface-averaged sublimation of an isothermal nucleus ($Q_m \cdot g(r)$). The solid curve is a least-squares fit of all observations to the function $Q_m \cdot f(r)$. The solid and dotted straight lines are results of fitting to a power law the pre- and post-perihelion observations, respectively.

The above mentioned standard model can be written as:

$$a_i = A \cdot g^*(r) \cdot C_i(\eta, I, \phi + \vartheta(t)), \quad i = 1, 2, 3,$$

where the direction cosines C_i of the non-gravitational acceleration depend on: the obliquity of the orbit plane to the nucleus equator, I, the solar longitude at perihelion, ϕ, the true anomaly, $\vartheta(t)$, of a comet and the lag angle, η, of the maximum outgassing behind the subsolar meridian.

Another type of the non-gravitational model is the model of forced precession of the spin axis of cometary nucleus proposed by Whipple and Sekanina (1979).

TABLE I

$g^*(r)$	α	m	n	k	r_o(AU)	Q_m(mol./s)
$g(r)$	0.111262	2.150	5.093	4.6142	2.808	$1.11 \cdot 10^{30}$
$g'(r)$	0.002726	2.100	3.200	3.9200	5.600	$1.11 \cdot 10^{30}$
$f(r)$	0.008860	2.828	3.761	3.7440	5.331	$7.8995 \cdot 10^{28} \pm 425$
		± 0.332	± 4.459	± 25.082	± 12.251	
$h(r)$	Pre-perihelion		$n = 3.39 \pm 0.08$			$9.732 \cdot 10^{30} \pm 1.061$
	Post-perihelion		3.87 ± 0.15			$9.407 \cdot 10^{30} \pm 1.063$

According to this model, nonlinear variations of the direction of the spin axis with time are caused by changes in the reaction force acting on the non-spherical nucleus. The orbital components of the non-gravitational acceleration are functions of time:

$$a_i = A \cdot f(r) \cdot C_i(\eta, I(t), \phi(t) + \vartheta(t)), \quad I = 1, 2, 3.$$

The parameters of the model are: A, η, I, ϕ, s and f_p. The last two parameters: the nucleus oblateness, $s = 1 - R_b/R_a$, and precessional factor, f_p, are hidden in the time dependences of the angles I and ϕ. They are related to the equatorial radius, R_a, of the nucleus and to its rotation period, P_r, by the expression: $\frac{P_r}{R_a} = \frac{4\pi f_p}{5s}$.

The parameters of all mentioned models were determined along with the six orbital elements $(T, q, e, \omega, \Omega, i)$ in an iterative process of orbit improvement.

4. Orbital Calculations and Results

The orbital motion of C/1995 O1 (Hale–Bopp) was investigated based on 3533 astrometric observations made between 1993 April 27 and 2001 August 13. The observations used were selected according to an objective data selection procedure (Bielicki and Sitarski, 1991). The pure gravitational orbit fits to the observations give an RMS of $1''.28$. However, the orbital solutions influenced by the non-gravitational effects give a better fit to the observations of the comet, with a mean residual $1''.07$ and $1''.06$ for the models: $A_i \cdot g^*(r)$ and $A \cdot g^*(r) \cdot C_i$, respectively. The orbital elements and the non-gravitational parameters are listed in Table IV and Table II.

The non-gravitational acceleration can be discussed in terms of some of the physical parameters of the cometary nucleus. The values of the parameter A de-

TABLE II

Non-gravitational parameters obtained from the orbital solution for two types of models: $a_i = A_i \cdot g^*(r)$ and $a_i = A \cdot g^*(r) \cdot C_i(\eta, I, \phi + \vartheta)$, where $g^*(r)$ represents functions: $g(r)$, $f(r)$ or $h(r)$ (Epoch: 1993 April 3.0 E.T.). The nucleus mass M was estimated based on appropriate values of the parameter A

	$A_i \cdot g(r)$	$A_i \cdot f(r)$	$A_i \cdot h(r)$
A_1 (10^{-8} AU/day^2)	1.5224 ± 0.0316	1.5678 ± 0.0331	1.5384 ± 0.0319
A_2 (10^{-8} AU/day^2)	0.1214 ± 0.0054	0.1248 ± 0.0057	0.1278 ± 0.0057
A_3 (10^{-8} AU/day^2)	-0.0191 ± 0.0118	-0.0206 ± 0.0121	-0.0208 ± 0.0126
Mean residual	1.″07	1.″07	1.″06
	$A \cdot g(r) \cdot C_i$	$A \cdot f(r) \cdot C_i$	$A \cdot h(r) \cdot C_i$
A (10^{-8} AU/day^2)	1.5085 ± 0.0311	1.5521 ± 0.0322	1.5209 ± 0.0311
η (deg)	8.16 ± 0.61	7.82 ± 0.51	8.94 ± 0.62
I_o (deg)	55.25 ± 3.05	53.50 ± 2.75	57.02 ± 2.65
ϕ_o (deg)	80.71 ± 5.49	88.05 ± 5.8	85.88 ± 4.90
Mean residual	1.″06	1.″06	1.″06
M (10^{14} kg)	2.466	2.142	2.346

rived from orbit determinations for models $A \cdot g^*(r) \cdot C_i$ were used to calculate the nucleus' mass, M (see Table II) from:

$$M = \frac{mvQ_m}{A\alpha},$$

where m is the molecular mass of water and v is the average ejection speed of the molecules. Putting $v = 0.16 + 0.1q$ km/s as a function of perihelion distance q(AU) (Rickman et al., 1987), one would estimate $v = 0.25$ km/s for Comet Hale–Bopp. The radius of an ice sphere of mass $M \approx 2.3 \cdot 10^{14}$ kg and of density $0.1 \cdot 10^{12} \leq \rho \leq 1.0 \cdot 10^{12}$ kg/km^3 is $8.2 \leq R_{min} \leq 3.8$ km. This is about a factor of 4 to 5 less than the best estimate of the radius for Comet Hale–Bopp. However, the comet's nucleus is not active over its entire surface and the contribution of sublimation from the grain halo surrounding the nucleus is unknown. Assuming the radius of the nucleus as equal to 20 km and taking into account the above mentioned limits for an ice sphere, one would expect that the activity level, f_a, of the comet is $17\% \leq f_a \leq 4\%$.

It was impossible to determine f_p and s along with the other parameters of the forced precession model. To derive reasonable values for all 12 parameters of the model a longer time span of observations is needed. Thus we calculated a set of values for the precessional factor, f_p, and oblateness, s, for a cometary nucleus of the radius between 10 km to 35 km (Weaver and Lamy, 1997) with the

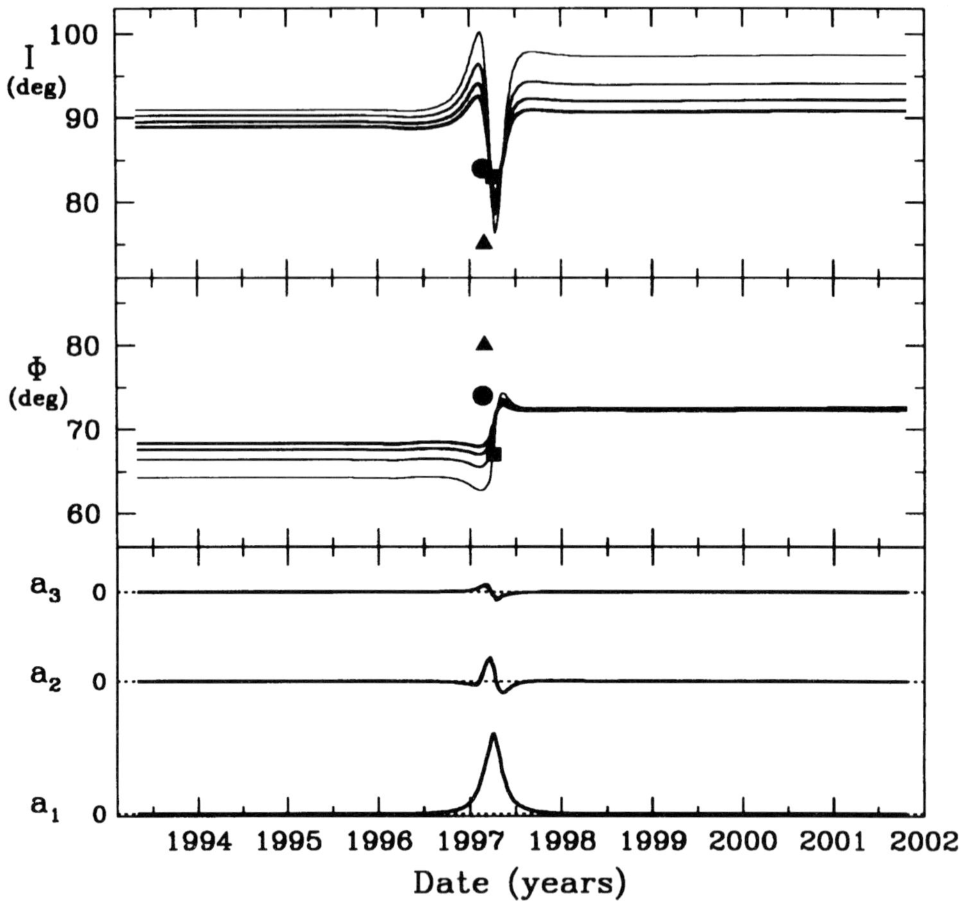

Figure 2. Time variation of the equatorial obliquity of nucleus, I, the solar longitude at perihelion, ϕ, and orbital components of the non-gravitational acceleration a_i derived from the forced precession model. The water outgassing from the comet nucleus was approximated by a function $Q_m \cdot f(r)$ (see Table I). An elongated nucleus ($s = -0.3$) with the rotation period $P_r = 11.34$ h was assumed. The curves from the thinnest to the thickest correspond to values of the nucleus radius as follows: 15 km, 20 km, 25 km and 30 km. Filled symbols denote the position of spin axis derived from observations of the arcs jets: square (Jorda et al., 1997), circle (Licardo et al., 1997) and triangle (Sekanina, 1998).

rotation period $P_r = 11.34$ hours (Jordan et al., 1997). Orbital calculations showed that precession models best fit the positional observations if the radius is larger than 15 km and the oblateness of the nucleus falls within the range from -0.3 to -0.5. The precession parameters derived for a nucleus of radius 20 km are shown in Table III. The changes of the spin axis orientation and the components of the non-gravitational acceleration due to forced precession are shown in Figure 2 for the models with a nucleus of different sizes ($15 \leq R \leq 30$ km) but the same oblateness ($s = -0.3$). The series of image sequences of the arc structures from February through May 1997 allowed the orientation of the spin axis to be obtained

TABLE III

Non-gravitational parameters derived from a forced precession model: $a_i = A \cdot f(r) \cdot C_i(\eta, I(t), \phi(t) + \vartheta(t))$. All orbital solutions fit to observations with an RMS of 1.″06 (Epoch: 2001 Oct.18.0 E.T.). The parameters f_p and s were assumed to be fixed and calculated for a nucleus of radius 20 km, rotating with the period 11.34 h

A 10^{-8} AU/day^2	η deg	I_o deg	Φ_o deg	f_p (10^5 AU/day)	s
1.4281 ± 0.0335	7.44 ± 0.88	89.89 ± 1.40	70.09 ± 1.51	-0.42187	-0.3
1.3617 ± 0.0263	8.16 ± 0.94	94.10 ± 1.22	72.52 ± 1.24	-0.56249	-0.4
1.3112 ± 0.0275	8.98 ± 1.02	97.45 ± 1.14	73.92 ± 1.16	-0.70311	-0.5

in equatorial coordinates. These data were recalculated to I and ϕ and are shown in Figure 2 with different symbols.

5. Dynamical Evolution

The long-term orbital evolution of Comet Hale–Bopp was calculated by Bailey et al. (1996). As the starting set of possible trajectories they took an ensemble of a few dozen heliocentric orbits for an epoch near the 1997 perihelion passage, derived from various arcs of pre-perihelion observations. They found that “the ensemble half-life for the comet to be captured or ejected is of the order of 0.5 Myr, in the backwards integrations and 1.2 Myr in the forward integrations”. The past and future motion of the comet was also discussed by Marsden (1997) on the basis of several orbit determinations, where the longest arc of observations included post-perihelion measurements up to the end of 1997.

Our sets of starting orbits for the integration of comet motion are constructed in a different way. Since we do not know the true orbit, we take a set of randomly selected orbits which all agree well with the observations used for the nominal orbit determination (for orbital elements of the nominal orbit see Table IV). According to Sitarski's procedure (1998), we randomly selected a set of 100 orbits well representing the positional observations. Thus we have a series of 101 current orbits in which the comet ‘could’ move. The ranges of randomly selected orbital parameters in the pure gravitational case and in the non-gravitational case are given in Table IV. These two sets of 101 orbits were evolved backwards and forwards up to 400 kyr. The equations of motion were numerically integrated in barycentric coordinates using the recurrent power series method (Sitarski, 1989) taking into account the perturbations by all nine planets. For the second set of orbits non-gravitational effects were included in the dynamical evolution. We assumed that non-gravitational accelerations accumulated over one period remain constant in

TABLE IV

Range of orbital elements around nominal orbit (Epoch: 1993 April 3.0 ET = JD2449080.5; Equinox: 2000.0) for randomly selected orbits in the pure gravitational case (set I) and in the non-gravitational case (set II). All 100 randomly selected orbits fit to 3533 positional observations (6931 equations) taken during the interval 1993 April 27 – 2001 August 13 with RMS of 1″.29 (set I) and 1″.07 (set II). Non-gravitational parameters A_1, A_2, A_3 are given in units of 10^{-8} AU/day^{-2}

Set I; pure gravitational case					
T	q	e	ω	Ω	i
19970401.63910	0.91811797	0.99652359	130°.40015	282°.43720	88°.98876
+0.00007	+0.00000070	+.00000056	+0°.00002	+0°.00002	+0°.00002
−0.00010	−0.00000054	−0.00000051	−0°.00002	−0°.00002	−0°.00002

Set II; non-gravitational case					
T	q	e	ω	Ω	i
19970401.63965	0.91806833	0.99648242	130°.40354	282°.43732	88°.98883
+0.00007	+0.00000215	+0.00000185	+0°.00018	+0°.00002	+0°.00004
−0.00009	−0.00000299	−0.00000185	−0°.00015	−0°.00001	−0°.00005
			A_1	A_2	A_3
			1.5224	0.1214	−0.01915
			+0.1208	+0.0136	+0.01785
			−0.0693	−0.0152	−0.03224

each revolution, even when a dramatic change in the orbit occurred. That implies that the non-gravitational parameters A_1, A_2, A_3 should be modified with the dynamical evolution of perihelion distance, q, and eccentricity, e. We obtained a simple function $\sim q^{2.78276} \cdot (1-e)^{-1.64026}$ for this modification. Without this assumption an unphysically high level of non-gravitational forces would be achieved when the orbital period or/and the perihelion distance decreases.

The results show all the possible types of dynamical evolution. The comet could be captured/ejected from the Oort cloud in the past/future 0.4 million years, or it could stay in a long-period orbit, or even experience a short-period stage, or a Sun-grazing transition. For example, the nominal orbit remained of long-period from the beginning to the end of integration, performing 210 revolutions around the Sun in the past and 62 in the future (the pure gravitational case). In the non-gravitational case the nominal orbit evolved quite differently. It experienced a short-period stage ($P < 200$ y) over 0.33 Myr in the past, and was ejected from the solar system after 0.32 Myr!

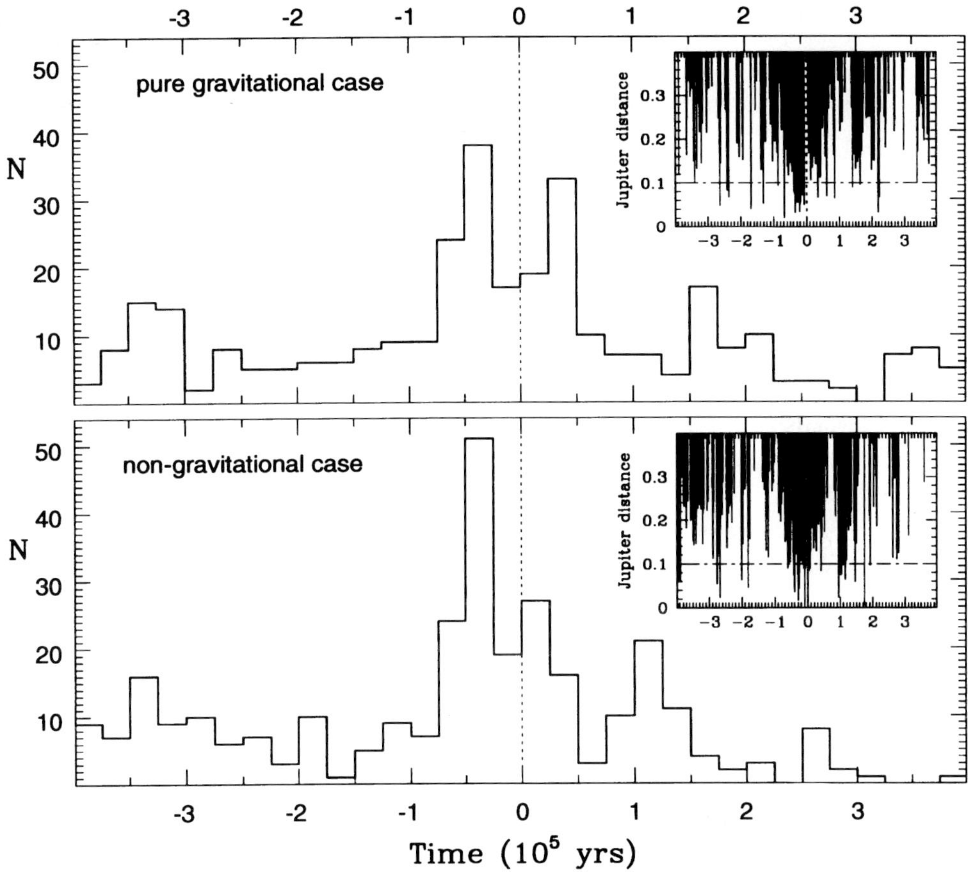

Figure 3. Distribution of all close approaches of the comet to Jupiter which appeared during the evolution of 101 orbits in the pure gravitational case and in the non-gravitational case, respectively. The evolution was performed backwards and forwards up to 400 kyr. The starting moment of integrations is shown by dotted vertical lines. The insets in both panels show the depths of all individual close encounters of the comet with Jupiter (closer than 0.33 AU).

The ascending node of the Comet Hale–Bopp's observed orbit (inclined almost perpendicular to the ecliptic) lies about 0.03 AU inside the orbit of Jupiter and thus the orbital evolution is dominated by Jovian perturbations. The distribution of all close comet encounters (closer than 0.33 AU) with Jupiter are shown in Figure 3 for pure gravitational and non-gravitational evolution. Some clustering of such encounters is visible in the past evolution up to 75 kyr, and a less significant clustering – in the future evolution up to 50 kyr.

The cumulative numbers of test objects from our sample captured in the past (or ejected in the future) over time-scales less than 50, 100, 200, 300, 400 kyr, respectively, are equal to 11 (10), 19 (17), 29 (28), 33 (34), 42 (36) for the pure gravitational evolution. The statistical results are different for non-gravitational integrations of a non-gravitational set of orbits. The corresponding cumulative

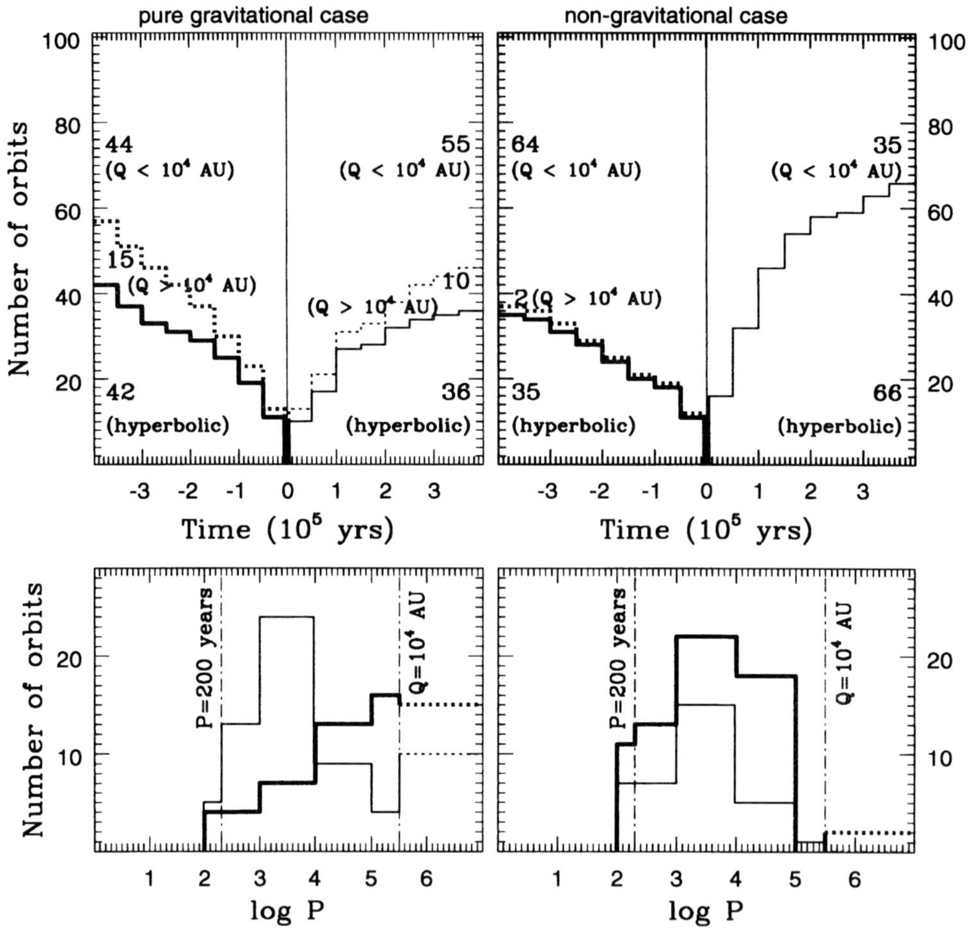

Figure 4. Upper panels: Cumulative number of orbits which evolved to: hyperbolic state (solid lines), long-period state with aphelion distance $Q \simeq 2a > 10^4$ AU (dashed lines) and $Q \simeq 2a < 10^4$ AU (rest of each sample), for the backwards (thick lines) and forwards integrations of 101 orbits up to 400 kyr. Lower panels: Number of orbits as a function of period for the past (thick lines) and the future subsets of orbits which evolved (after 400 kyr) to Halley state (20 y $< P <$ 200 y) or long-period state with $Q \simeq 2a < 10^4$ AU. Left side panels refer to pure gravitational evolution and right side panels show results for non-gravitational calculations.

numbers are then 11 (32), 18 (46), 28 (58), 34 (63), 35 (66) (see Figure 4). This suggests a larger ejection probability in the future than in the past (in contrast to Bailey et al. (1996) conclusion despite their longer integration intervals). We also found a few sungrazing orbits: 4 (8) in the past, and 3 (6) in the future for the pure gravitational case (non-gravitational case). This gives a similar probability of a sungrazing end-state of Hale–Bopp to Bailey et al. (1996) results (see their Table 7).

The mean number of past orbital revolutions is about four (six in the non-gravitational case) dozen for 42 (35) captured orbits, and a few hundred for the remaining 59 (76) orbits in the ensemble. Only one of 101 test particles in the non-gravitational case was captured just one revolution ago. Therefore it seems likely that Comet Hale–Bopp made several revolutions before coming to the inner part of the solar system.

6. Conclusions

- The orbital motion of C/1995 O1 (Hale–Bopp) evidently exhibits non-gravitational effects (compare the RMS in Table IV).
- The observed variations of water production rate with heliocentric distance do not follow the theoretical function $g(r)$ expected for an isothermal nucleus, but are very close to a model of outgassing restricted to the subsolar point on the nucleus surface (see Figure 1). That fact was previously pointed out by Weaver et al. (1999).
- The models of non-gravitational acceleration with various profiles of comet activity yield an estimate of the mass of the nucleus ($\approx 2.3 \cdot 10^{14}$ kg) and the lower limit of the radius (4–8 km). The presented forced precession models are consistent with the available observational data (see Figure 2) and indicate that the actual radius can be larger than 15 km.
- The evolutionary integrations show that the comet's motion is predictable only over a few orbital periods (less than 10 000 years). Non-gravitational effects play an essential role in the dynamical evolution of Comet Hale–Bopp. In particular, the non-gravitational future evolution gives a significantly higher probability of comet ejection from the Solar System than in the pure gravitational case (see Figure 4).

More advanced studies of the non-gravitational effects and the orbital evolution of Comet Hale–Bopp are in progress.

Acknowledgements

The authors would like to express their gratitude to Dr. Maciej Bzowski for his contributions to improve this paper. This work was supported by the Polish Committee for Scientific Research (the KBN grant 2.P03D.002.09).

References

Bailey, M. E., Emel'yanenko, V. V., Hahn, G. et al.: 1997, *M.N.R.A.S.* **281**, 916–924.
Bielicki, M. and Sitarski, G.: 1991, *Acta Astron.* **41**, 309–323.

Biver, N., Bockelée-Morvan, D., Colom, P., Crovisier, J., Germain B. et al.: 1997, *Earth Moon Planets* **78**, 5–11.
Colom, P., Gerard, E., Crovisier, J., Bockelée-Morvan, D., Biver, N., and Rauer, H.: 1997 *Earth Moon Planets* **78**, 37–43.
Combi, M. R., Reinard, A. A., Bertaux, J. L., and Quemerais, E.: 2000, *Icarus* **144**, 191–202.
Crovisier, J., Leech, K., Bockelée-Morvan, D., Lellouch, E. et al.: 1999, *ESA SP-427*, 137–140.
Dello Russo, N., Mumma, M. J., DiSanti, M. A. et al.: 2000, *Icarus* **143** 324–337.
Jorda, L., Rembor, K., Lecacheux, J., Colom, P., Colas, F., Frappa, E., and Lara, L. M.: 1997, *Earth Moon Planets* **77**, 167–180.
Licardo, J., Bellot Rubio, L. R., Casa, R., Gomez, A., Kidger, M. R., Sabalisk, N., Santos-Sanz, P. et al.: 1997, *Earth Moon Planets* **77**, 199–206.
Marsden, B. G.: 1997, *Earth Moon Planets* **79**, 3–15.
Marsden, B. G., Sekanina, Z., and Yeomans, D. K.: 1973, *Astron. J.* **78**, 211–225.
Rickman, H., Kamel, L., Festou, M. C., and Froeschle, Cl.: 1987, in *Diversity and Similarity of Comets*, ESA SP-278, 471–481.
Sekanina, Z.: 1988, *Astron. J.* **95**, 911–924.
Sekanina, Z.: 1998, *Astrophys. J.* **494**, L121–L124.
Sitarski, G.: 1989, *Acta Astron.* **39**, 345–349.
Sitarski, G.: 1998, *Acta Astron.* **48**, 547–561.
Stern, S. A., Colwell, W. B., Festou, M. C., Tamblyn, P. M., Parker, J. W., Slater, D. C., Weissman, P. R., and Paxton, L. J.: 1999, *Astron. J.* **118**, 1120–1125.
Weaver, H. A. and Lamy, P. L.: 1997, *Earth Moon Planets* **79**, 17–33.
Weaver, H. A., Feldman, P, D., A'Hearn, M. F., Arpigny, C., Brandt, J. C., and Stern, S. A.: 1999, *Icarus* **141**, 1–12.
Whipple, F. L. and Sekanina, Z.: 1979, *Astron. J.* **84**, 1894–1909.
Woods, T. N., Feldman, P. D., and Rottman, G. J.: 2000, *Icarus* **144**, 182–186.

73P/SCHWASSMANN–WACHMANN 3 – ONE ORBIT AFTER BREAK-UP: SEARCH FOR FRAGMENTS ⋆

HERMANN BOEHNHARDT, STUART HOLDSTOCK and OLIVIER HAINAUT
European Southern Observatory, Alonso de Cordova 3107, Santiago de Chile

GIAN PAOLO TOZZI
Osservatorio Astrofisico di Arcetri, Largo E. Fermi, I-50125 Firenze, Italy

STEFANO BENETTI
Astronomical Observatory of Padova, Vicolo dell' Osservatorio 5, 35122 Padova, Italy

JAVIER LICANDRO
Centro Galileo Galilei & Telescopio Nazionale Galileo TNG, P.O. Box 565, 38700 Santa Cruz de La Palma, Spain

(Received 15 February 2002; Accepted 19 May 2002)

Abstract. We report on our search campaign for the fragments of the Jupiter family comet and target of NASA's CONTOUR mission, Comet 73P/Schwassmann–Wachmann 3, one orbit revolution after the splitting of its nucleus in 1995 (Boehnhardt et al., 1995). Fragment C was found back with coma in November 1999 at 4 AU inbound and it continued to be active during the perihelion arc until at least December 2001 when we observed it last at 3.3 AU outbound. Fragment B was observed with coma between July and September 2001 when moving outbound from 2.35 to 2.75 AU. The search for other fragments including A, D (Boehnhardt et al., 1995) and E (Kodata et al., 2000a, b; Nakamura et al., 2000) in a search area of 15 $\times$ 4 arcmin starting at fragment C along direction of the extended radius vector was not successful in July and September 2001. The limiting magnitude of this search of $\sim$25 mag in R puts an upper limit on the radius of potential fragments of about 200 m (assuming albedo 0.04). The orbit deceleration parameter and the observed coma brightness of component C suggest that this object is the primary fragment that may contain a major part of the original nucleus.

Keywords: Comet 73P/Schwassmann–Wachmann 3, fragments, nucleus splitting

1. Motivation for the Search Program

Comet 73P/Schwassmann–Wachmann 3 (hereafter SW3) was discovered in 1930 by Arnold Schwassmann and Arno Arthur Wachmann (Hamburg Observatory, Germany), when its closest distance to Earth was only 0.06 AU. It has an orbital period

⋆ Based on observations performed at the Very Large Telescope VLT of the European Southern Observatory at Cerro La Silla in Chile and at the Italian Telescopio Nazionale Galileo (TNG) operated on the island of La Palma by the Centro Galileo Galilei of the CNAA (Consorzio Nazionale per l'Astronomia e l'Astrofisica) at the Spanish Observatorio del Roque de los Muchachos of the Instituto de Astrofisica de Canarias.

Earth, Moon and Planets **90:** 131–139, 2002.

of 5.3 years, and as such, is a short-period comet that belongs to the Jupiter family of comets.

The comet split into at least three pieces in 1995 (Boehnhardt et al., 1995). An analysis of this event can be found in Sekanina et al. (1996). Two of the fragments (B and C) were reobserved intact during the most recent return to the Sun together with a new fragment (E) not seen before (Kadota et al., 2000a, b; Nakamura et al., 2000).

The scientific interest in comet SW3 is very high mainly for two reasons: (1) While some of its fragments lasted only for a very limited period of time (Sekanina et al., 1996), some others (B and C) are still observable several years after the splitting event, thus giving the chance to study pristine material from the interior of the nucleus most likely now being exposed to sunlight for the first time after the formation period of the nucleus some 4.6 billion years ago. (2) In a few years time, NASA's CONTOUR mission will fly-by SW3 (along with two other comets) and will perform in-situ studies of this exceptional comet.

Here, we present ground-based observations of SW3 obtained during its past apparition (1999–2001). The main goals of the observations were (a) to recover as many fragments as possible, (b) to determine new astrometry for a refinement of the orbits and the splitting scenario of the components, and (c) to measure their coma activity. Here we present the results of part (a) and (b), the results of part (c) will be given in a separate paper elsewhere.

2. Observations at ESO VLT and at TNG

The observations were obtained in several observing runs at the 3.5 m Telescopio Nazionale Galileo (TNG) at La Palma/Spain and at the 8.2 m UT2 and UT4 telescopes of the Very Large Telescope (VLT) at Cerro Paranal/Chile. Observations were taken between November 1999 and December 2001, i.e. before and after the comet's perihelion passage on 29 January 2001. Observations were not possible around perihelion because the comet was close to the Sun. The 1999/2000 observations aimed for the detection of the fragments when inactive or when at low activity, the post-perihelion observations focused on the monitoring of the coma activity of the fragments, but allowed another search for faint components of the comet.

At the TNG, the Optical Imager Galileo (OIG) was used for imaging with Johnson-type BVRI filters. At the VLT SW3 was observed with the FORS2 instrument through Bessell-type VRI filters. Further information about the telescopes, instruments and CCDs used can be found on the web-sites of the respective observatories (TNG – http://www.tng.iac.es, VLT – http://www.eso.org/paranal). A summary of the observations, the sky conditions and the observing geometry of the comet is provided in Table I. The exposure time varied between 5 to 14min depending on the filter used thus keeping the trailing of stars below 5 arcsec.

TABLE I

A summary of the imaging observations, sky conditions and observing geometry of SW3

Date (UT) ΔT (days)	Observatory telescope + instrument	Filters	r (AU) Δ (AU)	E (deg) ϕ (deg)	Sky conditions Image quality
1999 November 10	ESO Paranal	R_S	4.03	95	photometric
−446	VLT UT2 + FORS2		3.81	14.2	1.8″
2000 March 1	La Palma	V_J	3.44	133	clear
−334	TNG + OIG		2.70	12.3	1.7–2.1″
2000 April 25	La Palma	$BVRI_J$	3.09	80	clear & thin cirrus
−279	TNG + OIG		3.12	18.7	1.5–2.0″
2001 July 20	ESO Paranal	$VRI_{B\&S}$	2.30	67	photometric
+172	VLT UT4 + FORS2		2.49	24.1	1.0–1.4″
2001 July 27	ESO Paranal	R_S	2.36	72	photometric
+179	VLT UT4 + FORS2		2.46	24.2	1.8–2.2″
2001 September 14	ESO Paranal	R_S	2.75	114	clear
+228	VLT UT4 + FORS2		2.17	19.3	1.0–1.4″
2001 December 9 + 10	ESO Paranal	$VRI_{B\&S}$	3.34	141–140	photometric
+314	VLT UT4 + FORS2		2.53–2.54	11.0	0.8–1.4″

Abbreviations: r = distance from the Sun, Δ = distance from Earth, ΔT = time to(−)/after(+) perihelion, ϕ = phase angle, E = elongation, subscript J = Johnson system (Golay, 1974), subscript B = Bessell system (Bessell, 1990), subscript S = special transmission range with central wavelength and equivalent width similar to Bessell R filter, but avoiding strong sky emission bands, see the FORS web page at http://www.eso.org/paranal for more details. The image quality as measured in the comet images is also listed.

Longer exposure times were applied for the search of the fainter components of the comet.

During the VLT exposures the comet was tracked using differential autoguiding at its predicted velocity in the sky. In July 2001 the elevation of the comet was low and the observations had to be performed at high airmass. At the TNG short integrations were obtained with the telescope in differential tracking mode (without guide probe control during the exposures). Small jitter offsets (of 20″ typical amplitude) were applied in between exposures of the comet.

The usual night and daytime calibration were taken: photometric standard stars (Landolt, 1992), sky flat-fields, biases.

3. Data Reduction

Data reduction of the images was done using ESO's MIDAS software package.

Basic data reduction: The basic data reduction consisted of bias subtraction, flat-field division, removal of cosmic rays, division by exposure time, coarse photometric calibration using the zero points of the night as calculated from the photometric standard star fields.

Astrometry: The CCD images were measured astrometrically using the software package of O. Hainaut – see http://sc6.sc.eso.org/ ohainaut/bin/pos1.cgi. The program determines first the astrometric constants of the images (field center, orientation, plate scale) iteratively and thereafter the astrometric position of an user selected object in image. For step (1) the astrometric reference stars were selected from the US Naval Observatory catalogue (USNO). The final solution was usually obtained with a set of at least 10 – in many cases more than 20 – reference stars approximately uniformly distributed over the field of view. For the plate scale solution in most cases a 5th order polynomial fit was applied separately in both cardinal directions of the image, a few measurements gave better results for 3rd order approximations. The residuals of the positions of the reference stars was better than $0.15''$. The same uncertainty applies for the position of the cometary fragments. The positions measured for the fragments B and C of SW3 are compiled in Table II.

4. Data Analysis and Results

The search for the fragments of comet SW3 was done in two steps: (1) By visual inspection and blinking of the images and (2) by shift-and-add of the comet images to reach greater exposure depth.

Search by visual inspection and blinking: the first step of the search was inspecting the images for objects of cometary nature at the predicted position of the fragments.

For the search of fainter fragments and to verify the ones found by visual inspection we examined the images also by blinking a set of two or more exposures. The respective images were carefully aligned to get a "frozen" display image of the star background while blinking the exposure series. Solar system bodies appear as moving objects that "jump" through the field of view in a systematic way following the time sequence of the exposures. Since differential tracking or autoguiding was applied during the exposures of the comet, the SW3 fragments have well-defined images while background objects appear trailed.

Since the comet position was close to the Ecliptic, several asteroid objects (in some FORS2 exposures more than 10 per field) show up as well in the images. These objects appear stellar (without coma and tail) and – in many cases – are slightly trailed, since their motion rate did not exactly match that of the comet.

Following the experience from previous split comet observations (Sekanina, 1982), the fragments should be located in the sector between the direction of the projected radius and the projected velocity vectors of the comet. A good guidance

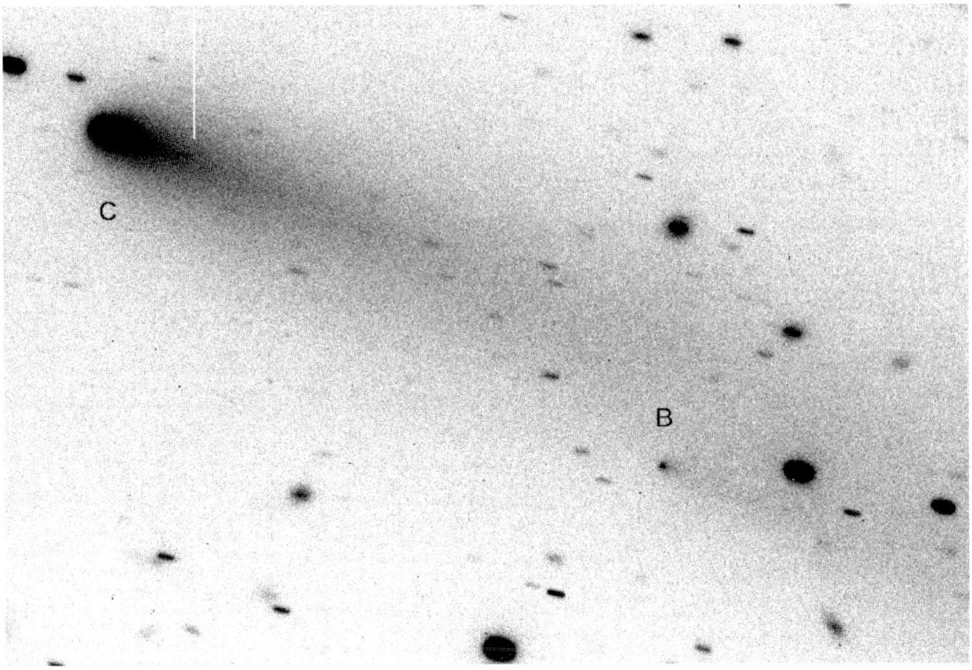

Figure 1. Fragments B and C of Comet 73P/Schwassmann–Wachmann 3. The image shows the comae and tails of fragments C and B as observed on 20 July 2001 with FORS2 through R_S filter. North is up and East is to the left. The field of view is 3. × 2.3 arcmin. Stars and galaxies appear trailed due to the telescope tracking of the comet. Close to fragment C a bad CCD column appears as a white vertical line.

was also given by the predicted position angle for the separation of the known fragments of SW3 in the sky.

Shift and add images: Systematic offsets with amplitude and direction of the predicted motion of the comet in the sky during the time interval between subsequent images were applied to the aligned images of the blinking search. Thereafter, the images were combined (a) by co-addition and (b) by median filtering. The latter was done only if more than three images per field, filter and night were available. The result images (both co-added and median filtered ones) were then again inspected for objects that appear untrailed and/or diffuse and of cometary nature. Great care was taken (by comparison with individual images or subsets of combined images of the exposure series) to identify galaxies in the images that mimic cometary appearance.

Fragments B and C: For all 7 observing runs we could easily identify fragment C by visual inspection of the images and confirm the detection by blinking. The same applies to fragment B for the images taken in July and September 2001. See Figure 1 for an example image of both fragments.

As verified a posteriori, fragment B was not in the 1999/2000 target fields since the telescope pointing was based on ephemerides that underestimated the separation distance of the fragments (even though this prediction was the best one available at that time). After identification the fragments' positions were measured astrometrically and the exact separation geometry was calculated when possible. The astrometry of the fragments was reported to the IAU Center for Minor Planets and the object identification was confirmed through orbit calculations.

Table II lists the astrometric positions of fragments B and C from our best images per night, Table III gives the separation geometry averaged over all available measurements of the fragments of a single night.

During the Dec. 2001 run we found an object of slightly fuzzy appearance and with constant distance to fragment C along the expected separation direction between B and C. One may speculate whether this object could have been fragment B. However, its calculated distance from fragment C was significantly larger than the one predicted. Therefore, the identification of this moving object as fragment B is – at present – uncertain.

If confirmed, the larger than expected offset between B and C in December 2001 implies that the non-gravitational deceleration of fragment B was altered significantly between September and December 2001. The optical appearance of fragment B in September 2001 and of the potential fragment B in December 2001 gives no indication that enhanced coma activity could be the reason for stronger non-gravitational forces.

Alternatively, the object could be a so far unidentified fragment of SW3 or just an unknown asteroid in the Main Asteroid Belt that happened to be at the expected position angle and moving with the same projected velocity as the comet in the sky.

Further observations are required to settle this identification issue for fragment B.

Other fragments: We also tried – repeatedly – to locate the reported fragments A&D and E (Boehnhardt et al., 1995; Nakamura and Kadota, 2000), in particular in September 2001 by taking longer exposures of 14 min each. The search area was About ± 5 arcmin around the predicted position of the fragment. Unfortunately, the fragments were not found. The limiting magnitude for a 5σ detection was 25 mag (R_S filter). From this and using Equation (1), we calculated the upper limit for the equivalent radius of the fragments A, D and E to be 0.2 km (assuming albedo 0.04).

$$R = r\sqrt{\frac{10^{0.4\times[(m_{\odot}-(m_{\rm obs}-\alpha\phi-5\log\Delta)]}}{A}}, \tag{1}$$

where r is the heliocentric distance to fragment (km); $m_{\odot}$, $m_{\rm obs}$ is the magnitude of Sun at 1 AU, limiting magnitude of fragment in same filter; α is the phase function coefficient, taken to be 0.035 (mag/deg); ϕ is the phase angle (deg); Δ is the geocentric distance to fragment (AU); and A is the albedo, taken to be 0.04.

We also searched our best images (20 July 2001: 5σ limit $\sim$ 24.5 mag in R; 14 September 2001: 5σ limit $\sim$ 25 mag in R_S) for other hitherto not yet discovered fragments of SW3, but could not find any within an area about $\pm 2'$ wide and $\sim 15'$ along the extended projected radius vector of fragment C. Hence, we conclude that

TABLE II
Astrometry of SW3 fragment's B and C

Date (UT)	Day (UT)	Fragment	Right ascension (hh mm ss.ss)	Declination (ddd mm ss.s)
2000 March	01.86672	C	07 58 42.84	+33 14 54.1
	01.87278	C	07 58 42.85	+33 14 52.9
	01.89061	C	07 58 42.27	+33 14 53.4
	01.89650	C	07 58 42.01	+33 14 53.8
	01.90241	C	07 58 41.64	+33 14 52.9
	01.90830	C	07 58 41.21	+33 14 53.0
	01.91426	C	07 58 41.04	+33 14 54.1
	01.92019	C	07 58 40.63	+33 14 53.5
	01.92613	C	07 58 40.38	+33 14 53.3
2000 April	25.93377	C	07 55 39.81	+31 08 59.3
	25.94319	C	07 55 40.13	+31 08 59.0
	25.95802	C	07 55 40.65	+31 08 55.1
	25.96752	C	07 55 41.03	+31 08 54.4
	25.98199	C	07 55 41.50	+31 08 49.3
	25.99145	C	07 55 41.98	+31 08 46.2
2001 July	20.35318	C	03 18 55.48	+10 39 20.1
	20.35738	C	03 18 55.73	+10 39 21.0
	20.36155	C	03 18 55.95	+10 39 21.8
	20.42598	C	03 18 55.96	+10 39 35.7
	20.43155	C	03 19 00.26	+10 39 36.3
	27.42622	C	03 25 48.38	+11 02 02.1
	27.43114	C	03 25 48.66	+11 02 03.0
2001 September	14.35951	C	03 43 01.55	+11 30 22.8
	14.36712	C	03 43 01.18	+11 30 23.3
2001 December	10.03111	C	02 26 27.97	+08 34 31.7
	10.08900	C	02 26 25.99	+08 34 33.2
2001 July	20.35318	B	03 18 47.68	+10 38 10.1
	20.35738	B	03 18 47.94	+10 38 11.1
	20.36155	B	03 18 48.20	+10 38 12.0
	20.39629	B	03 18 29.06	+10 35 34.4
	20.40118	B	03 18 50.61	+10 38 20.5
	20.42598	B	03 18 52.20	+10 38 25.9
	20.43155	B	03 18 52.52	+10 38 26.4
	27.39597	B	03 25 39.12	+11 00 47.5
	27.40660	B	03 25 39.66	+11 00 49.2
	27.41722	B	03 25 40.26	+11 00 50.7
2001 September	14.35951	B	03 42 51.97	+11 29 01.5
	14.36712	B	03 42 51.82	+11 29 00.8
2001 December	10.03111	B	02 26 16.09	+08 33 0.9
	10.08900	B	02 26 14.14	+08 33 1.9

The estimated accuracy of the astrometric positions of the comet is better than 0.15″, the time is given to 1 s accuracy.

TABLE III

Separation geometry of fragments B and C

Date (UT)		Separation B from C (arcsec)	Position angle of B from C (deg)
2001 July	20.39 ± 0.04	134.4 ± 0.2	238.5 ± 0.2
2001 September	14.36 ± 0.01	160.6 ± 0.2	239.0 ± 0.2
2001 December	10.09 ± 0.01	198.4 ± 0.2	242.5 ± 0.2

The table lists the total distance and position angle of fragment B from the main fragment C. The position angle is measured North over East. The identification of fragment B is uncertain for Dec. 2001 and the separation geometry should be taken with great care.

any fragment of SW3 present in the search area was smaller than 200–300 m in radius (again assuming albedo 0.04) and too faint to become safely detected in our images.

5. Conclusions

We could recover fragments B and C of Comet P73/Schwassmann–Wachmann 3 one orbit after the splitting event in 1995 (Boehnhardt et al., 1995). Considering also the reports on SW3 around perihelion (Kadota et al., 2000a, b; Nakamura and Kadota, 2000), fragment C was active – at least – between 4 AU pre- to 3.3 AU post-perihelion, fragment B between shortly before perihelion until at least 2.7 AU past closest approach to the Sun. A later detection of B further away post-perihelion is uncertain. If not confirmed, fragment B has declined dramatically in brightness between September and December 2001 and an upper limit of about 200–300 m (albedo 0.04) must be assumed for its radius. If confirmed, an unusual deceleration must have occurred to this fragment during the before mentioned time interval.

Fragments A, D and E could not be recovered post-perihelion in the search area around their expected positions. The non-detection puts an upper limit of about 200 m (albedo 0.04) for the radius of this fragment. A slightly larger radius limit (300 m) applies for any other SW3 fragment potentially present in the 15×4 arcmin wide search area of our images.

From the optical appearance of components B and C (the latter being much brighter), from the non-gravitational parameters of fragments B and C (B. Marsden, private communication) and from the non-detection of any other SW3 fragment in our images, we conclude that component C is the prime nucleus that remained from the splitting event in 1995. The small upper limits for the sizes of potential other fragments of SW3 and the high coma activity suggests that fragment C may contain the major mass of the original nucleus that was found to be of 1.3 km in radius (albedo 0.04; Boehnhardt et al., 1999). Therefore, fragment C may have a good chance to survive intact until the encounter with NASA's CONTOUR spacecraft during the next return of the comet in 2006. Since fragment C is a product of a recent splitting event, it may still display relatively fresh material from the interior of the nucleus providing the exceptional opportunity to explore through the spacecraft fly-by widely unaltered material from the formation period of the comet during the early phase of creation of our planetary system. However, since split comets have a certain tendency to split further (Sekanina, 1997), the still existing fragments of SW3 are at risk of further disintegration over the next revolutions around the Sun.

Acknowledgement

We like to acknowledge the support of M. T. Acevado, S. Brillant, M. Dahlem, C. Herrera, E. Pompei, M. Rojas, R. Scarpa, and A. Zarate from the VLT Science Operations Group during the service mode observations of the comet at ESO Paranal.

References

Bessell, M. S.: 1990, *PASP* **102**, 1181–1199.

Boehnhardt, H., Käufl, H. U., Keen, R., Camilleri, P., Carvajal, J., and Hale, A., 1995, 'Comet 73P/Schwassmann–Wachmann 3', *IAUC* 6274.

Boehnhardt, H., Rainer, N., Birkle, K., and Schwehm, G.: 1999, *A&A* **341**, 912–917.

Golay, M.: 1974, *Introduction to Astronomical Photometry*, Astrophys. Space Sci. Library 41, Reidel Publishers, Dordrecht.

Kadota, K. and Yoshida S.: 2000a, 'Comet 73P/Schwassmann–Wachmann 3', *IAUC* 7523.

Kadota, K., Jaeger, M., Galad, A., Koleny, P., Hale, A., and Yoshida S.: 2000b, 'Comet 73P/Schwassmann–Wachmann 3', *IAUC* 7534.

Landolt, A. U.: 1992, *AJ* **104**, 340–371.

Nakamura A., Kadota K., 2000, 'Comet 73P/Schwassmann–Wachmann 3', *IAUC* 7518.

Sekanina, Z.: 1997, *A&A* **318**, L5–L8.

Sekanina, Z.: 1982, in E. E. Wilkening (ed.), *Comets*, University of Arizona Press, pp. 251–287.

Sekanina, Z, Boehnhardt, H., Käufl, H. U., and Birkle, K.: 1996, JPL Cometary Sciences Group Preprint Series No. 183.

SPECTRA OF SPLIT COMET C/1999 S4 (LINEAR)

K. I. CHURYUMOV[1], I. V. LUK'YANYK[1], V. V. VLASSYUK[2] and N. V. BORISOV[2]
[1]*Astronomical Observatory, Kyiv National Shevchenko University, Kyiv, Ukraine;* [2]*Special Astrophysical Observatory of RAS, Russia*

(Received 5 March 2002; Accepted 12 July 2002)

Abstract. We obtained spectra of comet C/1999 S4 (LINEAR) with the UAGS spectrograph (long slit and CCD) installed on the 1-m Zeiss reflector of the SAO of the RAS (Northern Caucuses, Nizhny Arkhyz) on July 23/24, 26/27 and 27/28, 2000. On July 22/23, before the splitting of the cometary nucleus, several emission lines, such as C_2, C_3, CN, NH, CH, NH_2, CO^+, H_2O^+ were clearly identified in the spectra. The inspections of the CCD spectra obtained on July 27/28, 2000 reveals only very weak emission lines superimposed on the solar reflection spectrum. From analyzing the surface brightness profile of C_2 along the slit the velocity of separation of two secondary fragments (V = 10 km/h) and the energy of the fragment separation (E = 8.7×10^{15} erg) were estimated. A luminescence cometary continuum of 26% of the total continuum level is detected in the spectra of the comet at 5000 Å. Possible mechanisms of nucleus splitting are discussed.

Keywords: Comets, continuum, luminescence, nucleus, spectra, splitting

1. Observations and Reduction

The spectra of comet C/1999 S4 (LINEAR) were obtained with the UAGS spectrograph (long slit and CCD) installed on the 1-m Zeiss reflector of the SAO of the RAS (Northern Caucuses, Nizhny Arkhyz) on July 23/24, 26/27 and 27/28, 2000. Table I gives the technical details of the equipment used for the observations.

During 3 observating nights (July 23, 27 and 28, 2000) 12 spectra of the comet were obtained. The processing was done with the ESO-MIDAS standard reduction context "LONG". For flux calibration the standard star BD+28 4211 was used. Figure 1 gives an example of the spectra obtained on the 3 observing dates. To identify the cometary emission lines we used the catalog of the spectral lines in comet Brorsen–Metcalf (Brown et al., 1992). A number of emissions lines could be identified. On July 23, 2000, before splitting of the cometary nucleus, we identified among others the typical cometary emission lines of C_2, C_3, CN, NH, CH, NH_2, CO^+, H_2O^+. On that date the spectrum is a typical comet spectrum showing the spectral features which are characteristic for many other comets.

Earth, Moon and Planets **90:** 141–146, 2002.

TABLE I

Apparatus used for the observations of C/1999 S4 (LINEAR)

Collimator	A parabolic mirror, F/4 (F = 300 mm)
Camera	A Schmidt camera, F/0.9 (F = 110 mm)
Comparison spectrum	He + Ne + Ar spectral lamp
Slit length	140 arcsec
Image scale	0.41 arcsec/px
Wavelength range	4000–6000 Å
Gratings	651(8) grooves/mm
Linear dispersion	3.1 Å/px
TV-guide field	3 arcmin
CCD detector	ISD015A ("Electron" St-Petersburg), 530 × 580 px
Pixel size	18 × 24 mkm
Readout noise	12 electron

2. Results

Figure 1 shows that the intensity of the cometary emission bands changed very fast during the time span covered by the observations. It decreased considerably between 23 and 27 July and on 28 July the spectrum resembles practically completely the reflected solar spectrum without any emission lines.

The rather sharp decrease of the emission line fluxes is confirmed by a noticeable decrease of the comet visual magnitude. One of the authors (Churyumov) observed comet C/1999 S4 in Kyiv with binoculars and estimated its visual integral magnitudes on July 22.90 UT ($m_1 = 6.4^m$) and 23.90 UT ($m_1 = 6.3^m$). Together with the estimations of the integral visual magnitude made by M. Lehky (Chech Rep.) on July 31.85 UT ($m_1 = 8.5^m$) and D.Sergent (Australia) on Aug. 2.35 UT ($m_1 = 9.0^m$) it shows that the comet magnitude decreased by 2.5^m between July 22–Aug. 2, 2000. On Aug. 9 the comet brightness was $< 11^m - 12^m$.

Noticeable changes in the brightness profiles of some emission lines were also observed. Figure 2 shows the spatial profiles of the C_2 emission line on July 23, 27 and 28. On July 23 the profile is quite smooth, single-peaked and asymmetric (elongated in projected tail direction). On July 27 and 28 a double-peaked profile is present. These double-peaked C_2 profiles indicate the existence of a secondary maximum of brightness in the near nucleus region presumably representing a secondary fragment of the nucleus. The images of the comet obtained during this time show an elongated coma with no clear maximum of brightness. However, at least 16 fragments were detected in images of comet C/1999 S4 (LINEAR) taken on 5 August 2000 with the Hubble Space Telescope (HST) and on 6 August with the Very Large Telescope (VLT) (Weaver et al., 2001).

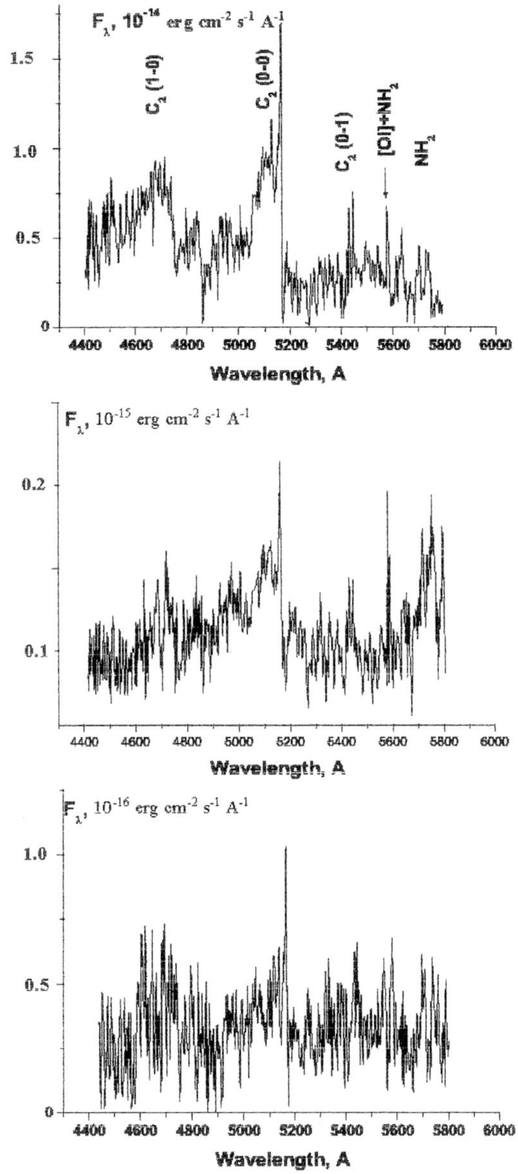

Figure 1. Evolution of the spectra of comet C/1999 S4 during July 23 to July 28, 2000.

Figure 2 shows that the distance between the brightness maxima changes between July 27 and 28. From the shift of the maxima we estimated the the velocity of fragments escaping in the near nucleus region to be about 10 km/h. Unfortunately, no confirmation measurement for this very high separation speed of the fragments is available.

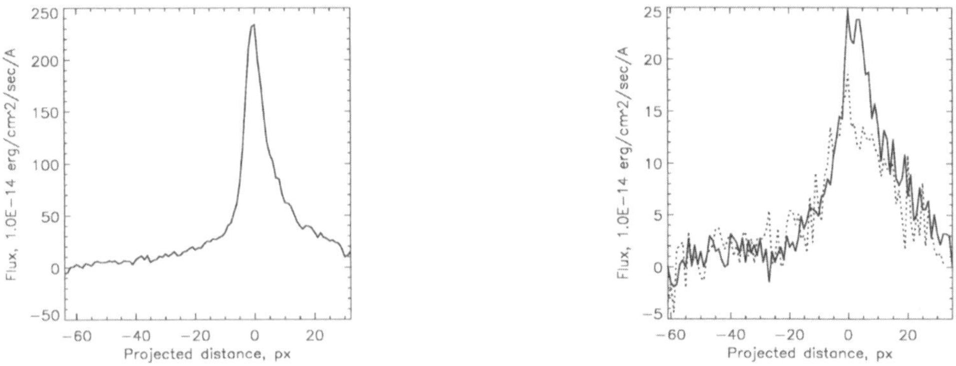

Figure 2. The surface brightness along the slit for the C_2 emission line on 23 July 2000 (left) and 27–28 July 2000 (right).

To estimate the size and possible mass of the nucleus we determined the absolute magnitude of the comet, H_0, from its light curve. We used the observations published in the International Comet Quarterly (2000) to produce the light curve shown in Figure 3 and to find its photometric parameters. After the exclusion of large errors using Tompson's rule the data were averaged over each day. The photometric parameters were calculated by Orlov's formula: $m = H_y + 5 \lg \Delta + 2.5n \lg r$, with Δ = geocentric distance and r = heliocentric distance. The photometric parameters of the comet before perighelion are $H_y = 8.79 \pm 0.03$ and $n = 6.7 \pm 0.1$. We used smaller intervals in the heliocentric distance than Yoshida (http://www.aerith.net/comet/catalog/1999S4/1999S4.html). Therefore the absolute magnitude of the comet obtained by us differs from that by Yoshida by approximately 1^m ($H_0 = 8.8$ – authors and $H_0 = 7.7$ –Yoshida).

Photographic observations of the comet made by T. Kryachko on the Pastoukhov mountain with the 40-cm Zeiss astrograph, and the analysis of the cometary light curve (Figure 3) indicate that the intense destruction of the comet nucleus began probably on July 23, 2000. The light curve shows that the comet had its maximum brightness on July 23, 2000, three days before the perihelion passage. Using the absolute magnitude of the comet ($H_0 = 8.8^m$), we estimated the approximate size of the comet nucleus by Whipple's formula (Whipple, 1987) to be R $\approx$ 1.7 km. Assuming that average density of the comet nucleus is close to 1 g/cm^3, we estimate the minimal kinetic energy needed for the escape of secondary fragments to be E $\approx 8.7 \times 10^{15}$ erg.

In the spectrum of comet C/1999 S4 we also detected the luminescence comet continuum which has non-solar origin. This effect is claimed to be present for a number of comets and may be caused by luminescence from cometary dust (Nazarchuk, 1987a). It was first described for comet Halley by Nazarchuk (Nazarchuk, 1987a, b) on the basis of decreasing of the Fraunhofer lines contrast. According to her technique the optical depth, τ_c, of the comet atmosphere is determined by

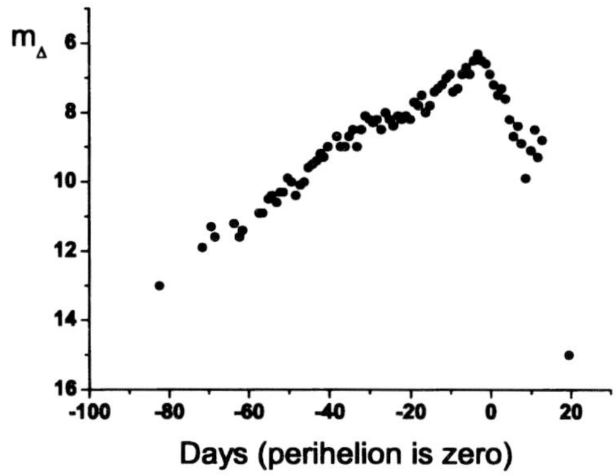

Figure 3. The light curve of comet C/1999 S4 (LINEAR).

scattering of the solar light. It is a cometary characteristic which is proportional to the ratio of the spectral intensities of the comet, I_c, and the Sun, $I_\odot$. In the presence of luminescence the optical depth is determined for this component by:

$$\tau_c(\lambda) = \frac{R_\odot^2}{4r_c^2} \frac{I_c(\lambda) - I_c^f(\lambda)}{I_\odot(\lambda)}, \tag{1}$$

with $R_\odot$ = the radius of the Sun, r = the heliocentric distance of the comet, $I_c^f(\lambda)$ = the intensity of the comet dust luminescence.

The determination of the additional component $I_c^f(\lambda)$ can be checked by the absence of traces of the Fraunhofer lines. For the given spectrum the level of the fluorescent continuum at the spectral range near λ 5000 Å is equal to 26% of the comet continuum.

Churyumov and Kleshchonok (2001) investigated spectra of three comets with the aim to determine the level of the non-solar-origin continuum in the spectral region 350–500 nm. Spectra of comets 24P/Schaumasse, C/1989 Y1 (Scoritchenko-George) and C/1995 O1 (Hale–Bopp) were observed with the 6-m BTA telescope and the long-slit spectrograph at the Special Astrophysical Observatory of the Russian Academy of Sciences. Spectra of comet Hale–Bopp (C/1995 O1) were also obtained with the 1 m Zeiss telescope and echelle-spectrometer of the SAO of RAS.

Analysis of the spectra of comet C/1999 S4 (LINEAR) also showed the presence of comet luminescence continuum (non-solar origin) with its maximum at 5000 Å, i.e. 26% of the total continuum level. A similar position of the maximum of the luminescence continuum was detected in the spectrum of comet C/2001 A2 (LINEAR) (see Lukyanyk et al., these proceedings). The number of comets studied by us showed that the maximum usually lies in the blue part of the spectra (at 4300 Å). The shift of the maximum of the luminescence continuum to the green spectral range (close to 5000 Å) is probably related to the presence of a special type of

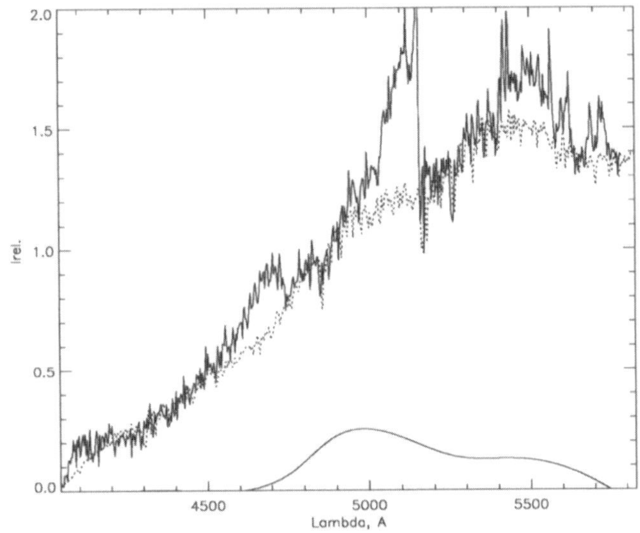

Figure 4. Detection of fluorescent continuum in the spectrum of comet C/1999 S4 (LINEAR) on July 23, 2000.

organic particles-luminofors in the coma that sublimated from the icy nucleus of comet C/1999 S4 (LINEAR).

3. Conclusion

The existence of a bright secondary nucleus in comet C/1999 S4, the high separation velocity of the fragments and the luminescence continuum of organic dust particles, needs to be confirmed by more and better observations.

References

Brown, M. E., Bouches, A. H., Spinrad, H., and Johns-Krull, C. M.: 1996, *Astrophys. J.* **112**, 1197.

Churyumov, K. I. and Kleshchonok, V. V.: 2001, *JENAM-01*, Astronomishe Gesellschaft, Abstract Series, Munich, p. 257.

Churyumov, K. I., Kleshchonok, V. V., and Vlassyuk, V. V.: 1997, *JENAM-97*, p. 70.

Nazarchuk, H. K.: 1987a, *Kometnyj Tsirkulyar* **372**, 2.

Nazarchuk, H. K.: 1987b, *Kometnyj Tsirkulyar* **377**, 2.

Weaver, H. A., Sekanina, Z., Toth, I., Delahodde, C. E., Hainaut, O. R., Lamy, P. L., Bauer, J. M., A'Hearn, M. F., Arpigny, C., Combi, M. R., Davies, J. K., Feldman, P. D., Festou, M. C., Hook, R., Jorda, L., Keesey, M. S. W., Lisse, C. M., Marsden, B. G., Meech, K. J., Tozzi, G. P., and West, R.: 2001, *Science* **292**(5520), 1329–1334.

Whipple, F. L.: 1987, *Astron. Astrophys.* **187**, 852–858.

SPLIT COMET C/2001 A2 (LINEAR) *

EMMANUEL JEHIN[1], HERMANN BOEHNHARDT[1], ZDENEK SEKANINA[2], XAVIER BONFILS[1], OLIVER SCHÜTZ[1], JEAN-LUC BEUZIT[3], MALVINA BILLERES[1], GORDON J. GARRADD[4], PIERRE LEISY[1], FRANCK MARCHIS[5], ANTONIO MÁS[6], LIVIA ORIGLIA[7], DANIEL SCARPA[6], DANIEL THOMAS[8] and GIAN PAOLO TOZZI[9]

[1]European Southern Observatory, Casilla 19001, Alonso de Cordova 3107, Vitacura, Santiago, Chile; [2]Jet Propulsion Laboratory, California Institute of Technology, Pasadena, California, USA; [3]Laboratoire d'Astrophysique, Observatoire de Grenoble, Université J. Fourier, BP 53, F-38041 Grenoble, France; [4]Loomberah, Australia; [5]University of California at Berkeley, Center for Adaptive Optics, 601 Campbell Hall, Berkeley, CA 94720, USA; [6]Observatorio Kappa Crucis, Montevideo, Uruguay; [7]Osservatorio Astronomico di Bologna, via Ranzani 1, I-40127 Bologna, Italy; [8]Universitäts-Sternwarte München, Scheinerstr. 1, D-41679 München, Germany; [9]Osservatorio Astrofisico di Arcetri, Largo E. Fermi, 50125 Firenze, Italy

(Received 8 April 2002; Accepted 3 July 2002)

Abstract. Comet C/2001 A2 experienced several splitting events during its 2001 perihelion passage. The first break-up event was observed in March 2001 (IAUC 7616). In this paper we report the first results of our extensive imaging and spectroscopic monitoring campaign with ESO telescopes over several weeks before and after the perihelion passage on May 25 2001.

Keywords: C/2001 A2 (LINEAR), comets, fragmentation

1. The Fragmentation History

A total of five new fragments were discovered in the course of this observing program and seen at least on three different dates (IAUC 7627, 7656 and Table I). During this campaign several telescopes/instruments were used in the La Silla and Paranal Observatories (NTT + EMMI & SUSI2, 3.6 + TIMMI2, 2.2 + WFI, UT3 and UT4 + test-camera, UT1 + FORS1). Beyond that we also collaborated with amateur astronomers from Uruguay and Australia.

The series of splitting events that we observed in this comet resembles the spill-off of smaller fragments from a larger nucleus (named B). The lifetime of the majority of the fragments appears to be short (orders of a few days to weeks). Examples are given in Figures 1a, b and c.

The astrometry of the fragments has been used (Sekanina et al., 2002) to estimate dates and dynamics of the separation process (Table I) by applying the extensively tested standard fragmentation model developed by Sekanina (1978, 1982).

* Based on observations carried out at the European Southern Observatory (La Silla and Paranal, Chile), program 267.C-5692.

TABLE I

Observational circumstances for companion nuclei and fragmentation solutions from primary nucleus B

Companion	A	G	C
First detection	Apr 30.1	May 21.4	May 17.0
Last detection	May 19.3	May 24.4	May 24.4
Number of measurements	15	3	4
Time of fragmentation	Mar 30.3	Apr 13.0	May 10.0
Related outburst (onset)	Mar 28.5	–	May 10.5
Relative nongravitational deceleration (units)[a]	16.3 ± 0.6	29.7 ± 1.0	53 ± 9
Companion	**D**	**E**	**F**
First detection	Jun 16.4	Jun 16.4	Jun 16.4
Last detection	Jun 28.4	Jun 27.4	Jun 19.4
Number of measurements	6	7	6
Time of fragmentation	Jun 7.4	Jun 9.8	Jun 11.3
Related outburst (onset)	Jun 5.0	Jun 5.0	Jun 5.0
Relative nongravitational deceleration (units)[a]	27.3 ± 1.7	54 ± 2	103 ± 13

[a] Units are 10^{-5} the solar gravitational acceleration.

The residuals are of the order of 0.5″ and are compatible with the measurements errors.

An interesting result of our study is a remarkable temporal correspondence between the onset of three of the outbursts (of a total of 4) and the fragmentation episodes that gave birth to fragments A, C and the sequence D, E and F. In the former case the rise time of the outburst was 8 days (compared to 3 days for the others) which corresponds well to a more complex splitting event. On the other side, the appearance of companion G is not linked to an outburst. And vice versa, after the outburst on July 12, 2001, no big fragment was discovered. The complete history of the nuclear fragmentation of C/2001 A2 and the link to the visual light curve is published in Sekanina et al. (2002). Hence, Comet C/2001A2 illustrates once again the complex and still unsolved problem of the outburst/fragmentation relationship.

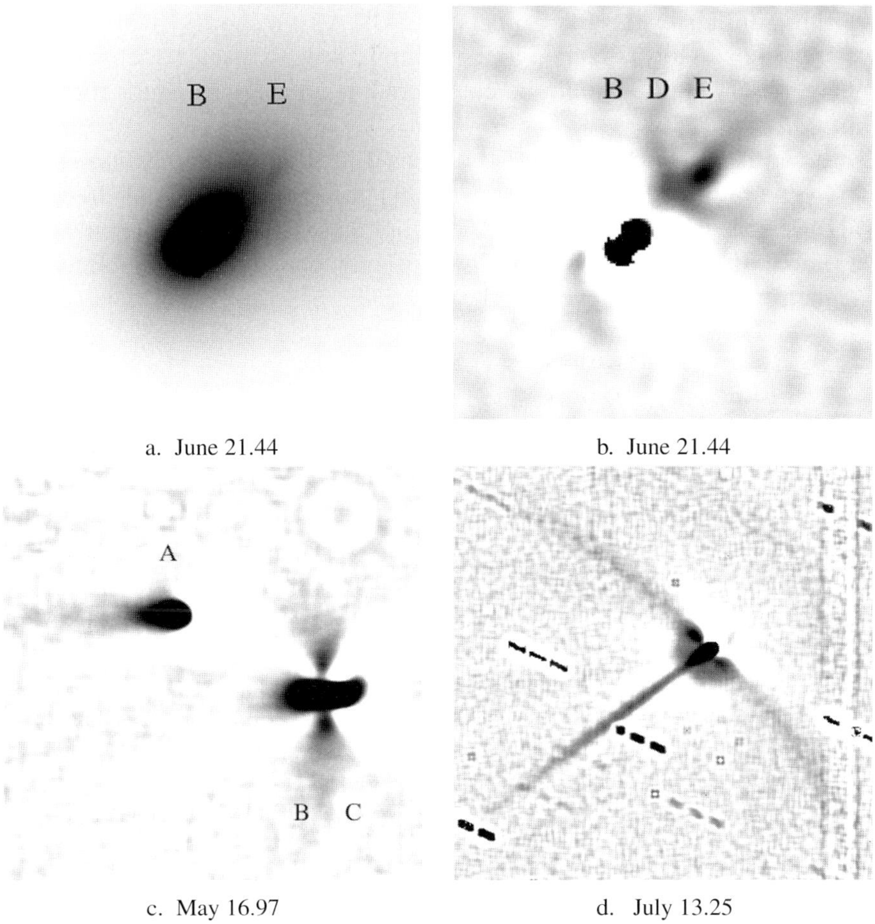

Figure 1. Fragments and arclets in the coma of Comet C/2001 A2 (LINEAR). (a) The E fragment is barely visible in this overexposed 40″ × 40″ R band NTT image. (b) A Laplace filter has been applied showing E (and its faint tail) very clearly. The image processing has also allowed the detection of D and a possible arclet between D and E. (c, d) Are Laplace-filtered R band images from UT4 + testcamera (30″ × 30″) and NTT + SUSI (2.5′ × 2.5′) revealing jets and wings. Nuclei B and C in image (c) are not resolved (the offset was only 1.6″) due to the box size used in the filtering. Companion A is visible with its own tail. The features seen in image (d) showed up July 13 and they were totally gone two days later.

2. Coma Features Associated with the Splitting Events

After structure enhancement on the computer using adaptive Laplace-filtering techniques (Boehnhardt et al., 1994 and references therein) we could successfully search in the comet images for arclet features similar to those found during the breakup of Comet C/1999 S4 (Farnham et al., 2001). Arclets (or coma wings) of short-term duration (see Figure 1c, during the separation of fragment C from the main component B) appeared perpendicular to the fragments' separation axis during the splitting events of Comets C/2001 A2 or shortly thereafter. Hence, arclets are – most likely – related to the splitting process itself (Boehnhardt, 2002) and, when observed, they could be considered as an early indicator for a nucleus break-up event.

Large and bright wings were also observed in R and V bands during the July outburst (Figure 1d.), but in this case no big fragment was found the following days and weeks. However, we should point out that the narrow dust spike observed at that time in the coma toward tail direction may contain many small unresolved fragments from a break-up event of the nucleus.

3. Production Rates

The second goal of this program was to obtain spectra of the different fragments. Except for the main fragment B, it was not possible to achieve due to their faintness and short lifetime. However, we managed to take long-slit low dispersion spectra of the comet (component B) one week before perihelion (May 19.0) with EMMI at the NTT as well as one and two months after perihelion (June 25.4, 27.4 and July 24.3) with FORS1 at the VLT. The long-slit was always oriented along and perpendicular to the direction of the extended radius vector of the comet projected on the sky. The spectra were processed in the usual way: bias and flatfield correction, cosmic ray removal, wavelength and response calibrations. The dust reflected sunlight in the emission bands of the detected molecules was subtracted by interpolating the continuum flux from wavelength regions free of gas emissions. From the flux measurements of the emission bands we have determined the gas production rates of the species C_3, C_2 and CN (and their ratios) of the main nucleus for, so far, the nights June 25 and July 24 (Table II). The comet was, respectively, at heliocentric distances of 0.99 AU and 1.36 AU and geocentric distances of 0.25 AU and 0.43 AU (1 arcsec corresponding to 184 km and 308 km at the distance of the comet). The determination of production rates involved the Haser model calculations using molecular g-factors and scale lengths from A'Hearn et al. (1995). For July 24 we have computed the production rates of the molecules in 10 numerical apertures along the slit that were symmetric to the nucleus position and ranging from 3500 km to 35000 km projected distance in the coma. The last row of Table II gives the standard deviation of the 10 measurements and shows that the production rates Q are nearly constant for different apertures.

On June 27 the water production rate was 2.5 10^{28} molecules/s (J. Crovisier, private communication). Hence, we estimate the following ratios of the carbon bearing molecules to hydroxyl ($\sim$ water): $C_2/OH = -2.088$, $C_3/OH = -3.244$ and $CN/OH = -2.014$. These abundance ratios indicate that C/2001 A2 belongs to the so-called "typical comets" group (A'Hearn et al., 1995), being a representative of the more carbon-rich objects in this group.

TABLE II
Log production rates (mol s^{-1}) and log abundance ratios

Date 2001 (UT)	$Q(C_2)$	$Q(C_3)$	Q(CN)	C_2/C_3	C_2/CN	C_3/CN
June 27.4	26.310	25.154	26.384	1.156	−0.073	−1.229
July 24.3	25.938	24.705	26.104	1.233	−0.166	−1.399
	±0.022	±0.039	±0.032	±0.056	±0.019	±0.073

References

A'Hearn, M. F., Millis, R. L., Schleicher, D. G., Osip, D. J., and Birch, P. V. 1995, *Icarus* **118**, 223–270.

Boehnhardt H.: 2002, 'Comet Splitting – Observations and Model Scenarios', these proceedings.

Boehnhardt H. and Birkle K.: 1994, *A&A* **107**, 101–120.

Farnham, T. L., Schleicher, D. G., Woodney, L. M., Birch, P. V., Eberhardy, C. A., and Levy L.: 2001, *Science* **292**, 1348–1354.

Hergenrother, C. W., Chamberlain, M., and Chamberlain, Y.: 2001, 'Comet C/2001 A2 (LINEAR)', *IAUC* 7616.

Jehin et al.: 2001, 'Comet C/2001 A2 (LINEAR)', *IAUC* 7627.

Schuetz et al.: 2001, 'Comet C/2001 A2 (LINEAR)', *IAUC* 7656.

Sekanina, Z.: 1978, *Icarus* **33**, 173–185.

Sekanina, Z.: 1982, 'The Problem of Split Comets in Review', in L. L. Wilkening (ed.), *Comets*, Univ. Arizona Press, Tucson, pp. 251–287.

Sekanina, Z., Jehin, E., Boehnhardt H., Bonfils, X., Schuetz O., and Thomas D.: 2002, *ApJ* **572**, 679–684.

ON THE 2001 SPLITTING OF THE NUCLEUS OF 51P/HARRINGTON

MARK R. KIDGER
Instituto de Astrofísica de Canarias, C/Vía Láctea s/n, E-38205 La Laguna, Tenerife, Spain
JOSÉ MANTECA
Observatorio de Begues, Begues, Barcelona, Spain

(Received 1 March 2002; Accepted 12 July 2002)

Abstract. The nucleus of 51P/Harrington suffered a major splitting event in Autumn 2001 discovered by amateur José Manteca. The light curve and separation of the fragments is examined based on data from the observers collaborating in the Spanish language "Cometas_Obs" mailing list, including a number of pre-announcement observations. There is some uncertainty in the epoch of the initiation of the split. Some evidence favours an earlier initiation than September 2001.

Keywords: Comets: 51P/Harrington, fragments

1. Introduction

Recently it has been recognised that the splitting of cometary nuclei is a relatively common event, both in periodic comets (e.g., 3D/Biela, 73P/Schwassmann–Wachmann 3, 141P/Machholz 2, D/1993 F1 Shoemaker–Levy 9) and in long-period objects (e.g., C/1965 S1 (Ikeya–Seki), C/1975 V1 (West), C/1996 B2 (Hyakutake), C/1999 S4 (LINEAR), C/2001 A2 (LINEAR)). A splitting, at least potentially gives valuable information on the structure and physics of the nucleus. Many types of splitting are observed; in many cases, such as Comet Hyakutake, the fragments are short-lived and probably represent small splinters from the main nucleus, in other cases the disintegration of the comet is total, either through tidal forces (D/1993 F1 (Shoemaker–Levy 9)) or for reasons currently not well understood (C/1999 S4 (LINEAR)). In some cases, such as 3D/Biela the fragments appear approximately equal in brightness and may be seen at more than one return.

2. 51P/Harrington

51P/Harrington is an object with a period of 6.77 years and perihelion outside the orbit of Mars (q = 1.57 AU). Discovered in 1953, the 2001 return was its 6th observed apparition. During the 1994 return (Scotti, 1994) detected two faint fragments "B" and "C" separated from the main nucleus by 23$'$, equivaten to ΔT = 0.285 days. A new splitting could be seen in observations on 2001 December 6.1 reported by (Manteca, 2001) with a separation of some 10-arcsec on an E–W line.

Earth, Moon and Planets **90:** 153–156, 2002.

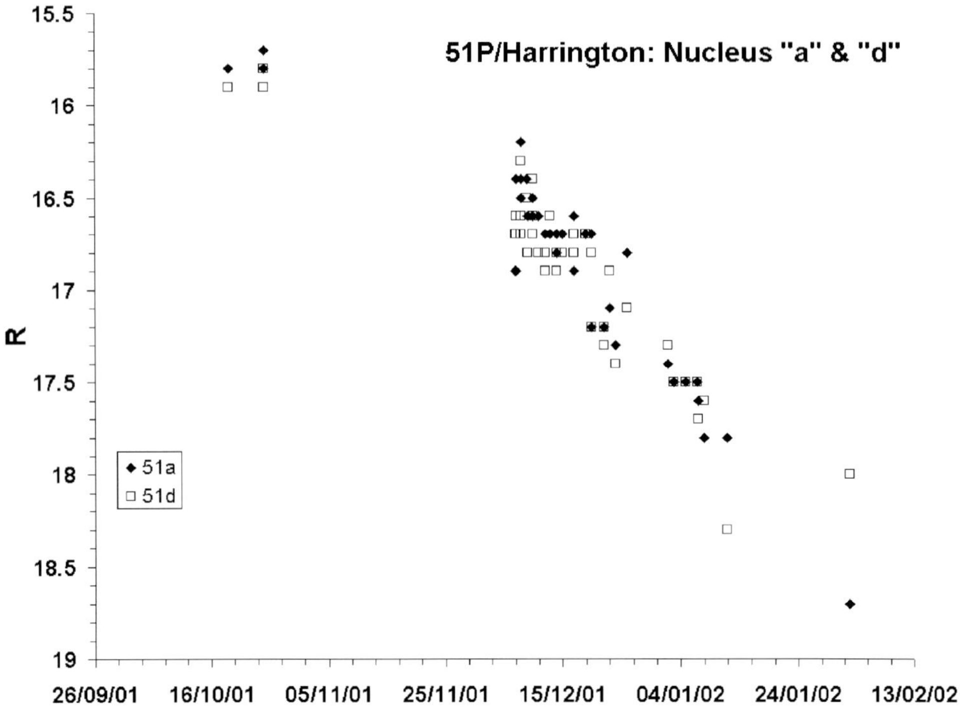

Figure 1. The light curve of fragments A (= "a") and D (= "d") of 51P/Harrington from the measures presented in the text. The data is taken in a 10 arcsec aperture and approximates to R.

This was confirmed by Naves and Campàs and by Ferrando (Naves et al., 2001). (Green, 2001) points out that this implies an effective difference in perihelion time of just 0.006 days. The magnitude of the two components was the same to within $\approx$25%, with *A* the brighter of the two. This contrasts to the nearly 12 magnitudes (factor of 5×10^4) difference between *A* and *B* & *C* in 1994.

On examination of archival images, it was found that the splitting was clearly visible in October, with only a small difference in the separation of components between October ($\approx$9 arcsec) and December ($\approx$12 arcsec). Manteca recorded the split nucleus on 2001 October 19.0 and 25.0 and Rodríguez at IAU station 458, on 2001 October 25.0 and November 18.0. There is even an indication that the innermost coma is elongated in an image by Manteca on 2001 September 24.1, suggesting that the separation at that epoch was at least 5 arcsec. Over the 3 months of data the variation in relative brightness of the fragments is minimal, with fragment *A* found to be a mean of 0.08 magnitudes brighter than *D* (see Figure 1).

One of the difficulties in analysis of the splitting event is that the mainly amateur observers who have data use CCDs with large pixel scales (the pixel scale for the observations described in (Manteca, 2001) is 2.5 arcsec/pixel, typical of all the observers whose data has been used here), thus making resolving the two nuclei difficult and accurate measurement of the separation of the nuclei complic-

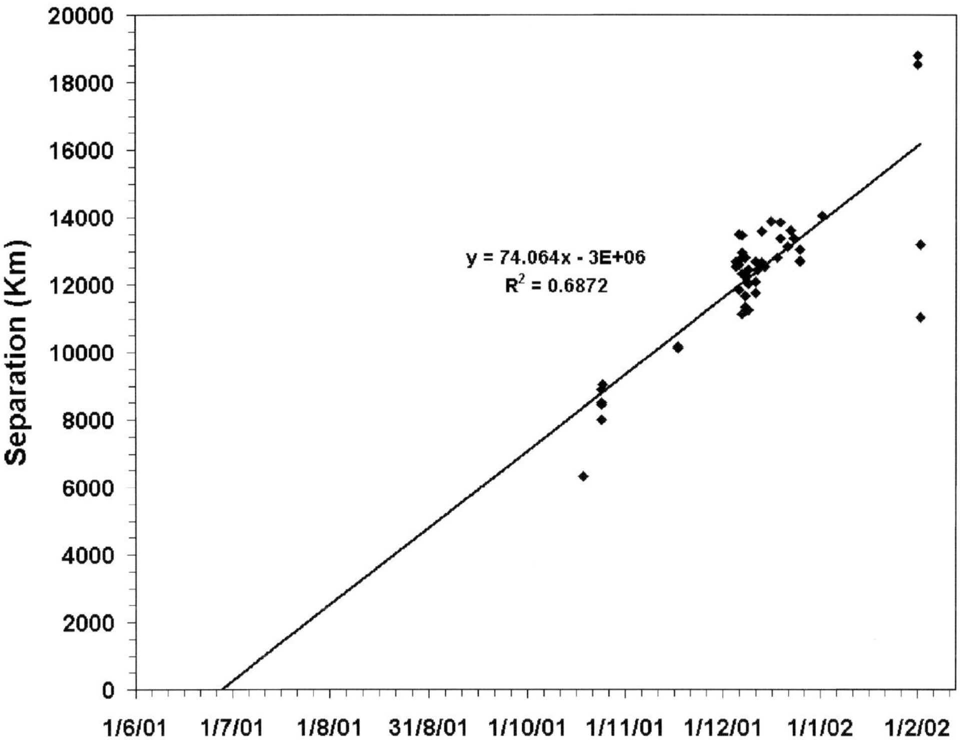

Figure 2. The separation of nuclei A and D of 51P/Harrington from the astrometry reported in the text.

ated. There are 73 sets of measurements reported to the *Cometas_Obs* mailing list by IAU stations 170, 213, 235, 458, 620 and 941. These observations form a substantial part of the entire MPC data archive for this splitting, but suffer from considerable dispersion for the reasons given previously. The mean rate of increase in separation of the components is found to be 0.041 arcsec/day, with a projected velocity measured from the linear separation of 68 km/day, or 0.78 m/s, suggesting a very low impulse splitting. However, many measures seem to underestimate the separation of the nuclei. Using a carefully selected sub-set of the 45 most internally consistent measures (see Figure 2) shows the mean rate of separation increases to 86 km/day (1.0 m/s) and gives a tentative date of intersection of the fragments of 2001 July 19, our best estimate for the date of splitting. This is somewhat earlier than the provisional value estimated by (Sekanina, 2001). Kadota (2001) notes that the comet increased in brightness by >2 mags. between 2001 August 22 and September 23, this is confirmed by photometry from Begues Observatory that shows a 3 magnitude brightening and would support Sekanina's original splitting date. In contrast though there is no evidence of a photometric event in July, in part due to lack of light curve coverage. Of course, a linear extrapolation is almost certainly not the best fit to the data due to the changing geometry, but the data in

Figure 2 require the initial projected expansion rate to be much higher to satisfy a September splitting date due to geometric effects, although the projected rate of expansion is consistent with a constant rate from mid-October.

3. Conclusions

Analysis of 73 sets of astrometry of the splitting of 51P/Harrington between 2001 October 19 and 2002 January 7 gives a mean rate of separation of the sub-nuclei of 1.0 m/s and a projected date of separation of 2001 July 19. This is inconsistent with the photometric signature of light curve that suggests a splitting in late August or early September. The reason for this discrepancy is uncertain, but the dispersion in the data may prove too large to permit detailed modelling of the circumstances of the fragmentation.

References

Green, D.: 2001, *IAUC* 7769.
Kadota, K.: 2001, *IAUC* 7773.
Manteca, J.: 2001, *IAUC* 7769.
Naves, R., Campàs, M., and Ferrando, R.: 2001, *IAUC* 7769.
Scotti, J.V.: 1994, *IAUC* 6089.
Sekanina, Z.: 2001, *IAUC* 7773.

THE BREAKUP OF C/1999 S4 (LINEAR), DAYS 0–10

MARK R. KIDGER
Instituto de Astrofísica de Canarias, E-38205 La Laguna, Tenerife, Spain

(Received 7 March 2002; Accepted 16 July 2002)

Abstract. The evolution of the morphology of C/1999 S4 (LINEAR) is examined from a series of images taken from shortly before the disruption of the nucleus until 10 days afterwards. This is combined with light curve data to provide a unique documentation of the early evolution of the disruption event. Neither images from the 1-m Jacobus Kapteyn Telescope nor the 2.5-m Isaac Newton Telescope at the Roque de los Muchachos Observatory (La Palma, Canary Islands, Spain) show no evidence of bright sub-nuclei, although the presence of a well-defined stable lance-point structure in the head of the comet indicates that a dust and gas producing source remained active in this region. The centre of brightness of the coma moved in the anti-solar direction at a few tens of metres per second after disruption indicating that it was a mainly dust structure. The contrast in the fragmentation history of comets such as C/1999 S4, C/2001 A2 and 141P/Machholz 2 suggests that there is a wide variation in nucleus properties from highly unstable and loosely bound rubble piles to relatively consolidated conglomerates.

Keywords: C/1999 S4 (LINEAR), comets, nucleus splitting

1. Introduction

C/1999 S4 (LINEAR) was discovered on 1999 September 27 at a distance of r = 4.27AU. Given its brightness at discovery and relatively small perihelion distance significant interest was generated in this object because of the possibility that it would become naked-eye visible some 10 months after discovery, allowing significant preparation of observations.

The comet was dynamically new in the Oort Cloud sense with $1/a = +0.000024 \pm 0.000001$ against the "typical" value for an Oort Cloud comet of $1/a = +0.000040$ (Marsden, private communication). Such comets are generally rich in volatiles and follow a rather large power law brightening rate before water switch-on at around 3AU. The brightening rate of C/1999 S4 was though similar to somewhat evolved objects such as C/1995 O1 (Hale–Bopp), rather than to dynamically new objects such as C/1989 (Austin) or C/1973 E1 (Kohoutek) that may follow a 5th or 6th power law initially. Another somewhat anomalous feature of C/1999 S4 was the unusually large non-gravitational parameters (see Table I).

Note though that Weaver (2000), Weaver et al. (2001) reported that the comet was strongly depleted in low temperature volatiles such as CO compared to dynamically "fairly" new comets such as C/1996 B2 (Hyakutake) of C/1995 O1 (Hale–Bopp). Such a situation is somewhat unusual given that dynamically new

Earth, Moon and Planets **90:** 157–165, 2002.

TABLE I

Non-gravitational terms and light curve brightening for C/1999 S4 and three recent, bright comets. Note the unusual size of the parameters for C/1999 S4 despite the lack of any observed jet activity. The brightening rate was typical of a dynamically evolved object rather than a new one in the Oort Cloud sense

Comet	A1	A2	Brightening	References for A1 and A2
C/1995 O1	1.27	0.1144	$m \propto 10.0 \log r$	(Marsden, 2001)
C/1996 B2	2.56	0.0485	$m \propto 8.7 \log r$	(http://www.aerith.net/comet/catalog/1996B2/1996B2.html)
C/1999 S4	9.50	−2.11	$m \propto 8.0 \log r$	(Marsden, 2002)
C/2001 A2	0.47	−0.3558	$m \propto 7.5 \log r$	(Marsden, 2001)

objects are usually found to be richer in low-temperature volatiles than more evolved objects. (Mumma et al., 2001) suggest that the explanation for this depletion is that the comet was formed at a much lower heliocentric distance (~5-10AU) than for other comets such as 1P/Halley, C/1995 O1 (Hale–Bopp) and C/1996 B2 (Hyakutake) that are thought to have formed at >20AU from the Sun.

2. Fragmentation History

Weaver (2000) reported a factor of 7 variation in flux in 2.2 days in observations with the Hubble Space Telescope, coincident with the observation of the separation of a small fragment on 2000 July 5–7. The aperture used was 0.15 arcsec, equivalent to 93 km at the comet. Observations taken on 2000 June 28th and July 1st suggest that there was an important colour and morphological change in the comet's inner coma between the two nights (Schulz and Stüwe, 2002). Schleicher (2000) also found evidence of significant variability in 1999 December.

Figure 1 shows CCD monitoring of the inner coma of C/1999 S4 (LINEAR) in late February and early March 1999 by Spanish amateur astronomers (see Kidger, 2002; Kidger and Manteca, 2002 for details of the observational method used). We see strong evidence of a major outburst in the light curve with an amplitude of 1 magnitude and a rise time of a few days. After maximum the flux returned to close to pre-outburst level in ~48 h. At Δ=3.01 AU, the physical radius of the aperture was 11 000 km at the comet and thus the decay time only implies velocities within the aperture of ~60 m/s. We suggest that this event has the signature of a minor fragmentation event similar to, but probably larger than that observed by Weaver (2000). There was a similar, bit less well observed event in late January 2001. Such events and their repeated occurrence in the light curve suggest that the comet was probably splitting off small fragments constantly during practically all its inbound

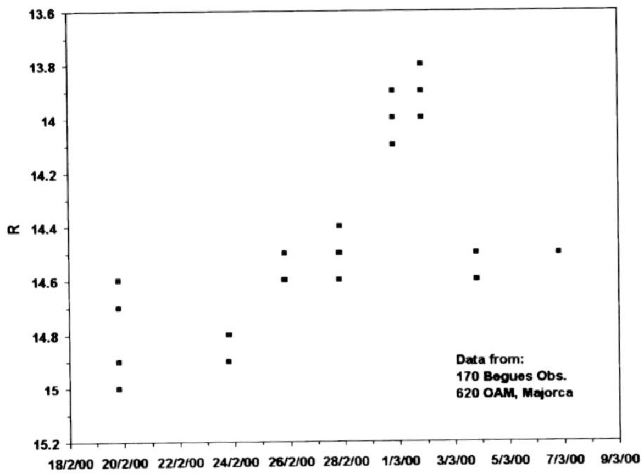

Figure 1. The light curve of C/1999 S4 (LINEAR) taken from Spanish CCD data between 2000 February 18 and March 06. Clear evidence is seen of a major outburst with an amplitude of ≈ 1 magnitude.

leg. This is consistent with the observation of 4 major outbursts by the SWAN camera of SOHO between late May and August 2001 (Mäkinen et al., 2000).

3. Outburst to Disintegration

Systematic observation of C/1999 S4 (LINEAR) through its perihelion passage was carried out with the 1-m Jacobus Kapteyn Telescope (JKT) and 2.5-m Isaac Newton Telescope (INT) at the Roque de los Muchachos Observatory (La Palma, Canary Islands, Spain). Observations were carried out in photometric conditions on 2001 July 23, 24, 25, 26, 27, 28 and 30 (no observations were possible on the 29th due to the evacuation of the observatory caused by a nearby severe forest fire) with the JKT and on 2001 Jul 31 and Aug. 1 with the INT. The JKT used the standard Cassegrain CCD camera with a field of 11 arcmin and the INT the Wide Field Camera with a field of half a degree. Standard UBVRIZ filters were used. Details of the filters and photometric system are given in González-Pérez et al. (2001).

The comet was in outburst on 2001 July 23.9 with a well-defined plasma tail (Figure 2). The surface brightness of both coma and tail was sharply reduced in July 24.9; despite increasing the exposure from 2 to 10 s the size of the coma is greatly reduced and the tail has almost disappeared (Figure 3). On July 25.9 the first images were taken in strong twilight. It was at once evident that the comet had suffered a severe disruption: the inner coma was strongly elongated in a manner reminiscent of the early post-discovery images of D/1993 F1 (Shoemaker–Levy 9) and its brightness had suffered a further sharp decline. On successive nights the surface brightness continued to decrease rapidly.

Figure 6 shows the decline in the surface brightness of the inner coma which, on nights from July 25.9 onwards drifted away from the nominal position of the

Figure 2. C/1999 S4 (LINEAR) observed with the JKT on 2000 July 23.9 in U. In this and all figures North is to the top and East to the left.

Figure 3. C/1999 S4 (LINEAR) observed with the JKT on 2000 July 24.9 in R.

Figure 4. C/1999 S4 (LINEAR) observed with the JKT on 2000 July 25.9 in R.

Figure 5. C/1999 S4 (LINEAR) observed with the JKT on 2000 July 26.9 in R.

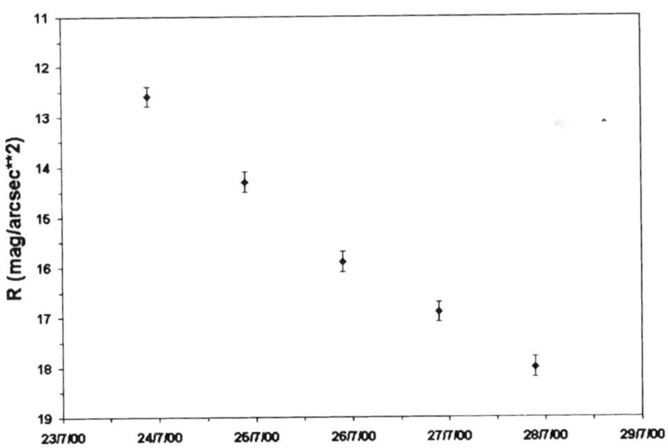

Figure 6. The variation of peak sufrace brightness of the coma of C/1999 S4 (LINEAR) during its disruption.

comet. This is measured as the average intensity of the pixels on the flat central plateau of the inner coma and converted to a surface brightness in magnitudes per square arcsecond after subtraction of the background level close to the comet. We see that the peak surface brightness of the coma dropped by more than 2 orders of magnitude in 4 days.

Figure 7. A deep exposure of the coma of C/1999 S4 (LINEAR) with the INT on 2000 August 01.9. No evidence of sub-nuclei is seen.

4. Sub-nuclei

The surface brightness of the expanding cloud showed a flat maximum with no evidence of sub-nuclei, in contrast to C/1993 F1 (Shoemaker–Levy). This conclusion was supported by the observations of Licandro et al. (2000) who took deep integrations in JHK with the 3.6-m Telescopio Nazionale Galileo (TNG) on 2000 July 26.9 and 27.9, which also show no evidence of sub-nuclei. A deep exposure with the INT on 2000 Aug. 1.9 (Figure 7) shows no discrete sub-nuclei down to B = 22.5, implying $H_V > 25.0$ for solar colours. The brightest sub-nuclei reported by Weaver et al. (2001) on 2000 August 5 and 6 had $H_V \approx 27$. Their rapid disappearance suggested a rapid rate of fading, although the INT, TNG and to a lesser extent JKT results suggest that even the brightest fragments were never very much brighter than when detected by the HST. $H_V \approx 27$ implies a diameter of 27-m for a rocky fragment of albedo 4%. Active, subliming fragments could be much smaller than this limit, whilst very dark fragments would be larger. The INT upper limit implies a maximum diameter of 70-m for rocky fragments of 4% albedo. Note though that the presence of a stable "lance point" structure at the head of the coma (seen most clearly in Figure 7) shows that a stable source of dust and gas emission was present at the nominal position of the comet. This led to the successful prediction (Kidger, 2000) that small sub-nuclei would be found in this position, as was later confirmed by Weaver et al. (2001).

5. The Zero Point of Expansion

Measurement of the expansion rate of the expanding cloud is difficult due to the rapid drop in surface brightness and the difficulty in measuring its centre accurately in the later JKT frames. However, combination of the JKT and INT data gives a mean expansion rate of 10.2 arcsec/day from 2000 July 26.9 to August 01.9

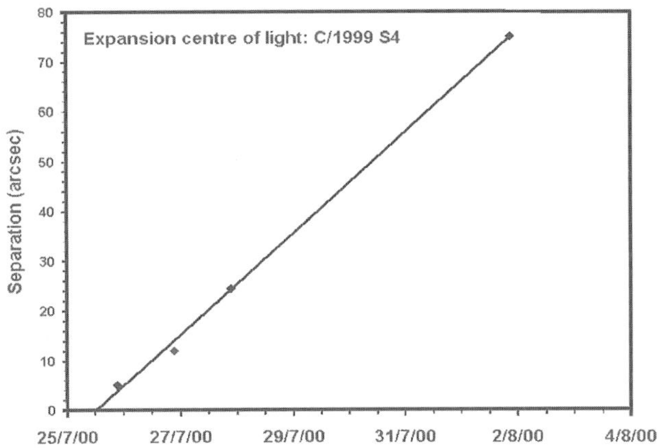

Figure 8. The rate of expansion of the peak of sufrace brightness (magnitudes per square arcsec) in R of the coma of C/1999 S4 (LINEAR) during its disruption.

(Figure 8). This leads to a nominal initiation date for the expansion of 2000 July 25.5. Assuming that the expansion is in the anti-solar direction, the deprojected expansion velocity of the centre of brightness is 77 m/s – consistent with the cloud being composed of dust.

6. Discussion

A review of cometary splittings is given by (Boehnhardt, 2002). Some comets (e.g., 51P/Harrington, 73P/Schwassmann–Wachmann 3) seem to be able to survive major and even repeated fragmentation events and may even develop into stable nearly co-orbital independent comets (e.g., 42P/Neujmin 3 and 53P/Van Biesbroeck that survive for many revolutions). Others (e.g., D/1993 F1 (Shoemaker–Levy 9)) suffer catastrophic disruption, but do not disintegrate completely, maintaining consolidated and apparently largely stable sub-nuclei. A third category (e.g., C/1996 B2 (Hyakutake)) suffered minor fragmentation(s) similar to the event observed in C/1999 S4 (LINEAR) in early July 2000, but without further disruption. These objects provide strong support for the rubble-pile model of the nucleus, which exists in three main variants that mainly differ in the strength and type of bonding forces. In contrast to the well-bound cometary nuclei that are able to survive major splittings, C/1999 S4 (LINEAR) seems to be more akin to a loosely bound dust ball and may be related to the class of comets that apparently disappear without explanation, an example of which is C/1997 N1 (Tabur), which was a 10th magnitude object that disappeared during the August 1997 Full Moon.

The properties shown by C/1999 S4 (LINEAR): depletion in low-temperature volatiles, fragmentation, very large non-gravitational forces, disintegration with minimal remnant, etc. suggest that at least at the time of its finally disintegration it was an unusually poorly consolidated object with loosely bound blocks of *relatively* small size.

Although observations up to 2000 June show that the comet was completely normal (Peschke et al., 2002), other observations demonstrate that the disruption of the nucleus was not met with the major outburst of water emission that would be expected with virgin sub-surface volatiles being released (Mäkinen et al., 2000; Farnham et al., 2001). This suggests that the mechanism for the disruption may simply have been the exhaustion of the icy "glue" that maintained the integrity of the nucleus. (Mäkinen et al., 2000) suggest that 90% of the water in the nucleus had sublimed before the terminal outburst on 2000 July 21.

Estimates of the size of the original nucleus vary widely. The very large non-gravitation terms and sudden disintegration into very small fragments suggest that the original nucleus was small. Farnham et al. (2001) suggest a lower limit of 0.44 km. A significantly larger value is given by Peschke et al. (2002), but Bockelée-Morvan et al. (2001) give a diameter from 0.2–0.6 km from outgassing and the mass budget. Fernández et al. (1999) finds a lower limit to the size of Jupiter family comets of ~600-m and suggests that below this limit cometary nuclei are not stable, consistent with the fate of C/1999 S4 (LINEAR). It seems that the best estimate for the diameter of the nucleus before disintegration was in the range from 0.4–0.6 km.

A suitable model for the nucleus of C/1999 S4 (LINEAR) at the moment that it disintegrated might be the tradition Spanish Christmas sweet, the pólveron. These appear to the uninitiated to be solid, but often break apart embarrassingly into dust in one's lap when bitten. The model is especially apt for those pólverons covered with crushed almond, or seeds, that represent the rocky fragments that survived the disintegration of the nucleus.

7. Conclusions

We have discussed the disintegration of the nucleus of C/1999 S4 (LINEAR). A complete disruption of the nucleus into an expanding dust cloud was detected on 2000 July 25.9. Extrapolation of the expansion of this cloud to its origin suggests that the expansion started around 2000 July 25.5, although evidence suggests that the complete disruption of the nucleus initiated between July 18th and 22nd. The surface brightness of the expanding cloud decreased by more than a factor of 3 per day. No evidence is found, either in these observations, or in other observations taken at the same time with alternative techniques, of bright sub-nuclei that may have evolved into the mini-comets reported by Weaver et al. (2001).

References

Bockelée-Morvan, D., Biver, N., Moreno, R., Colom, P., Crovisier, J., Gérard, É., Henry, F., Lis, D. C., Matthews, H., Weaver, H. A., Womack, M., and Festou, M. C.: 2001, *Science* **292**, 1339–1343.

Boehnhardt, H.: 2002, these proceedings.

Farnham, T. L., Schleicher, D., Woodney, L. M., Birch, P. V., Eberhardy, C. A., and Levy, L.: 2001, *Science* **292**, 1348–1354.

Fernández, J. A., Tancredi, G., Rickman, H., and Licandro, J.: 1998, *A&A* **352**, 327–340.

González-Pérez, J. N., Kidger, M. R., and Martín-Luis, F.: 2001, *AJ* **122**, 2055–2098.

Kidger, M. R.: 2000, *IAUC* 7474.

Kidger, M. R.: 2002, these proceedings.

Kidger, M. R. and Manteca, J.: 2002, these proceedings.

Licandro J., Tessicini, G, Pérez, I., and Hidalgo, S.: 2000, *IAUC* 7468.

Mäkinen, J., Teemu, T., Bertaux, J. L., Combi, M., and Quémerais, E.: 2001, *Science* **292**, 1326–1328.

Mumma, M. J., Dello Russo, N., DiSanti, M. A., Magee-Sauer, K., Novak, R. E., Brittain, S., Rettig, T., McLean, I. S., Reuter, D. C., and Xu, Li-H.: 2001, *Science* **292**, 1334–1339.

Marsden, B. G.: 2000, *MPEC* 2000–2007.

Marsden, B. G.: 2001, *MPC* 45447.

Marsden, B. G.: 2002, *MPC* 44030.

Peschke, S. B., Lisse, C. M., Fernandez, Y. R., Ressler, M., Stickel, M., Kaminski, C., and Golish, B.: 2000, *BAAS* **32**, 36.07.

Schleicher, D. G.: 2000, *IAUC* 7342.

Schulz, R. and Stüwe, J. A.: 2002, these proceedings.

Weaver, H.: 2000, *IAUC* 7461.

Weaver H., Sekanina, Z., Toth, I., Delahodde, C. E., Hainaut, O. R., Lamy, P. L., Bauer, J. M., A'Hearn, M. F., Arpigny, C., Combi, M. R., Davies, J. K., Feldman, P. D., Festou, M. C., Hook, M. C., Jorda, L., Kessey, M. S. W., Lisse, C. M., Marsden, B. G., Meech, K. J., Tozzi, G. P., and West, R.: 2001, *Science* **292**, 1329–1334.

COMPARATIVE STUDY OF THE DUST POLARIMETRIC PROPERTIES IN SPLIT AND NORMAL COMETS

NIKOLAI KISELEV
Institute of Astronomy, Kharkiv National University, 35 Sumskaya Street, 61022 Kharkiv, Ukraine

KLAUS JOCKERS
Max–Planck–Institut für Aeronomie, D–37191 Katlenburg–Lindau, Germany

VERA ROSENBUSH
Main Astronomical Observatory of the National Academy of Sciences, 27 Zabolotny Street, 03680 Kyiv, Ukraine

(Received 7 March 2002; Accepted 21 June 2002)

Abstract. Our polarimetric database contains six comets, C/1975 V1 (West), 16P/Brooks 2, C/1988 A1 (Liller), D/1996 Q1 (Tabur), C/1999 S4 (LINEAR), and C/2001 A2 (LINEAR), which can be related to the group of split comets. Comets West, S4 (LINEAR) and A2 (LINEAR) were observed during splitting. We compare the polarimetric measurements of the dust particles in these comets, sometimes together with available photometric and colorimetric data, with those in normal comets. We conclude that there is no significant evidence for differences of polarization between tidally split comets (e.g., Brooks 2), dissipating comets (e.g., Tabur), non-tidally split comets (e.g., West) and normal comets. The total disintegration of Comet S4 (LINEAR), however, did produce significant changes in the observed properties of dust.

Keywords: Comets, dust, polarization, split comets

1. Introduction

For several years, we have carried out polarimetric observations of comets which disintegrated to a different degree: C/1975 V1 (West), 16P/Brooks 2, D/1996 Q1 (Tabur), C/1988 A1 (Liller), C/1999 S4 (LINEAR), and C/2001 A2 (LINEAR). Our data set contains also the results of polarimetric observations of many comets, which do not belong to the group of split comets. Most of these "normal" comets are dusty comets. Their individual curves of polarization are very similar. The phase dependence of polarization of this group of dusty comets can be represented by a single synthetic curve, which is well described by the following trigonometric fit (Lumme and Muinonen, 1993)

$$P(\alpha) = b(\sin\alpha)^{c1}(\cos\alpha/2)^{c2}\sin(\alpha - \alpha_o), \qquad (1)$$

where the coefficients b, $c1$, $c2$, α_o are the free parameters. For the fit the dusty comets 67P/Churuymov–Gerasimenko, 22P/Kopff, P/1983 V1 (Hartley–IRAS), 27P/Giacobini–Zinner, 1P/Halley, C/1987 P1 (Bradfield), C/1990 K1 (Levy),

Earth, Moon and Planets **90:** 167–176, 2002.

TABLE I

Coefficients of the trigonometric fit for the blue and red continuum filters

λ [nm]	b	$c1$	$c2$	α_o
484/6	26.832	0.740	0.260	20.9
684/9	31.130	0.797	0.357	21.1

4P/Faye, C/1991 T2 (Shoemaker–Levy), C/1996 B2 (Hyakutake), and C/1999 U1 (LINEAR) were used in the continuum filters λ484/6 nm, λ684/9 nm and close to them (Chernova et al., 1993, 1994; Kiselev and Velichko, 1998; Kiselev et al., 2000a). The coefficients of the trigonometric fit for the blue and red continuum region are given in Table I.

Comet Hyakutake, which is accounted above as "normal" comets, may actually belong to the group of split comets. On March 26.2, 1996, two trailing condensations were observed at distances of about 1700 km and 3500 km from the nucleus by Harris et al. (1997). Both condensations moved along the tail with a velocity of about 19 m/sec. Therefore, they were included in our diaphragm (6622 km) during the polarimetric observations on March 25.0 and 26.0 (Kiselev and Velichko, 1998). Nevertheless, there were no significant variations in polarization of the comet at that time. A similar fragment was seen in the polarimetric image by Jockers on March 30, 1996. However, the brightness of the fragment was too weak to measure its degree of polarization. Polarimetric observations of comets near the maximum of polarization are still rare. Therefore, we have included the polarization measurements of Comet Hyakutake into the synthetic phase dependence of polarization for normal dusty comets. Using the trigonometric fit the polarization degree of split comets can be compared with the synthetic curve for normal comets. In this way we can draw conclusions on a possible difference in their dust properties.

2. Comet C/1975 V1 (West)

During February 19.1–March 6.5, 1976, close to the perihelion passage (Sekanina, 1976, 1982), the nucleus of Comet West broke up into four fragments A, B, C, D.

Polarimetric observations of this comet were carried out by Kiselev and Chernova (1978) from March 6.0 to July 4.8, 1976. Before March 21, 1976, all four fragments of the nucleus of Comet West were observed in the 88 arcsec diaphragm of the polarimeter. After this date, only fragments A and D were centered in the diaphragm. Polarimetry of the single fragments A, B, and D in the V filter was made by Narizhnaia et al. (1977). Michalsky (1981) measured the polarization

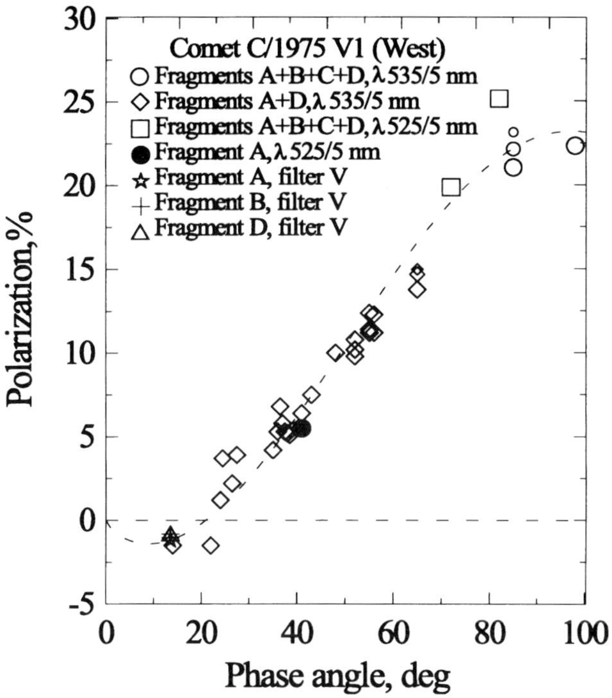

Figure 1. The phase-angle dependence of polarization for fragments of Comet West in comparison with the trigonometric fit to the phase curve for normal comets (blue continuum filter λ484/6 nm).

of Comet West with an aperture of 27 arcsec, in which nucleus A and D were observed. All above-mentioned polarimetric data for Comet West are presented in Figure 1. The employed filters are indicated.

In Figure 1 the data points of polarization degree of Comet West measured in different diaphragms scatter significantly. This is not related to the location of the fragments in the diaphragms of different size. A dependence of polarization on the size of the diaphragm was observed for different dusty comets and was explained by an evolution of dust particle properties after their release from the cometary nucleus (Kolokolova et al., 2001). Therefore one can conclude that there is no difference in polarization of different fragments of the Comet West nucleus and other dusty comets within the accuracy of the polarization measurements. On the other hand, Rosenbush (1986) have found spectral differences in fragments A and D. Tarashchuk (1980) has noted a difference of the color indexes B–V and U–B of fragments A, B, and D. According to these authors the spectral and color distinctions were caused by the difference in gas production of the fragments but not by intrinsic properties of their dust. Thus, neither spectrophotometric nor polarimetric data have revealed significant differences in the dust properties of individual fragments of Comet West as compared to normal dusty comets.

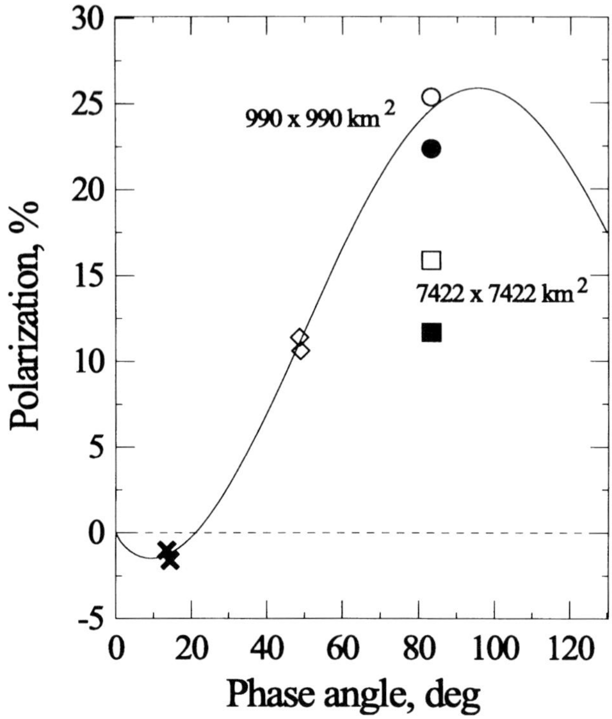

Figure 2. Polarization degree of Comet Brooks 2 (crosses), gas-rich Comet Tabur (circles) and dusty Comet Liller (diamonds) in comparison with trigonometric fit for dusty comets in the red continuum filter λ684/9 nm (solid line). Open and filled circles and squares denote the corrected (Kiselev et al., 2001a) and observed (Jockers, 1997) polarization of Comet Tabur, respectively, in different diaphragms: 990 × 990 km^2 and 7422 × 7422 km^2. For Comet Liller, the aperture used corresponds to the coma area of 39875 km.

3. Comet 16P/Brooks 2

Comet Brooks 2 belongs to the group of tidally split comets (Sekanina, 1997). After a close approach to Jupiter in July 1886, the comet broke up into several components, from which only the brightest fragment was observed in 1889. Since 1896 only the principal nucleus A has been visible and is still surviving today. Unique polarimetric observations of Comet Brooks 2 were carried out by Jockers and Chernova (unpublished data) at the 1.23-m telescope of the Calar Alto Observatory on December 3 and 5, 1994 at phase angles 13.5° and 14.4°, respectively. The focal reducer of the Max Planck Institute for Aeronomy with the four-beam Wollaston prism assembly and the wideband red filter λ694/79 nm was used for the observations (Jockers et al., 2001b). Comparison of the polarization degree of Comet Brooks 2 with the trigonometric fit for dusty comets is shown in Figure 2. One can see that there is no difference between the polarization of Comet Brooks 2 and normal comets.

4. Comets D/1996 Q1 (Tabur) and C/1988 A1 (Liller)

The similarity of orbital parameters of comets D/1996 Q1 (Tabur) and C/1988 A1 (Liller) suggests that these comets were fragments of a common parent body. However, these fragments were very different according to their gas-to-dust ratio (Kawakita et al., 1997). Comet Tabur was a gas-rich comet, while Comet Liller was a dusty one. In this respect a comparison of polarization for both comets is very important for the study of the diversity of secondary nuclei of the split parent body. Comet Tabur belongs to the so-called dissipating comets (Sekanina, 1997). Its brightness decreased catastrophically in late October 1996. Since neither outbursts nor disruption of the nucleus were observed, one can assume that the sharp decrease of comet brightness could be caused by a deactivation of the nucleus (Kiselev et al., 2001a). According to Fulle et al. (1998) the deactivation of the cometary nucleus was caused by seasonal effects, or by the complete end of the nucleus surface activity (possibly caused either by nucleus mantling or by exhaustion of the ice reservoirs).

Polarimetric observations of Comet Tabur were carried out on October 10, 1996, just before the dissipation of the comet (Jockers, 1997). A comparison of polarization degree of comets Tabur and Liller with normal dusty comets is shown in Figure 2. One can see that the polarization of Comet Tabur strongly depends on aperture size. The polarization in the near nucleus region corrected for the NH_2 contamination is similar to that for dusty comets, including Comet Liller. This may imply that the dust particles freshly emitted by the nucleus of a gas-rich comet have the same polarimetric properties as the dust particles in a dusty comet. It is necessary to emphasize that in the large aperture even the corrected polarization remains still small. The possible reasons of this are discussed by (Kiselev et al., 2001a; Rosenbush et al., 2002).

5. Comet C/1999 S4 (LINEAR)

This comet fragmented several times, as manifested by a series of sporadic outbursts in June and July before the complete disintegration of the nucleus around July 23, 2000. In Figure 3 we present the results of aperture polarimetry of Comet S4 in the green continuum filter $\lambda 526/5$ nm (Kiselev et al., 2000b). The polarization values are considerably larger than those for normal comets in the continuum filter $\lambda 484/6$ nm (dashed line). However, they are in a good agreement with the fit obtained for the red filter $\lambda 684/9$ nm of the normal comets (solid line). For most of the comets the polarization of dust usually is higher in the red spectral region as compared to the blue one. The spectral dependence of polarization of Comet S4 was very unusual. Most of the time the degree of polarization in the red filter was lower than in the green and the blue spectral range (Kiselev et al., 2001b).

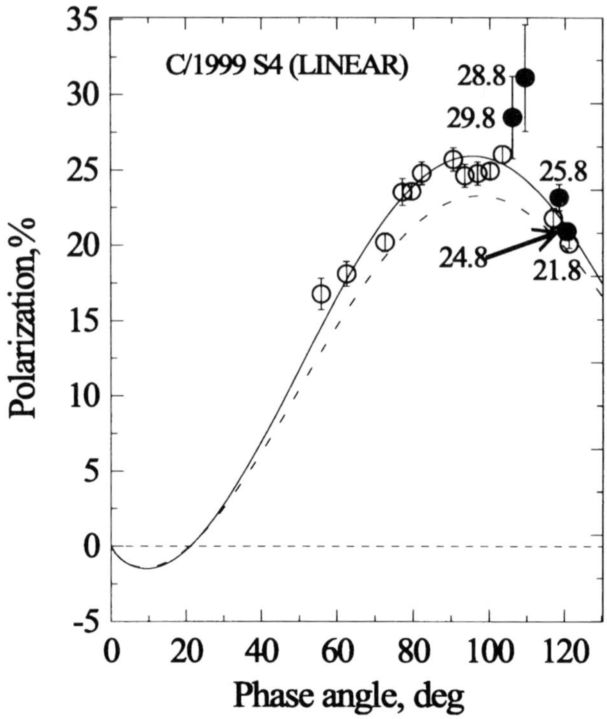

Figure 3. Polarization degree of Comet S4 (LINEAR) versus phase angle. The polarization of the comet measured in the green continuum filter λ526/5 nm (open circles) is in a good agreement with trigonometric fit for normal dusty comets in the filter λ684/9 nm (solid line). The trigonometric fit for normal comets in the filter λ484/6 nm is shown (dashed line). After the final fragmentation, the degree of polarization of the comet became significantly higher (filled circles).

This anomalous spectral dependence of polarization was also detected in Comet 21P/Giacobini–Zinner (Kiselev et al., 2000a). One more feature of S4 is that this comet as well as Giacobini–Zinner belongs to the taxonomic class of carbon chain-depleted comets (Farnham et al., 2001). Moreover, Comet S4 had a strong continuum with very high color gradient, greater than 20% per 100 nm (T. Bonev et al., private communication).

As it is seen in Figure 3, there were no features in the polarization curve before the total disintegration of the comet. After its final fragmentation, however, a dramatic increase in polarization was observed The cometary coma faded rapidly between July 25 and 28. At this time the degree of polarization increased up to 31.1% on July 28.8 and up to 29.5% on July 29.8.

We suggest that during the final disintegration of the comet the properties of the dust particles changed. In particular, fresh dust in the form of small particles could have been ejected and this may have resulted in a considerable increase of polarization (Jockers et al., 2001a).

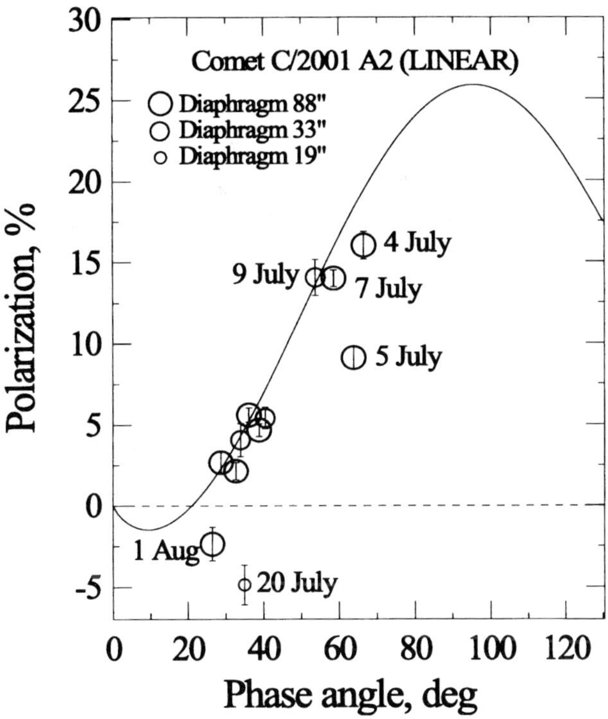

Figure 4. Polarization degree of Comet A2 (LINEAR) versus phase angle. Polarization measured in the red continuum filter λ713/6 nm as compared with the trigonometric fit for normal comets in the filter λ684/9 nm (solid line).

6. Comet C/2001 A2 (LINEAR)

Comet C/2001 A2 (LINEAR) showed considerable variability during its approach to the Sun. This was caused by a partial fragmentation of the nucleus. The breakup of the nucleus into the fragments A and B took place on March 29.9, 2001. A separation of additional components C, D, E, and F from nucleus B occurred on May 11, June 3.5, June 9.5, and June 11.3, respectively (Hergenrother et al., 2001; Broughton, 2001; Jehin et al., 2001; Schuetz et al., 2001; Sekanina, 2001).

We observed the brightest fragment of the comet – nucleus B (Rosenbush et al., 2002). The dust-equivalent A(α)fρ production rates and the color of the dust were calculated according to Farnham et al. (2000). The gas-to-dust ratio expressed by the equivalent width of the C_2 emission was large, $W(C_2) \approx 1000$ Å, for a coma size of 20400–33800 km. Therefore, Comet A2 (LINEAR) was a gas-rich comet. The dust in the comet was blue relative to the Sun in the wavelength range λ345–526 nm (UC–GC: -25% per 100 nm) and red in the wavelength rangeλ526–713 nm (GC–RC: 22% per 100 nm). The blue color of dust is more often observed for gas–rich comets. However, this effect can be due to the gas contamination in the UC continuum filter (A'Hearn et al., 1995). Therefore the different colors of Comet

A2 (LINEAR) in the blue and red wavelength regions can be due to the differential gas contamination in the continuum filters. A procedure developed by Farnham et al. (2000) allows to remove the contamination caused by the main emissions, such as C_2 and C_3. However, according to Arpigny (1995), there are a lot of weak unidentified lines in high resolution spectra of comets. For gas-rich comets such unidentified emissions will look as pseudo-continuum. Therefore the observed color as well as polarization of gas-rich comets may be significantly distorted. It is important to note that comet C/A2 was more dusty ($W(C_2) = 562$ Å) for the smaller diaphragm (area of coma 7700 km) (Rosenbush et al., 2002). In the smaller diaphragm the UC–GC color is redder (UC–GC = -8.4% per 100 nm), while the GC–RC color did not change with respect to the value in the large diaphragm. It is possible that the gas contribution was incompletely removed in Comet A2 (LINEAR), in particular in the UC filter. In such a case the manifestation of the gas contamination should be larger in the larger aperture. However, the intrinsic properties of dust may also change with distance from the nucleus (Kolokolova et al., 2001). Therefore, the origin of the blue colors of the gas-rich comets still remains unknown.

The comparison of polarization of Comet A2 in the filter λ713/6 nm with the trigonometric fit for dusty comets observed in the red continuum filter λ684/9 nm is shown in Figure 4. As in the case of Comet Tabur, the polarization degree of A2 strongly depended on the aperture size. Nevertheless, the polarization measured in the narrow-band filter and the near nucleus region of the coma was similar to that for the normal comets.

Along with this, an unexpected change of the polarization plane was observed in Comet A2 on July 19.9 and 30.9 at phase angles 35° and 26.5°, respectively. Usually, the position angle of the polarization plane is perpendicular to the scattering plane (the polarization is positive) for phase angles larger than 22°. We cannot explain these results by observational errors and a change of dust properties in Comet A2 (LINEAR) during its splitting cannot be ruled out as an explanation for this observation.

7. Conclusions

We have compared the observed properties of dust in several comets, which disintegrated to a different degree. The process of partial fragmentation of comets, such as West, Brooks 2, Liller, Tabur, and A2 (LINEAR), not always causes changes in polarization. Disintegration of Comet S4 (LINEAR) represents an example of an extreme disintegration of a comet. In this case significant changes of the scattering properties of dust particles were found.

Acknowledgements

This work was partially supported by a Max Planck fellowship and a grant of the DAAD.

References

A'Hearn, M. F., Millis, R. L., Schleicher, D. G., Osip, D. J., and Birch, P. V.: 1995, *Icarus* **118**, 223–270.

Arpigny, C.: 1995, in A. J. Sauval, R. Blomme, and N. Grevesse (eds.), *ASP Conference Series* **81**, 362–382.

Broughton, J.: 2001, *IAUC* 7625.

Chernova, G. P., Ahmedzyanov, M., and Kiselev, N. N.: 1994, *Planet. Space Sci.* **42**, 623–625.

Chernova, G. P., Kiselev, N. N., and Jockers, K.: 1993, *Icarus* **103**, 144–158.

Farnham, T. L., Schleicher, D. G., and A'Hearn, M. F.: 2000, *Icarus* **147**, 180–204.

Farnham, T. L., Schleicher, D. G., Woodney, L. M., Birch, P. V., Eberhardy, C. A., and Levy, L.: 2001, *Science* **292**, 1348–1353.

Fulle, M., Mikuz, H., Nonino, M., and Bosio, S.: 1998, *Icarus* **134**, 235–248.

Harris, W. M., Combi, M. R., Honeycutt, R. K., Mueller, B. E. A., and Scherb, F.: 1997, *Science* **277**, 676–681.

Hergenrother, C. W., Chamberlain, M., and Chamberlain, Y.: 2001, *IAUC* 7616.

Jehin, E. et al.: 2001, *IAUC* 7627.

Jockers, K.: 1997, *Earth Moon Planets* **79**, 221–245.

Jockers, K., Bonev, T., Delva, M., Kiselev, N. N., and Petrova, E.: 2001a, *Astronomische Gesellschaft. Abstract Series* **18**, 139.

Jockers, K., Credner, T., Bonev, T., Kiselev, N. N., Korsun, P., Kulik, I., Rosenbush, V. K., Andrienko, A., Karpov, N., Sergeev, A., and Tarady, V.: 2001b, *Kinematics and Physics of Celestial Bodies, Supplement* **17**, 13–18.

Kawakita, H., Furusho, R., Fujii, M., and Watanabe, J.: 1997, *Publ. Astron. Soc. Japan* **49**, L41–L44.

Kiselev, N. N. and Chernova, G. P.: 1978, *Sov. Astron. J.* **55**, 1064–1071.

Kiselev, N. N. and Velichko, F. P.: 1998, *Icarus* **133**, 286–292.

Kiselev, N. N., Jockers, K., Rosenbush, V. K., and Korsun, P. P.: 2001a, *Solar System Res.* **35**, 480–495.

Kiselev N. N., Jockers K., and Rosenbush V. K.: 2001b, in *Proceedings of the NATO Advanced Research Workshop on the Optics of Cosmic Dust*, 16–19 November 2001, Bratislava, Slovak Republic, pp. 47–48.

Kiselev, N. N., Jockers, K., Rosenbush, V. K., Velichko, F. P., Bonev, T., and Karpov N.: 2000a, *Planet. Space Sci.* **48**, 1005–1009.

Kiselev, N. N., Velichko, F. P., and Velichko, S. F.: 2000b, in *Proceedings of the International Conference "Fourth Vsekhsvyatsky Readings. Modern Problems of Physics and Dynamics of the Solar System"*. 4–10 October 2000, Kyiv, Ukraine, Kyiv University, pp. 127–131.

Kolokolova, L. O., Jockers, K., Gustafson, B. A. S., and Lichtenberg, G.: 2001, *J. Geophys. Res.* **106**, 10113–10127.

Lumme, K. and Muinonen, K.: 1993, in *IAU Symposium 160. Asteroids, Comets, Meteors 1993*, 14–18 June 1993, Belgirate, Italy. LPI 810, Houston, p. 194.

Michalsky, J. J.: 1981, *Icarus* **47**, 388–396.

Narizhnaia, N. V., Shakhovskoj, N. M., and Efimov, Yu. S.: 1977, *Astron. Circ.* No 963, 1–2.

Rosenbush, V. K.: 1986, *Icarus* **66**, 230–240.

Rosenbush, V., Kiselev, N., and Velichko, S.: 2002, these proceedings.

Schuetz, O., et al.: 2001, *IAUC* 7656.
Sekanina, Z.: 1976, *Sky Tel.* **51**, 386–393.
Sekanina, Z.: 1982, in L. L. Wilkening (ed.), *Comets*, University of Arizona, Tucson, pp. 251–287.
Sekanina, Z.: 1997, *Astron. Astrophys.* **318**, L5–L8.
Sekanina, Z.: 2001, *IAUC* 7630, 7656.
Tarashchuk, V. P.: 1980, *Probl. kosmich. phiz.* **15**, 79–92.

OBSERVATIONS OF COMET C/2001 A2 (LINEAR) WITH THE MULTIPUPIL FIBER SPECTROGRAPH

IGOR V. LUK'YANYK* and KLIM I. CHURYUMOV

Astronomical Observatory of Kyiv Shevchenko National University, Observatorna Str., 3, Kyiv, 04053, Ukraine

(Received 1 March 2002; Accepted 6 July 2002)

Abstract. We present preliminary results of observations of the spectrum of comet C/2001 A2 (LINEAR) obtained with the Multipupil Fiber Spectrograph, installed in the Prime Focus of the 6-m Telescope of the SAO of the RAS (Northern Caucasus, Nizhny Arkhyz), on August 11/12, 12/13 and 14/15, 2001. High resolution spectra of the comet were obtained in the range from 3800–5200 Å. The emission lines in the spectrum of comet C/2001 A2 (LINEAR) on Aug. 12, 2001 were identified. The fluorescent continuum in the spectrum of comet C/2001 A2 (LINEAR) on Aug. 12, 2001 is detected.

Keywords: Comets, Continuum, C/2001 A2 (LINEAR), Luminescence, Nucleus, Spectra, Splitting

1. Instrument and Observations

Comet C/2001 A2 (LINEAR) was observed with the Multipupil Fiber Spectrograph (the Spectrograph is intended for 2D spectrophotometry of extended objects), installed in the Prime Focus of the 6-m Telescope of the SAO of the RAS (Northern Caucasus, Nizhny Arkhyz), on August 11/12, 12/13 and 14/15, 2001. The main optical components of the spectrograph are: Enlarger lens, multi-fibre array, wide field spectrograph. Objects in the focal plane are enlarged by the enlager lens and projected onto multi-fibre array. The entrance lens of each fibre forms an image of the pupil of the telescope illuminated by the object. The fibres (numerical aperture 0.2 and core diameter 200 mmk) image the light onto different parts of the long slit of the spectrograph. A separate fibre acts as night sky probe (22 sq. arcsecs) and is placed at a distance of 4.5 arcmin from the field center. Geometric distortion is calibrated via low-order Fabry-Perrot marks. Table I lists the main opto-mechanical parameters of the spectrograph.

The spectrograph is mounted on the universal prime focus adapter which has a TV guiding system with 2 arcmin field of view for pointing the object and two moveable probes for offset guiding with 40 arcsec field of view each. Limiting magnitude for pointing mode is near $V \sim 20$ at 2 arcsecs seeing, and $V \sim 18$ on the guide probes. Low speed automatic guiding system helps the observer to follow the object during the exposure. The spectrograph and universal prime focus

* E-mail: iluk@observ.univ.kiev.ua

Earth, Moon and Planets **90:** 177–183, 2002.

TABLE I

The spectrograph parameters

Spatial sampling	Resolution element 0.5 × 0.5″, 0.75 × 0.75″, 1 × 1″
Field of view	16 × 16 resolved elements
Spectral range	3600–9600 Å
Dispersion	1.35–5 Å/px, realized by a set of gratings of 300–1200 g/mm
Resolution	FWHM 2.5 px over the entire field of view
Collimator	Off-axis Maksutov f = 400 mm, f/3.5 field of view 10 deg
Collimated beam	100 mm
Camera	Schmidt-Cassegrain f = 150 mm, f/1.2, field of view 14 deg
Detector	CCD TK 1024 1024 × 1024px, 24 × 24 mm/px, readout noise $3e^-$
Calibration	Geometric distortion, 3-D flat-field, comparison spectrum Ar-Ne-He, spectrophotometric standards
Total through-put	4–5% per 5000 Å(telescope + spectrograph + detector)
Flextures	10 mm at zenith distances between 10 and 70 deg
Scattered light	< 2%
Recorded data cube	16 × 18 × 1024 elements

adapter are operated with remote control. Data acquisition performed by a PC/AT 586 under LIMA software.

2. Results

Cometary gas emission: High resolution spectra of comet C/2001 A2 (LINEAR) were obtained in the wavelength range from 3800–5200 Å. The catalog of the spectral lines see in Comet Brorsen–Metcalf (Brown et al., 1996) is used for the identification of the emission lines in the spectra of the comet. The results of the identification of emission lines are given in Table II.*

Observations using the Multipupil Fiber Spectrograph allowed us to build detailed maps of the distribution of monochromatic surfaces brightness of the coma in cometary emission lines. Figure 1 shows isophote maps of the C_2 and C_3 emission lines.

Dust continuum: In order to detect the dust continuum we compared the integrated continuum in the comet spectrum with a solar spectrum which was calculated by convolving a high dispersion reference spectrum of the Sun with a Gaussian of width of the spectral resolution of our spectrograph. Incomplete match of the line widths between object and solar comparison spectrum leads to pseudo-emission

* Notes: λ_{obs} – observational wave length, I – the absolute flux, λ_{cat} – laboratory wave length, Mol. – molecule, VT – vibrational transition, RT – rotational transition.

TABLE II

Identification of the emission lines in the spectrum of comet C/2001 A2 (LINEAR) on Aug. 12, 2001

λ_{obs}, Å	I, 10^{-14} $\frac{erg}{cm^2\ sec\ Å}$	λ_{cat} ,Å	Mol.	VT and RT
3869.08	11.4	3869.17	CN	(0,0) R(08)
3881.15	18.8	3881.30	CN	(0,0) P(14)
3991.15	3.4	3990.96	C_3	(0,2,0)-(0,0,0) R(06)
4020.66	5.8	4020.68	C_3	(0,1,0)-(0,1,0) Qdc(17)
4042.13	7.0	4042.18	C_3	(0,1,0)-(0,3,0) Rc(11)
4052.86	9.3	4052.79	C_3	(0,0,0)-(0,0,0) P(08)
4075.66	5.9	4075.53	C_3	(0,0,0)-(0,2,0) P(10)
4099.80	4.4	4099.77	C_3	(0,0,0)-(0,4,0) Q(08)
4196.38	3.8	4196.40	CN	(1,2) P(16)
4199.07	3.6	4199.10	CN	(0,1) R(10)
4203.09	3.5	4202.90	CN	(0,1) R(06)
4213.82	5.1	4214.02	CN	(0,1) P(11)
4239.30	3.1	4239.41	CH	(0,0) Q(2)
4334.54	4.2	4334.65	CH	(0,0) P{1cd}(4)
4341.25	3.3	4341.42	C_2	(2,0) R2(30)
4381.50	4.0	4381.52	C_2	(2,0) P1(27)+P3(26)
4537.10	5.6	4537.09	C_2	(2,1) R1(78)+R2(77)+ R3(76)
4543.80	5.7	4543.79	C_2	(2,1) R1(76)+R2(75)+ R3(74)
4577.34	6.4	4577.48	C_2	(1,0) R1(70)+R2(69)
4588.07	6.4	4588.09	C_2	(2,1) R1(62)+R2(61)
4588.07	6.4	4588.17	C_2	(2,1) R3(60)
4593.44	6.2	4593.46	C_2	(3,2) R1(56)+R2(55)
4597.46	5.9	4597.53	C_2	(1,0) R1(64)+R2(63)
4606.85	6.3	4606.81	C_2	(1,0) R1(61)+R2(60)
4610.88	6.7	4610.89	C_2	(2,1) P1(91)+P2(90)+ P3(89)
4622.95	7.0	4623.00	C_2	(3,2) R1(44)+R2(43)
4629.65	7.0	4629.68	C_2	(3,2) R1(41)+R2(40)
4663.19	8.2	4662.99	C_2	(4,3) R1(15)
4677.95	8.8	4677.82	C_2	(1,0) R3(33)
4680.63	8.7	4680.59	C_2	(1,0) P3(66)
4683.31	9.2	4683.26	C_2	(4,3) P2(10)
4687.34	8.9	4687.31	C_2	(2,1) R2(18)
4696.73	10.1	4696.69	C_2	(1,0) P3(58)
4703.43	7.8	4703.50	C_2	(1,0) R1(22)
4707.46	8.4	4707.45	C_2	(2,1) P2(36)+R2(01)

TABLE 2

Continued

λ_{obs}, Å	I, 10^{-14} $\frac{erg}{cm^2 \sec Å}$	λ_{cat} ,Å	Mol.	VT and RT
4714.16	10.6	4714.21	C_2	(2,1) P3(09)+P3(23)
4722.21	7.9	4722.21	C_2	(1,0) P3(41)
4736.96	10.3	4736.96	C_2	(1,0) P2(16)+P2(17)+ P3(13)+P3(18)
4795.99	7.2	4795.93	CH	(0,1) Q(2)
4916.71	7.3	4916.85	C_2	(0,0) R1(75)+R2(74)
4948.91	7.3	4949.00	C_2	(0,0) P1(99)+P2(98)+ P3(97)
4951.59	7.5	4951.52	C_2	(0,0) R3(68)
4954.27	8.2	4954.19	C_2	(0,0) P1(98)+P2(97)+ P3(96)
4960.98	7.3	4960.94	C_2	(0,0) R3(66)
4971.71	7.2	4971.54	C_2	(1,1) R3(57)
4975.73	7.2	4975.66	C_2	(1,1) R3(56)
5006.59	7.5	5006.52	C_2	(1,1) R3(48)
5009.27	7.7	5009.15	C_2	(2,2) R1(41)+R2(40)
5014.64	7.4	5014.49	NH_2	(0,13,0) 321-431
5017.32	7.1	5017.18	C_2	(1,1) R3(45)
5034.76	7.9	5034.90	C_2	(1,1) P1(71)+P2(70)+ P3(69)
5037.44	7.4	5037.48	C_2	(1,1) R3(39)
5046.83	8.4	5046.93	C_2	(1,1) R1(38)+R2(37)
5064.26	8.9	5064.37	C_2	(1,1) R1(32)
5070.97	9.4	5070.78	C_2	(3,3) P3(12)+P3(16)
5077.68	9.3	5077.61	C_2	(1,1) R3(25)
5083.04	8.6	5083.05	C_2	(0,0) R1(37)+R2(36)
5088.42	9.2	5088.43	C_2	(2,2) P1(33)+P2(32)
5093.78	9.6	5093.83	C_2	(1,1) P3(47)
5097.81	9.7	5097.71	C_2	(1,1) P1(47)+P2(46)+ R3(16)
5103.17	10.4	5103.08	C_2	(1,1) R2(14)
5113.90	9.5	5113.72	C_2	(1,1) R1(08)
5116.58	9.5	5116.66	C_2	(0,0) R1(25)
5120.61	9.6	5120.64	C_2	(0,0) P1(52)+P2(51)
5136.70	8.9	5136.66	C_2	(0,0) R3(14)
5156.82	10.7	5156.69	C_2	(0,0) P3(29)
5164.87	17.1	5164.90	C_2	(0,0) P1(13)+P1(17)+ P2(11)+P2(16)+ P3(10)+P3(16)
5199.75	5.1	5199.80	NH_2	(1,8,0) 1 1 0 - 0 0 0

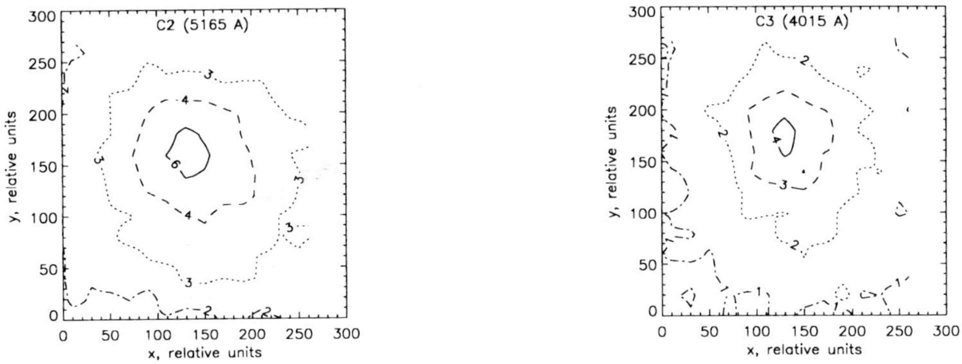

Figure 1. Isophote maps of the coma in C_2 (5165 Å) and C_3 (4050 Å) monochromatic images taken on Aug. 12, 2001 (emission line flux = isophote label to be multiplied by $10^{-14}\frac{\text{erg}}{\text{cm}^2\ \text{sec Å}}$). The field of view is $16 \times 16''$.

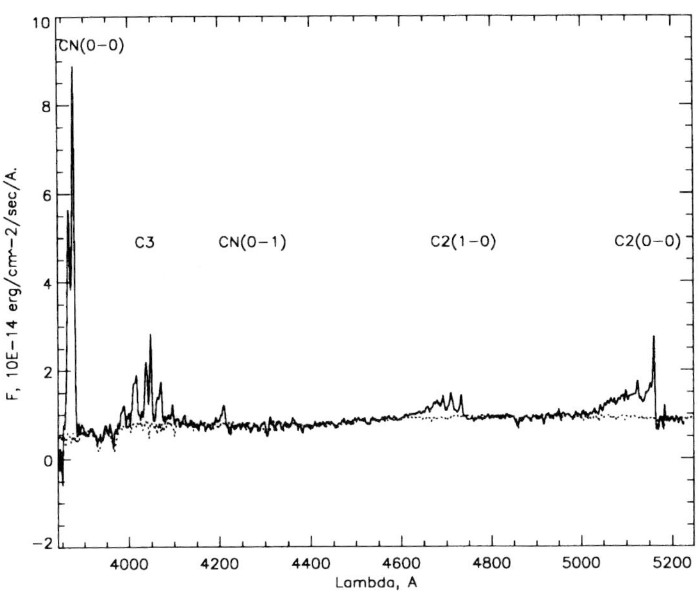

Figure 2. Detection of luminescence continuum in the spectrum of comet C/2000 A2 (LINEAR) on August 12, 2000 (the dot curve is the solar spectrum)

lines located at wavelengths of the absorption lines of the Fraunhofer spectrum, when subtracting both spectra from each other.

The resulting spectra is interpreted as a combination of two components: the solar reflected and a fluorescent continuum from the comet. According to Nazarchuk (1987a, b) we assume that the latter radiation detected here is due to the luminescence of cometary dust. Figure 2 shows how the comet continuum is de-

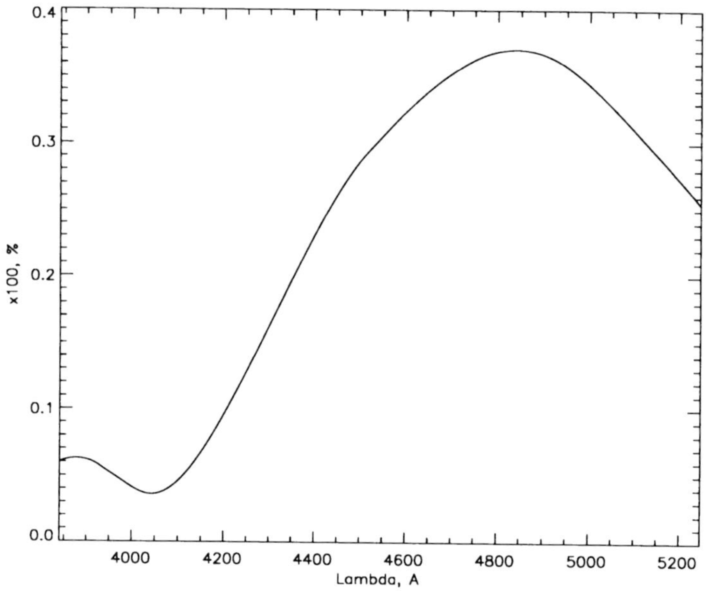

Figure 3. The luminescence continuum in the spectrum of comet C/2000 A2 (LINEAR) on August 12, 2000.

termined with regard to the level of the luminescent dust continuum. The level of the cometary continuum I_c is determined by the formula

$$I_c = k_s \cdot I_s + I_f,$$

where I_s is the level of the solar spectrum, k_s the proportional coefficient, and I_f the fluorescent continuum level. The values k_s and I_f were calculated by a least squares fit for a small part of the cometary spectrum. The relative flux of solar continuum and luminescent cometary continuum at 4800A were determined to be 63 and 37 percent, respectively (this is the maximum of the cometary luminescent continuum). We believe that we have detected a cometary continuum of non-solar origin in the spectral range 3600–5200 Å. This effect was observed in a several comets and is attributed to luminescence of cometary dust (Churyumov and Kleshchonok, 2001; Churyumov et al., 1997a, b), maybe from cometary CHON particles.

Figure 3 shows the result of detection of the luminescence continuum in the spectrum of comet C/2001 A2 (LINEAR) on Aug. 12, 2001. It appears that the maximum is shifted to 4800 Å, a value similar to the one found for Comet C/1999 S4 (LINEAR) (Churyumov et al., 2002).

3. Conclusions

The total number of the detected emission lines in the spectrum of Comet C/2001 A2 (LINEAR), obtained Aug. 12, 2001 is 250. We identified emission lines from C_2, C_3, CN, CH, CH^+, NH_2 gas. The luminescent continuum is detected at 37% of the total level of the continuum at 4800 Å in the spectrum of comet C/2001 A2 (LINEAR) on Aug. 12, 2001.

References

Brown, M. E., Bouches, A. H., Spinrad, H., and Johns-Krull, C. M.: 1996, 'A High-Resolution Catalog of Cometary Emission Lines, *Astrophys. J.* **112**, 1197.

Churyumov, K. I. and Kleshchonok, V. V.: 2001, *JENAM-01*, Astronomishe Gesellschaft. Abstract Series, Munich, p. 257.

Churyumov, K. I., Kleshchonok, V. V., Mussaev, F. A., Bikmaev, I. F., and Galazudinov G. H.: 1997a, *Earth Moon Planets* **78**, 105–110.

Churyumov, K. I., Kleshchonok, V. V., and Vlassyuk, V. V.: 1997b, *Earth Moon Planets* **78**, 111–117.

Churyumov, K. I., Luk'yanik, I. V., Vlassyuk, V. V., and Borisov, N. V.: 2002, these proceedings.

Nazarchuk, H. K.: 1987a, *Kometnyj Tsirkulyar* **372**, 2.

Nazarchuk, H. K.: 1987b, *Kometnyj Tsirkulyar* **377**, 2.

LONG-TERM AMATEUR CCD MONITORING OF COMET C/1999 S4 (LINEAR)

JUAN RODRÍGUEZ, SALVADOR SÁNCHEZ, ANTONIO GARCÍA, MANOLO BLASCO and MIQUEL A. VILLALONGA

Observatori Astronòmic de Mallorca, Cami de l'observatori s/n, E-07144 Costitx, Mallorca, Spain

(Received 18 March 2002; Accepted 3 July 2002)

Abstract. The results from the analysis of the extensive CCD photometry that we have obtained during the long-term monitoring (from its discovery until its break-up) of comet C/1999 S4 (LINEAR) are presented here. Our observations show that the comet did not always behave as expected: it showed activity ahead of time and in the end became fainter instead of brighter as it approached perihelion. A sequence of images and a lightcurve are shown.

Keywords: Comets: C/1999 S4 (LINEAR), comets: long-term photometry

1. Introduction

The study of comets at the Observatori Astronòmic de Mallorca (OAM),* began with the passage of the comets C/1996 B2 (Hyakutake) and C/1995 O1 (Hale–Bopp). Thereafter, brightness and size estimates of many more comets such as 21P/Giacobini–Zinner, C/1998 U3 (Jäger), C/1998 U5 (LINEAR), P/1998 W1 (Spahr), C/1999 H1 (Lee), and C/1999 L3 (LINEAR) were carried out.

On the 27th of September 1999, the Lincoln Near Earth Asteroid Research (LINEAR) team in New Mexico discovered (MPEC 1999-T03, IAUC 7267) comet C/1999 S4 (LINEAR), hereafter S4. At the time of its discovery it was about 4.0 AU** away from Earth and appeared as a relatively bright object of $V \sim 17$ mag. Thus at the time it looked as if it could become one of the most spectacular comets since comet C/1995 O1 (Hale–Bopp).

This comet seemed ideally suited for a long-term monitoring campaign with CCD cameras on small diameter telescopes $\Phi < 0.5$ m.

In the following we describe the data, how it was obtained, reduced and analysed. Furthermore we describe the evolution of the comet, discuss the results of the analysis of the CCD photometry and compare it to the results obtained by other groups.

* Minor Planet Center (MPC) code 620.

** 1 Astronomical Unit (AU) = 149.6×10^6 km.

Earth, Moon and Planets **90:** 185–194, 2002.

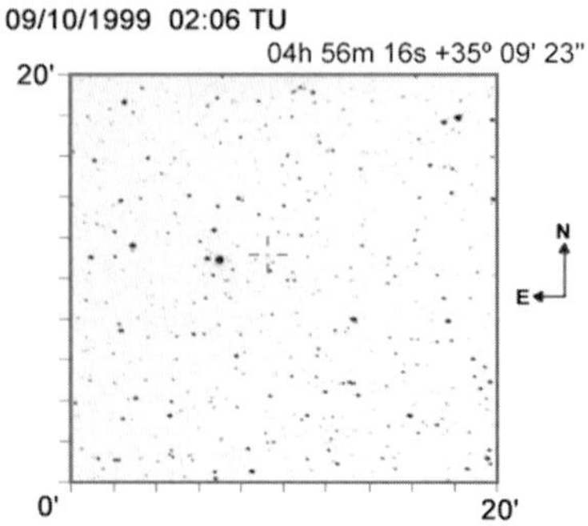

Figure 1. First image of C/1999 S4 (LINEAR) obtained on Oct. 9, 1999 from the OAM.

2. CCD Observations

During most of the time that comet C/1999 S4 (LINEAR) was visible from the OAM we observed it using several different combinations of telescope and CCD camera (see the journal of observations and Table I).

We used 0.30 m and 0.35 m and Schmidt-Cassegrain telescopes with respective aperture ratios of f/3.0 and f/3.2. Our main workhorse for wide field observations in this campaign was a 0.41 m Schmidt telescope with an aperture ratio of f/2.0. We used commercial, Peltier cooled CCD cameras from Santa Barbara Instruments (SBIG): The ST-7E (765 × 510 pixels, KAF0401E) and the ST-8E (1530 × 1020 pixels, KAF1602E) Camera. The blue enhanced CCD chip pixels are $9\mu \times 9\mu$ for both cameras (see combinations of observing equipment in Table II). Comet S4 was observed on 34 nights over the 10-month time-span from its discovery until its break-up.

At times two telescopes were observing the comet simultaneously. All of the images were obtained without filter and using integration times between 10 and 300 sec. All images were dark and bias subtracted as well as flat-fielded. Both astrometry and photometry of comet S4, was performed using the programs "PISES ATLAS-PRISM 98", with the "GUIDE 8" database.

This program allows for a simple, individual adjustment of the extraction boxes for series of reference stars (S1, S2, S3, ...). For the brightness determinations of the inner coma of the comet we used between 3 and 6 reference stars, in our case this is fairly easy due to the large field of view 45′–60′ in our images. The sky background was determined close to each reference star using the same extraction boxes. We obtain the photometric calibration by using the Tycho II catalogue V-filter magnitudes for the reference stars. With these we get a zero point C which relates the amount of counts of a given region to its magnitude. To check the quality of the photometric calibration we measure each reference star again, recompute

TABLE I
Journal of observations

Date	Estimate Time (UT)	Mag.	Error	Coma size (′)	Tail size (′)	Tail PA (°)	Delta (AU)	r (AU)	Equipo utilizado
09/10/1999	1:24	15.4	0.05	0.4	0.0	0	3.6390	4.2000	1
06/12/1999	20:10	15.1	0.10	0.6	1.4	105	2.6384	3.5495	5
07/12/1999	23:26	15.7	0.02	0.6	0.7	105	2.6331	3.5365	1
08/12/1999	0:16	15.7	0.03	0.6	0.6	105	2.6330	3.5361	3
12/12/1999	2:07	15.4	0.06	0.3	0.6	95	2.6187	3.4893	2
06/01/2000	20:14	14.5	0.20	0.5	1.2	75	2.6702	3.1884	5
13/01/2000	23:33	14.5	0.05	0.7	1.2	75	2.7148	3.1032	8
14/01/2000	18:43	15.2	0.04	0.6	1.3	74	2.7202	3.0936	5
03/02/2000	19:27	14.9	0.08	0.6	1.9	73	2.8670	2.8502	5
05/02/2000	20:33	15.1	0.20	0.3	0.6	57	2.8816	2.8249	5
10/02/2000	18:30	14.8	0.05	0.5	1.6	68	2.9150	2.7640	4
19/02/2000	19:02	14.7	0.05	0.5	1.3	70	2.9680	2.6511	9
23/02/2000	18:43	14.7	0.04	0.6	2.2	68	2.9871	2.6007	9
25/02/2000	19:18	14.8	0.06	0.6	1.9	68	2.9957	2.5750	9
27/02/2000	19:11	15.0	0.03	0.6	1.9	70	3.0032	2.5496	9
29/02/2000	18:57	14.4	0.04	0.6	1.9	66	3.0099	2.5242	9
01/03/2000	19:07	14.7	0.09	0.6	1.8	66	3.0129	2.5114	8
03/03/2000	19:17	14.5	0.07	0.5	3.1	66	3.0183	2.4857	8
06/03/2000	19:02	14.3	0.04	0.6	2.2	65	3.0244	2.4472	8
07/03/2000	19:34	14.4	0.08	0.5	2.7	62	3.0260	2.4340	8
08/03/2000	19:14	14.5	0.13	0.5	2.0	63	3.0272	2.4213	8
02/06/2000	2:35	10.9	0.05	1.1	0.0	0	1.9411	1.2636	6
03/06/2000	2:53	11.2	0.06	1.7	1.8	287	1.9125	1.2499	8
04/06/2000	2:40	11.2	0.04	2.0	2.9	285	1.8842	1.2365	8
08/06/2000	2:42	11.1	0.04	0.7	1.1	280	1.7663	1.1828	6
09/06/2000	1:54	10.3	0.06	2.7	2.7	285	1.7370	1.1700	5
16/06/2000	2:11	10.5	0.02	2.7	2.7	274	1.5145	1.0791	8
29/06/2000	2:55	10.5	0.06	2.0	7.7	271	1.0612	0.9263	8
01/07/2000	2:02	10.3	0.09	2.7	10.4	271	0.9897	0.9061	8
05/07/2000	1:50	9.8	0.02	4.3	9.3	270	0.8436	0.8679	8
08/07/2000	2:01	9.5	0.11	4.4	16.2	273	0.7345	0.8422	8
14/07/2000	18:38	8.7	0.16	4.8	22.0	253	0.5107	0.7967	7
21/07/2000	21:30	6.8	0.30	6.4	22.0	27	0.3746	0.7700	8
22/07/2000	21:23	6.8	0.23	7.0	22.0	36	0.3725	0.7680	7
24/07/2000	21:18	6.5	0.27	9.0	27.0	60	0.3830	0.7660	8
25/07/2000	20:40	7.1	0.11	8.7	30.0	60	0.3956	0.7650	8

TABLE I
Continued

Date	Estimate Time (UT)	Estimate Mag.	Estimate Error	Coma size (′)	Tail size (′)	Tail PA (°)	Delta (AU)	r (AU)	Equipo utilizado
27/07/2000	21:29	7.8	0.16	6.3	32.0	84	0.4333	0.7660	8
28/07/2000	20:44	7.4	0.24	4.5	24.0	86	0.4565	0.7670	8
29/07/2000	20:22	7.9	0.12	2.8	16.0	91	0.4829	0.7680	8
30/07/2000	20:25	8.0	0.16	3.0	18.3	92	0.5116	0.7700	8
31/07/2000	20:20	8.6	0.08	3.0	18.3	96	0.5418	0.7730	8
04/08/2000	20:15	9.0	0.21	1.9	11.5	100	0.6757	0.7880	8

TABLE II
Combinations of observing equipment used

Equipe number	CCD Camera Type	CCD Camera Chip size (mm)	CCD Camera Pixel size (μ)	CCD Camera Pixels number (x, y)	Telescope Type	Telescope Diameter (m)	Telescope f/	Telescope Pixel scale ″/pixel
1	ST-8E	13.77 × 9.18	9	1530 × 1020	S/C	0.35	3.30	1.55
2	ST-8E	13.77 × 9.18	9	1530 × 1020	S/C	0.35	4.60	1.16
3	ST-8E	13.77 × 9.18	9	1530 × 1020	S/C	0.35	6.30	1.68
4	ST-7E	6.88 × 4.59	9	765 × 510	S/C	0.35	3.30	1.55
5	ST-7E	6.88 × 4.59	9	765 × 510	S/C	0.30	3.30	2.06
6	ST-7E	6.88 × 4.59	9	765 × 510	S/C	0.30	6.30	0.88
7	ST-9E	10.24 × 10.24	20	512 × 512	S/C	0.30	3.00	4.58
8	ST-8E	13.77 × 9.18	9	1530 × 1020	Schmidt	0.41	2.00	2.32
9	ST-7E	6.88 × 4.59	9	765 × 510	Schmidt	0.41	2.00	2.32

its brightness and compare it to the original value in the catalogues. We find no deviations greater than 0.05 mag.

$$m_2 = -2.5\log(S2) + (m1 + 2.5\log(S1)), \tag{1}$$

$$C = m1 + 2.5\log(S1), \tag{2}$$

$$m = C - 2.5\log(S). \tag{3}$$

It is not easy to measure the brightness of diffuse objects such as comets as they have no well defined edges. To measure the brightness of the inner coma we

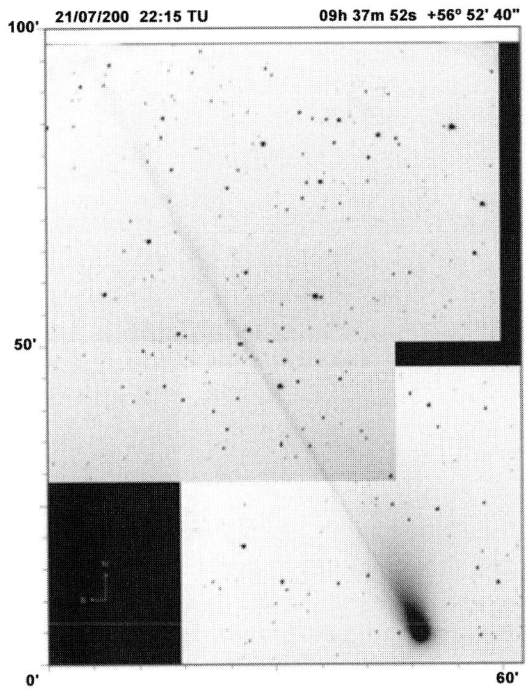

Figure 2. Composite of 3 images of C/1999 S4 (LINEAR) observed on July 21, 2000.

used a projected square extraction region with each side 30,000 km long. This corresponds to extraction boxes with sides from 0.17′ to 1.85′ thus covering the inner coma region from the beginning of our observations.

The total error for the inner coma measurements taking into account all error sources such as, correct dark and flatfield corrections, the errors of V-magnitudes in the Tycho 2 catalogue, and the fact that no filters were used, adds up to a maximum error of 0.3 magnitudes (see Table II, column 4).

In the end a total of 122 detailed measurements of the comet parameters such as its coma brightness as well as "whenever possible" the size and orientation of the coma were obtained from these observations, thus making it possible to analyse the behavior and to study the evolution of S4 during the complete observation period.

3. Analysis

From the observations described in the previous section we can now determine the absolute magnitude m_0^b (b – Broadband) and the brightening coefficient n.

The evolution of the apparent magnitude m_1 of a comet is described as a function of comet-Earth distance Δ and comet-Sun distance r, both in AU's:

$$m_1 = m_0 + 5 \log \Delta + 2.5n \log r. \tag{4}$$

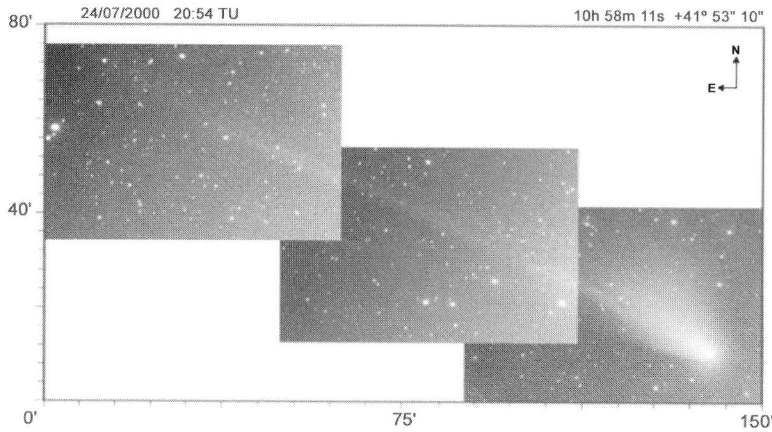

Figure 3. Composite of 3 images of C/1999 S4 (LINEAR) observed on July 24, 2000.

The parameters in this equation are the absolute magnitude m_0 of the comet and the brightening coefficient n. m_0 is defined as the absolute magnitude of the comet if it were situated at both 1 AU from the Earth and the Sun. n denotes the photometric parameter that describes the evolution of the comet brightness as a function of its distance to the sun.

The absolute magnitude m_0 and the brightening coefficient n where determined using a regression method. This method yield the correlation coefficient (this value approaches 1 the better the data are matched).

3.1. First observing period: Oct. 8, 1999, to Mar. 3, 2000

For this first observing period of comet S4 we obtain $m_0 = 10.0$ and $n = 2.1$ from fits to our photometry. We find that the value of n is quite low when compared with some periodic comets such as: 21P/Giacobini–Zinner with $m_0 = 8.3$ and $n = 11.4$ and 93P/Lovas with $m_0 = 9.3$ and $n = 13.8$ or some newer comets such as: C/1998 P1 (Williams) with $m_0 = 5.1$ and $n = 15.5$ and C/1999 L3 (LINEAR) with $m_0 = 6.0$ and $n = 19.0$ (all values here from Shanklin (1999)).

This low value for n means that the increase in brightness of the comet as a function of time is mainly due to the comet approaching us. During this time comet S4 did not show any real activity that could had led to an increase in brightness. This could be explained by the comet having a very small nucleus and a lack of volatiles.

3.2. Second observing period: June 2, 2000 to July 26, 2000

For this observing period we get $m_0 = 8.9$ and $n = 3.7$. The new value for m_0 indicates that the comet is at least 3 times brighter than during the first observing period. This value is compatible with the expected decrease of activity after the comet having lost most of its volatile ices during the first phase. A value of $n = 3.7$,

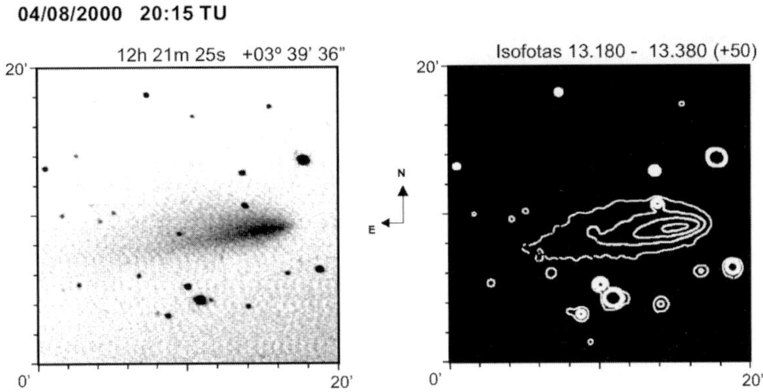

Figure 4. Image of C/1999 S4 (LINEAR) after break-up of the comet nucleus obtained on Aug. 4, 2000 from the OAM. Right: CCD image. Left: The contours illustrate in more detail the abnormal elongation of the inner coma of the comet.

indicates that the increase of brightness of the comet is not only due to its gradual approach to Earth, but also due to additional activity.

3.3. The CCD Lightcurve for All Observations

In Figure 5 the long-term lightcurve of comet S4 determined from our CCD observations are shown. The solid line in Figure 5 represents the best fit of the apparent magnitude equation to the complete data set:

$$m_1 = 9.5 + 5 \log \Delta + 6.34 \log r, \tag{5}$$

which gives an absolute magnitude $m_0 = 9.5$, a brightening coefficient $n = 2.5$ and a correlation coefficient of 0.88.

In Figure 6 we show the evolution of the comet brightness as a function of its distance to the Sun. Here one can clearly see that the comet did not behave as expected; it showed several anomalies, the first one between 2.5 AU and 1.2 AU where there seems to be a sudden increase of activity that led to a brightening and subsequent dimming (see Figure 5, around June 2000). Finally when it is at its perihelion at about 0.7 AU, at its maximum approximation to the Sun, it suddenly becomes fainter, most likely due to the break-up of its nucleus. Even though one would usually expect it to brighten, because when the nucleus breaks apart a large quantity of gases should be set free leading to a short but sharp increase in comet activity which was not detected here.

During the night of July 24 and 25 first indications of the fragmentation of the nucleus were detected (see Figure 4) the inner coma region had shifted with respect to where it should have been and started rapidly to become fainter.

Our observed brightness function for comet S4 is compared in Figure 7 to the brightness functions obtained from other data sets, i.e., from the Japanese

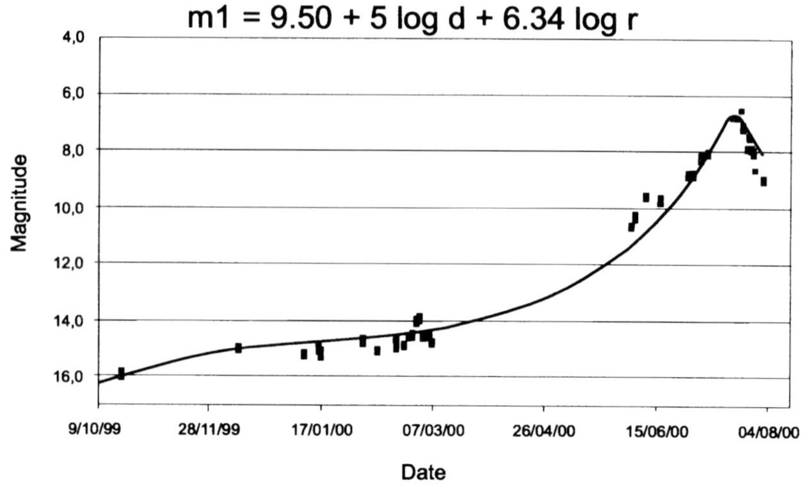

Figure 5. Observed magnitude of C/1999 S4 (LINEAR) as a function of time. The points represent the observed magnitudes obtained from our photometry at the OAM see Table I. The solid line represents the best fit of the brightness function to the OAM data.

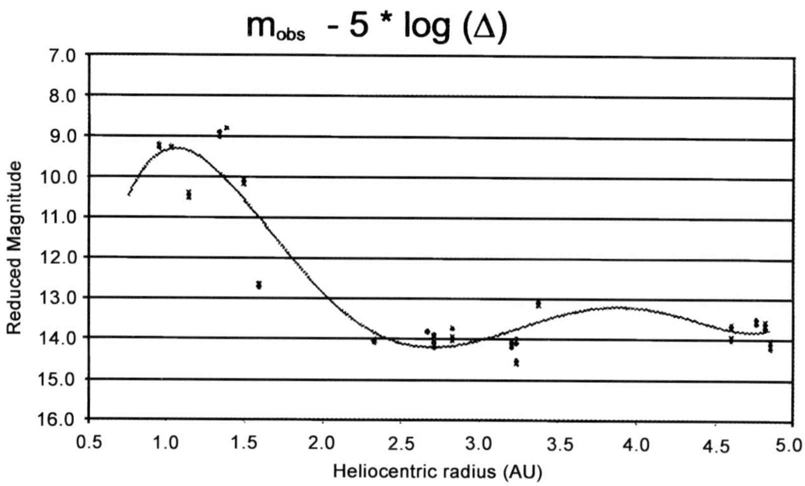

Figure 6. Reduced magnitude of C/1999 S4 (LINEAR) as a function of distance to the Sun.

observations (Seiichi Yoshida),* Spanish visual observations[3] and Spanish CCD observations.** The trend in all lightcurves is the same. The offsets are due to differences in the effective band passes of the naked eye and CCDs cameras.

* http://aerith.net/index.html
** http://www.iac.es/galeria/mrk/

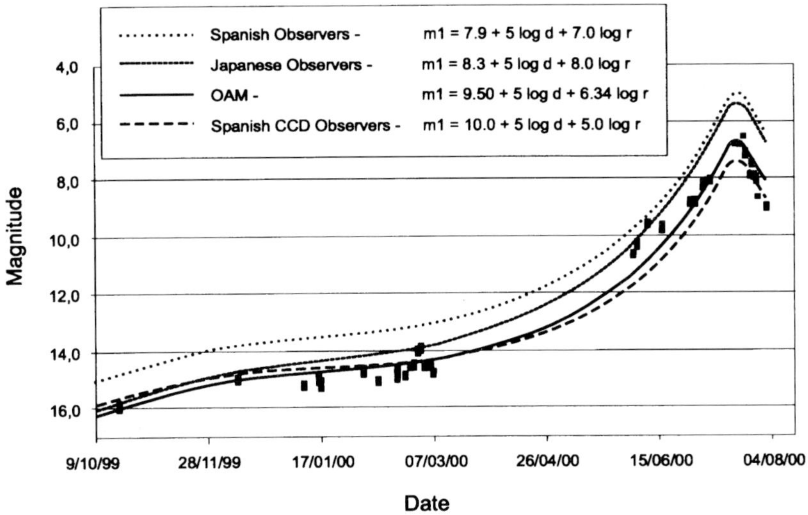

Figure 7. Observed magnitude of C/1999 S4 (LINEAR) as a function of time. The points represent the observed magnitudes obtained from our photometry at the OAM (see Table I). The solid line represents the best fit of the brightness function to the OAM data. The observed magnitude functions fitted to the data obtained by other groups are overplotted: the dotted, short-dash, and long-dash correspond, respectively, to the Japanese, Spanish visual, and Spanish CCD observers.

4. Conclusions

Despite the absolute errors of the up to 0.3 mag for the individual observations our large sample of data permits us to generate a very good long-term light curve for C/1999 S4 (LINEAR), i.e., the brightness variation as a function of the comet's distance to the Sun and the Earth is much larger than the errors of the individual observations (relative error in the several percent range).

Our long-term monitoring observations of C/1999 S4 from just after its discovery until its break-up have shown that its behaviour was irregular. This was most apparent shortly before its perihelion where the break-up of the nucleus began. At that time we thought that we had observational problems but actually the comet became fainter instead of brightening further. The lightcurves we obtain are comparable with those obtained by other observing groups and found to be in general agreement with each other.

Our last image of comet S4 on Aug. 4, 2000 (see Figure 4) was taken a few days after the final break-up of the comet and just one day before this catastrophic event was detected in simultaneous Hubble Space Telecope (HST) and University of Hawaii 2.2-m telescope observations* The HST images clearly show the break-up of the nucleus into a large number of cometesimals That are no smaller than 30 m. On Aug. 6, 2000, further images** were taken using the European Southern

* http://oposite.stsci.edu/pubinfo/PR/2000/27/index.html
** http://www.eso.org/outreach/press-rel/pr-2000/phot-20-00.html

Observatory (ESO) Very Large Telescope (VLT) at Paranal in Chile. These images show that the extended tail ends in a formation of small ice fragments of its nucleus. Comparing these images with those taken with HST the day before, we see that the fragment distribution has changed. From that we infer that it still was very active between the HST and VLT observations.

We would like to conclude by saying that long term monitoring and tracking of comets with CCD cameras on small telescopes is a very useful method to keep track of how comets evolve and behave as the comets swing by the Sun.

Acknowledgements

We would like to thank Mark Kidger of the Instituto de Astrofìsica de Canarias (IAC) for the kind and motivating support he has given during this long-term observation of C/1999 S4. We would also like to thank Teresa Sanchez (OAM) for translating the article and Vadim Burwitz (MPE) for typesetting the manuscript in LaTeX. Many thanks to the International Comet Quarterly (ICQ) and and the Minor Planet Center (MPC) for the feedback, in the form of forecasts on the behaviour and morphology of this comet, to the images and data we provided them.

References

Kidger, M.: 2000, July 27, *IAUC* 7467.

Shanklin, J.: 1999, 'The Comet's Tale', *Newsletter of the Comet Section of the British Astronomical Association* **6**(1), (Issue 11). http://www.ast.cam.ac.uk/~jds/tail11.pdf

THE DUST COMA OF COMET C/1999 S4 (LINEAR) *

RITA SCHULZ
Research and Scientific Support Department of ESA, ESTEC, Postbus 299, 2200 AG Noordwijk, The Netherlands

JOACHIM A. STÜWE
Leiden Observatory, P.O. Box 9513, 2300 RA Leiden, The Netherlands

(Received 27 February 2002; Accepted 21 June 2002)

Abstract. Comet C/1999 S4 (LINEAR) showed a very special behaviour between 28 June and 1 July 2000. Optical observations of the dust coma in two distinct continuum bands revealed that it changed morphologically as well as in colour. The two-dimensional coma morphology indicates a splitting of the nucleus which probably occurred shortly before the observations of 28 June 2000. The distribution of the dust particles in sun and tail direction reflected by the slopes of the radial profiles indicate the presence of a considerable amount of disintegrating dust particles in the sunward hemisphere and an overabundance of dust, reflecting at 440 nm, within the first 18,000 km of the dust tail. The spatial profiles of the (BC–RC) colour index in sun direction are distinctly different on 28 June and 1 July, indicating the production of a large amount of particles observable in blue continuum after 28 June.

Keywords: Dust coma, split comets

1. Introduction

One of the most spectacular cometary splitting events was the complete disruption of the nucleus of comet C/1999 S4 (LINEAR), a probably dynamically new comet discovered at $r = 4.3$ AU on 27 September 1999 by the **Li**ncoln **N**ear **E**arth **A**steroid (LINEAR) programme. Evidence of the nucleus' disruption was first detected during observations between 23 and 27 July 2000 (Kidger, 2000). Optical observations of the comet's debris taken with HST and VLT on 5 and 6 August 2000 revealed that apart from the bright (dust) tail dominating its visual appearance, more than a dozen individual fragments were spread around the predicted position of the original nucleus (Weaver et al., 2001). All these fragments have faded rapidly and were no longer detected on further VLT images obtained on 9 and 14 August 2000.

From the position angle of the tail axis the time of the outburst producing the dust tail (which is commonly referred to as the time of disruption) was determined to 22.8 ± 0.2 July 2000 (Weaver et al., 2001). However, the cross correlation of two pairs of fragments detected on 5 and 6 August 2000, and subsequent estimation of their separation times indicated that these fragments separated already some time

* Based on observations at Wendelstein Observatory.

Earth, Moon and Planets **90:** 195–203, 2002.

TABLE I
Imaging observations of C/1999 S4 (LINEAR)

Date	UT	Filter	τ_{exp} (s)	Airm.	r (AU)	Δ (AU)	Phase
28.06.2000	00:43	RC	900	2.292	0.94	1.10	59°.2
28.06.2000	01:01	RC	900	2.106	0.94	1.10	59°.2
28.06.2000	01:37	BC	1200	1.789	0.94	1.10	59°.3
28.06.2000	01:59	BC	1200	1.650	0.94	1.10	59°.3
07.01.2000	00:45	BC	1200	2.096	0.91	0.99	64°.6
07.01.2000	02:10	RC	900	1.537	0.91	0.99	64°.7

between late June and mid-July 2000 (Weaver et al., 2001). Indeed it turned out that the comet had splitting events already before its final disintegration was observed (Weaver et al., 2001). Here, we report on imaging observations of comet C/1999 S4 (LINEAR) which indicate a splitting of the nucleus on 28 June 2000.

2. Observations and Data Reduction

C/1999 S4 (LINEAR) was observed with the **Mon**ochromatic **I**mage **Ca**mera (MONICA) mounted on the 0.8-m telescope at Wendelstein Observatory, Germany, on 28 June and 1 July 2000. Images (field of view: 3′.1 × 4′.7, pixel size: 0″.5) were taken in blue continuum (BC) and red continuum (RC) with two special narrow-band filters. The band passes of these filters cover regions of the cometary spectrum known to be free of gaseous emission. A summary of the observations is given in Table I. Table II lists the basic properties of the two narrow-band filters. In addition to the cometary observations we took images of the spectrophotometric standard star BD +33 2642 in both filters at different air masses for calibrational purposes. These images showed that, although the sky appeared to be clear during the observations on 28 June, some thin cirrus must have been present. The observations of 1 July were effected by thin clouds occasionally moving into the field of view. The brightness values presented in this paper are therefore too low on the absolute scale and the apparent decrease of the comet brightness between 28 June and 1 July may be due to clouds. However, all relative values, such as profile shapes and spatial colour variations are not affected by weather conditions. The tracking errors of the telescope, assessed from the length and shape of the star trails, was determined to be below 1″ for the longest exposure time, i.e., 1200 s.

All images were bias subtracted and flat-fielded. The air-mass coefficient, k_λ, of the filters (Table II) can be determined from the images of the spectrophotometric standard star, as the slope of a [log(ADU/s) vs. Air mass] plot. The absolute values for the flux of BD +33 2642 as a function of wavelength were taken from Oke (1990). By relating these flux values convolved with the appropriate filter curve to the count rates in the corresponding images of the standard star, we were able to determine a count rate to flux conversion factor for each filter (Table II). These were then used for a flux calibration of the comet images. The flux tables for the Sun (Colina et al., 1996) and for Vega (Hayes, 1985) can further be used to establish the zero-point fluxes for a conversion of fluxes to magnitudes, since for Vega all colour indices should be zero *per def.*, which then enable the computation of the (BC–RC) colour indices. The C.I. of the sun was then computed to: $(BC\text{–}RC)_{\odot} = 1.13 \pm 0.01$.

Since the comet did not cover the entire CCD, the amount of night sky flux could be determined from the images directly by computing the average over the whole frame excluding objects. This procedure neglects possible variations of the sky over the frames, but since the sky in the frames was uniform within a margin of a few percent, this appears to be a justifiable approach.

TABLE II

Properties of narrow-band filters

Filter	λ_{eff} (nm)	W_{λ} (nm)	k_{λ} (mag/Airm.)	Flux/cps. 10^{-16} $\frac{erg/(cm^2\ s)}{ADU/s}$
BC	444.0	2.6	1.62 ± 0.41	22.6 ± 10.0
RC	685.6	7.2	0.18 ± 0.03	51.5 ± 18.0

3. Analysis and Results

3.1. Two-dimensional analysis

Figure 1 shows the dust coma of Comet C/1999 S4 (LINEAR) as it appeared in the blue and red continuum on 28 June and 1 July 2000. It is obvious that the coma morphology has changed significantly within the three days between both observing runs. On 1 July 2000 the dust coma is so strongly elongated already in the inner part that it is very difficult to determine an optocenter in these images. (The optocenter is defined as the brightest spot in the coma.) On 28 June 2000 the images appear to be rather "ordinary", on the first view, showing an elliptical coma and the dust tail. However, the immediate vicinity of the brightest pixel is already somewhat displaced towards the south east with respect to the extended rest of the coma. The optocenter is at very different positions on both observing dates. Altogether, this might suggest that shortly before the observations of 28

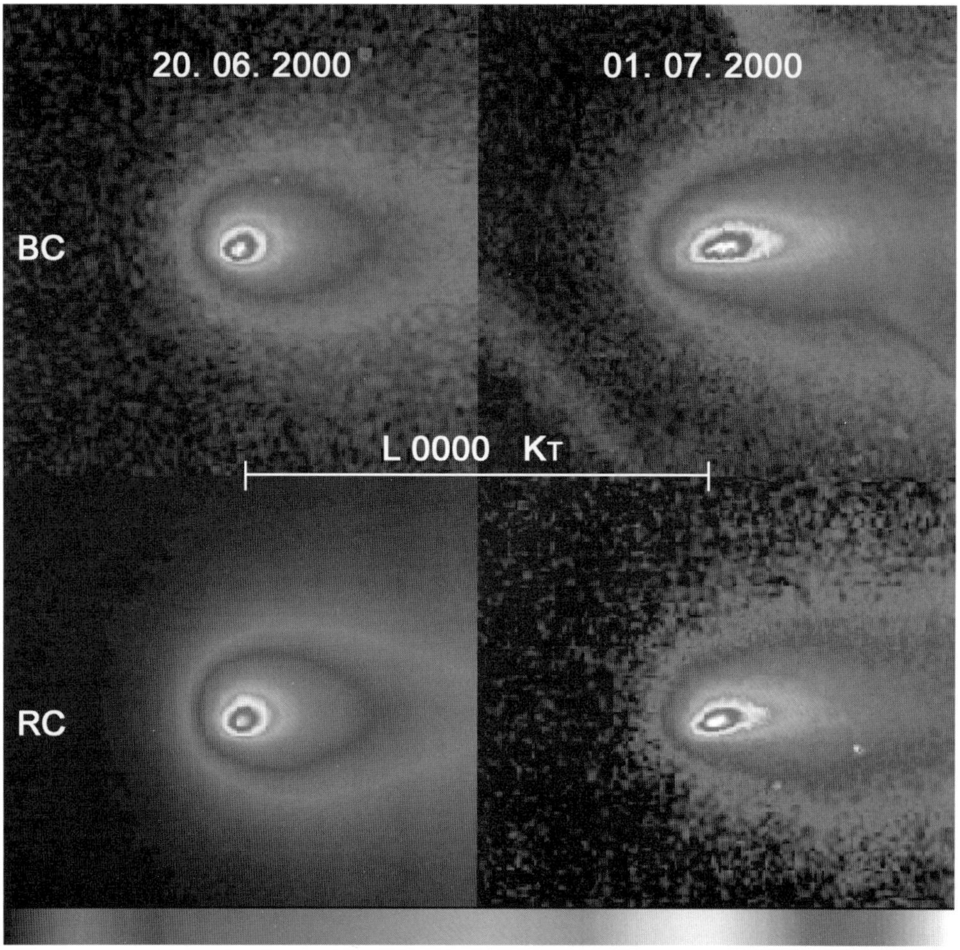

Figure 1. Comet C/1999 S4 (LINEAR) imaged in blue continuum (BC) and red continuum (RC). On 1 July 2000, the coma is much more elongated than on 28 June 2000. The Sun is to the left in all images.

June were obtained, the nucleus or a highly active fragment started to drift with respect to the rest of the coma that still moves undisturbed along the original orbital path of C/1999 S4 (LINEAR). As the undisturbed coma was released prior to the disruptive event it represents older material. The more recently released dust forms a new tail arranged along the path of the active fragment, which is different from the original orbital path of C/1999 S4 (LINEAR). Consequently, the displacement is more pronounced in the images of 1 July 2000.

3.2. Coma Profiles

Spatial coma profiles were obtained from the images of 28 June and 1 July 2000. The profiles shown in Figure 2 represent the radial distribution of the dust as seen in blue and red continuum in the sunward hemisphere and in tail direction. (The five innermost points of the profiles, $\approx 2\farcs5$, reflecting the seeing and uncertainties owing to tracking errors, are not plotted.) The sunward profiles were obtained by azimuthal averaging over the sunward hemisphere (whereby the brightest pixel of each image was assumed to contain the nucleus). The profiles in tail direction were averaged over $\pm$ 2 CCD rows, centered along the symmetry axis of the coma, which was determined as follows. First, all images were oriented along the direction of the projected extended radius vector of the comet, with the tail pointing along the (positive) x-axis. Then a first estimate of the "center" row (in other words, symmetry axis) of the coma by eye was assumed and the image was mirrored along this row, i.e., the y-axis was "flipped" upside down. A "reference" image was computed by taking the minimum of the original and flipped images for each pixel position (x, y) and this reference was subtracted from the original. The resulting residual images show quite distinct artefacts, which enable a new estimate of the symmetry axis and the whole procedure was repeated until the undisturbed coma vanished completely from the residual images leaving only the "displaced" material.

From Figure 2 it is evident that the profiles in tail direction are much flatter on 1 July than on 28 June, whereas the sunward profiles only slightly differ on both dates. To quantify the shapes of the profiles we have fitted them by a ρ^{-m} law, which results in straight lines with certain slopes, m, in the double logarithmic representation chosen for Figure 2. The slopes and minimum distances to which a good fit could be obtained are given in Table III. It turned out that almost all profiles were well fitted by this law down to projected nucleus distances between 3,700 km and 6,000 km. The tailward profile of 1 July in blue continuum could, however, be fitted by a ρ^{-m} law only at distances greater than 18,000 km. Inside this distance the BC brightness profile is very flat indicating that the number density of the dust particles represented by this colour are not diluting with r^{-2}. This can only be the case if they are continuously replenished.

Another peculiarity is that the slopes resulting for the sunward profiles are all much higher than expected. In a steady-state coma the decrease of brightness should follow a ρ^{-m} law with $1 \leq m \leq 1.5$ (Jewitt and Meech, 1987), which is indeed the case for the outer parts of the tailward profiles. In the sunward hemisphere, however, the brightness decreases at a much higher rate. The determined slopes are of order of 2.3 on 28 June and 2.7 on 1 July. Such high slopes are unexplainable at steady-state conditions indicating that the dust grains seen in the sunward hemisphere are further disintegrating while moving away from the nucleus. The sunward profiles in Figure 2 extend to a nucleus distance of about 20,000 km. Thus, most of the dust in these profiles has been ejected from the nucleus within less than 1 day, if a dust velocity of about 0.5 km/s is assumed.

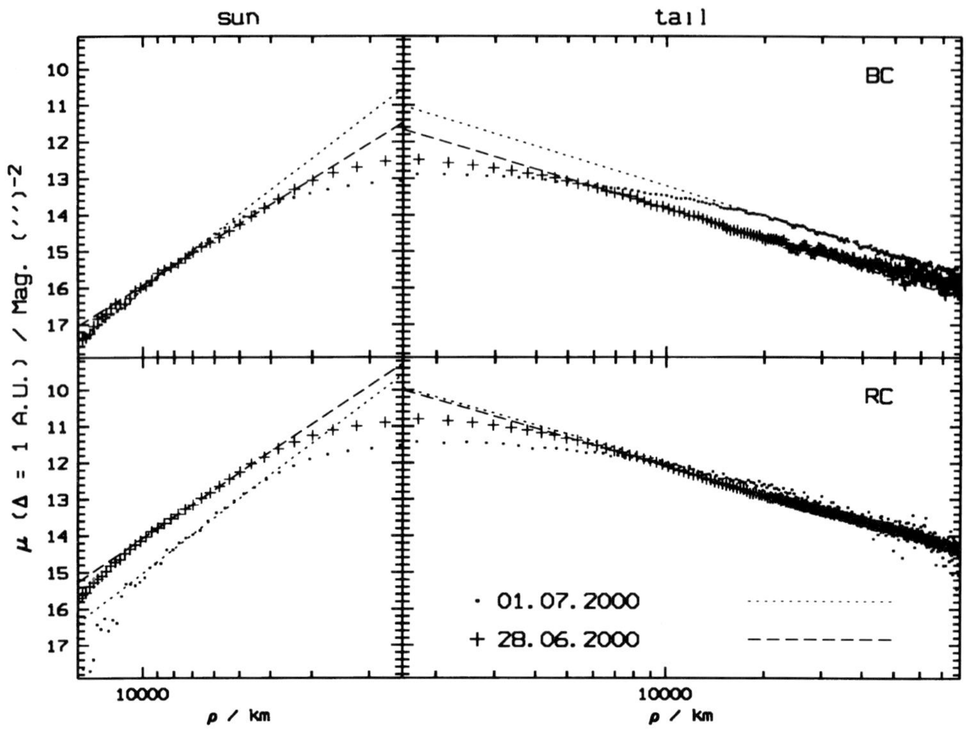

Figure 2. Spatial profiles in sun and tail direction with straight lines fitted to them. The surface brightness, μ is plotted as a function of the projected distance to the nucleus. The sunward profiles were azimuthally averaged over the sunward hemisphere. The profiles in tail direction were centered along the symmetry axis of the coma (see text for details).

TABLE III

Slopes of straight line-fits to profiles in Figure 2

Date	Filter	m_{Tail}	ρ_{min}	m_{Sun}	ρ_{min}
28.06.2000	BC	1.08 ± 0.02	5500	2.22 ± 0.03	4000
	RC	1.05 ± 0.01	4000	2.40 ± 0.02	4000
01.07.2000	BC	1.09 ± 0.01	18000	2.75 ± 0.04	5500
	RC	1.02 ± 0.02	6000	2.72 ± 0.06	3700

This indicates that the amount of *unstable* dust increased between 28 June and 1 July. A detailed modelling of the dust distribution is required to quantify this effect and further investigations are planned for the future.

3.3. Dust Colour

Figure 3 shows the (BC–RC) colour index profiles in sun and tail direction determined directly from the profiles of Figure 2. In tail direction the dust colour is as expected, redder than the Sun and constant along the profiles. In sun direction the dust colour profiles have very different shapes. On 28 June the (BC–RC) colour index increases with distance from the nucleus peaking at 4,000–5,000 km. On 1 July it shows the opposite behaviour, decreasing with distance to the nucleus reaching its minimum at 6,000–7,000 km. In both cases the distance of the maximum/minimum of the curve coincides with the distance from which on straight lines could well be fitted to the brightness profiles of Figure 2. Further in, the coma is obviously dominated by non-steady-state conditions. We therefore refrain from speculating about the particle and colour distribution inside these boundaries. Although the error of the colour index is rather large (see error bar depicted in Figure 3) the difference in the dust colour between 28 June and 1 July is still pronounced if the worst case error is assumed. For 28 June there is a rather straightforward explanation for the colour index profile. If dust particles continue to disintegrate while moving away from the nucleus, larger (redder) particles should predominantly be present at closer distances while the region further away from the nucleus should exhibit more smaller (blue) particles. In this scenario the vertex of the curve (reddest value) indicates the border between the regions more dominated by the red or blue particles. Note, that the value of the (BC–RC) colour index is generally higher than in tail direction ($\sim$0.3 at the peak position) indicating that there are more red particles in sun than in tail direction on this date.

The shape of the colour profile on 1 July indicates that more small (blue) than large (red) particles must be present in the sun than in the tail direction. The decrease of the (BC–RC) colour index towards "bluer" values could, in principle, be produced by gas emission bands. However, since we used narrow-band continuum filters of which the gas contamination is minimal, only a huge outburst of gas with emission bands in the BC filter could have caused such a pronounced effect. We therefore argue that the decrease of the (BC–RC) colour index towards "bluer" values reflects the production of additional particles that reflect light in this wavelength range. The position of the minimum would thus represent the peak of "blue" particle production, hence define the lifetime of the respective disintegrating grains. The latter need not necessarily be visible in the optical, but could be very large particles that might be observable at IR wavelength.

3.4. Discussion and Conclusions

The analysis of the imaging observations of Comet C/1999 S4 (LINEAR) from 28 June and 1 July 2000 resulted in clear evidence for a distinct difference between the dust coma morphology on both dates. The change observed in the two-dimensional coma morphology (Figure 1) suggests that a disruptive event took place shortly before the observations of 28 June 2000. This presumption is further supported

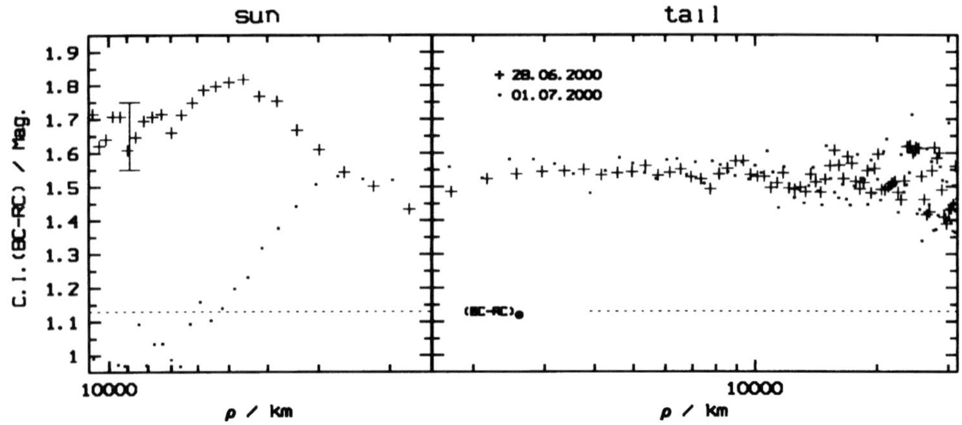

Figure 3. Spatial colour index profiles in sun and tail direction. The BC–RC colour index is plotted as a function of the projected distance to the nucleus. The profiles were constructed from the brightness profiles shown in Figure 2.

by the radial distribution of dust particles in sun and tail direction and by the different variations of the spatial colour distribution within the coma. The fact that the tailward profile obtained in the blue continuum is very flat out to a distance of about 18,000 km indicates the presence of an additional source of dust which may be one or more small fragments. The low resolution of the images does, unfortunately, not allow the detection of such fragments in our observations. However, HST observations reveal that fragmentation of the nucleus was already on-going well before the complete disintegration of the comet at around 23 July 2000. At least one fragment at ~460 km projected distance tailward from the nucleus was detected on 7 July 2000 by HST (Weaver et al., 2001). With a typical separation velocity of 0.5–1 m/s (cf. Sekanina, 1982) this fragment would have been close to the nucleus at the time our observations took place.

The brightness profiles reflecting the distribution of dust as a function of distance to the nucleus are very different in sun and in tail direction and also vary significantly between both dates of observation. The slopes of the sunward profiles indicate the presence of a considerable amount of disintegrating dust particles on both dates, which is, however, increasing between 28 June and 1 July. The existence of such particles is also confirmed by the difference between the radial distribution of the (BC–RC) colour in sun and in tail direction and the distinctly different shape of the sunward colour profiles on 28 June and 1 July.

All these results are consistent with the assumption that a splitting of the nucleus occurred a short time before the observation of 28 June 2000, which produced at least one major fragment and probably a number of small fragments drifting in the approximate tail direction. The fragments remained active at least until 1 July 2000 producing further disintegrating dust particles. A splitting of the nucleus around this time would also be consistent with the finding that two pairs of fragments detected on 5 and 6 August 2000 separated already sometime between late June and mid-July 2000 (Weaver et al., 2001).

Acknowledgements

We thank the Wendelstein Observatory for the granting observing time to our programme and C. Ries and O. Bärnbantner for taking the observations in service mode.

References

Colina, L., Bohlin, R. C., and Castelli, F.: 1996, *Astron. J.* **112**, 307–315.

Hayes, D. S.: 1985, *Proc. of IAU Symp. 111*, 225.

Jewitt, D. G. and Meech, K. J.: 1987, *Astrophys. J.* **317**, 992–1001.

Kidger, M.: 2000, *IAUC* 7467.

Oke, J. B.: 1990, *Astron. J.* **99**, 1621–1631.

Sekanina Z.: 1982, in L. L. Wilkening (ed.), Tucson, pp. 251–287.

Weaver, H. A., Sekanina, Z., Toth, I., Delahodde, C. E., Hainaut, O. R., Lamy, P. L., Bauer, J. M., A'Hearn, M. F., Arpigny, C., Combi, M. R., Davies, J. K., Feldman, P. D., Festou, M. C., Hook, R., Jorda, L., Keesey, M. S. W., Lisse, C. M., Marsden, B. G., Meech, K. J., Tozzi, G. P., and West, R.: 2001, *Science* **292**, 1329–1333.

FAST VARIATIONS OF SPECTRA OF COMET HALLEY

SERGIY BORYSENKO, HALYNA NAZARCHUK and LEONID SHULMAN
Main Astronomical Observatory of the National Academy of Sciences of Ukraine, 27 Akad. Zabolotny Str., 03680, Kyiv, Ukraine

(Received 28 February 2002; Accepted 12 July 2002)

Abstract. Results of the analysis of the Comet Halley spectrophotometry, which has been carried out by H.K. Nazarchuk in 1985 with the TV scanner of the 6-meter telescope (SAO, Russia) are presented. Time variations in the intensities of the CN, CH, C_2 and NH_2 bands were investigated using the series of spectra obtained in the spectral region $\lambda\lambda = 410 \div 510$ nm. The auto-correlation functions for all the bands, the cross-correlation functions for C_2 lines and their Fourier transformations are calculated to determine the frequencies and periods of the variations. A possibility is considered that high-amplitude fast variations of spectral lines in Comet Halley are caused by solar flares. The daily numbers of solar spots and proton fluxes with energies of more then 1 MeV are compared with the spectral variations of these lines. It is shown that in the observation periods the comet was projected onto an active region of the Sun, but, among all the kinds of solar activity, mainly the solar proton flux with energies less than 4 MeV coincides in time with fast intensity variations in the spectral lines. An algorithm of cross-correlation analysis of discrete sampling series with gaps is built, and average cross-correlation coefficients are calculated.

Keywords: Comet Halley, fast spectral variations, luminescent continuum in comets, solar proton flux, spectroscopy

1. Introduction

To check if a cometary nucleus is chemically homogeneous one can use series of spectra with sufficient spectral and time resolution. High-resolution spectral series of comet Halley were obtained by H. K. Nazarchuk in November–December 1985 with 6-m telescope of Special Astrophysical Observatory (Russia). These spectra are processed to get the physical parameters of the nucleus. The results of this analysis are presented in this paper.

2. Observations and Data Reduction

The TV-scanner which is mounted in the Nasmyth's focus of the 6-meter telescope was used as a spectral device. The applied diffraction grating gave an inverse dispersion of 50 Å/mm. The spectrum on the inlet window of the image intensifier used was 18 mm long. All spectra were obtained by counting the photons in each of 1024 spectral channels. Thus, the width of the channel was about 0.9 Å and,

Earth, Moon and Planets **90:** 205–215, 2002.

therefore, the width of the spectral range, which the TV-scanner could cover in one exposition, was about 900–980 Å. Four exposures are needed to cover the whole sensitivity range of the input photocathode (3320–6600 Å) with a spectral resolution $\lesssim 2$ Å.

The TV-scanner is designed especially for spectroscopy of stars, galaxies and other objects of very small angular size. Originally the spectrograph has a long (>40 arcsec) slit. The slit was covered by the special mask, which cut from the slit two rectangular diaphragms. The height of each slit was equal to 4 arcsec. So we had two channels separated by 40 arcsec and every exposure gave one-dimensional spectra of two different points of the comet atmosphere.

Four pointing modes of the diaphragms were used:

(1) N – the nucleus and the region at 40 arcsec from it;
(2) C – two regions of the coma each at 20 arcsec from the nucleus between them;
(3) NN – near-nuclear region at 4 arcsec from the nucleus and the point in the atmosphere at the nucleocentric distance 36 arcsec;
(4) S – one of the diaphragms was accompanying the occulted star, while another was tracing the coma. The log of the observations is given in Table I.

The measured spectra are sampled by the TV-scanner in an array of 1024 channels (16-bit integer). A Neon gas spectrum was taken before every set of the observations in order to connect a spectral channel number with the wavelength. The dispersion curves were obtained for both slits. The effect of nonhomogeneity of the photocathode of the image intensifier was eliminated by flat-field correction. The spectrophotometric standard stars Feige 15 and Feige 34 were chosen as the reference stars for spectral calibration. The primary processing removed the effect of geomagnetic field on the image intensifier and atmospheric extinction.

After the dispersion curve was obtained all the comet spectra were linearized, i.e. reduced to an uniform wavelength grid. The reference stars were reduced in the same manner. On the next step the comet spectra were reduced to relative intensities by dividing them by the observed reference star spectrum and then multiplying the result by the published distribution of energy in the stellar spectrum. Although this distribution is known in absolute flux units, an absolute flux calibration of the comet spectra is still impossible, because only an unknown part of the reference star image penetrated in the slit of the spectrograph.

3. Analysis

Variations of intensity of some C_2 lines are shown in Figure 1. Here we used the results of the N and NN observing modes. Assuming that each spectral channel registered a random Poisson process one can estimate the statistical error of the number of counted photons n as $\sqrt{n}$. As we can see from Fig. 1 the mean square error does not exceed 5 percent for the nuclear region and 20 percent for the more distant coma.

TABLE I

Series of cometary spectra. Observers were: A – Afanasyev, V. L., L – Lipovetskiy, V. D., S – Shulman, L. M., N – Nazarchuk, H. K.

Days	Range	Method	Observers
November 11	V	N(1), C(1)	A, L, S
November 12	V	N(1), C(1)	A, L, S
	B	N(1), C(1)	
	G	N(1), C(1)	
	R	N(1), C(1)	
	R	N(4)	
November 13	B	N(29), S(7)	A, L, S, N
November 14	VB	N(13)	A, L, S, N
November 15	VB	N(36), S(7)	A, L, S, N
December 8	R	N(1), NN(1), C(1)	N
	G	N(1), NN(1), C(1)	
	B	N(1), NN(1), C(1)	
	V	C(1)	
December 9	V	N(1), NN(1), C(1)	N
	B	N(1), NN(1), C(1)	
	R	NN(31)	
December 10	R	NN(17)	N
	B	NN(15)	
December 11	G	NN(27)	N
December 12	R	NN(34)	N

The autocorrelation function $K(\tau)$ of a continuous function $F(t)$ is defined by the relationship (Bendat and Piersol, 1980):

$$K(\tau) = \int_{-\infty}^{+\infty} F(t)F(t+\tau)dt. \quad (1)$$

A sequence of non-equidistant samples A obtained in the moments t_i may be considered as a continuous function

$$F(t) = \sum_{i=1}^{N} A(t)\delta(t - t_i). \quad (2)$$

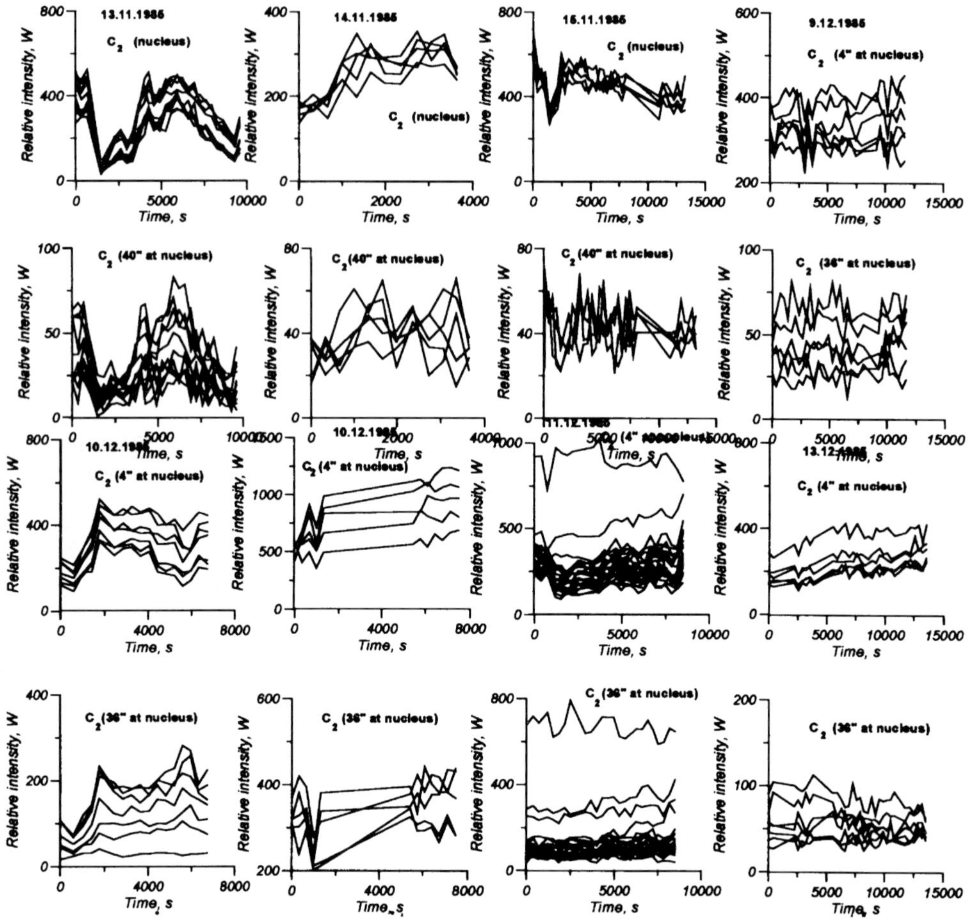

Figure 1. Variations of intensity for some spectral lines of the C_2 bands in November–December 1985.

After substitution this expression into (1) and the integration one can have

$$K(\tau) = \sum_{i=1}^{N} \sum_{j=1}^{N} A(t_i) A(t_j) \delta(t_j - t_i + \tau) = \frac{1}{T} \sum_{k=1}^{N} C_k \delta(t - \tau_k), \tag{3}$$

where $\tau_k = t_j - t_i$ is the time shift and $C_k = A_i A_j$ are the auto-correlation functions. This expression can be used to cross-correlate function by the substitutions $\tau_k = |t_{1j} - t_{2i}|$, $C_k = A_{1j} A_{2i}$ (Borysenko, 2000).

Equation (3) gives a discrete non-equidistant series of $K(\tau_k)$. Before the Fourier analysis is performed, this series was smoothed by convolution with a Gaussian kernel and transformed into an equidistant series $K_s(i\Delta t)$ with a time shift step Δt.

$$K_s(i\Delta t) = C \sum_{j=i-n_{\max}}^{i+n_{\max}} K(\tau_j)\exp(-(\tau_j - i\Delta t)^2)/d, \tag{4}$$

where $i = 0, 1, 2 \ldots N$; $d = 2\sigma^2$; $C = 1/(\sigma\sqrt{\pi})$. Parameter σ controls the smoothing (we used $\sigma = 170$ and $\Delta t = 75$ s), which decreases the effect of sampling frequency. The number of terms in the sum (4) is restricted by the exponent that goes to zero with increasing $n_{\max}$. The result of the smoothing is shown in Figure 2. One can see rather small residual variations on the sampling frequency. After the Fourier transformation we have found two periods of variations: the first is about 20–40 min (variations of type I) and the second is between 1 and 2 h (variations of type II). The type II variations were not present every observing night. Variations of type I are short-periodic and have a low amplitude. It may be caused by active processes in the cometary nucleus. Variations of type II have high amplitudes and long durations (Borysenko et al., 2000).

4. Method of Trapecia

Our observations are series of 13–36 exposures per night, separated by few minutes. So it is necessary to check the results of the correlation analysis using the spectral window of data (Terebizh, 1992; Deeming, 1975). In this case a cross-correlation function is defined as

$$K_{1,2}(\tau) = \int_{t_{\min}}^{t_{\max}} k_1(t)k_2(t+\tau)dt, \tag{5}$$

where

$$k_{1,2}(t) = \sum_{i=1}^{n} \frac{N_i^{(1,2)}}{\Delta t_i^{(1,2)}} E(t_{ib}, t_{ie}) \tag{6}$$

and

$$E_i(t) = \begin{cases} 1, & \text{if } \; t_{i,b} < t < t_{i,e} \\ 0, & \text{if } \; t\bar{\in}(t_{i,b}, t_{i,e}) \end{cases}. \tag{7}$$

So, each exposure is described by a rectangle of unit height. Finally, the cross-correlation function is

$$K_{1,2}(\tau) = \sum_{i=1}^{n}\sum_{j=1}^{n} \frac{N_i^{(1)}N_j^{(2)}}{\Delta t_i \Delta t_j} \int_{t_{\min}}^{t_{\max}} E(t_{ib}, t_{ie}, t)E(t_{ib}, t_{ie}, t+\tau)dt \tag{8}$$

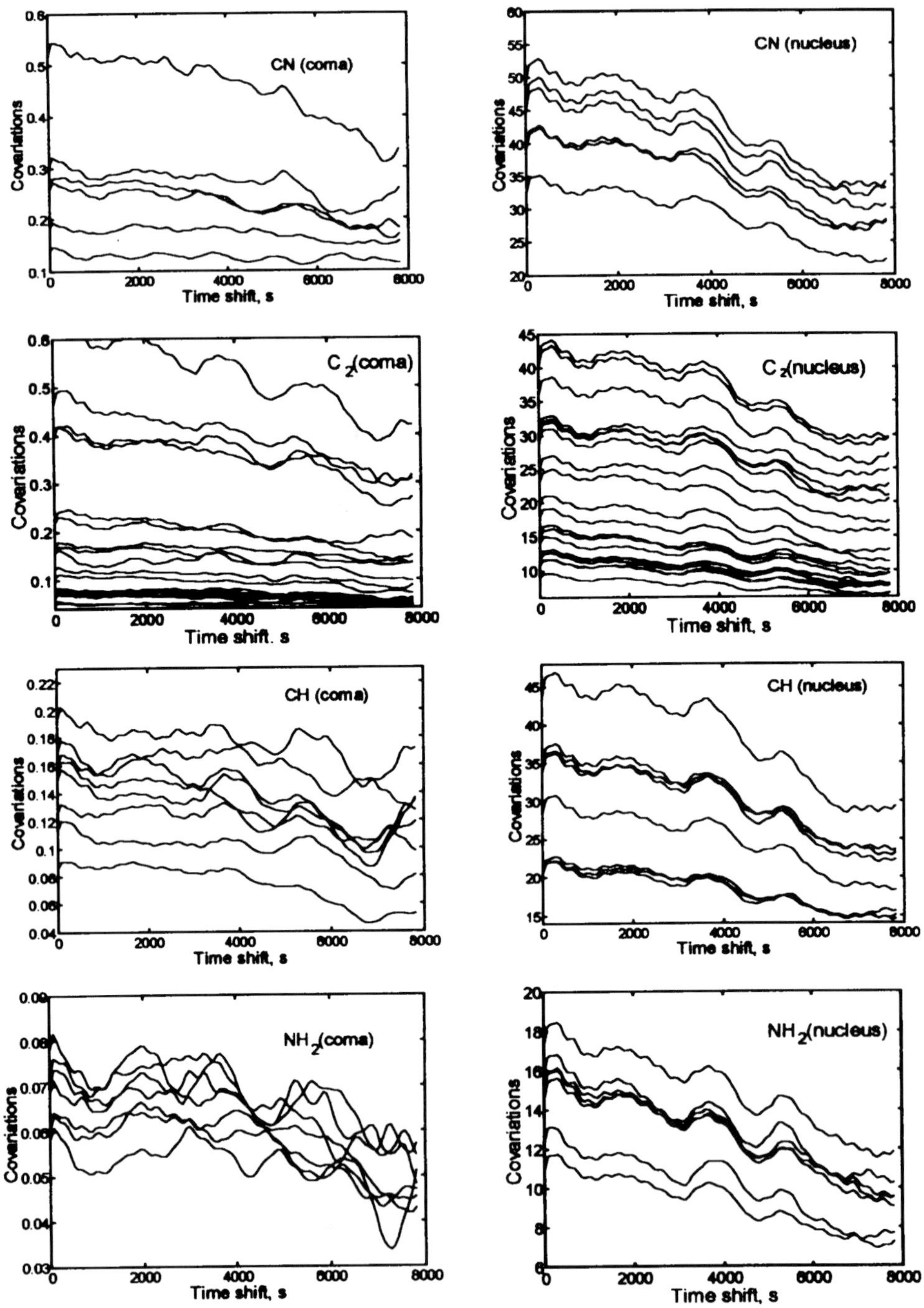

Figure 2. Auto-correlation functions.

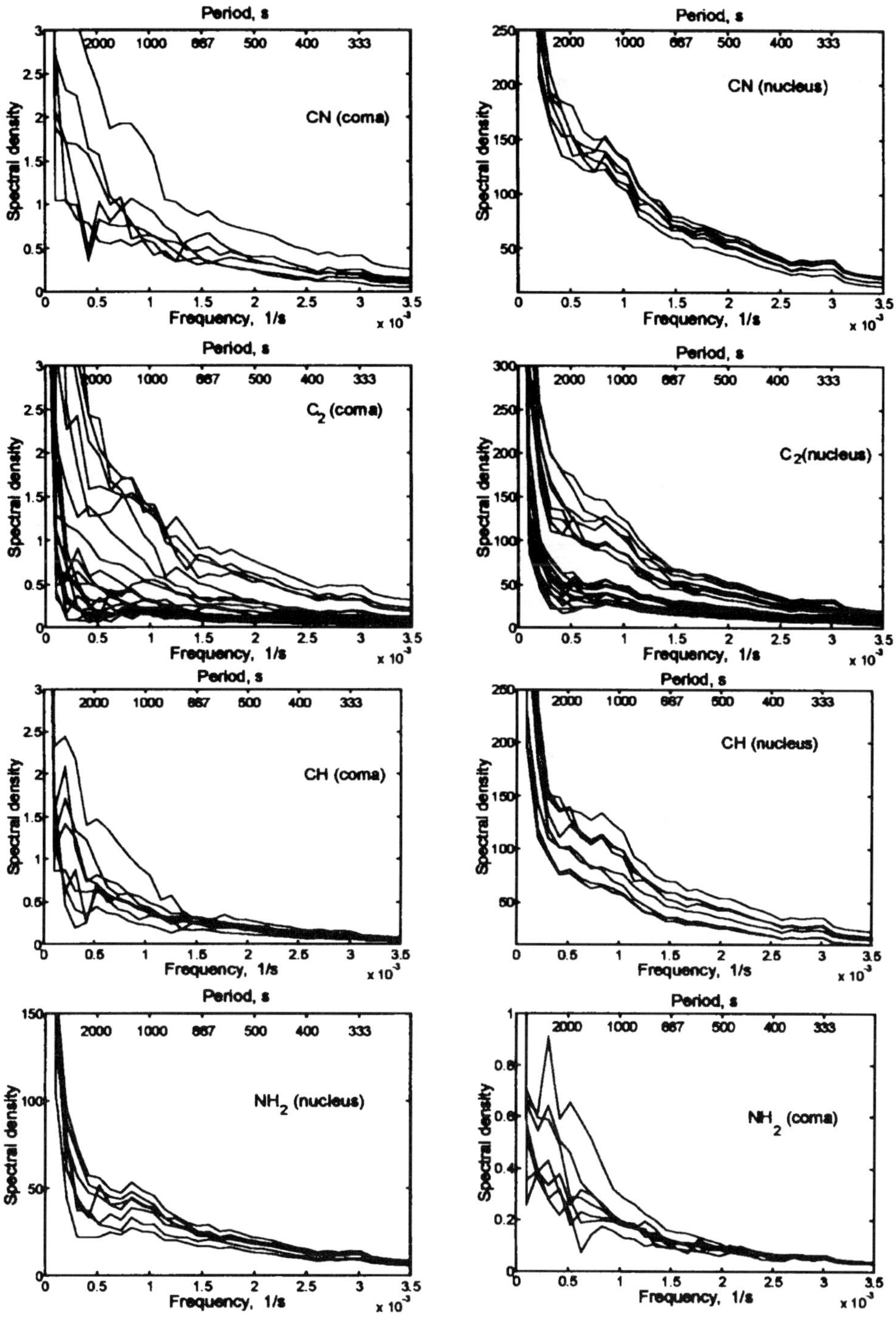

Figure 3. Statistical power spectra.

or

$$K_{1,2}(\tau) = \sum_{i=1}^{n}\sum_{j=1}^{n} \frac{N_i^{(1)} N_j^{(2)}}{\max(\Delta t_i, \Delta t_j)} Trap_{ij}(\tau), \tag{9}$$

where

$$Trap_{ij}(\tau) = \begin{cases} 0, & \text{if } \tau < \tau_{\min,ij} \\ \dfrac{\tau - \tau_{\min,ij}}{\min(\Delta t_i, \Delta t_j)} & \text{if } \tau_{\min,ij} < \tau < \tau_{\min,ij} + \min(\Delta t_i, \Delta t_j) \\ 1, & \text{if } \tau_{\min,ij} + \min(\Delta t_i, \Delta t_j) < \tau < \tau_{\max,ij} \\ & \quad - \min(\Delta t_i \Delta t_j) \\ \dfrac{\tau_{\max,ij} - \tau}{\min(\Delta t_i, \Delta t_j)} & \text{if } \tau_{\max,ij} - \min(\Delta t_i, \Delta t_j < \tau < \tau_{\max,ij} \\ 0, & \text{if } \tau > \tau_{\max,ij} \end{cases} \tag{10}$$

$$\tau_{\min,ij} = t_{jb} - t_{ie}, \qquad \tau_{\max,ij} = t_{je} - t_{ib}. \tag{11}$$

The correlation coefficient is:

$$\tilde{K}_{1,2}(\tau) = \frac{K_{1,2}(0) - \langle k_1 \rangle \langle k_2 \rangle T_S}{\sigma_1 \sigma_2}, \tag{12}$$

where

$$\sigma_{1,2} = \sqrt{\sum_{i=1}^{n} \frac{(N_i^{(1,2)})^2}{\Delta t_i} - \langle k_{1,2} \rangle^2 T_S}. \tag{13}$$

We obtained correlational coefficients of 0.7–0.9 for the nuclear region and 0.1–0.8 for the more distant coma. The large amplitude and long period variations show the highest correlation. Sometimes (see Figure 4) fast variations are well synchronized (indicating chemical homogeneity of the nucleus), whereas on some days they are not, looking like a "white noise".

5. Solar Activity and Fast Variations of Intensity of the Cometary Spectrum

We considered the following parameters of solar activity: (1) The number of solar spots, (2) the mean daily solar proton fluxes with energies more then 1 MeV (as measured by the IMP8 spacecraft). The solar activity was very low at the time of our observations. The number of spots as well as the Wolf's number is a rather general index which cannot explain many details. But there was a maximum of the flux of protons with energies of 1–4 MeV) on November 11 and December 5 (Figure 5). The projection of the cometary location onto the solar surface shows

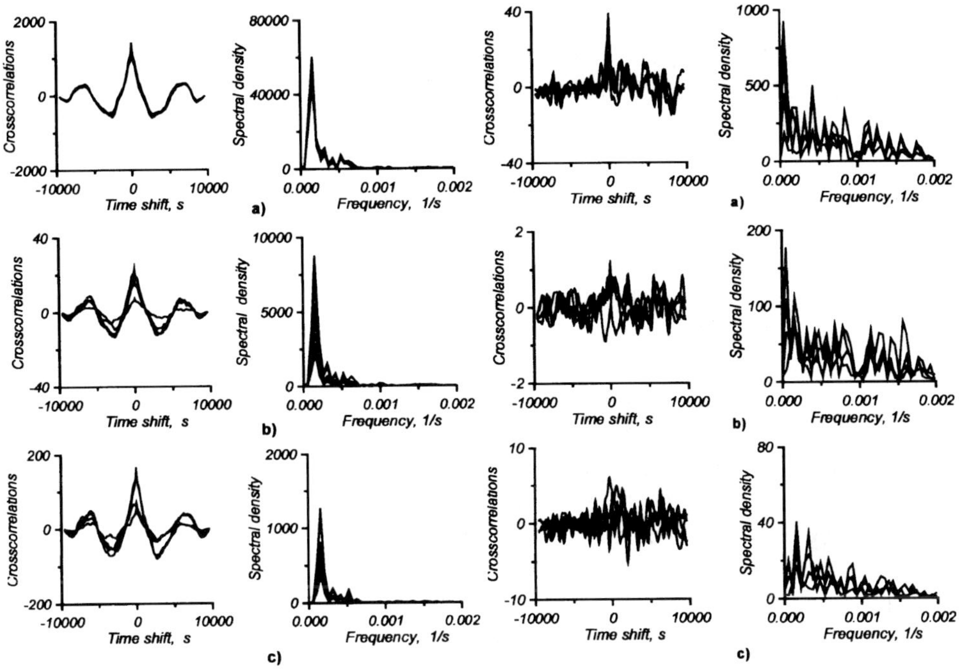

Figure 4. Cross-correlation functions and Fourier transformations for II type variations of some C_2 lines: (a) For nucleus; (b) for coma; (c) between nucleus and coma.

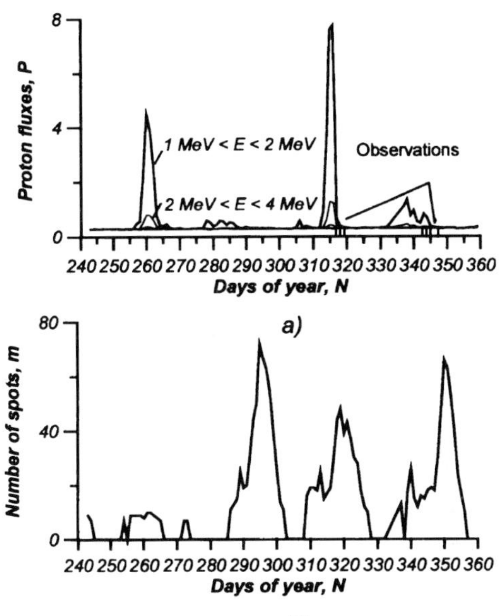

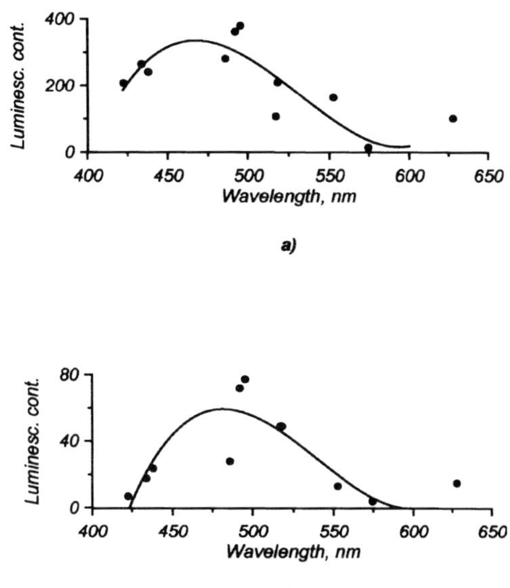

Figure 5. Solar proton fluxes (a) and numbers of solar spots (b).

good agreement with locations of active regions and coronal transients. Secondary emissions of electrons and photons from cometary dust caused by solar hard radiation may be one of the reasons for the type II variations (Borysenko and Shulman, 2001).

6. Cometary Luminescent Continuum

An interesting phenomenon was discovered by Nazarchuk (1987a, b) in the spectra of Comet Halley. This feature can be observed only with a telescope which can provide high spatial resolution. Comets have no own sources of optical radiation. Their spectra are the result of scattering of the solar radiation by cometary matter, i.e., by dust and gas molecules. The molecules are exited from the lowest quantum level and they emit through the resonance fluorescence.

One must divide a cometary spectrum by the solar one to get the optical thickness of the cometary atmosphere as a function of the wave length. When we divided our cometary spectra by the solar one, all Fraunhofer lines appear as "emission" features. Since these "emissions" are not real, we conclude that an observed cometary continuum is a sum of solar spectrum scattered by cometary dust (classic cometary continuum) and an additional continuum S_l (spectrum of luminescence), which is produced by absorbing of solar photons with further reemission in another spectral regions. The total cometary spectrum is

$$S_{\rm com} = \Omega I_0^{\lambda} \tau_{\lambda} + S_l(\lambda), \tag{14}$$

where the luminescent continuum is given by the relationship

$$S_l(\lambda) = \Omega \int \frac{\lambda'}{\lambda} I_0(\lambda')\tau(\lambda')w(\lambda' \rightarrow \lambda)d\lambda', \tag{15}$$

where Ω is the space angle of the Sun as seen from the comet, $w(\lambda' \rightarrow \lambda)$ is the probability of emission of a photon with the wavelength λ after absorbing of a photon with the wavelength λ'. Thus, the optical thickness of cometary atmosphere is:

$$\tau_{\lambda} = \frac{S_{\rm com}(\lambda) - S_l(\lambda)}{\Omega I_0^{\lambda}}. \tag{16}$$

We suppose that S_l is the emission of cometary particles excited by solar ultraviolet radiation. The level of the luminescent continuum S_l can be determined only in vicinity of an isolated Fraunhoffer line. The level of the S_l is assumed to be correct if the τ_{λ} function given by the Equation (16) shows neither emission nor absorption at this wave length.

The problem of luminescent continuum in comets needs a more detailed analysis best to be done with high-resolution spectra. After the luminescent continuum was found in the Comet Halley spectra, it was detected in other comets (Churyumov, Kleshchonok, 2001). It is interesting that at least one comet (C\1995 Y1 Hyakutake) did not have the luminescent continuum.

Acknowledgement

This work was supported by the State Basic Research Foundation of Ministry for Education and Science of Ukraine.

References

Bendat, J. S. and Piersol, A. G.: 1980, *Engineering Applications of Correlation and Spectral Analysis*, A Wiley-Interscience Publ., John Wiley and Sons, New York, 312 pp. (in Russian).

Borysenko, S. A.: 2000, *Kinematics and Physics of Celestial Bodies*, Supplement Series **3**, 256–258.

Borysenko S. A. and Shulman L. M.: 2001, *Kinematics and Physics of Celestial Bodies* **17**, 389–402 (in Ukrainian).

Borysenko, S. A., Nazarchuk, H. K., and Shulman, L. M.: 2000, *Kinematics and Physics of Celestial Bodies* **16**, 32–39 (in Ukrainian).

Churyumov, K. I. and Kleshchonok, V. V.: 2001, Astronomishe Gesellschaft, Abstract Series, *JENAM 2001*, Munich, **18**, 257.

Deeming, T. J.: 1975, *Astrophys. Space Sci.* **36**, 137–158.

Nazarchuk, H. K.: 1987, *Cometary Circular*, 372 (in Russian).

Nazarchuk, H. K.: 1987, *Cometary Circular*, 377 (in Russian).

Otnes R. K. and Enochson, L.: 1978, *Applied Time Series Analysis*, Volume 1. *Basic Techniques*, A Wiley-Interscience Publ., John Wiley and Sons, New York, Chichester, Brisbane, Toronto, 428 pp. (in Russian).

Terebizh V. Yu.: 1992, *Analysis of Time Series in Astrophysics*, Nauka, Moscow (in Russian).

C/1995 O1 HALE–BOPP: SHORT AND LONG DISTANCE ACTIVITY FROM A THEORETICAL MODEL

MARIA TERESA CAPRIA, ANGIOLETTA CORADINI and MARIA CRISTINA DE SANCTIS
CNR, IASF, Area di Ricerca di Tor Vergata, v. del Fosso del Cavaliere, 100, 00133 Rome, Italy

(Received 28 February 2002; Accepted 26 May 2002)

Abstract. A theoretical model of the nucleus thermal evolution and differentiation is used to simulate the evolution along the orbit of the gas and dust emissions of comet Hale–Bopp. The model was already applied to this comet (Capria et al., 2000b): At that time only the results of the observations obtained shortly after the perihelion were available. Now much more data have been published and we present more refined and complete results about the production rates of gas and dust along the orbit and the internal stratigraphy. The results of our model on long distance activity and its explanation are also presented.

Keywords: C/1995 O1 Hale–Bopp, comet, nucleus

1. Introduction

A well tested model (Coradini et al., 1997a, b; Capria et al., 2000a, b, 2001, 2002; De Sanctis et al., 1999, 2000a, b) of nucleus thermal evolution and differentiation was used to reproduce the activity of comet C/1995 O1 Hale–Bopp. The model was already applied to this comet (Capria et al., 2000b), and a good simulation of the activity of the comet in the range of few astronomical units around perihelion was obtained. This time we want to follow the comet far from perihelion, using better determined initial parameters, to see if our model is able to predict the long distance activity.

The 1997 passage of comet Hale–Bopp through the inner Solar System offered the astronomical community an unprecedented opportunity to follow a comet for a long part of its orbit. When discovered at a distance of 7 AU from the Sun the comet was already active. The earlier activity was attributed to the sublimation of CO (Jewitt et al., 1996; Biver et al., 1996); at 3–4 AU before perihelion water production began to be higher than CO production. At perihelion, dust and gas production was enormous. Dello Russo et al. (2000) derived the following heliocentric dependence between 0.93 and 1.49 AU: $Q_{H_2O} = (8.35 \pm 0.13) \times 10^{30}[R_h^{-1.88\pm0.18)}]$ molecules s^{-1}. For native CO, the following heliocentric dependence between 4.1 and 2.02 AU was found (Di Santi et al., 2001): $Q_{CO} = (1.06 \pm 0.44) \times 10^{30}[R_h^{-1.76\pm0.26)}]$ molecules s^{-1}. A dust to ice mass ratio of 5.1 ± 1.2 was obtained (Dello Russo et al., 2000) within a heliocentric distance of 1.5 AU. The

Earth, Moon and Planets **90:** 217–225, 2002.

comet was still very active at 13 AU after perihelion, when it was imaged at La Silla Observatory in March 2001.

Several papers about the interpretation of the activity of this comet using nucleus theoretical models were published in these years. One of the firsts was the paper by Prialnik (1997), published before the arrival of the comet at perihelion. The model nucleus is made by amorphous water ice, dust and a fraction of 5 per cent in mass of CO gas trapped in amorphous ice and released during the transition. The author predicts that CO emission, coming only from a subsurface layer, should not continue to rise until perihelion, and that should become quite flat. In the paper by Flammer et al. (1998), the nucleus is composed by a homogeneous grainy admixture of dust, H_2O-clathrate grains and grains of CO. "Guest" molecules of CO are trapped in the clathrate (5 per cent). The authors predict outgassing values slightly lower post-perihelion, and conclude that the production rate curve of the comet can be explained by the mantled nature of a chemically differentiated nucleus. Their CO production curve is flat. In the model by Enzian (1999) solid CO is assumed to be both trapped in amorphous water ice and as an independent ice phase. The author obtains a good agreement with observations and explains high water production near 3 AU pre-perihelion with the sublimation from icy grains in the inner coma. In the paper published by Kuehrt (1999), it is suggested that the early rise of water sublimation can be explained with the contribution of seasonal effects.

2. The Model

Only a short description of the nucleus model and of the physical processes accounted for will be given in the following; for more details the reader is referred to Capria et al. (2001); De Sanctis et al. (1999 and the references therein).

The model is one-dimensional. The spherical nucleus, porous and initially homogeneous, is composed of ices (water and CO, in this application) and a refractory component; a fraction of CO can be trapped in the amorphous matrix and released during the transition to crystalline phase. The refractory material is described as spherical grains with given initial size distribution and physical properties.

Energy and mass conservation is expressed by the following system of coupled equations, solved for the whole nucleus:

$$\rho c \frac{\partial T}{\partial t} = \nabla[k \cdot \nabla T] + \sum_{i=1}^{n} Q_i + Q_{am-cr}, \tag{1a}$$

$$\frac{1}{RT}\frac{\partial P_i}{\partial t} = \nabla[G_i \cdot \nabla P_i] + Q\prime_i \quad i = 1, n, \tag{1b}$$

where Q_i are the energies exchanged by the solid matrix in the sublimation and recondensation of the ices, Q_{am-cr} is the heat released during the transition of water ice from amorphous to crystalline form, R is the gas constant, P_i the partial pressure of component i, G_i its diffusion coefficient, and Q'_i is the gas source term due to sublimation-recondensation processes. In the case of the CO diffusion equation, the gas source term due to the release of trapped CO, $Q_{CO_{tr}}$, must be added.

Thermal conductivity of the bulk material is computed by means of Russel's formula:

$$k = \frac{k_s[\psi^{2/3}k_p + (1-\psi^{2/3})k_s]}{k_s[\psi - \psi^{2/3} + 1] - k_p\psi^{2/3}[\psi^{1/3} - 1]}, \qquad (2)$$

where ψ is the porosity, k_s is the conductivity of the solid phase and k_p is the effective conductivity due to radiative energy transfer across the pores (expressed as $4\epsilon\sigma r_p T^3$, ϵ being the thermal emissivity, σ the Stefan-Boltzmann constant and r_p the mean pore radius).

Gas diffusion coefficients are computed on the basis of the mean free path of the molecules in the pore system; the model accounts for three different diffusion regimes: Knudsen, viscous and a transition one. It can be noted that under usual conditions diffusion regime is Knudsen's, so we can assume that each gas flows independently from the others.

The temperature on the surface is obtained by a balance between the solar energy reaching the surface, the energy re-emitted in the infrared, the heat conducted to the interior and the energy used to sublimate surface ices. Due to the rising temperature, ices start to sublimate, beginning from the more volatile ones, and the initially homogeneous nucleus differentiates giving rise to a layered structure in which the boundary between different layers is a sublimation front.

When the ices near the surface of the nucleus begin to sublimate, the refractory particles become free and can be dragged by the escaping gas, so that they can be blown off or they can accumulate on the surface to form a crust. To determine how many particles can be blown off and how many can be accumulated on the surface, the balance between the drag of the outflowing gas and the gravitational attraction force, corrected by the centrifugal force, is determined, obtaining for each distribution a critical radius that represents the radius of the largest particle that can leave the comet. Surface erosion due to ice sublimation, particles ejection and crust compaction is taken into account.

2.1. Hale–Bopp Physical Parameters

The model depends on many initial parameters that have been chosen according to the measured values, when available, or to parameters commonly used in the literature of cometary nuclei modelling. Some of these initial parameters are listed in the Table I. In the model CO is present both as an ice and as a trapped gas: Half of the trapped CO is released when the transition amorphous-crystalline occurs, and half when the water sublimation starts, as observed during the laboratory experiments (Bar-Nun and Owen, 1998).

TABLE I
Initial parameters

Albedo	0.04
Dust/ice	4.0
Porosity	0.7
CO/H_2O ice	0.001
CO/H_2O trapped	0.1
Pore radius	0.0001 m
Initial ice phase	Amorphous
Rotation period	11.4 h
Initial temperature	20 K

TABLE II
Dust distribution parameters

Heat conduction coefficient	3 W/K/m
Size distribution	5 size classes from 10^{-7}–10^{-3} m
Density	1000 kg/m^3

In terms of dust emission, it is open to debate if Hale–Bopp is a comet anomalous or not. It is certain, in any case, that it was particularly dusty (Kidger, 1997). The high dust-to-ice mass ratio is chosen according to values derived from near-perihelion observations (Dello Russo et al., 2000). The values chosen to simulate the refractory component are listed in the Table II. In this application only one grains distribution was considered.

3. Discussion of the Results

The model comet was followed for two revolutions, because Hale–Bopp is not a new comet and already passed through the inner Solar System. At the beginning of the orbit we are considering the surface layers of the comet are already differentiated, that means depleted of supervolatiles and amorphous ice.

The results obtained with the model described above are in reasonable agreement with the results of observations. In Figure 1 gas production rates of water and CO from the nucleus are plotted in a range of 10 AU before and after perihelion, along with measured production rates from various observations. It should be noticed that the water flux is well reproduced around perihelion, while at a distance beyond 4 AU before perihelion the model is not able to simulate the observed flux.

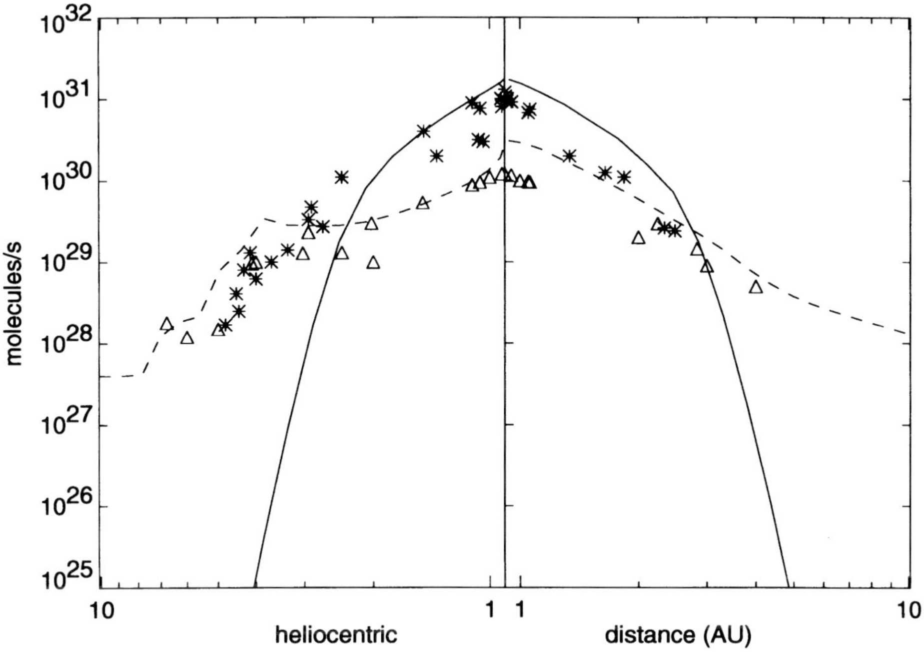

Figure 1. Gas flux from 10 AU before to 10 AU after the perihelion. Continuous line represents computed water production, while dashed line represents computed CO production. Asterisks and triangles represent, respectively, water and CO production from various published observations

The early rise of water sublimation before 4 AU is probably due to a distributed source of icy grains in the inner coma: near-infrared absorption features of water ice were detected at 7 AU (Davies et al., 1997).

The CO production rate from a direct source seems to be well reproduced by the model (in Figure 1 only the measured values attributed to a direct source are shown by the triangles). It should be remembered that in the assumptions of this model CO is present, in the nucleus, both as an ice and as a gas released during the phase change of amorphous ice. The release of CO trapped molecules determines the sharp increase at 4–5 AU pre-perihelion, corresponding to the onset of the transition of amorphous water ice. Direct sublimation from ice explains the flux of CO at longer distances, when the surface and internal temperatures do not allow for water sublimation. It should be noticed also that a source of CO should be found close to the surface, otherwise a perihelion production peak would not exist. In the model, a contribution to this peak is given by the fact that some of the trapped gas is released only at water ice sublimation.

The fact that the gas production from the nucleus can be well simulated without introducing unusual parameters means that the exceptional activity of this comet is due mainly to the great size of its nucleus.

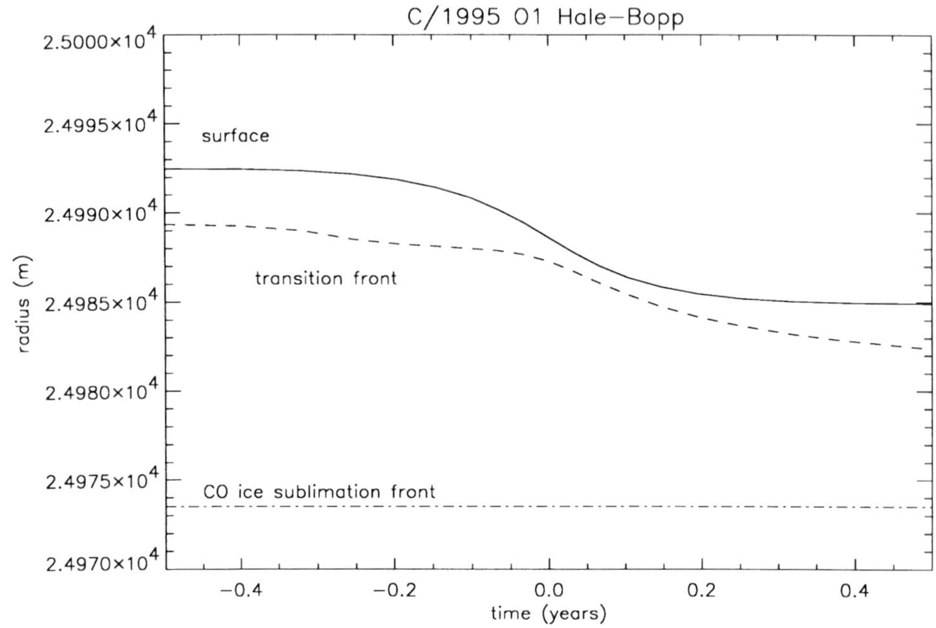

Figure 2. Stratigraphy of the comet around perihelion.

The stratigraphy of this comet, shown in Figure 2, is quite different from those normally obtained for Jupiter family comets and should be typical of long period comets. Jupiter family and short period comets are usually much more differentiated, because they spend most of their orbit not far from the Sun. Long period comets, instead, spend most of their lifetime very far from the Sun, and when they enter the inner Solar System they pass near to the Sun at a high speed: They remain exposed to high temperatures for a short period of time. At each passage they undergo a strong ablation of the surface, and transition and sublimation fronts remain close to the surface and follow the nucleus erosion. The amorphous ice transition front is close to the surface, explaining the peak in the CO gas production at perihelion. Most of the interior of the comet should be still pristine, or it has experienced only the changes due to the presence of radioactive elements in the refractory component.

3.1. Long distance activity

From the analysis of earlier images it seems that comet Hale–Bopp was already active at about 13 AU pre-perihelion and probably even farther, at 18–20 AU pre-perihelion (Cartwright et al., 1999). After the perihelion passage, the comet was still clearly active at the distance of 13 AU, when it was imaged at La Silla Observatory with the WFI camera on the MPG/ESO 2.2 m telescope: at that moment, in the pictures, the whole comet had an estimated size of 2 millions of km and exhibited a broad, fan-shaped tail.

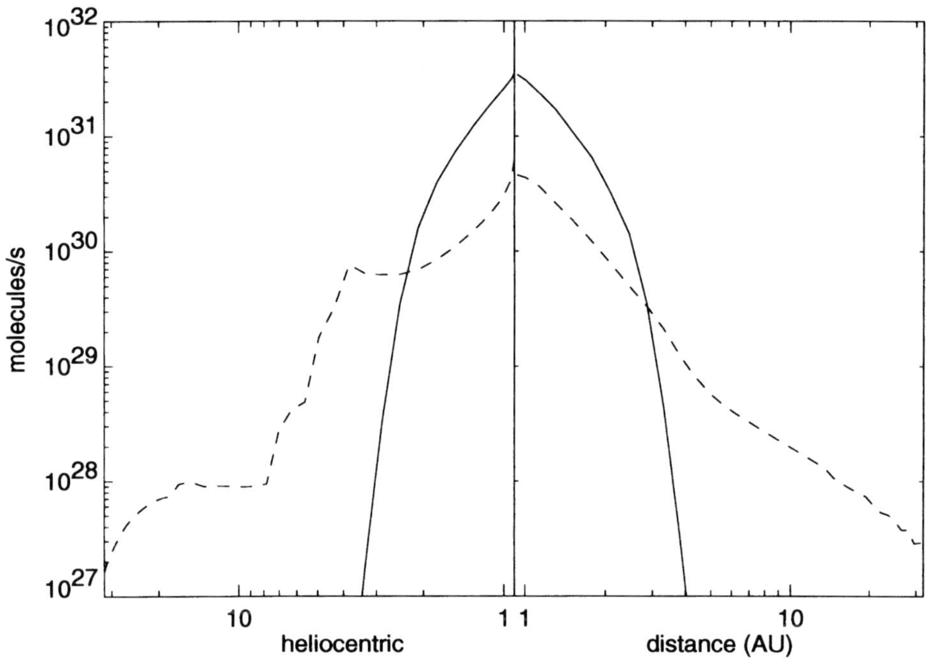

Figure 3. Long distance activity: From 30 AU pre to 30 AU post-perihelion. Water: Continuous line, CO: dashed line.

Activity at this distance can be explained only by the sublimation of supervolatile ices, because the temperatures are too low for the transition from amorphous to crystalline ice phase. In Figure 3 the activity of comet Hale–Bopp is shown to a distance of 30 AU pre- and post-perihelion.

In the plot, CO is sublimating from the nucleus along the whole range represented in the figure; the flux should be able to entrain dust particles and sustain a coma, at least at a distance between 20 and 10 AU.

Hale–Bopp's dust emission between 10 AU pre-perihelion and 10 AU post-perihelion is shown in Figure 4. The peak at 5 AU corresponds to the beginning of the phase change of amorphous ice, when the total CO flux becomes stronger due to the contribution of released gas (see Figure 1).

Dust emission is driven mainly by H_2O flux between 3 AU pre- and 3 AU post-perihelion, while at higher heliocentric distances it should be driven by supervolatiles such as CO. This is clearly possible, due to the high CO abundance in Hale–Bopp and other comets.

In the model, dust flux is well reproduced when the driver of the activity is water ice subliming mainly from the surface, while when the driver is a gas coming from sub-surface layers the dust flux is well simulated only when this flux is very strong. We are working on an improvement of our model that should better simulate the dust flux far from perihelion. It should be noticed, anyway, that the output of this

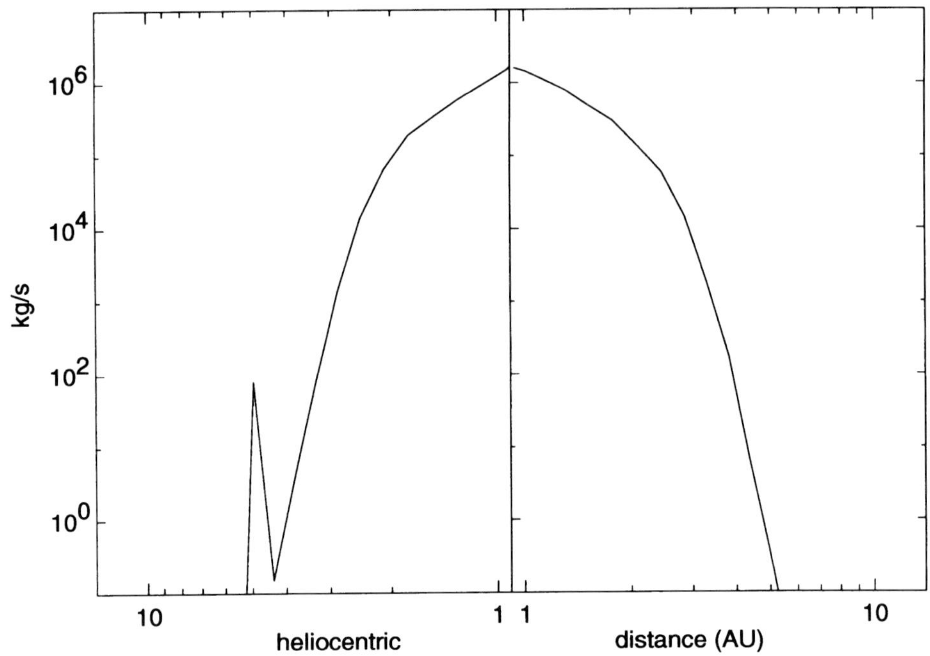

Figure 4. Dust flux between 10 AU pre-perihelion and 10 AU post-perihelion.

nucleus model, plotted in Figure 4, is the instantaneous dust flux. The number of particles that are present in the coma at any moment should be higher than this instantaneous dust flux: in a real comet, the coma is formed not only by the particles emitted in that moment from the surface, but also from the particles emitted earlier.

4. Conclusion

Our model is able to reproduce the Hale–Bopp gas flux coming from a direct source, including the long distance emission of CO and the increase of CO flux at 4–5 AU pre-perihelion. To explain the observed long distance activity of the comet it is needed to assume that CO ice is still present under the surface: in this case, its flux is able to entrain some dust and sustain a coma around the nucleus also at large heliocentric distances.

The model does not reproduce very well the dust flux at higher distances, when the driver of activity are the supervolatiles. This does not mean that new physical processes should be introduced, but that our simulation of dust flux should be improved.

From our results the interior of the comet should not be very differentiated, and the original structure and composition could be found close to the surface. This behaviour is mainly a consequence of the orbital parameters (very elliptical orbit): due to the strong ablation, the CO sublimation front and the phase transition front remain close to the surface and move inward more or less at the same rate of the nucleus erosion. It is not necessary to introduce an extremely low thermal conductivity to obtain such internal profile.

In our opinion, the exceptional activity of comet Hale–Bopp was due only to its great size, because our model reproduced quite well the observations whithout using unusual initial parameters or introducing new physical processes. The success of this simulation constitutes a good test of the ability of our model to obtain reasonable results.

References

Bar-Nun, A. and Owen, T.: 1998, in *Solar System Ices, 27–30 March 1995, Toulouse, France*, Kluwer Academic Publishers, Dordrecht, Astrophysics and Space Science Library Series 227, pp. 353–366.

Biver, N., Rauer, H., Despois, D., Moreno, R., Paubert, G., Bockelée-Morvan, D., Colom, P., Crovisier, J., Gerard, E., and Jorda, L.: 1996, *Nature* **380**, 137–139.

Capria, M. T., Coradini, A., De Sanctis, M. C., and Blecka, M. I.: 2001, *Planet. Space Sci.* **49**, 907–918.

Capria, M. T., Coradini, A., and De Sanctis, M. C.: 2002, *Adv. Space Res.* in press.

Capria, M. T., Coradini, A., De Sanctis, M. C., and Orosei, R.: 2000a, *AJ* **119**, 3112–3118.

Capria, M. T., Coradini, A., De Sanctis, M. C., and Orosei, R.: 2000b, *A&A* **357**, 359–366.

Cartwright, I. M., Fitzsimmons, A., and Pollacco, D. L.: 1999, *Earth Moon Planets* **78**, 127–127.

Coradini, A., Capaccioni, F., Capria, M. T., De Sanctis, M. C., Espinasse, S., Orosei, R., Salomone, M., and Federico, C.: 1997a, *Icarus* **129**, 317–336.

Coradini, A., Capaccioni, F., Capria, M. T., De Sanctis, M. C., Espinasse, S., Orosei, R., Salomone, M., and Federico, C.: 1997b, *Icarus* **129**, 337–347.

Davies, J. K., Roush, T. L., Cruikshank, D. P., Bartholomew, M. J., Geballe, T. R., Owen, T., and de Bergh, C.: 1997, *Icarus* **127**, 238–245.

Dello Russo, C., Mumma, M. J., Di Santi, M. A., Magee-Sauer, K., Novak, R., and Rettig, T. W.: 2000, *Icarus* **143**, 324–337.

De Sanctis, M. C., Capaccioni, F., Capria, M. T., Coradini, A., Federico, C., Orosei, R., and Salomone, M.: 1999, *Planet. Space Sci.* **47**, 855–872.

De Sanctis, M. C., Capria, M. T., Coradini, A., and Orosei, R.: 2000a, *AJ* **120**, 1571–1578.

De Sanctis, M. C., Capria, M. T., and Coradini, A.: 2000b, *AJ* **121**, 2792–2799.

Di Santi, M. A., Mumma, M. J., dello Russo, N., and Magee-Sauer, K.: 2001, *Icarus* **153**, 361–390.

Enzian, A.: 1999, *Space Sci. Rev.* **90**, 131–139.

Flammer, K. R., Mendis, D. A., and Houpis, H. L. F.: 1998, *ApJ* **494**, 822.

Jewitt, D. and Matthews, H.: 1999, *AJ* **117**, 1056–1062.

Jewitt, D., Senay, M., and Matthews, H.: 1996, *Science* **271**, 1110–1113.

Kidger, M.: 1997, *Earth Moon Planets* **79**, 79–102.

Kuehrt, E.: 1999, *Space Sci. Rev.* **90**, 75–82.

Prialnik, D.: 1997, *ApJ* **478**, L107.

CHALLENGING A PARADIGM: DO WE NEED ACTIVE AND INACTIVE AREAS TO ACCOUNT FOR NEAR-NUCLEAR JET ACTIVITY?

J.-F. CRIFO
Service d'Aéronomie du CNRS, BP 3, F 91371 Verrières le Buisson Cedex, France
E-mail: crifo@aerov.jussieu.fr

A. V. RODIONOV
Central Research Institute on Machine Building (TsNIIMASCH), Pionyerskaya St., 4, Korolev, Moscow Region, 141070, Russia

K. SZEGŐ
KFKI Research Institute for Particles and Nuclear Physics, PO Box 49, H-1525 Budapest, Hungary

M. FULLE
Osservatorio Astronomico di Trieste, Via Tiepolo, 11, I-34131 Trieste, Italy

(Received 7 March 2002; Accepted 29 June 2002)

Abstract. We briefly describe an advanced 3D gas dynamical model developed for the simulation of the environment of active cometary nuclei. The model can handle realistic nucleus shapes and alternative physical models for the gas and dust production mechanism. The inner gas coma structure is computed by solving self-consistently (a) near to the surface the Boltzman Equation (b) outside of it, Euler or Navier-Stokes equations. The dust distribution is computed from multifluid "zero-temperature" Euler equations, extrapolated with the help of a Keplerian fountain model. The evolution of the coma during the nucleus orbital and spin motion, is computed as a succession of quasi-steady solutions. Earlier versions of the model using simple, "paedagogic" nuclei have demonstrated that the surface orography and the surface inhomogeneity contribute similarly to structuring the near-nucleus gas and dust coma, casting a shadow on the automatic attribution of such structures to "active areas". The model was recently applied to comet P/Halley, for which the nucleus shape is available. In the companion paper of this volume, we show that most near-nucleus dust structures observed during the 1986 Halley flybys are reproduced, assuming that the nucleus is strictly homogeneous. Here, we investigate the effect of shape perturbations and homogeneity perturbations. We show that the near nucleus gas coma structure is robust vis-a-vis such effects. In particular, a random distribution of active and inactive areas would *not* affect considerably this structure, suggesting that such areas, even if present, could not be easily identified on images of the coma.

Keywords: Comets, dust, gas, P/Halley, structure

1. Introduction

While a wealth of coma observations surely carry clues of the nucleus surface properties, the theoretical foundations of the inferences from these data to nuclei properties, found in the cometary literature, are extremely tenuous. This is true, in first place, for the widely accepted paradigm according to which the structures

Earth, Moon and Planets **90:** 227–238, 2002.

observed in the (dust) coma are due to patterns of "active" and "inactive" areas at the nucleus surface. Inconsistencies in this assumption were already pointed-out (Whipple, 1982) before 3-D gasdynamic simulations of a nucleus activity – which are the only way by which robust conclusions can be reached – were at hand. The first (and unique, prior to our work) such simulation (Kitamura, 1990) was prompted by the 1986 Halley flyby observations: the structure of a dust coma was for the first time revealed on a geometrical scale comparable to the nucleus size. From the point of view of numerical simulation, on such a scale time-dependent effects can be neglected, while the very large scale images offered by ground-based observations do require a fully time-dependent model (allowing for the nucleus rotation and orbital motion). Kitamura's work robustly proved that brightness maxima near a spherical nucleus with active and inactive areas do not trace maxima in surface flux, but the interaction regions between the active areas. An even more severe suspicion was cast on the prevailing paradigm when we demonstrated that the surface orography of a non-spherical nucleus creates similar jet-jet interactions and resulting structures in the near-nucleus coma. Thus, while structures due to active/inactive areas *may* exist in *some* comae, those due to the orography *must* be present in *all* comae (Crifo and Rodionov, 1999).

The next step towards a robust understanding of the nucleus activity and coma formation appeared to us to be the handling of the correct shape of the nucleus, and the quantitative comparison between the ab-initio computed Halley near-nucleus coma, and the 1986 flyby images. It is this effort which we describe (partially) in this and the companion paper of Szegő et al. This effort does not exhaust the topic, as it will remain to address to the large scale observations of the Halley's coma, using a time-dependent model. This is the goal of our effort of the forthcoming years.

The formation of a coma by sublimation under vacuum (or by outgassing) is similar to underexpansion of molecular jets studied at the laboratory or in the industry: not only the dominant molecule, H_2O, but even the temperature and pressure ranges are identical. Modelling the nucleus activity therefore amounts to adapting standard, existing, modelling techniques. Four difficulties however appear: (1) The physicochemical nature of the nucleus is purely a matter of speculations. One is therefore free to adopt his preferred views, but it will be important to identify which coma properties depend critically upon these assumptions, and which ones do not. (2) It is necessary to deal with another poorly defined component: "cometary dust". Here, let us observe that the aerodynamical behavior of a solid is primarily dependent upon its shape – as do its optical properties, which determine its distant motion. So, one has to represent the dust in an arbitrary manner, and, again, to discuss the sensitivity of the results upon the assumptions. (3) The coma is submitted to solar illumination: both the chemical implications (cf. Schmidt and Wegmann, 1980) and the thermodynamic implications (molecular excitation, Bockelée-Morvan, 1987) have been treated in a fairly advanced way, but only in 1-D, which renders the results unusable in the innermost coma. (4) The initial boundary conditions (i.e., the nucleus surface) are quite complicated. This is a serious difficulty, even now.

The present model combines in a modular way a nucleus shape model, a dynamical nucleus rotation model, a nucleus gas and dust production model, a physicochemical coma model and a coma gas and dust outflow model. The gas and dust production model is usually that of Whipple (1950) – but applied to nuclei with realistic shapes. Other production models can be (and have been) used to deal e.g. with the distant gas production by subsurface sublimation. The physicochemical coma model is that of (Crifo, 1995), usually (and hereafter) simplified by omitting distributed production and chemistry. The outer dust coma Keplerian model is described in (Fulle and Crifo, 1999). The model can be used to compute the net nucleus mass loss, net recoil force, and net sublimation torque, hence to compute the evolution of the spin motion (Szegő et al., 2001) and – this also is planned – of the nucleus orbit.

The present model spatial resolution at the nucleus surface is suitable for dealing with data such as those acquired on P/Halley in 1986, and on P/Borrelly last year. Here, and in the companion paper (Szegő et al., 2002), we address the first set. Our basic full model is described in all details in Rodionov et al. (2002). Here, we will only emphasize the specific assumptions used for the P/Halley application. Then, we will return on the shape determination, in view of its critical role. The agreement between model results and flyby observations will be discussed in the companion paper. Here, we will investigate the dependence of the results upon perturbations of the nucleus shape, and of the nucleus homogeneity.

2. Outline of the Basic Full Model

2.1. The Shape and Size Model of Comet Halley

Two VEGA spacecrafts flew by comet Halley, VEGA 1 on March 6, 1986, and VEGA 2 on March 9, 1986, and took images with a nucleus-pointed telescope with 26.4×39.6 arc min. field of view.*

During the VEGA 1 flyby sixty-three 128×128 pixel images were taken sequentially through three different colour filters; the first image of this set was taken from a distance of 51,290 km, before closest approach, the last one from 42,585 km, after closest approach. During the VEGA 2 flyby eleven large size (512×512 pixel) images were taken through four different colour filters; the first image of this set was taken from a distance of 29,539 km, before closest approach, the last one from 42,567 km, after closest approach. The one-pixel resolution changed from 850 to 133 m (in the horizontal direction), the phase angle varied from 102° to 57°.

* We call the reader's attention to the fact that the pixels were not rectangular, therefore if one retrieves VEGA image data from data centres, he should be careful to check whether the image is pixel size-corrected or not!

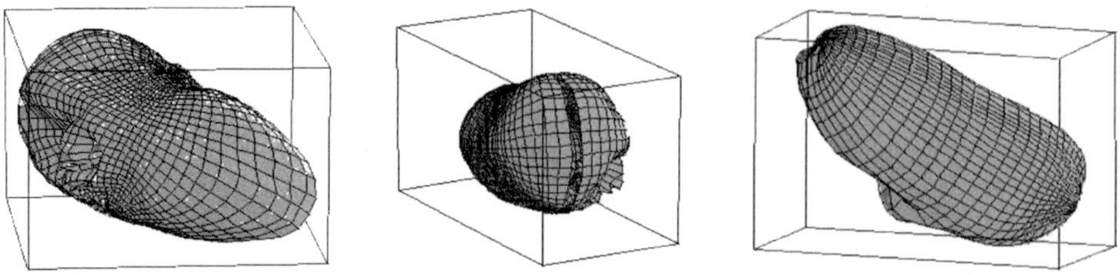

Figure 1. The shape of Halley's nucleus seen from three of the 1986 flyby view directions.

The raw data were corrected using standard image processing techniques. One of the difficult problems was to get rid of a strong, quasi-coherent periodic noise (the signal-to-noise ratio was not greater than 32:1 on the raw pictures). The VEGA 1 camera produced severely out-of-focus images – the point spread function was estimated to be about 10–12 pixels wide; besides it, all measured intensity values were shifted by an offset. The manner in which these problems were overcome is described in Merenyi et al. (1990) and in Szegő et al. (1995).

The VEGA 1 image set was used to generate the 3-D model of the nucleus, in two steps: first the main characterics were determined, using a simple starting shape, then the nucleus was defined as a polyhedron of 31 × 39 vertices and triangular faces, to be able to adjust easily its shape. It was a critical problem to decide about the orientation of the longest axis. Two models were created for the two possible options, and comparing the resulting appearance to the images, it was concluded that during the VEGA 1 encounter the big end pointed toward the camera. The refinement of the model was a long and tedious job, requiring: (1) To derive brightness features and contours from the real data and adjust the polyhedron shape accordingly (in favourable cases bright features on a certain image appeared as silhouettes on some other images), and (2) to compare the brightness distribution between the model and the data. In doing this, the scattering law $I_0(\cos\alpha_i)^k(\cos\alpha_e)^{k-1}$ was used with $k = 0.7$. The final shape agreement with all VEGA 1 images is shown in (Szegő et al., 1995).

The VEGA 2 and GIOTTO images (Keller et al., 1995) were used with caution in the 3-D reconstruction, because their use required the adoption of a nucleus rotation model, and we did not want to produce a model-dependent shape. We assessed the orientation of the long axis separately on those images (for GIOTTO we had two solutions because only one 2-D view is available), and assessed the possible rolls about the long axis to obtain compatibility with the images.

The 3-D shape of Halley, shown in Figure 1, has overall dimensions are 7.2 × 7.22 × 15.3 km; this determination is accurate to no better than 0.5 km. As the VEGA-1 images covered about 160° field of view only, the shape near the small end of the nucleus is especially inaccurate. Yet, one will notice the presence of a large, irregular "cavity" of which only the edge appears on the GIOTTO image. On the other hand, the small-size details discussed by (Keller et al., 1995) are not present here, due to the insufficient model resolution.

The shape derived above exhibits a significant topographic noise; to filter it out, the shape was expanded into spherical harmonics and the expansion truncated at some adequate degree. In fact, this procedure also offered the opportunity to bring controlled perturbations to the shape, for the sake of investigating their effect on the coma, as described below.

2.2. The Nucleus Surface Model

The present model can accommodate a wealth of alternative physical nucleus gas and dust production models (NGDPM). Whatever model is adopted must in a sense be considered arbitrary, since we have strictly no direct information on the structure of any cometary nucleus. It is therefore important to find out whether the computed coma structure will or not be greatly affected by a change in NGDPM. Unfortunately, none of the many NGDPM which have been or are being developed, e.g., Podolak and Prialnik (1996), Enzian and Klinger (1999), Skorov et al. (2001) lends itself to the insertion in a 3-D code: This is because all authors assume free vacuum outflow, while, in reality, the surface outflow occurs *into the coma*, which is by no means equivalent. We expect that it should be possible to develop suitable versions of these models in the future. The basic Whipple model (of which the posterior ones are sophisticated variants) lends itself to insertion in a 3-D code, because "normal" sublimation into an atmosphere has been studied long since inside the gasdynamic community. Also this basic paradigm has only one free parameter, hence it recommends itself as the optimum initial choice. On the other hand, the main effect of changing the NGDPM is potentially to alter the profile of gas production over the surface. This, we have simulated crudely in two separate ways: (1) By introducing arbitrary analytic gas production profiles (Crifo et al., 1999), or (2) by introducing random patterns of surface icy area fractions (see below). In both cases significant changes are induced in the coma, but its general structure is not strongly affected, owing to the dominance of the orography (see below). In passing, let us also mention that we did not find a significant coma structure dependence upon which molecule is assumed dominant in the gas production (Crifo et al., 1999).

Thus, the results presented in this volume were obtained assuming solar-driven surface sublimation of "normal" ice with imbedded inert dust. The essential free parameter is the surface icy fraction f, which is defined at each surface element as the ratio of the area of exposed ice in the element, to that of exposed dust. It is evident that the sublimating flux of pure ice is reduced in proportion to f as inert dust typically does not emit water vapour.* If the nucleus is assumed homogeneous, f is constant, otherwise it is position-dependent. A surface sublimation energy budget is solved at each point, neglecting internal thermal conduction and coma light scattering effects. A secondary parameter κ is introduced, representing an

* However, this was ignored in the post-1986 estimates of "the minimum area needed to produce the observed gas production", which surely contributed to bias the image interpretations made at that times.

internal heat flux to the shadowed surface elements, to maintain a finite temperature there; the absolute value of this coefficient was found to play a negligible role in the dayside coma. An essential point is that in these equations the net surface gas flux depends upon the downstream gas flow: accordingly, all surface element budget equations are coupled to the gas flow equations and solved together with them.

2.3. The Gasdynamic Model

The gasdynamic approach is the most efficient way to compute a fluid motion, but provides correct results only to the extent that the gas velocity distribution function is Maxwellian (in a frame co-moving with the gas), or nearly so. In the present case, a strict Maxwellian prevails in all regions of interest, making it possible to use Euler equations (EE), with however the following essential exception. In the immediate vicinity of the sublimating surface, the emitted molecules (having $V_\perp > 0$) are distributed according to a "half-Maxwellian" (i.e., a Maxwellian from which the part $V_\perp < 0$ is absent) – resulting from the solid-state process of molecular emission. If the molecules moving towards the surface were distributed according to the other half of the Maxwellian, the net flux would be zero. This cannot be the case, and therefore the resulting velocity distribution is not a Maxwellian one. Furthermore, the $V_\perp < 0$ distribution surely depends upon the downstream coma conditions: if it is extremely rarefied, this component does not exist; at the other extreme (frost deposition) this component is dominant. Naturally, after a certain number of collisional mean free paths (m.f.p.), the Maxwellian shape is re-established. Thus, in a disequilibrium layer with thickness of several m.f.p., adjacent to the surface, the surface molecular distribution adjusts itself to the surrounding coma conditions. In our model, this region is treated by solving the fundamental Gaskinetic equation – the Boltzman equation (BE) – by an analytical method, assuming that local plane-parallel conditions hold. For a Halley-size comet at 1 AU, the m.f.p. is of the order of 1 m or less close to the surface, so that the preceding assumption is fairly well satisfied for a nucleus model smooth on a 100 m scale. Thus, analytical expressions relate the flux of $V_\perp < 0$ molecules at the surface to the flow Mach number at the top of the disequilibrium layer, and the initial temperature at this point to the ice temperature. In this way, the initial boundary conditions for solving the EE are coupled to the surface budget equations at each surface element.

In principle, the EE should incorporate terms allowing for the perturbations of the gas flow by the gas-dust interaction. However, for the present conditions, i.e., for the dust mass spectrum inferred from the 1986 in-situ dust detector data, we have checked that these pertubations are unimportant, so that the computation of the gas coma can be done first, and the resulting dust coma computed independently.

2.4. Dust outflow model

For the dust component, a fluid regime is not established. For instance, the initial number density of micron-size dust particles is about 10–100 cm^{-3} in the present case, corresponding to a free-space isotropic m.f.p. on the order of 3–30 km, but, in reality, the gas drag coaligns the grain motions, eliminating the collisions. Thus, the problem is that of a classical collision-less effusion in an external field of force (the aerodynamic force field). This effusion obeys, *in geometrically simple cases*, zero-temperature EE (with a term representing the aerodynamic force); one set of equations must be used by dust species (size and shape): this makes-up the so-called multifluid model used in many 1-D coma models. We use it in the results presented here, even though the geometry is not simple. It can be shown that the results are, in this case, approximate only – we return on this in the companion paper. It should be reminded, here, that another, impossible to circumvent, source of modelling uncertainty is the total lack of information concerning the dust grain shape: Simple examples (Crifo and Rodionov, 1999) indicate that shape dispersion will result in grain trajectory spreading. Finally, as illustrated in Crifo and Rodionov (1999), grains with the same shape, but different masses, assume not radically different, but significantly spread trajectories. Fortunately, we will see that the coma dust structures can only result from gas structures, so that these computed dust structures should not be too severely dependent upon the dust model used.

3. Results

Model computations were made of H_2O and 0.9 μm radius dust grains distributions around P/Halley nucleus at the time of the three 1986 flybys. An overview comparison of the results with the observations made during the flybys is done in the companion paper (Szegő et al., 2002). It was found, as expected, that the dust coma structures are the direct consequence of gas coma structures, both being formed in the immediate vicinity of the nucleus. Here, we wish to demonstrate briefly that these gas structures are robust versus perturbations in nucleus shape and homogenity.

To analyze how significant is the dependence of the gas outflow on topography, we calculated the gas and dust outflow for smoothed nucleus shapes obtained by retaining less and less terms in the spherical harmonic expansion of the original shape. Figure 2 shows excerpts from the results: the strongest filtered shape corresponds to harmonic 1 plus only 1/2 of the amplitude of harmonic 2, and does not create much coma structuring. But already with harmonics 1 + 2 one sees the subdivison of the sunward coma characteristic of the Giotto images, and which suggested “two distinct active areas”. With harmonics 1–20, the coma is fairly structured, and a detailed analysis shows that it contains most of the azimuthal gradients (designated as “filaments”) of the GIOTTO image, at the right position.

That the dust coma image exhibits the structure of the computed gas coma is an overwhelming evidence that the dust structures have nothing to do with "jets" but simply trace the gas flow – dust seeding in flows is a well know gas flow diagnostic technique.

To investigate the coma dependence upon surface inhomogeneity, we placed a random pattern of "gaussian" active areas – i.e., having a Gaussian distribution in f with angular distance from a peak value f_M; both the position of the peak, the Gaussian FWHM, and f_M were generated randomly (see Figure 4). Figure 3 compares the resulting computed coma gas density with that for a uniform surface (constant f). While some differences exist between the two structures, one can say that the effect is very minor: clearly, the gross features of the shape are the predominant factors which govern the coma structure.

4. Discussion and Conclusions

We have presented here the outline of a new 3-D coma fluid dynamic model, fully described in (Rodionov et al., 2002). Using this model, it was possible to show that the structured gas environment of the nucleus is robustly rooted in the overall nucleus shape: small topography details produce only small coma details. Most instructive is the fact that random variations of the surface homogeneity also do not appear to influence the coma structure considerably. Perhaps it will be possible, under extreme assumptions such as limiting the gas production to a few small, highly contrasted active areas, to produce substantial changes in coma structure, but most likely at the cost of ruining the agreement with the observations.

The observed near-nucleus gas coma structure is already not so badly reproduced using a low harmonic approximation to the shape, and a uniform nucleus composition; it is only to recover the finer details that a more accurate shape representation is required. In a simplified manner one could state that the coma structure is the consequence, essentially, of the "basic" form of the nucleus (lower degree spherical harmonics), not of the topographic or compositional quirks of the surface. Surely, this is the deep reason of our ability to interpret the coma successfully, despite our imperfect information on the nucleus shape.

There is an obviously very hard to accept implication of the present model results: to interpret coma structures, the shape of the nucleus is needed. Perhaps in the future it will be possible to guess the critical components of this shape from coma observations. We are not sure. Tentatively, we suggest that flyby and rendezvous missions to comets supplemented by (not replaced by) ground-based observations will, on the long term, be recognized as the only means of advancing our understanding of the nucleus activity and structure.

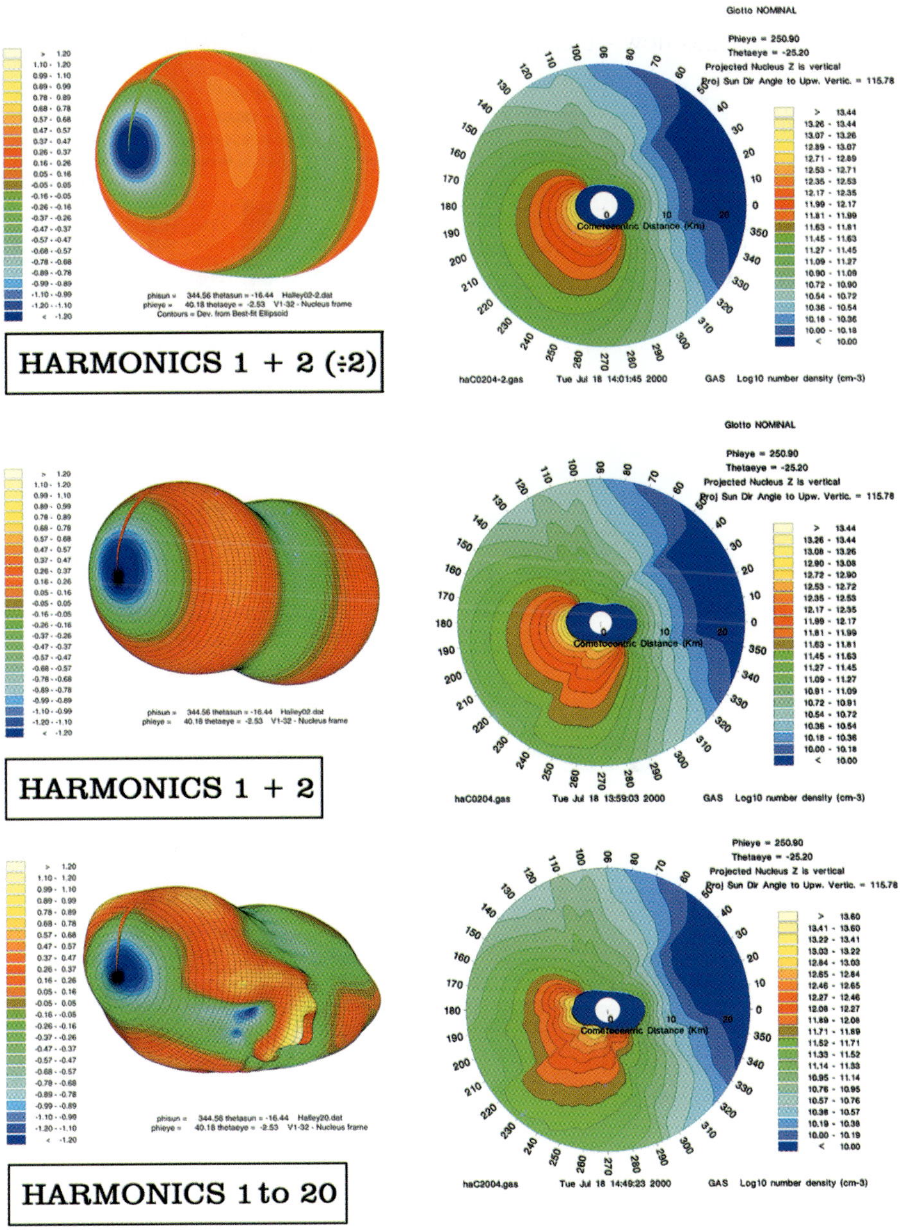

Figure 2. Influence of the shape on the gas coma density. *Left panels:* the smoothed nucleus as would been seen from the VEGA-1 (image 32) direction; *right panels:* $\log_{10}$(number density, cm^{-3}) in the GIOTTO image plane. Top: Only the harmonic 1, plus one half of harmonic 2 are retained; middle: Only harmonics 1 and 2 are retained; bottom: Harmonics 1 to 20 are retained (i.e., the shape filtering is negligible).

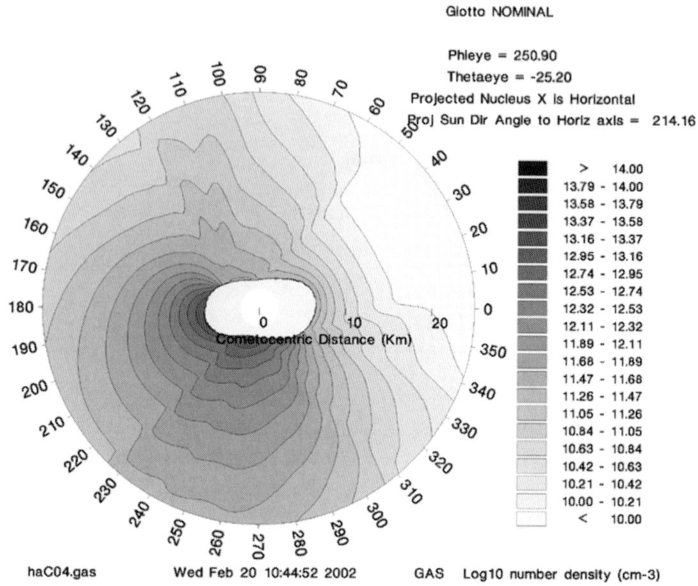

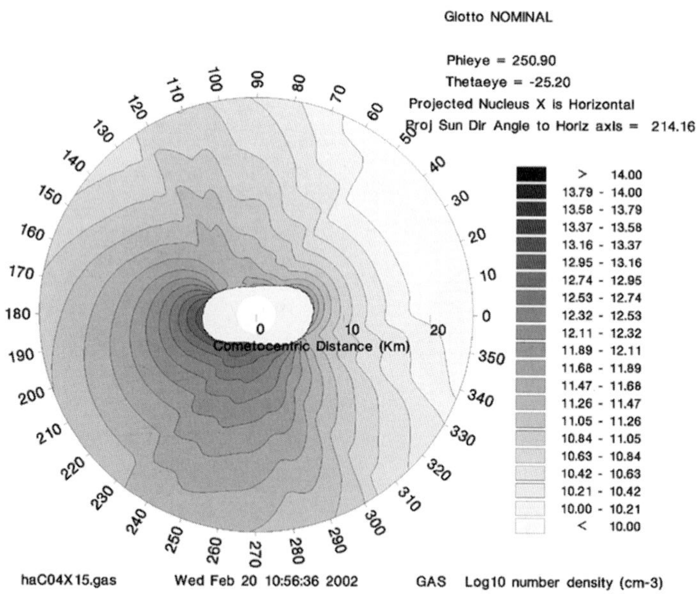

Figure 3. Top panel: $\log_{10}$(gas number density) for a homogeneous (constant area fraction f) nucleus; Bottom panel: $\log_{10}$(gas number density) for the same nucleus, but with the random pattern of f shown on Figure 4. In both cases, the shape smoothed by removal of harmonics > 10 is used.

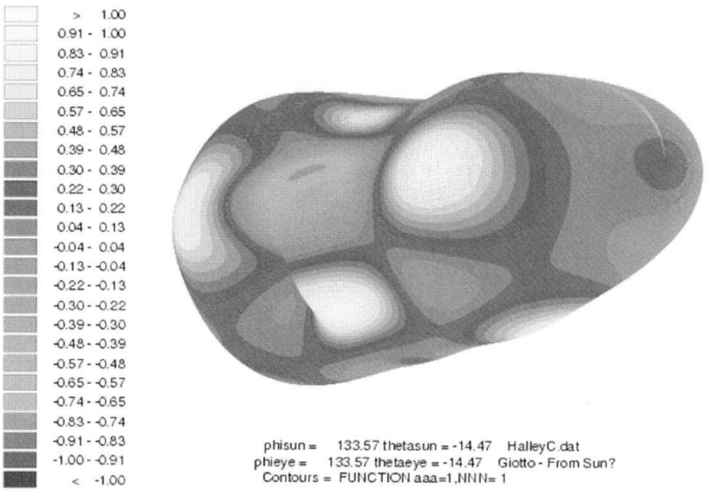

Figure 4. Random Gaussian active area pattern superimposed on Halley nucleus' shape. The icy area fraction f is related to the isocontour value X by the relation $f = (1 + X)/2$.

Acknowledgements

The work of J.-F.C. and A.V.R. was supported by French CNRS/INSU and CNES/CT contracts. The work of K. Sz. was funded by the Hungarian OTKA grant T-32634.

References

Bockelée-Morvan, D.: 1987, *Astron. Astrophys.* **181**, 169–181.

Crifo, J.-F.: 1995, *Astrophys. J.* **445**, 470–488.

Crifo, J.-F. and Rodionov, A. V.: 1999, *Planet. Space Sci.* **47**, 797–826.

Crifo, J.-F., Rodionov, A. V., and Bockelée-Morvan, D.: 1999, *Icarus* **138**, 85–106.

Enzian, A., Klinger, J., and Schwehm, G.: 1999, *Icarus* **138**, 74–84.

Fulle, M. and Crifo, J.-F.: 1999, *Astron. Astrophys.* **347**, 1009–1028.

Keller, H. U., Curdt, W., Kramm, J. R., and Thomas, N.: 1995, *ESA SP-1127*, Vol. 1.

Kitamura, Y.: 1990, *Icarus* **72**, 555–567.

Merenyi, E., Foldy, L., Szegő, K., Toth, I., and Kondor, A.: 1990, *Icarus* **54**, 388–405.

Podolak, M. and Prialnik, D.: 1996, *Planet. Space Sci.* **44**, 655–664.

Rodionov, A. V., Crifo, J.-F., Szegő, K., Lagerros, J., and Fulle, M.: 2002, *Planet. Space Sci.*, submitted.

Schmidt, H. U. and Wegmann, R.: 1980, *Comp. Phys. Com.* **19**, 309–326.

Skorov, Yu. V., Kömle, N. I., Keller, H. U., Kargl, G., and Markiewicz, W. J.: 2001, *Icarus* **153**, 180–196.

Szegő, K., Crifo, J.-F., Foldy, L., Lagerros, J. S. V., and Rodionov, A. V.: 2001, *Astron. Astrophys.* **370**, L35–L38.

Szegő, K., Crifo, J.-F., Rodionov, A. V., and Fulle, M.: 2002, these proceedings.

Szegő, K., Sagdeev, R. Z., Whipple, F. L., Abergel, A., Bertaux, J. L., Merenyi, E., Szalai, S., and Varhalmi, V.: 1995, *ESA SP-1127*, Vol. 2.

Whipple, F. L.: 1950, *Astrophys. J.* **111**, 375–394.
Whipple, F. L: 1982, 'Cometary Nucleus and Active Regions', in T. I. Gombosi (ed.), *Cometary Exploration*, Vol. 1, Hung. Acad. Sci. Budapest, pp. 95–110.

EVOLUTION OF THE ROTATIONAL STATE OF IRREGULAR COMETARY NUCLEI

PEDRO J. GUTIÉRREZ, JOSÉ L. ORTIZ, RAFAEL RODRIGO and JOSÉ J. LÓPEZ-MORENO
Instituto de Astrofísica de Andalucía, CSIC Aptd 3004, 18080 Granada, Spain

LAURENT JORDA
Laboratoire d'Astrophysique de Marseille, Site Pereisc, 13376 Marseille Cedex 12. France

(Received 1 March 2002; Accepted 26 May 2002)

Abstract. A simplified thermal model has been used to calculate the non-gravitational forces acting on small irregular nuclei in the orbit of Comet 46P/Wirtanen. The torque of the non-gravitational force has been calculated and the Euler equations have been solved in order to investigate the rotational evolution of several irregular nuclei during a single orbital step. Several initial spin axis orientations and activity patterns on their surfaces have been considered. The nuclei considered have a mean radius of 1 km and their inertia moments have been calculated assuming a homogeneous bulk density of 500 kg/m^3. In all the simulations, the initial spin period is 6h and the nuclei initially rotate around their shortest axis. Under these assumptions, significant changes in the angular momentum and in spin period have been obtained in all the simulations, but the nucleus is found to practically remain in its spin state of lowest energy during the entire orbital period.

1. Introduction

Since Whipple (1950)'s historical article, it is well known that the non-gravitational forces due to sublimation of ices can affect the orbital and rotational cometary motion. Since then, theoretical modeling of the evolution of the cometary rotational state considering the effect of the non-gravitational force has become of great interest. These type of models can help to improve future observational searches for the rotational parameters of comets which are essential for a correct interpretation of many kinds of observations.

Several theoretical studies about the evolution of the rotational state of Halley-like comets have been published during the last decades (e.g., Peale and Lissauer, 1989; Julian, 1988; Samarasinha and Belton, 1995; etc.). However, they all assume that the nucleus is ellipsoidal because of the lack of reliable observational constraints on nucleus shapes. Nevertheless, like all bodies smaller than a few hundred kilometers, comets are likely very irregular bodies. Taking this into account, we built a model to study the temporal evolution of the rotational state of a comet in which the nucleus is described as an irregular body. This model can be used in the future, when accurate nucleus shapes will be measured, for instance at the beginning of the Rosetta mission (Schwehm and Schulz, 1999).

Earth, Moon and Planets **90:** 239–247, 2002.

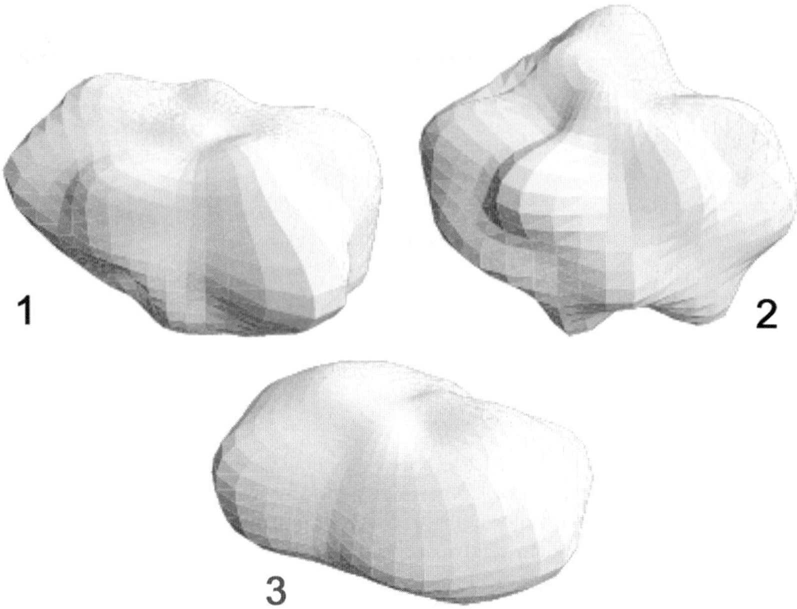

Figure 1. Three Gaussian random shapes used for the simulation. The moments of inertia $(I_x:I_y:I_z)$ of these objects are (normalized by mass·radius^2): (1) 0.32:0.56:0.61, (2) 0.40:0.51:0.59, (3) 0.42:0.64:0.70.

In this article, the evolution of the rotational state of small irregular cometary nuclei in the orbit of Comet 46P/Wirtanen during one orbital period is studied. The non-gravitational force is calculated using a simplified version of the thermal model described by Gutiérrez et al. (2001a). This model is similar to that presented by Keller et al. (2000) but here, more initial conditions are considered.

2. Comet Nucleus and Non-Gravitational Torque

In this study, the irregular surfaces of the nuclei have been defined using the Gaussian random shapes developed by Peltoniemi et al. (1989) and Muinonen (1998). The key idea of this method is to describe the surface radii as a number of lognormal random variables which depend on the spherical coordinates. Figure 1 shows three Gaussian random shapes with different moments of inertia. The moments of inertia, the direction of the principal axes of inertia and the position of the center of mass of these irregular shapes have been computed assuming that the nucleus has a density uniformly equal to 500 kg/m^3.

Regarding the activity pattern of the simulated comets, we have considered two cases: uniform activity across the surface, and a mixture of active and inactive regions. In this model, the active region of the nucleus consists of crystalline water

Figure 2. Distributions of active regions. In these figures, white regions consist of water ice and black regions represent the non-sublimating inactive areas.

ice. The active areas have been randomly distributed using the lognormal statistics. Several lognormal distributions in spherical coordinates have been considered. Only the surface points characterized by a value for the random pattern greater than a defined value are assumed to be active, while the rest of the surface is considered inactive.

For computational purposes, the nucleus surface is divided into 2320 cells (triangles). In order to calculate the net flux of momentum leaving the nucleus and the torque of the non-gravitational force, it is first necessary to compute the local surface temperature of each active surface cell by solving the energy balance equation:

$$\frac{F_0(1 - A_i)c_i}{r_h^2} \cos(\theta_i) = \epsilon_i \sigma T_i^4 + LZ(T_i), \tag{1}$$

where F is the solar constant, A the Bond albedo, r_h the heliocentric distance, θ the solar zenith angle, c is 1 if the cell is illuminated and 0 otherwise. The parameter ϵ is the infrared emissivity, σ Stefan–Boltzmann's constant, L the latent heat and T the temperature of the active cell number i. In this equation, heat conduction into the nucleus have been neglected. In principle, this is a reasonable asumption because heat diffusion is important from a energetic point of view only at large heliocentric distances where the non-gravitational force and torques are small. The water sublimation rate, Z, is given by the Hertz–Knudsen relationship, corrected by a factor $(1-\alpha)$, where $\alpha = 0.25$ to consider the recondensation of backscattered sublimated molecules (Crifo, 1987):

$$Z = (1 - \alpha)Z_{\mathrm{HK}} = (1 - \alpha)\frac{p_v(T)}{\sqrt{2\pi mkT}}, \tag{2}$$

where, m is the mass of water molecules, k is Boltzmann's constant, and $p_v(T)$ is the vapor pressure (Fanale and Salvail, 1984). For the other physical parameters involved in the energy balance equation, we have considered that the albedo and emissivity of the nuclei are equal to 0.05 and 0.95 respectively, and that the latent heat is constant and equal to 48600 J/mol.

When the surface temperature or the flux of molecules leaving the surface has been computed, it becomes possible to calculate the non-gravitational force. In this study, the torque of the non-gravitational force has been approximated by the relationship:

$$N_{\text{nongrav}} = -m \sum_{i=1}^{N} Z_{\text{HK}} S_i \eta v_g \mathbf{r_i} \times \mathbf{n_i}, \tag{3}$$

where the summation is over the N illuminated active cells, S_i is the area of the cell number i, $\mathbf{r}$ is the radial vector joining the center of mass of the body and the center of the cell, $\mathbf{n}$ is the vector normal to the surface, v_g is the initial gas velocity, given by the known expression:

$$v_g = \sqrt{8kT/\pi m} \tag{4}$$

and $\eta = 0.5$, in Equation (3), represents the momentum transfer coefficient (Crifo, 1987; Rickman and Froeschlé, 1983).

In the present status of the program, the nucleus is considered as a rigid body and the erosion due to the sublimation of water ice is neglected. Under this assumption, the Euler equations can be used to study the temporal evolution of the spin state.

3. Simulations and Results

A total of 36 simulations using the shapes of Figure 1, scaled to a mean radius of 1 km, and the active fractions of Figure 2 have been performed on the orbit of Comet 46P/Wirtanen (Marsden and Williams, 1999). Three different initial spin axis orientations have been considered: (1) $I = 0°$, Φ=90°, (2) $I = 90°$, Φ=90°, and (3) $I = 45°$, Φ=60°, where I is the obliquity of the spin axis, and Φ its the argument at perihelion (Sekanina, 1981). The initial rotation period is 6 h in all the simulations. Initially, the nucleus rotates in the minimum energy state. Starting from aphelion, the non-gravitational force and the torque acting on the nucleus are calculated at each time step. When the torque is calculated, the Euler equations are numerically solved (using a Bulirsch–Stoer algorithm) to calculate the orientation of the body in space. The details of the numerical scheme and the tests we have performed to check the integrator are described in a previous work (Gutiérrez, 2001b). In this section, some results of these simulations are shown.

It is known that if the moments of inertia verify $I_x \leq I_y \leq I_z$, the quantity $L^2/2E$ verifies $I_x \leq L^2/2E \leq I_z$, where L is the angular momentum and E the rotational energy. Taking this into account, the quantity $(1 - L^2/2EI_z)/(1 - I_x/I_z)$ is an adequate variable to study the evolution of the rotational excitation level. This quantity is equal to zero when the nucleus is rotating around the axis with the highest moment of inertia (rotation in the minimum energy state), and to one when the nucleus rotates around the axis with the lowest moment of inertia (rotation in

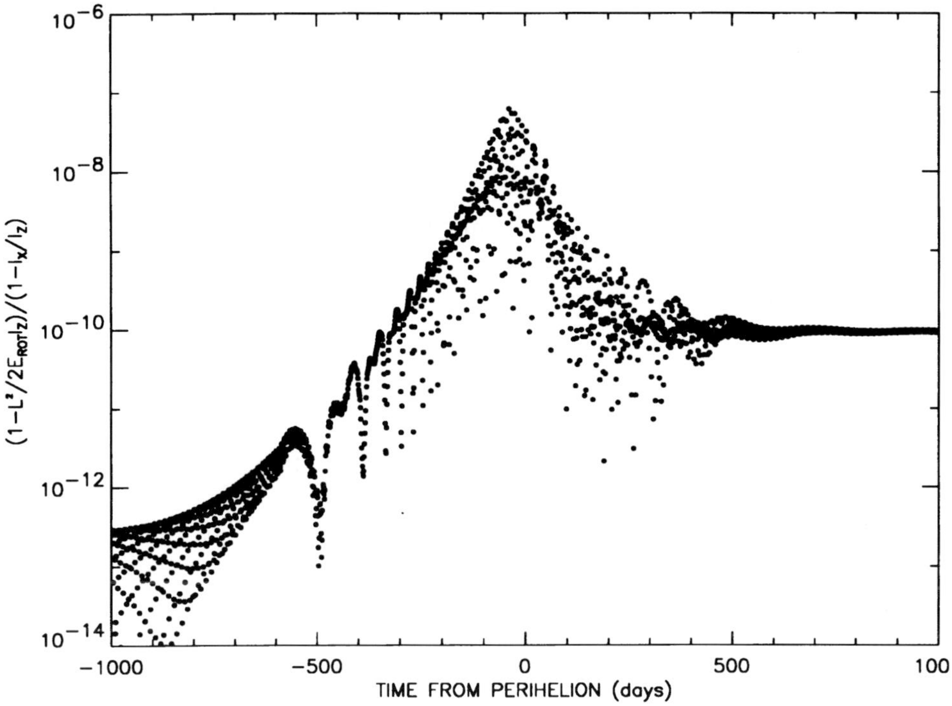

Figure 3. Rotational excitation level for the object 1 when the initial spin axis orientation is $I = 45°$, $\Phi = 60°$. Dots represent the instanteneous value of $(1 - L^2/2EI_z)/(1 - I_x/I_z)$.

the maximum energy state). Hereafter, the quantity $(1 - L^2/2EI_z)/(1 - I_x/I_z)$ is called "rotational excitation level".

Figure 3 shows the temporal evolution of rotational excitation for the uniformly active object 1 (Figure 1) when the initial spin axis orientation is ($I = 45°$, $\Phi = 60°$). In this case, the nucleus spins up after perihelion. In Figure 3, it can be noticed that the excitation level remains very small for the whole orbital period. Only a small excitation takes place around perihelion. After perihelion, a small excitation remains but this excitation level is so low that we can consider that the nucleus remains in a principal axis spin state. This result is also obtained for the other objects when they are uniformly active for the same initial conditions, which confirms the results of Keller et al. (2000).

Although the excitation level does not change significantly, the torque gives raise to important changes in the rotation period and in the spin axis orientation. This can be seen in Figures 4 and 5, respectively. In Figure 4, changes in the rotation periods for the three irregular uniformly active objects and for two initial spin axis orientations are plotted. It shows that the rotation period can either increase or decrease around perihelion, depending on the object and on the initial spin axis orientation. A decrease is observed for instance for object 2 when the spin axis is

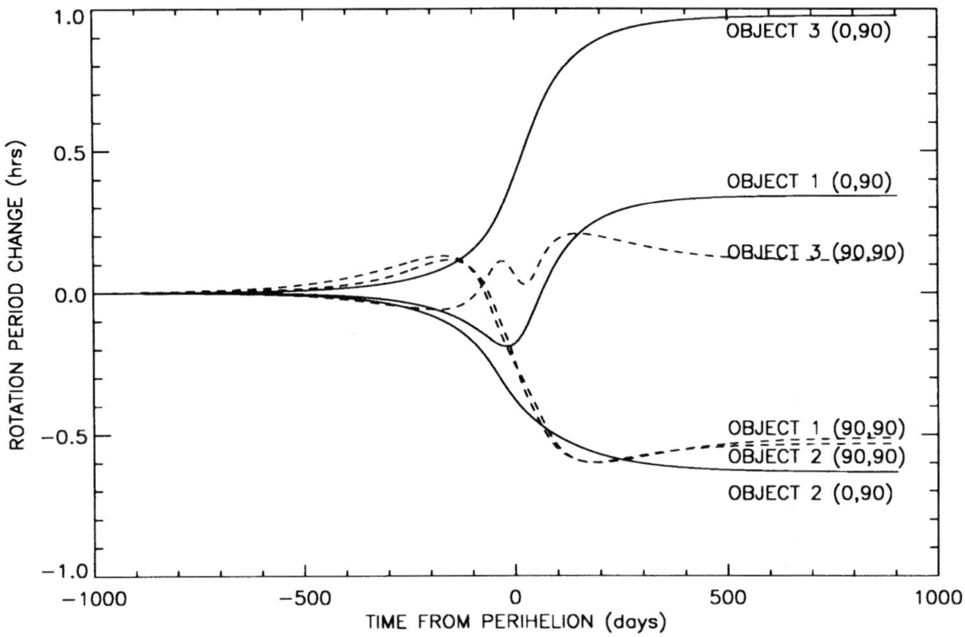

Figure 4. Change of the spin periods for the 3 objects of Figure 1 when the initial spin axis orientations are ($I = 0°$, $\Phi = 90°$) and ($I = 90°$, $\Phi = 90°$).

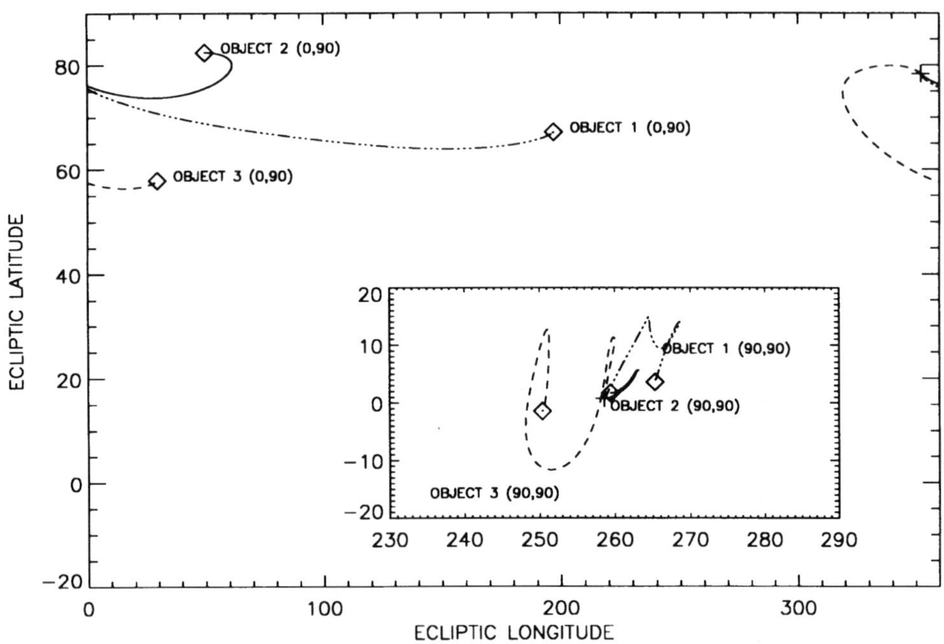

Figure 5. Changes of the spin axis orientations for the 3 objects shown if Figure 1 when the initial spin axis orientations are ($I = 0°$, $\Phi = 90°$) and ($I = 90°$, $\Phi = 90°$). The symbol + represents the initial spin axis orientation and the symbol ◇ represents the final spin axis orientation.

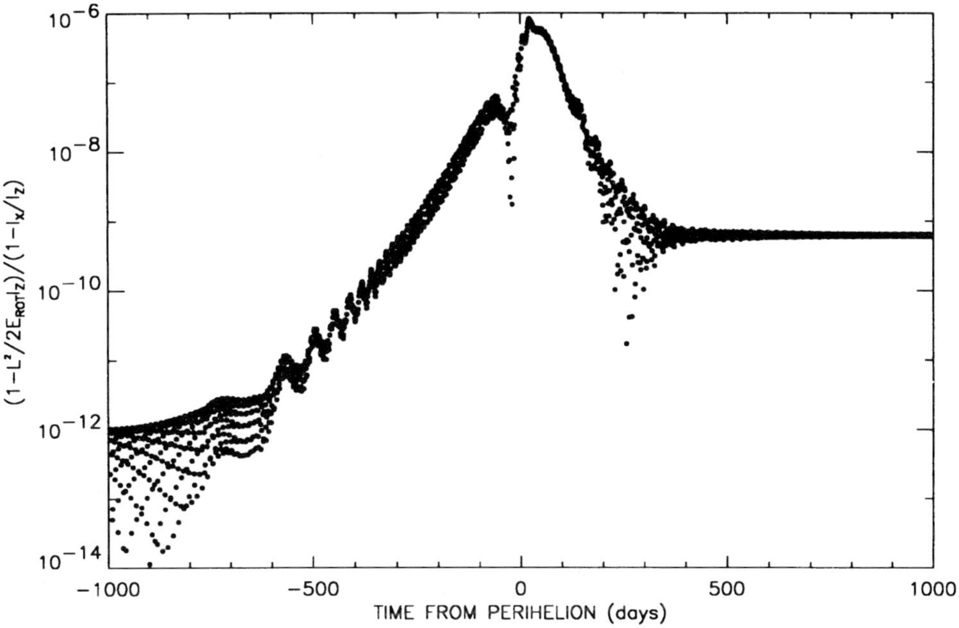

Figure 6. Excitation level for the object 1 when the initial spin axis orientation is $I = 45°$, $\Phi = 60°$ and only the 30 % of its surface is active (distribution A of Figure 2). Dots represent the instanteneous value of $(1 - L^2/2EI_z)/(1 - I_x/I_z)$.

perpendicular to the orbital plane. A more complicated behavior is also observed for object 3 with the spin axis initially in the orbital plane.

Figure 5 shows the evolution of the spin axis in ecliptic coordinates during an orbital period for the 6 cases shown in Figure 4. This Figure shows that large changes in the spin axis orientation (and, therefore, in the angular momentum orientation) can occur in a single perihelion passage. As expected, the magnitude of these changes depends on the shape and on the initial spin axis orientation. The rate of change of the spin axis orientation for all these cases, i.e., when the nucleus remains rotating around its shortest axis of inertia, has a mean value close to perihelion between 0.05 and 0.25 deg/day. Further away from perihelion ($r_h >$ 2.0 AU), the rate of change of the spin axis orientation is always lower than 0.001 deg/day. These values are comparable to those found by Keller et al. (2000). The other initial spin axis orientation shows a similar behavior, i.e., important changes in the rotation period and in the spin axis orientation.

Non-uniform active nuclei also show a similar behavior. The evolution of the excitation level of object 1 with an active region representing only 30% of its surface (distribution A in Figure 2) is shown in Figure 6. The initial spin axis orientation is ($I = 45°$, $\Phi = 60°$). The other spin axis orientations, and the other distributions of active regions show the same behavior, i.e., the excitation level

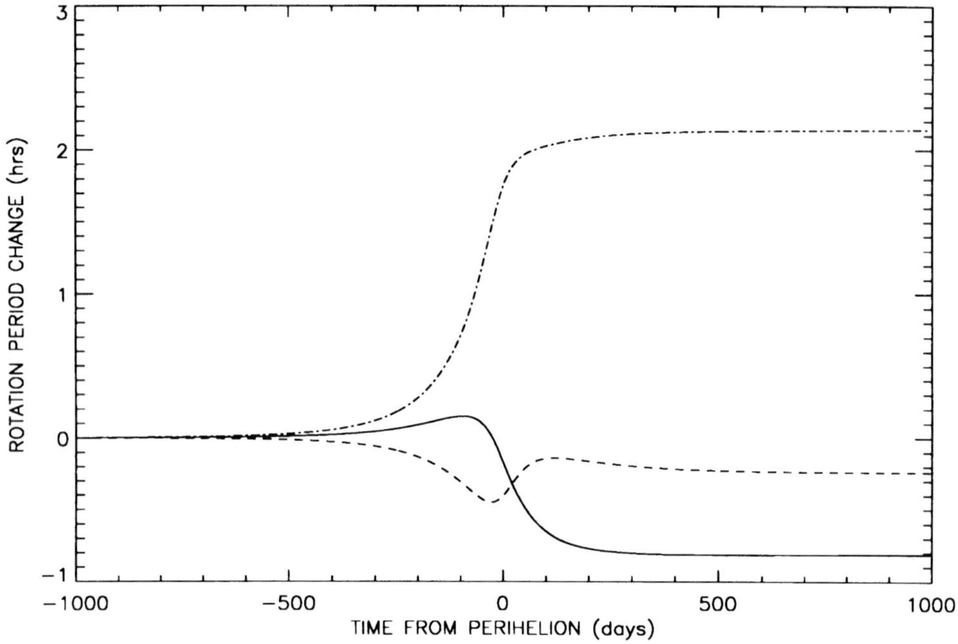

Figure 7. Change of the spin period of the object 1 with the 3 distributions of active area shown in Figure 2 when the initial spin axis orientation is $I = 45°$, $\Phi = 60°$. Solid: Distribution A of Figure 2, dotted-dashed: Distribution B, dashed: Distribution C.

remains negligible for the whole orbital period as in the cases of homogeneously active nuclei.

Significant changes in the rotation period and in the spin axis orientation have also been obtained when the nucleus is partially active. The changes in the rotation period for the object 1 with the three active area distributions displayed in Figure 2 are shown in Figure 7.

Taking all the simulations into account, we have obtained that, if the nuclei are initially rotating in the lowest energy state, half of the simulated nuclei spin up and the other half slow down in an orbital period. In principle, it is reasonable to think that the accelerated nuclei will remain in the lowest energy level for a long time because the excitation in a single perihelion passage is very low. We want to stress that the result could be different for different initial spin periods. In order to check this, long term simulations during more than 10 orbital periods have been done for several nuclei rotating with an initial period of 6–48 h (Gutiérrez, 2001b; Jorda and Gutiérrez, 2002). The final results will be presented in details in a forthcoming article. Different nucleus physical parameters (size, density) could also produce different results: a non-principal axis spin state would be easier to get for smaller and lighter nuclei.

4. Summary and Conclusions

Several simulations to study the rotational state evolution during a single orbital step of different irregular small nuclei in the orbit of Comet 46P/Wirtanen have been performed. The mean radii of the simulated nuclei are equal to 1 km and a bulk density of 500 kg/m^3 has been assumed in order to calculate their moments of inertia. In all the simulations, the nucleus initially rotates around its shortest axis of inertia with a spin period of 6h. *Under these assumptions*, significant changes in the rotation period (accelerations and decelerations) and in the spin axis orientation are observed in all the simulations. In half of the simulations, the nucleus spins up by a significant amount which shows that this mechanism is a plausible cause for the splitting of cometary nuclei. Nevertheless, for these fast rotating nuclei, the "rotational excitation level" defined in Section 3 does *not* change significantly during a single orbital period. This result is different (but not contradictory) with what Samarasinha et al. (1996) found for a symmetric prolate body. However, it shows the importance of exploring the whole range of physical parameters involved before getting to a reliable conclusion regarding the rotational excitation of short-period comets.

References

Crifo, J. F.: 1987, *Astron. Astrophys.* **187**, 438–450.
Fanale, F. P. and Salvail J. R.: 1984, *Icarus* **60**, 476–511.
Gutiérrez, P. J.: 2001, 'Modelo tridimensional de núcleos cometarios: producción de agua y evolución del estado rotacional', Ph. D. Dissertation, Universidad de Granada.
Gutiérrez, P. J., Ortiz, J. L., Rodrigo, R., and López-Moreno, J. J.: 2001, *Astron. Astrophys.* **355**, 326–336.
Jorda, L. and Gutiérrez, P.: 2002, 'Rotational Properties of Cometary Nuclei', these proceedings.
Julian, W. H.: 1988, *Icarus* **74**, 377–382.
Keller, H. U., Jorda, L., Rickman, H., and Thomas, N.: 2000, *Bull. Amer. Astron. Soc.* **32**, 1062.
Marsden, B. G. and Williams, G. V.: 1999, *Cataloge of Cometary Orbits*, 13th edn., IAU Minor Planet Center, Smithsonian Astrophysical Observatory, Cambridge.
Muinonen, K.: 1998, *Astron. Astrophys.* **332**, 1087–1098.
Peale, S. J. and Lissauer, J. L.: 1989, *Icarus* **79**, 396–430.
Peltoniemi, J. I., Lummne, K. ,Muinonen, K. et al.: 1989, *Appl. Opt.* **28**, 4088–4095.
Rickman, H. and Froeschlé, C.: 1982, in *Cometary Exploration I; Proceedings of the International Conference*, Nov. 15–19, 1982, Budapest, Hungary, pp. 75–84.
Samaransinha, N. H. and Belton, M. J. S.: 1995, *Icarus* **116**, 340–358.
Samaransinha, N. H., Mueller, B. E. A., and Belton, M. J. S.: 1996, *Planet. Space Sci.* **44**, 275–281.
Schwehm, G. and Schulz, R.: 1999, *Space Sci. Rev.* **90**, 313–319.
Sekanina, Z.: 1981, *Annu. Rev. Earth Planet. Sci.* **9**, 113–145.
Whipple, F. L.: 1950, *Astrophys. J.* **111**, 375–394.

A MODEL OF AN ACTIVE REGION ON THE SURFACE OF A COMETARY NUCLEUS

ALEKSANDRA IVANOVA and LEONID SHULMAN
Main Astronomical Observatory of the National Academy of Sciences of Ukraine, 27 Akad. Zabolotny Str., 03680, Kyiv, Ukraine

(Received 27 February 2002; Accepted 13 July 2002)

Abstract. An active region on the surface of a cometary nucleus is considered as a conic hole in the surface dust mantle with icy bottom and dusty side-walls. This conic structure should concentrate solar energy onto the bottom and therefore enhance sublimation. Preliminary results of the calculation of this effect are given. The temperature distribution at the bottom of the crater is calculated for different sets of its geometrical parameters. Effects of intensified sublimation depending on the geometrical parameters are considered for the specific case when a single active region is located exactly at the pole of the nucleus and the pole is directed to the Sun.

Keywords: Active region, cometary nucleus, craters, sublimation rate

1. Introduction

After the successful space missions to comets we know that cometary nuclei are non-spherical and their surfaces are inhomogeneous. It is necessary to replace the classic idealized models of a cometary nucleus by more realistic ones. Observations, laboratory simulation (see for example Kajmakov and Sharkov, 1957; Ibadinov, 1971; Ibadinov et al., 1991; Grün et al., 1991), and theoretical calculations (Shulman, 1972; Prialnik and Bar-Nun, 1988), as well as the space missions show, that nuclei of comets are coated by dust layers. This mantle prevents the ice of a nucleus from being heated directly by solar radiation. Therefore, the outgassing of a cometary nucleus is often localized in several active areas. The mantle is formed when the greatest dust particles are accumulated on the surface because gravity of the nucleus is greater than the gas drag force which pushes the grains away from the surface (Dobrovolsky, 1961; Shulman, 1972). It is assumed that an active region on a nucleus appears after a local crack in the dust mantle. This process is almost not investigated theoretically and it is not considered here. The subject of the present paper is to describe the heat transfer and sublimation in an active region which has been created before by some unknown mechanism.

Earth, Moon and Planets **90:** 249–257, 2002.

2. Structure and Composition of the Surface of a Cometary Nucleus

The classic icy model of a nucleus was developed by Whipple (1950, 1951) who considered a cometary nucleus as a conglomerate of frozen gases and dust particles. This model was improved many times (Shulman, 1987) afterward. An idealized spherically symmetric model of a cometary nucleus does not take into account the macroscopic non-uniformity of their surfaces. Many of cometary nuclei are covered by dust mantles which attenuate both the heat flux to ice and the gas flow outward because both effects are "diffusion" (heat conduction and gas flow) processes through the pores in the dust. As it was first shown by Shulman (1972) the molecular flux by diffusion is considerably less than that from the bare icy surface. Therefore, practically the whole atmosphere of a comet is formed by sublimation from active areas.

A simple model of an active region that looks like a conic hole (crater) in the dust mantle of a nucleus is proposed here. The crater has bare ice at the bottom and dust on its side walls. This model differs of that considered by Colwell et al. (1998) who assumed that the whole surface of the crater is icy. The simplest case is considered when the active area is exactly on the pole of the cometary nucleus and the axis of its rotation is directed toward the Sun. This case is not unrealistic, as it became clear after the DS-1 mission that the nucleus of Comet Borrelly has its active regions near its pole.

In this simple case there is no effect of axial rotation on the heating and sublimation. Nevertheless such an approach is good enough to check whether the crater structure of an active region can amplify the sublimation rate.

3. Heat Balance on the Walls of the Active Area

The formation and evolution of a cometary atmosphere is defined by the thermal conditions and chemical composition of the nucleus. There are many papers where the temperature of cometary nuclei was calculated theoretically – Dobrovolsky (1961), Ibadinov (1971), Markovich (1963), Colwell (1999), Weissman and Kieffer (1981) – and measured experimentally in laboratory simulation – Kajmakov and Sharkov (1957). The nucleus temperature depends on the absorbed solar energy, the chemical composition and the structure of the nucleus. The calculation of the temperature of a cometary nucleus has been carried out first by Dobrovolsky (1961) and thereafter by Markovich (1963). The heat transfer equation can be written as:

$$\rho_n C \frac{\partial T}{\partial t} + div(K grad T) = S(\mathbf{r}, t), \qquad (1)$$

where T ($\mathbf{r}$, t) is the temperature as a function of time t and position r. ρ_n is the density of the nucleus, C is the thermal capacity per unit mass, K is the heat conductivity, S denotes contributions by intrinsic energy sources. The solution of Equation (1) needs a boundary condition. If the crater structure is absent, the boundary condition is the equation of heat balance on the surface of the nucleus, i.e.,

$$\frac{q(1-A)\cos z}{r^2} = \varepsilon\sigma T^4 - (K\nabla T)_{R(\boldsymbol{r},t)=0} + ZL, \tag{2}$$

where q is the solar constant, z is the zenith distance of the Sun, A is the albedo of the surface of the nucleus, ε is its emissivity, σ is the Stefan–Boltzmann constant, Z is the sublimation flux per unit surface, L is the energy of sublimation per molecule, and $R(\boldsymbol{r}, t) = 0$ defines the equation of the surface of the nucleus. We assume that the active region is in heat equilibrium. It means that the temperature does not depend on time, and the conductive flux of energy into the nucleus is neglected. The temperature distribution is calculated taking into account that solar radiation that reaches bare ice (the bottom of the crater) in two ways: directly from the Sun and by the re-emission – in the infrared – from the walls of the crater. Thus, ice at the bottom receives more energy per unit area than in the case when the crater is absent.

To solve the integral equations of energy balance numerically, a discrete grid was designed. Both the bottom and side walls of the cone were divided into small elements of equal surface area. For the conic crater model, the energy balance equation for the ice bottom can be written as follows:

$$\begin{aligned}\frac{q(1-A_{\mathrm{i}ce})\cos z}{r^2_{\mathrm{C}om}} &= (1-A_{\mathrm{i}ce})\sigma T^4_{\mathrm{i}ce} + L\frac{\mathrm{e}^{A-B/T_{\mathrm{i}ce}}}{\sqrt{2\pi mkT_{\mathrm{i}ce}}} \\ &\quad +(1-A_{\mathrm{d}ust})\frac{\sigma}{\pi}\int T^4_{\mathrm{d}ust}\frac{\cos\beta_1\cos\beta_2}{l^2}\mathrm{d}^2s,\end{aligned} \tag{3}$$

where $r_{\mathrm{C}om}$ is the heliocentric distance of the comet, $A_{\mathrm{i}ce}$ the albedo of the ice surface, $T_{\mathrm{i}ce}$ and $T_{\mathrm{d}ust}$ are the temperatures of the ice and the dust, A and B are the constants for the approximation of the saturated vapour pressure as a function of the temperature: $P_s = \exp(A - B/T_{\mathrm{i}ce})$, l is the distance between a point at the icy bottom and another one at the dusty walls of the cone, β_1 and β_2 are the angles between the connecting line of both points and the normal vectors of the respective surface elements. Equation (3) has to be integrated over the entire surface of the cone (d^2s as differential of area). In our specific case the Sun is in zenith, therefore $\cos z = 1$ on the icy bottom of the cone. The balance of energy on the dust walls is given by the relationship:

$$\begin{aligned}\frac{q(1-A_{\mathrm{d}ust})\cos z}{r^2_{\mathrm{C}om}} &= (1-A_{\mathrm{d}ust})\sigma T^4_{\mathrm{d}ust} \\ &\quad +(1-A_{\mathrm{d}ust})\frac{\sigma}{\pi}\int T^4_{\mathrm{d}ust}\frac{\cos\beta_1\cos\beta_2}{l^2}\mathrm{d}^2s.\end{aligned} \tag{4}$$

In this case $\cos z = \sin \varkappa$, where $\varkappa$ is the half vertex angle of the cone. Here we did not consider heat exchange between gas inside the cone and its dusty side walls.

4. Numerical Analysis

A discrete grid was constructed for the numerical solution of the system of integral Equations (3) and (4). Both the bottom and the lateral surface of the cone were divided into small elements of equal areas. The geometry of an active area is determined by the following three quantities: The bottom radius r of the crater, its height h and half vertex angle $\varkappa$. Subscript s indicates the properties of an side wall element, subscript b refers to the icy bottom. One can rewrite the integral Equations (3) and (4) as a system of the nonlinear algebraic equations:

$$\frac{q(1-A_{\mathrm{ice}})}{r^2_{\mathrm{Com}}} = (1-A_{\mathrm{ice}})\sigma T^4_{bi} + L\frac{P_s(T_{bi})}{\sqrt{2\pi k T_{bi}}} + \sum_{j=1}^{N_s}(1-A_{\mathrm{ice}})\sigma\frac{T^4_{sj}}{\pi}K_{ij}, \tag{5}$$

for the ice bottom and

$$\frac{q\sin\varkappa}{r^2_{\mathrm{Com}}} = \sigma T^4_{si} - \sum_{j\neq i}^{N_s}\sigma\frac{T^4_{sj}}{\pi}K_{ij} + 2\alpha k(T_{si}-T_{\mathrm{gas}}), \tag{6}$$

for the lateral surface. The last term on the right side of this equation approximates the heat exchange between the mineral substance of the side wall and the gas inside the cone which has the temperature T_{gas}. It is assumed that the gas receives the energy from the wall, proportional to the temperature difference between the gas and the wall. The efficiency of heat exchange is determined by coefficient α.

In Equations (5) and (6) K_{ij} is the matrix given by the relationship:

$$K_{ij} = \frac{\cos\beta_i\cos\beta_j}{l^2_{ij}}\frac{\Delta S_j}{\Delta S_i}, \tag{7}$$

where ΔS is the area of a surface element. $\cos\beta$ is a dot product of two unit vectors and is given by the relationship

$$\cos\beta_i = \mathbf{n}_i\cdot\mathbf{l}^0_{ij}, \tag{8}$$

where $\mathbf{n}$ is a unit vector of the normal to the center of the ith element of the side wall:

$$\mathbf{n} = \{n_{xi}, n_{yi}, n_{zi}\}, \tag{9}$$

and

$$\mathbf{l}^0_{ij} = \frac{\{x_{bi}-x_{sj}, y_{bi}-y_{sj}, z_{bi}-z_{sj}\}}{l_{ij}}. \tag{10}$$

The length l of the vector is

$$l_{ij} = \sqrt{(x_{bi} - x_{sj})^2 + (y_{bi} - y_{sj})^2 + (z_{bi} - z_{sj})^2}, \tag{11}$$

where x, y, z are the coordinates of two points: one on the side wall (subscript sj) and the second at the bottom (subscript bi) of the crater.

Using the cylindrical symmetry one can reduce this two-dimensional problem to one-dimensional one. So it is possible to calculate the distribution of the temperature over the meridian cross section of a conic crater. The algebraic systems (5) and (6) can be solved by an iteration method (Shulman, 1987). For the iteration procedure the initial non-linear system should be transformed into a special form where the next approximation of the temperature is being calculated by substituting the previous approximation in the right sides of the equations. This kind of iteration procedure converges if the functions in the right sides of the transformed equations give so-called contracting projection, i.e., each following approximation must reduce the errors of the previous one. The iteration is terminated when the difference between two subsequent approximations is less than a given minimum accuracy. One can write:

$$T_{si} = \{\frac{1}{\sigma}[\frac{q \sin \varkappa}{r^2_{\text{Com}}} + \sum_{j \neq i} \frac{\sigma T^4_{sj}}{\pi} \frac{\cos \beta_i \cos \beta_j}{l^2_{ij}} \Delta S - 2\alpha k (T_{si} - T_i)]\}^{\frac{1}{4}}, \tag{12}$$

$$\begin{aligned} T_{bi} &= B/\{A - \log\{\frac{\sqrt{2\pi m k T_{bi}}}{L}[\frac{q(1 - A_{\text{ice}})}{r^2_{\text{Com}}} - \sigma T^4_{bi} \\ &\quad + \sum_j \frac{\sigma T^4_{sj}(1 - A_{\text{ice}})}{\pi} \cdot \frac{\cos \beta_i \cos \beta_j}{l^2_{ij}} \Delta S]\}\}. \end{aligned} \tag{13}$$

These algebraic equations were used to calculate the distribution of the temperature. From practical experience, the iteration procedure converges very rapidly.

5. Results

We have calculated the distribution of the temperature on the meridian cross-sections of different craters. The geometric parameters of the craters are given in Table I.

The temperature distributions for different parameter sets of the crater are plotted in Figure 1. The sublimation flux was calculated using the Knudsen–Hertz formula

$$Z = \frac{p_n(T)}{\sqrt{2\pi m k T}}, \tag{14}$$

TABLE I

Geometric parameters for different model of active areas on the surface of a comet

r, cm	90	130	150	160	190	210	220	230	240	250
h, cm	300	320	330	340	350	360	370	400	450	500
$\varkappa$, degree	30	32	34	36	38	42	47	54	56	60

where $p_s(T)$ is the pressure of saturated vapour at a given temperature of the ice. The total flux of water vapour from the bottom of the crater is given by the relationship:

$$Q = 2\pi \int_0^r \frac{r e^{A-B/T(r)}}{\sqrt{2\pi mkT(r)}} dr \approx n_b \cdot \Delta S \cdot \sum_{i=1}^{n_b} \frac{e^{A-B/T_i(r)}}{\sqrt{2\pi mkT_i(r)}}. \tag{15}$$

The values $A = 31.085$, $B = 6120.3$ K were taken for the approximation of the saturation pressure of water vapour. The sum in the right side of Equation (15) is taken over all bottom elements of the cone.

If the crater is absent, then the gas production of the part of the nucleus with the area πr^2 should be

$$Q = \pi r^2 \frac{e^{A-B/T}}{\sqrt{2\pi mkT}}. \tag{16}$$

Here, the temperature is calculated for the given heliocentric distance via Equation (13) (the double sum in the right side – the additional flux of energy from the walls – is omitted). The amplification factor G of the sublimation rate is

$$G = Q/Q_0. \tag{17}$$

This quantity has been calculated for each model and the results are shown in Figures 2–3 for the different craters. The dependence of Q on the heliocentric distance of the comet is shown on Figure 3 (right panel).

6. Conclusions

The results of our modeling show that a crater structure can amplify the sublimation rates by a factor ≤ 3. It is easy to understand that in more general case (i.e., non-polar position of the active area) the cometary nucleus will generate gas-and-dust jets when the active areas are rotating into the day side of the nucleus. One can suspect that large dust particles are accumulated on the bottom of the crater forming

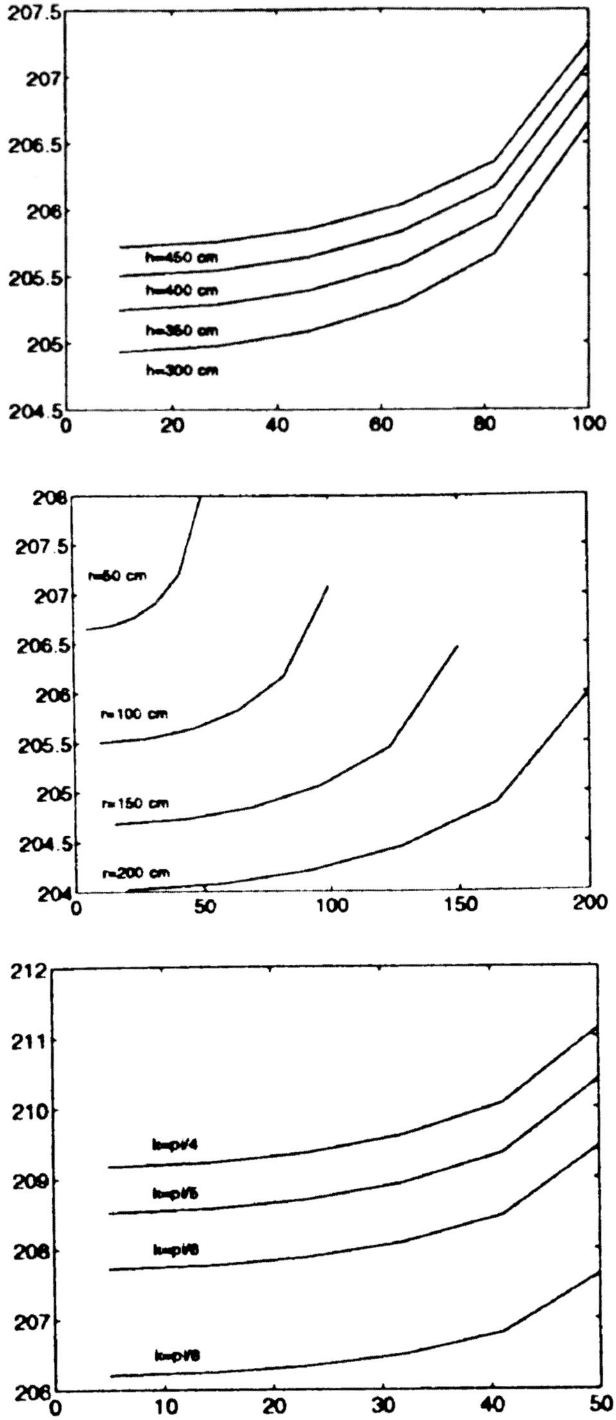

Figure 1. The temperature distribution $T(r_b)$ for different depth of the cone (300–450 cm, 50 cm steps; top panel), different bottom radii (50–200 cm, steps 50 cm; middle panel) and different vertex angles ($\pi/4,5,6,8$; bottom panel).

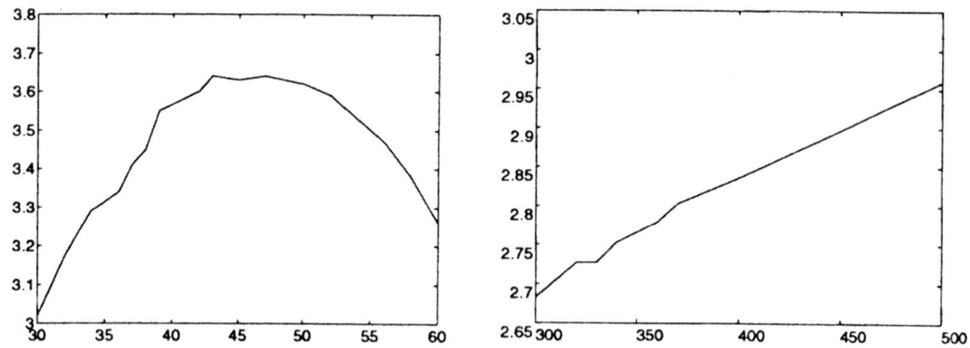

Figure 2. Amplification of the sublimation rate as a function of the vertex angle of the cone (left panel) and of the depth of the cone (right panel).

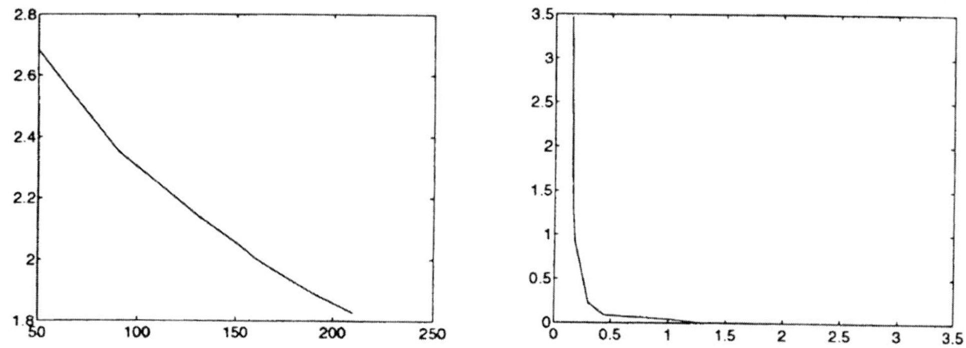

Figure 3. Amplification of sublimation rate as a function of the bottom radius of the cone (left panel) and of the heliocentric distance of the comet (right panel).

a mantle there. It is interesting to note that the temperature of ice at the bottom of the crater is increasing from center to the edge. Therefore, an initially plain bottom of the crater becomes convex and potentially a massive icy fragment could be released from the bottom of the cone. For this reason, a spontaneous outburst of sublimation can occur when the dust mantle on the bottom of the crater will be blown off.

Acknowledgement

This work was supported by the State Basic Research Foundation of Ministry for Education and Science of Ukraine.

References

Colwell, J. E., Bruce, M. J., Bradford, J. S., and Stern, S. A.: 1999, *Icarus* **85**, 205–215.

Crifo, J. F. and Rodionov, A. V.: 1998, Invited at the *IAU Symp. 168*, *PHSP*, submitted.
Dobrovolsky, O. V.: 1961, *Non-Stationary Processes in Comets and Solar Activity*, Dushanbe, 193 pp. (in Russian).
Grün, E., Bar-Nun, A., Benkhoff, J. et al.: 1991, in *Comets in the Post-Halley Era*, Vol. 1, Kluwer Academic Publishers, Dordrecht, pp. 277–297.
Ibadinov, H. I.: 1971, 'Investigation of Dust Component of Comets', Autoref. of the PhD Thesis, Dushanbe (in Russian).
Ibadinov, H. I., Rakhmonov, A. A., and Bjasso, A. S.: 1991, in *Comets in the Post-Halley Era*, Vol. 1, Kluwer Academic Publishers, Dordrecht, pp. 299–311.
Julian, W. H., Samarasinha, N. H., and Belton M. J. S.: 1999, *NOAO* **859**, 26.
Kajmakov E. A. and Sharkov V. I.: 1957, *Comets and Meteors*, No. 15, 21–25 (in Russian).
Klinger, J.: 1980, *Science*, pp. 260–271.
Markovich, M. Z.: 1963, 'Temperature of Cometary Nuclei', Autoref. of the PhD Thesis, Leningrad (in Russian).
Prialnik, D. and Bar-Nun, A.: 1988, *Icarus* **74**, 272–283.
Shulman, L. M.: 1972, *Dynamics of Cometary Atmospheres. Neutral Gas*, Kyiv, 242 pp. (in Russian).
Shulman, L. M.: 1987, *Nuclei of Comets*, Moskow, 230 pp. (in Russian).
Weissman, P. R. and Kieffer, H. H.: 1981, *Astrophysics*, p. 390.
Whipple, F. L.: 1950, *Astrophysics* **47**, 375–394.
Whipple, F. L.: 1951, *Astrophysics* **113**, 464–474.

SPANISH MONITORING OF COMETS: MAKING SENSE OF AMATEUR PHOTOMETRIC DATA

MARK R. KIDGER
Instituto de Astrofísica de Canarias, E-38205 La Laguna, Tenerife, Spain

(Received 11 March 2002; Accepted 12 July 2002)

Abstract. Amateur astronomers have always represented an important observing group in cometary astronomy. Much of our knowledge of cometary light curves has come from amateur data, initially in the form of total visual magnitude estimates and now increasingly in the form of CCD observations. The increasingly widespread use by amateur astronomers of CCD cameras of excellent sensitivity and good cosmetic quality has revolutionised astrometry, allowing far more intensive astrometric monitoring of comets to be carried out down to magnitude 18 and fainter, with a corresponding enormous increase in the quality of calculated orbits. Although amateur CCD photometry is extensively available in the Internet, its use has been less widespread. The reason is the lack of standardisation in the way that this data is taken that leads to amateur CCD light curves having enormous dispersion. All amateur CCD photometry is aperture photometry, but it is seen that neither does it represent well the equivalent of m1 (total visual magnitude), even with a large aperture, nor is it close to the definition of m_2 (the nucleus magnitude). The problem is examined using data from the Spanish Comet Observers Group archives to show that by careful standardisation of data acquisition amateur CCD data can produce high quality, well-sampled and physically meaningful light curves. Examples are presented of the results for recent comets including 19P/Borrelly, 51P/Harrington, C/2001 TU80 (LINEAR-NEAT), C/2000 WM1 (LINEAR) & C/2001 A2 (LINEAR).

Keywords: 19P/Borrelly, 51P/Harrington, C/2000 WM1 (LINEAR), C/2001 A2 (LINEAR), C/2001 TU80 (LINEAR-NEAT), observational techniques, photometry

1. Introduction

Over the last few years there has been an increasing use of CCD cameras by amateur astronomers to monitor comets on a regular basis. A CCD camera attached typically a 20 or 25-cm Schmidt/Cassegrain telescope is the preferred configuration (although an increasing number of amateur astronomers use telescopes of 30-cm, 40-cm and even as large as 60-cm). This gives a typical image scale of approximately 2 arcsec/pixel and a field that may be half a degree or larger. Even from a moderately light-polluted urban site a 20-cm telescope + CCD is capable of detecting comets of magnitude 18 with sufficient signal-to-noise-ratio (s/n) to make usable astrometry and photometry.

Much of the amateur effort has gone into taking astrometry. Over the last 5 years the quantity of astrometry of even relatively faint comets reported to the Minor Planet Center (MPC) has increased many fold. Although a significant part of this increased database is due to professional observatories such as LINEAR, a

Earth, Moon and Planets **90:** 259–268, 2002.

large part is due to amateur observers. An example of this is seen by comparing the astrometric coverage of a bright pre-CCD comet (27P/Crommelin) or 23P/Brorsen-Metcalf, with recent comets (see Table I).

2. The *T/N* Problem

At present all CCD photometric data is submitted to the Minor Planet Center Archive or to bodies such as the International Comet Quarterly as either Total magnitude (T, or m_1), or as nuclear magnitude (N, or m_2). These designations date back to the pioneering work of Roemer (see e.g., Jeffers et al., 1954; Roemer, 1956) when the first serious attempts were made to measure systematically the brightness of the cometary nucleus in her programme of observations from the Flagstaff station of the US Naval Observatory from 1950–1976. It became necessary to have a system to distinguish between the integrated magnitude of the comet, as measured by visual or photographic observers and the magnitude of the central condensation of the coma. These magnitudes, generally referred to as m_1 and m_2 have been used for some 50 years, proving durable and generally unambiguous for most of that time.

With the increase in the amount of photometry reported using CCDs, starting in the 1980s and by the end of the 1990s spreading to include most amateur astrometry too, the simple separation into nuclear and total magnitudes has become increasingly difficult. A review of some of the difficulties has been presented by (Tancredi et al., 2000) in the context of calculating nucleus magnitudes for comets, while (Kamel, 1992) discusses the problem of standardisation of cometary magnitudes.

In particular, with CCD detectors a magnitude that is reported as T will, except in the case of very faint comets, rarely include the whole coma as it would be seen by a visual observer. In most cases the "total" magnitude reported by a CCD observer will be fainter by some unspecified amount than the magnitude reported by a visual observer, dependent on exposure time, degree of condensation of the coma, diameter, etc. Similarly, a magnitude reported as N will rarely correspond to the bare nucleus of the comet, except for distant and inactive objects. In some cases the magnitude reported is the result of a coma subtraction process, in others it is the magnitude within an unspecified aperture around the nucleus. Converting CCD data to N or T is, at present, impossible given the traditional observing and reporting proceedures.

With the widespread use of CCDs by amateur astronomers the quantity of high quality astrometry being received by the MPC has increased at an extremely rapid rate. However, the photometry of comets (but not asteroids) associated with that astrometry is of such a diverse nature that it is difficult or impossible to interpret. We thus risk throwing away what is a huge resource for cometary studies. At present the MPC is studying the reporting of comet magnitudes and it may be anticipated

TABLE I

The total number of astrometric observations of selected comets in the MPC database. In the final column we give the mean number of points per apparition (total number of points divided by apparitions covered) for periodic comets and the number of points *per month* of observation for non-periodic objects. Note that even the 15th magnitude C/2001 TU80 has more astrometry in 3 months than the naked-eye pre-CCD comet 23P/Brorsen–Metcalf in its 5 observed apparitions. Most, although not all of these points have associated photometry, providing a huge potential resource for studying light curves

Comet	Apparition/period covered	Maximum brightness (mag)	MPC data	Data/apparition (/month)
23P/Brorsen–Metcalf	1988	5 (1988)	86	29
27P/Crommelin	1985	8 (1985)	279	56
C/1999 S3 (LINEAR)	99/09/24–00/02/11	12	598	(133)
C/1999 S4 (LINEAR)	99/09/27–00/07/22	6	1070	(107)
C/2001 A2 (LINEAR)	01/01/03–01/11/18	3	1599	(152)
C/2001 TU80 (LINEAR-NEAT)	01/10/12–01/12/28	15	93	(37)
C/2001 X1 (LINEAR)	01/12/13–01/12/28	14	108	(210)

that a change from a pure N, T system for reporting photometry may be made in the future to make allowance for CCD photometric data.

3. Treating the Problem

For photometry of comets to be genuinely valuable, standardisation of both data acquisition and reduction is required. As the amateur observing community is highly diverse, it is extremely difficult to standardise fully. Most observers do not have the resources to purchase standard filter sets, or to measure the colour transformations or second-order extinction corrections that would be necessary to homologate fully their measures. However, such effects are very much of second order when compared, for example, to the consequences of variable aperture, or even using unsuitable star catalogues for photometry.

Here we will show that the dispersion of amateur CCD photometry of comets can be reduced by 1–2 orders of magnitude by simply defining an appropriate standard aperture for the measurements to be reported, a standard photometric system and a standard reference catalogue. In some cases the best that can be obtained is an *approximation* to the standard that has been defined, but even that is sufficient to improve greatly the consistency of data.

4. The "Cometas_Obs" Data Archive

The proving ground for the standardised photometric method has been the Spanish language "Cometas_Obs" mailing list. Started in Spring 2001 to improve coordination of observations of C/2001 A2 (LINEAR) between Spanish observers, this list has more than 70 active observers in at least 7 countries subscribed. This mailing list was in turn based on a group of observers who had submitted CCD privately to the author of this paper for some years. All data submitted to the "Cometas_Obs" list is archived and CCD observers are encouraged to submit their data to the MPC, obtaining the necessary IAU/MPC site code. Observers are encouraged to adopt standard observing and reduction practices where possible and new observers are assisted by more experienced ones.

It was evident that even though data were always separated by the m_1 or m_2 criterion for analisis that this alone was insufficient (see Figure 1). After discussion it was decided to adopt a series of criteria to standardise the photometry that was being made:

- Where possible observations should be taken through an R filter or, in its absence, treated as R-band photometry.
- A 10 arcsec aperture should be used for all photometry reported to the MPC, although observations in other *specified* apertures are encouraged.

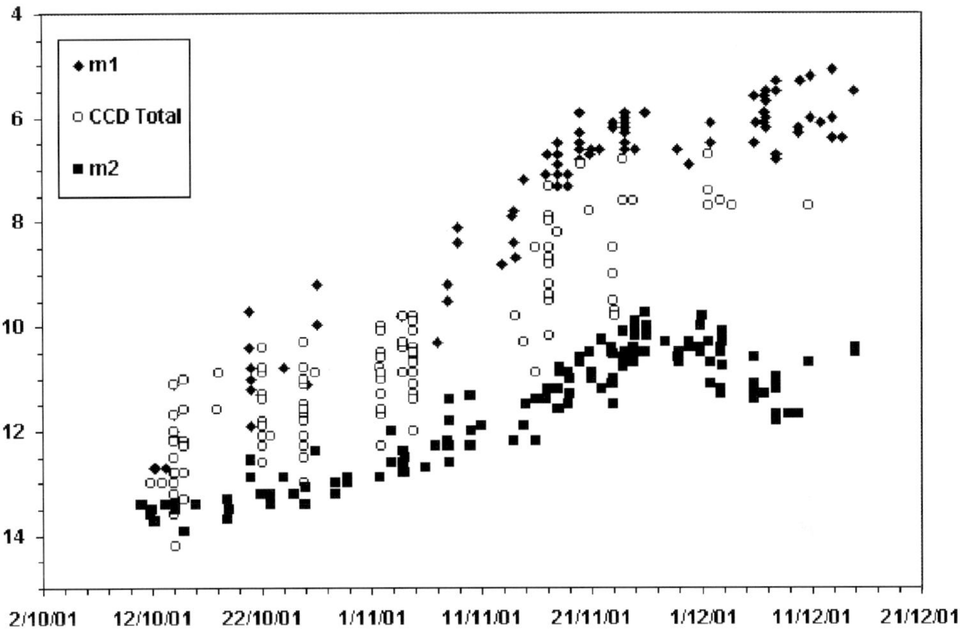

Figure 1. The light curve of C/2000 WM1 (LINEAR) taken from the "Cometas_Obs" database. Data are plotted as "m_1" (visual total magnitudes), "m_2" (CCD photometry with a 10 arcsecond aperture) and "CCD" (all other apertures). Note how the "CCD" points fill the locus between the visual and standard aperture CCD data.

– Due to the large number of fainter comets that may be observed with CCDs it was decided to use the USNO A2.0 catalogue as the standard reference.

Some unfiltered CCDs available on the market are found to be closer to V than to R but, in the majority of cases the approximation to R is best. This is also coherent with the use of the USNO A2.0 catalogue that gives B and R magnitudes for stars. The choice of this catalogue was forced as it is the only one that is widely available that covers the full sky to a faint magnitude with reasonable photometric reliability (of the order of 0.2 magnitudes, with the photometry being linked at the bright end to the Hipparcos/Tycho scale), although obviously it is not suited to high-precision professional photometry.

Note that a 10 arcsec aperture is several times the typical seeing, but small enough to represent the activity of the nucleus, although the contribution of the nucleus to the flux in the aperture is negligible. At $\Delta = 1$ AU this aperture corresponds to a radius of 7250 km, thus giving a typical crossing time of only a few hours. Given the typical seeing conditions and pixel sizes of amateur CCD data a smaller aperture was not regarded as practical. A larger aperture increases the uncertainty due to the sky background and the danger that faint field stars will adversely affect the results. An obvious criticism of this method though is the lack of *physical* meaning of a standard 10 arcsec aperture or any other standard size

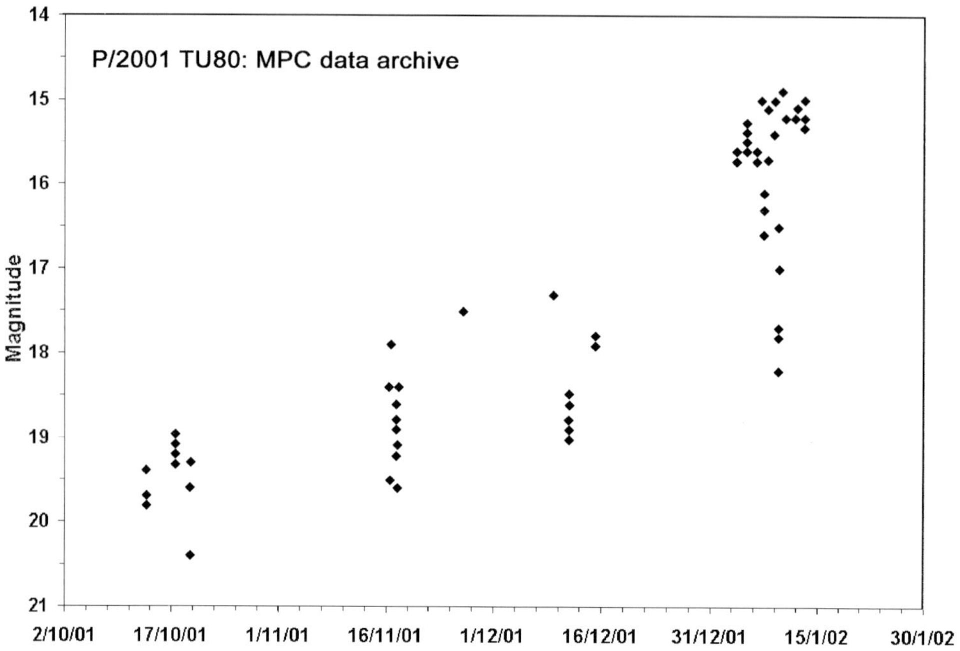

Figure 2. The light curve of C/2001 TU80 (LINEAR-NEAT) taken from the MPC database. A large dispersion can be seen in the data.

defined. For this reason it is only proposed as a measure that allows photometry to be carried out conveniently and consistently and not as a definitive solution to the N, T problem for CCD observers.

5. Results

Many places are available on the Internet where amateur and professional photometry of comets are available, in many cases the data are presented as light curves. Such lightcurves are a valuable aid, both to the observer when planning observations, and in the detailed study of a particular comet. Unfortunately, the non-separation of data by type and widely differing data acquisition methods leads to a huge level of dispersion. In the case of C/2000 WM1 (LINEAR) even at an early epoch when all data was taken by CCD we find 2 magnitudes of dispersion between observers and at later epochs 5 or more.

In Figures 1 and 3 we compare photometry of C/2001 TU80 (LINEAR-NEAT) from the MPC database and from the "Cometas_Obs" mailing list taken in a standardised way. Note that the data shown in Figure 3 forms a significant subset of that shown in Figure 2. The total dispersion in the data reduces from ≈ 3 magnitudes to ≈ 0.5 magnitudes, with $\sigma = 0.70$ and $\sigma = 0.35$ mags respectively, most of the

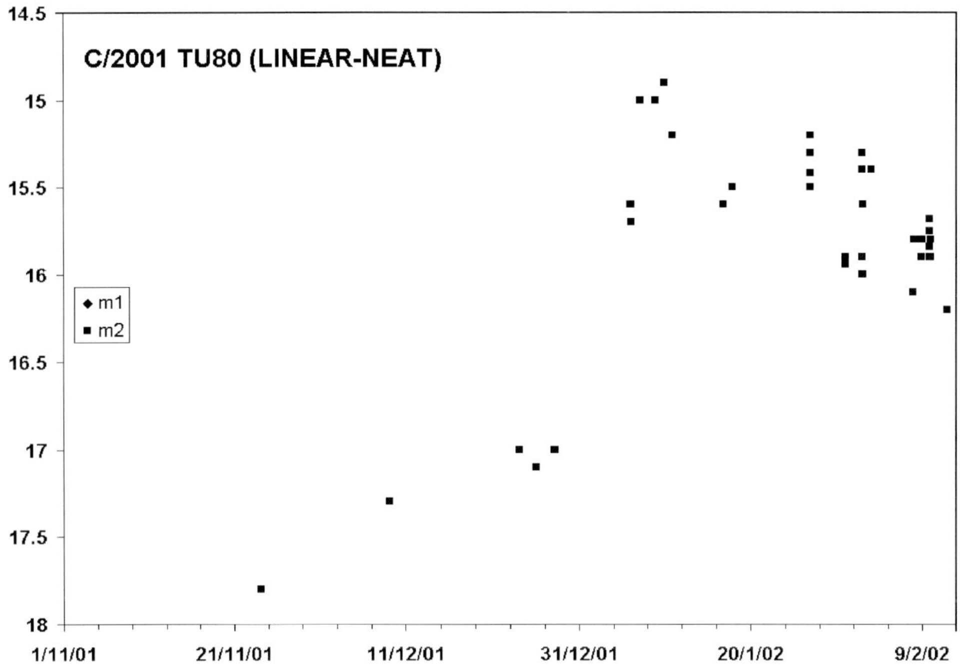

Figure 3. The light curve of C/2001 TU80 (LINEAR-NEAT) taken from the "Cometas_Obs" database. Most of the dispersion is due to some active observers working with unfiltered CCDs that are close to V rather than to the more standard R.

latter being due to the split in the data into magnitudes measured as *V* and *R* that are not separated here.

An example of the improvement of the light curve quality when standardised data is used is found in the case of 51P/Harrington. The light curve taken using the standardised method (see Figure 4) shows a standard deviation $\sigma = 0.17$ magnitudes. In contrast, the data taken from the MPC database (see Figure 5) shows a very much larger dispersion ($\sigma = 1.04$ mags).

Given the much reduced dispersion in the data it is possible to detect light curve structures on a finer scale than otherwise. The "Cometas_Obs" database has over 700 observations of C/2001 A2 (LINEAR). It is obvious from the data that the rate of decline in brightness showed significant variations between mid- July and November 2001 with a suspicion of a posible periodicity of 25 ± 4 days.

It should be noted though that CCD photometry continues to suffer problems in one particular case. When the comet to be measured is particularly diffuse and the central condensation is poorly defined, the dispersion in the data remains much larger than is desirable. A particular case of this is seen in the case of C/2000 SV74 (LINEAR), or in 19P/Borrelly (see Figure 6). Here the dispersion is approximately double what it would be in a more favourable case of a comet with a highly condensed inner coma. Even in this case though, apart from a few points in August

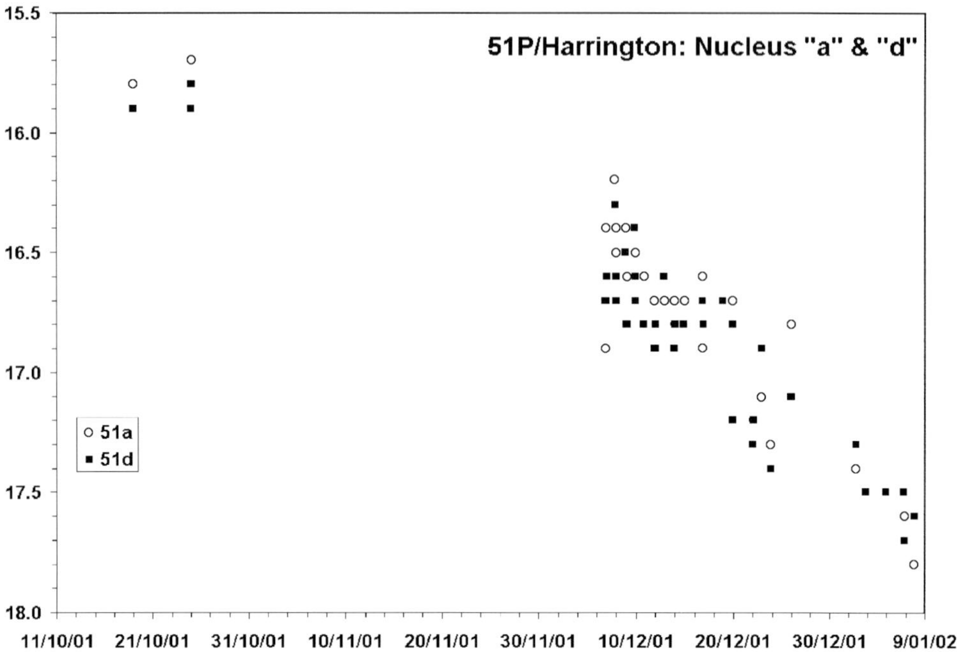

Figure 4. The light curve of nuclei *a* and *d* of 51P/Harrington taken from the "Cometas_Obs" database. The standard deviation of the light curve is <0.2 magnitudes.

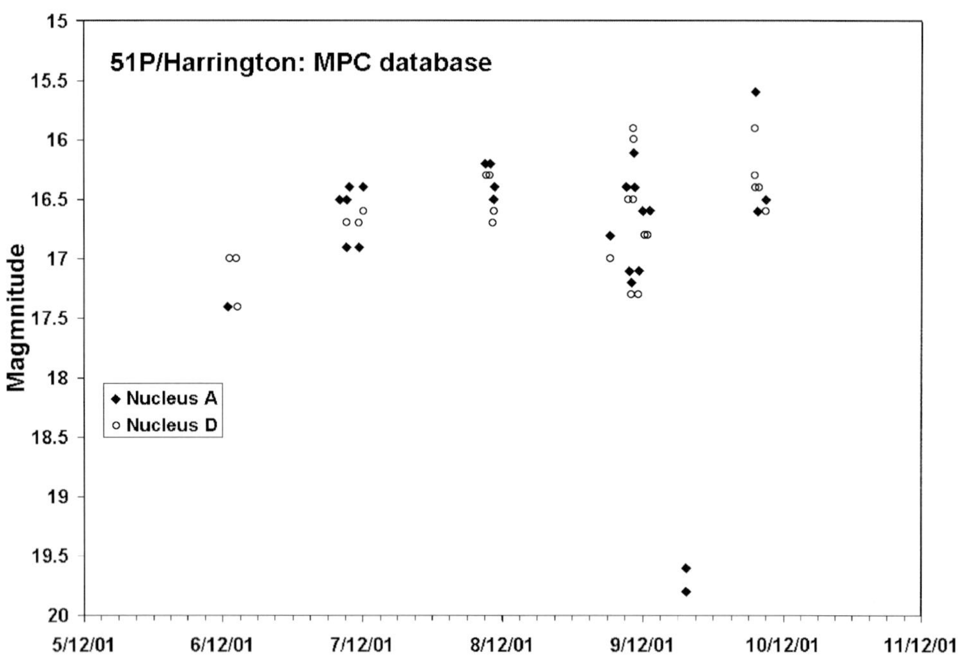

Figure 5. The light curve of nuclei *a* and *d* of 51P/Harrington taken from the MPC database. Even ignoring the two outliers that represent genuine nucleus magnitudes the dispersion of the data is 1.5 magnitudes.

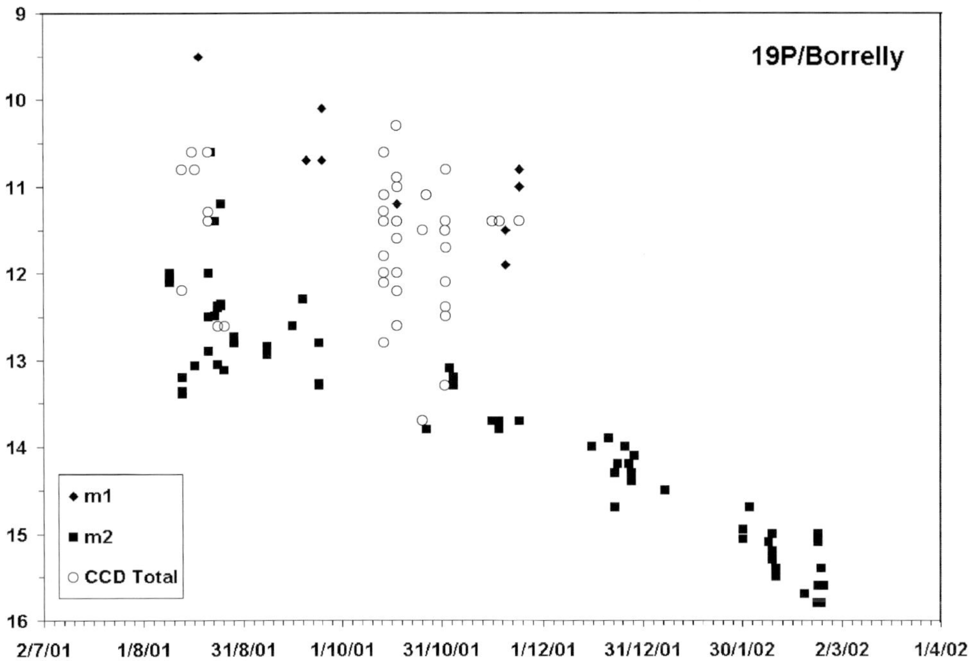

Figure 6. The light curve of 19P/Borrelly taken from the "Cometas_Obs" database. This much less centrally condensed comet shows the limitations of a relatively small aperture for photometry.

2001, the data is relatively well behaved. Photometrically we see that 19P/Borrelly showed a very simple light curve with no unusual activity observed, either at the time of the Deep Space 1 encounter, or at any other epoch between 2001 August and 2002 January.

6. Discussion

Ideally, if we are attempting to standardise cometary photometry, we might want a standard aperture that represents a constant physical radius at the comet, but this is impractical for two principal reasons:

1. Many comets are covered over a wide range of r and thus Δ. An aperture adapted to photometry of a distant comet will become inconveniently large when the comet is closer to the Earth.
2. During the initial period, usually a few days, when the comet's orbit is unknown (i.e., no orbit has been published) or, for very distant objects, highly uncertain it is impossible to use a standard aperture. In some cases photometry would have to be remeasured as the orbit is refined. This would cause significant inconvenience to observers and databasing problems as "old" photometry becomes incorrect and has to be either revised or removed completely.

A possible alternative, suggested in consultation with the MPC (B. G. Marsden, private communication), that has only been attempted in a limited way to date is to define, for example, a minimum of three apertures of differing sizes (e.g., 10, 30 and 60-arcseconds) and, assuming an exponential coma brightness law, interpolate to a given physical aperture. Although this method is seen to be promising in the trials made to date, due to the time-consuming nature of this process, such a refinement would presumably need to be incorporated into astrometry packages, allowing the calculations to be made automatically, if it is to become widely accepted in the astronomical community. Whilst such options appear difficult or impossible we suggest that the compromise proposed here is probably the best available to observers at present.

7. Conclusions

We have shown that amateur CCD photometry of comets is a potentially highly valuable resource for the study of their light curves and, from there, the production rates of different species. We have suggested a series of measures to standardise CCD photometry of comets which, although not ideal, at least make a considerable improvement in the quality and consistency of the light curves that can be produced from the data. A possible extension of this standardisation is suggested, although this is probably an impracticable step at present as it would require collaboration from the manufacturers of the most commonly used astrometric software to adapt their products to the needs of the comet-observing community. By taking appropriate measures the potential of the enormous amount of high-quality CCD photometry that is available can be realised.

References

Jeffers, H. M., Vasilevskis, S., and Roemer, E.: 1954, *AJ* **59**, 305.
Kamel, L.: 1992, *A&AS* **92**, 85.
Roemer, E.: 1956, *AJ* **61**, 391.
Tancredi, G., Fernandez, J. A., Rickman, H., and Licandro, J.: 2000, *A&AS* **146**, 73.

MEASURING THERMO-MECHANICAL PROPERTIES OF COMETARY SURFACES: *IN SITU* METHODS

NORBERT I. KÖMLE, GÜNTER KARGL
Space Research Institute, Austrian Academy of Sciences, Schmiedlstrasse 6, A-8042 Graz, Austria
KARSTEN SEIFERLIN
Institut für Planetologie, Westfälische Wilhelms-Universität Münster, Wilhelm-Klemm-Strasse 10, D-48149 Münster, Germany
WOJIECH MARCZEWSKI
Space Research Centre, ul. Bartycka 18A, Warsaw 00-716, Poland

(Received 9 March 2002; Accepted 20 May 2002)

Abstract. Thermal and mechanical properties of cometary ices are closely associated with each other. Both are largely determined by the texture (porosity, grain size distribution, grade of sintering) of the material. The surface probe of the Rosetta mission to comet 46P/Wirtanen (Rosetta Lander) will for the first time measure these thermo-mechanical key parameters *in situ*, using a hammering device, a drill, and anchors to be shot into the ground by pyrotechnical means. Several of these components are associated to the experiment MUPUS (*MUlti-PUrpose-Sensors for surface and sub-surface science*). The development of this instrument has now reached a mature state, as the flight model is already delivered and integrated with the spacecraft. We describe the main aspects of the experiment, outline the evaluation methods, and show representative results from test measurements.

Keywords: Comets, Rosetta Lander, MUPUS

1. Introduction

Cometary nuclei are thought to be composed of a complex mixture of ices with different volatility, organic compounds, and dust particles. Apart from the chemical composition, the physical behaviour of this material in response to irradiation by the Sun is highly dependent on the texture which develops by the thermal history of the near-surface layers (Mekler and Prialnik, 1991; Espinasse, 1991; Kührt and Keller, 1994; Seiferlin et al., 1995; Kossacki et al., 1997; Capria et al., 2002). Texture as well as bulk porosity control key material parameters like the effective thermal conductivity and the material strength. Both may vary over a large range (orders of magnitude) in comets and therefore comet modellers are frequently faced with the problem to choose reasonable values in order to make realistic predictions.

Up to now the only way to determine values is to perform laboratory tests on comet analogue samples in a cryogenic vacuum environment. Such experiments have been performed extensively in the frame of the KOSI (comet simulation) program in Germany (Grün et al., 1991) and in other smaller laboratories (e.g., Kömle et al., 1996). However, while the results obtained from these experiments were certainly helpful for cometary modeling, they nevertheless should be used with caution. As the real texture of cometary near surface material is still not known, measured parameters derived from laboratory tests with analogue samples could be arbitrarily wrong. Only *in situ* measurements on a cometary surface (probing at least several decimetres into depth and best at many places!) can provide really reliable parameter values.

A first step into this direction will be undertaken by the experiment MUPUS as well as by several other experiments to be flown aboard the *Rosetta Lander* spacecraft. *Rosetta Lander* is a major part of ESA's cornerstone mission ROSETTA to comet 46P/Wirtanen. It will land on the surface of the nucleus and perform chemical and physical investigations at the landing site. (See Biele et al., 2002, for a description of the other *Rosetta Lander* experiments.) This paper focuses on MUPUS, describes its main goals and components, and the development work done from the early beginnings to present.

2. Thermal and Mechanical Properties: Key Parameters for the Evolution of Cometary Activity

Since the only or at least the by far dominating energy source of comets is incident sunlight, the most interesting phenomena occur where the available energy is distributed among various processes at the surface or at shallow depths below the surface. These layers are in principle accessible, and are studied by MUPUS. Here, energy is consumed by re-radiation in the IR, warm-up of near-surface layers, sublimation of water ice and other species, chemical reactions, and finally sintering and Oswald ripening of ice grains. Sublimation processes are responsible for a number of interesting and characteristic features of comets, like (a) inward flow of gas and, as a consequence, chemical differentiation and densification; (b) outward flow of gas and trapped or dragged dust particles; and (c) dehydration and total dessication of the uppermost layer, which is turned into a dry dust mantle of low cohesion.

All of the above processes modify the structure of the uppermost layers, and leave the material at the very surface in an unknown, but certainly not pristine state. Instruments like evolved gas analysers, gas chromatographs or mass spectrometers have access only to more or less modified material. The activity of comets, characterised by the formation of a coma with gas and dust emissions from the surface, is only a consequence of the processes occurring in the shallow subsurface layers. The composition of the coma strongly depends on the depth of the sublimation fronts of various molecular species, and almost never reflects the bulk composition of the nucleus.

Keeping this in mind, it is evident that remote sensing data from the *Rosetta Orbiter* and from Earth-bound observatories – and to some extent even data recorded by lander instruments *in situ* – can only be interpreted correctly if the energy balance of the comet is understood. Two key parameters to be measured as a function of depth and time are the temperature profile and the thermal conductivity. The MUPUS penetrator aims to measure these two quantities.

The third parameter to be measured is the mechanical strength of the material and its variation with depth. Such variations of the material strength are predicted by theoretical models describing the thermal history in the near-surface layers (see references given in the Introduction), because the sintering processes mentioned above usually lead to cementation of originally loose icy particles. This often results in an increase of the material strength by orders of magnitude within very short distances, and thus in the formation of strong local ice crusts. Especially the ices more volatile than water ice (e.g., CO, CO_2) tend to form extremely hard crusts, as was demonstrated in laboratory experiments (Kossacki et al., 1997).

Another possibility is the formation of cohesive non-volatile crusts with low permeability to gases subliming below, which might form due to glueing of non-volatile mineral particles by organic macromolecules (Kömle et al., 1996). It is well-known that the structure of such crusts strongly influences both the thermal behaviour and the gas and dust output from a cometary surface – in other words – the "cometary activity". Knowledge of the strength distribution is at least one important item to understand this phenomenon.

In order to measure the strength profile MUPUS makes use of another subsystem of the *Rosetta Lander*, namely the anchoring projectile that will be shot into the cometary soil after landing in order to ensure save operation of various instruments and to avoid that gas drag – due to sublimation of surface ices and/or torques caused by mechanical operation of instruments (drill, penetrator, APX, cameras, etc.) disturb the stability of the lander as a whole. For this purpose a shock accelerometer is implemented into the tips of the anchors. It will measure the deceleration history in the soil, from impact at the surface until the anchoring projectile comes to rest in a particular depth. To obtain a strength profile from the measured deceleration signal is not a trivial task. It demands careful calibration, an exact knowledge of the anchor geometry, and an appropriate theoretical model. Several theoretical approaches to solve the "penetration equation" for a cylindrical body with a conical tip have recently been published (Kömle et al., 2001; Johnson, 2001; Frew et al., 2001). The inversion of the full penetration equation (as given, e.g., by Johnson, 2001) depends – apart from the strength parameter – also on the dynamic drag and on skin friction along the walls of the penetrator/anchor. Therefore for a sound quantitative interpretation of the deceleration signal in terms of material strength additional laboratory experiments with well characterized samples will be necessary. On the other hand, differences in strength, if they are substantial, will be much easier to be detected. Because, as outlined in the Introduction, such discontinuities in strength are expected at phase boundaries, there is a high probability to find this layering with the dynamic penetrometry experiment described here. Additional information concerning the strength distribution with depth might be obtained from the operation of the MUPUS penetrator, simply by counting the number of hammer strokes necessary to insert it to a certain depth in the soil.

3. The Experiment MUPUS

MUPUS consists of several components, shown in Figure 1. The main part is the so-called penetrator (MUPUS-PEN), a thin hollow tube of glass fiber with a titanium cone at the tip and a hammering device at the upper end. It is shown in stowed (b) and in deployed position (c). (a) shows the first (main) anchoring projectile along its path through the various soil layers. Atop of the hammering device a front-end electronics including an analog-to-digital converter is mounted, which is designed to survive under cold conditions. The interior wall of the penetrator tube is covered by a kapton foil partially coated with a thin film (0.5 micrometer) electrically conducting layer of titanium. Titanium is used instead of the more common platinum temperature resistors for technological reasons (Gregorczyk et al., 1999). Figure 2 shows development models both of the penetrator (thermal probe) with the temperature sensors and of the harpoon/anchoring device.

The thermal probe can be used in two different modes: a "passive" mode, and an "active" mode. In the passive mode, the 16 temperature sensors are used like conventional RTD-type sensors (e.g., PT 100), and the result of such a measurement is a temperature profile given by 16 data points located along a vertical line, if the probe is deployed correctly. Each value is averaged over the length of the corresponding sensor (1 cm at the top to 4 cm at the maximum depth of 32 cm). These data sets are typically obtained every few (1 to 15) min.

In the active measurement mode, the 16 sensors are employed as heaters by operating them with a rather high average power. At the same time, the heated sensors are still used as temperature sensors. This allows us to measure the self heating of each sensor. The observed temperature increase is a function of the known average dissipated power and of the unknown thermal diffusivity of the medium that encompasses the cylindrical probe at the level of the heated sensor.

The concept described above is very similar to the commonly used "line heat source method" (LHS), also known as "hot wire method", "needle probe" and "thermal conductivity probe" (see e.g., Seiferlin et al., 1996). The main difference is that MUPUS uses torus-shaped heated segments, while the LHS employs an "infinitely" (or sufficiently) long and thin cylindrical heat source, e.g., a wire or needle. The LHS has the big advantage that – because of approximations that are a consequence of cylindrical geometry – the thermal *conductivity* can be determined directly, while in the MUPUS case only the thermal *diffusivity* is measured; the *conductivity* can only be computed with *a priori* information about density and specific heat. The LHS cannot be used with MUPUS because several conditions cannot be met. The two most important conditions that are violated are: (a) *Homogeneous medium* and (b) *isothermal medium*. The result of an LHS measurement

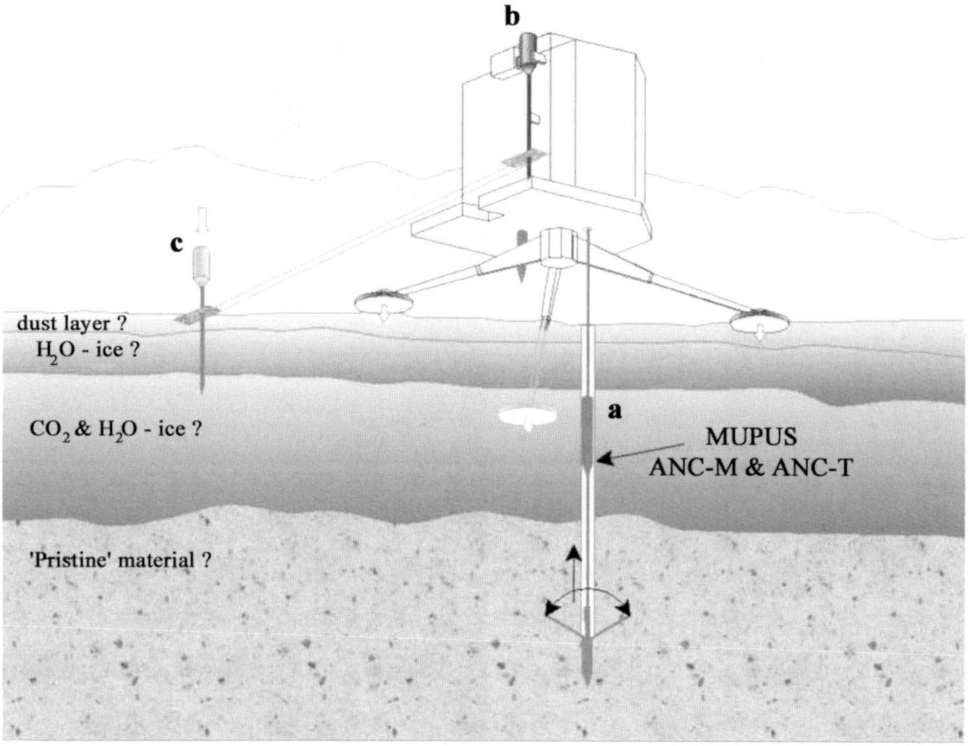

Figure 1. Components of MUPUS mounted on the Rosetta Lander: (a) Anchor during and after harpoon shot; (b) and (c): Penetrator in stowed and deployed position, respectively. The box to the left of (b) represents the thermal radiation sensor TM. The vertical cross section of the comet ground shows the expected "layering" as described in the text.

would be only *one* value for the average thermal conductivity of the uppermost 32 cm. In the MUPUS case, we heat each sensor separately. Thus we are able to measure the thermal diffusivity and derive the thermal conductivity at 16 different depth levels.

In combination with the temperature profile we can determine the heat flow across 15 depth levels (i.e., at the boundary between two adjacent sensors), and finally derive the energy balance for each of the 16 depth levels as a function of time. Depending on the duration of the whole mission, we hope to observe the depth level (and possible progression penetration) of sublimation fronts of water ice, CO–ice and other species and possibly even follow the "motion" of these discontinuities. We also hope to see layers of different thermal conductivity, indicating different structures like (a) dry dust mantle (low thermal conductivity, low cohesion), (b) hard sintered crust and (c) soft and rather weakly modified layers.

An important supplement to the subsurface temperature measurements performed by the sensors in the penetrator is MUPUS-TM, shown as a box beside position (b) in Figure 1, but not discussed in detail here. This sensor measures the

a b c

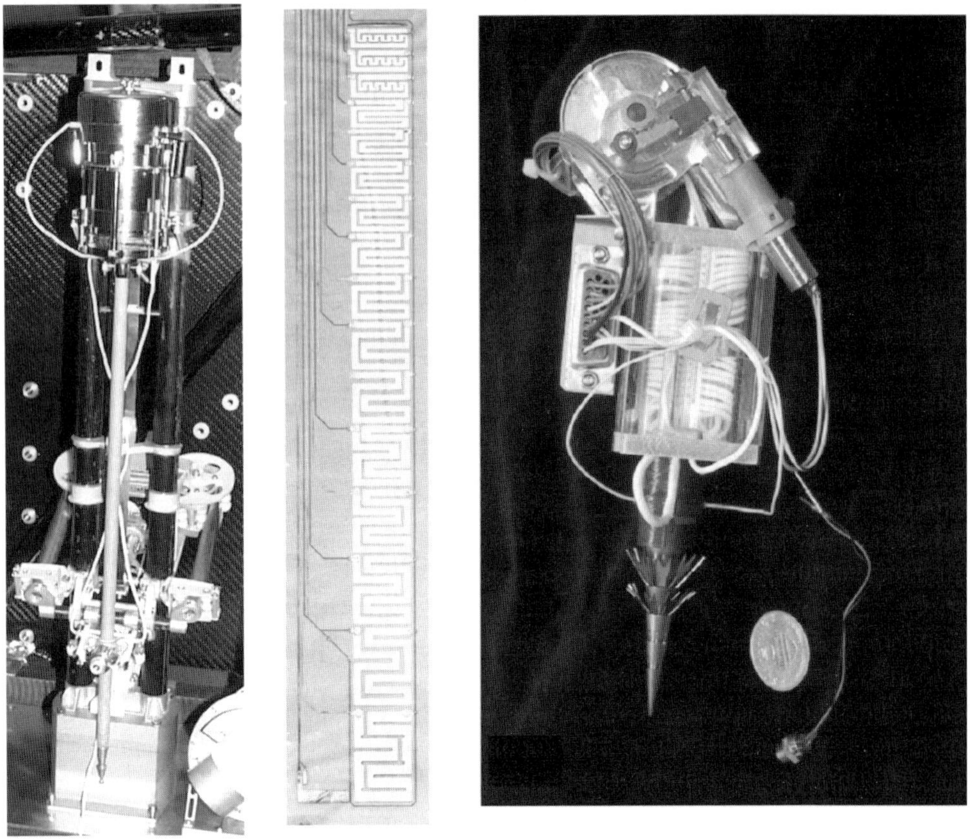

Figure 2. Main components of the MUPUS experiment: (a) Penetrator with tube and hammering device, (b) the temperature sensors integrated in the tube of the penetrator, consisting of a kapton foil carrying 16 titanium conductors in a row each between 1 and 4 cm long, 3 cm wide and $< 1\ \mu$m thick. Copper conductors on the same kapton carrier replace conventional cabling to the electronics. The kapton foil is rolled and glued to the inner wall of the penetrator tube. (c) Anchoring device with integrated shock accelerometer; the accelerometer is shown separately together with a "historical" 5 ATS coin to have an impression of the scale.

temperature of the surface itself by detecting the IR-flux within its field of view. This measurement is of particular importance if it turns out that a thin dust mantle covers the cometary ice, because it is known from theoretical models that in this case a very steep temperature gradient should exist at the surface, due to the low conductivity of the dust layer (see e.g., Steiner and Kömle, 1992).

The third component of MUPUS (MUPUS-ANC, Figure 2c), are sensors implemented in the two anchors of the *Rosetta Lander* spacecraft. Two types of sensors are incorporated: ANC-T is a temperature sensor monitoring the local temperature at the final "resting place" of the anchor, once it has been shot into the cometary soil. Thus it provides an additional temperature measurement, laterally displaced from the measurements of the MUPUS-PEN sensors, and probably from a larger depth. The second sensor is a miniature shock accelerometer monitoring the acceleration and deceleration of the anchoring projectile while it is fired into the ground. The acceleration will be performed pyrotechnically within a few milliseconds to reach an impact speed of approximately 50 ms^{-1}. A peak acceleration value of about 90000 ms^{-2} (i.e., ca. 9000 g) is expected. After a very short free flight period the anchor starts to penetrate the soil and is decelerated by the resistance of the cometary material. The main challenge concerning the selection of an appropriate sensor is the high dynamic range it has to cover. On the one hand it has to survive the high amplitude acceleration phase – but on the other hand it should be able to resolve the probably much smaller deceleration signal expected when the anchor penetrates soft porous ice layers with low cohesion. The chosen sensor type compromises these stringent requirements. It allows to measure the full acceleration phase (which is also important for calibration purposes) while still giving a reasonable signal for sintered porous ices, e.g., the typical water ice crusts observed in comet analogue samples.

4. Field and Laboratory Tests

During the several years development phase for the MUPUS experiment and the anchors extensive tests were performed both with the penetrator and with the anchors. The penetrator can be operated in several modes, which differ (a) in the power of a single hammer stroke and (b) in the stroke frequency. In most penetrator insertion tests the weight of the penetrator was compensated for, in order to take into account the zero gravity conditions at the comet. The tests can be roughly subdivided in two categories:

1. Thermal conductivity and temperature profile measurements after insertion of the probe into different samples. Temperature measurements could be done in a continuous mode, while active thermal conductivity measurements should be repeated from time to time, in order to monitor structural changes of the material that might occur as a consequence of ice sublimation and sintering.
2. Strength measurements obtained by firing the anchor into different materials and interpreting the deceleration signal recorded by the shock accelerometer.

In the following we present selected results, in order to demonstrate the ability of the probe to fulfill its tasks at the surface of comet 46P/Wirtanen.

4.1. Temperature Profiles

Figure 3 shows two test measurements performed with the MUPUS penetrator. The black horizontal lines mark the sensor boundaries (the 16 sensors are closely packed and leave only a thin gap between them). Temperatures are interpolated between the centers of two adjacent sensors, thus inverting the process of averaging performed implicitly by the sensors themselves. This data processing of course suppresses short-wavelength features in the temperature profile that would violate Nyquist's theorem, e.g., local anomalies small compared to the sensor length, but reproduces the long–wavelength temperature profile quite well, as theoretical considerations and numerical simulations show.

The probe structure, including the metal conductors from the sensors to the electronics, transfers some heat. This inevitable effect disturbs the temperature profile to be studied. However, the MUPUS penetrator makes use of several design features to minimize this problem: for example, no conventional cables are needed; instead, very thin copper conductors are printed onto the kapton carrier that holds the titanium sensors. The extremely small copper cross sections reduce the heat flow compared to even thin cables by orders of magnitude. The probe structure itself is hollow, again minimizing the cross section, and is made of a glass fiber composite with rather low thermal conductivity.

Figure 3a shows an example data set, measured with a MUPUS–type thermal probe and a conventional data logger in a field test in Poland. The duration of the test was about two days, starting in the morning of a rather cloudy day, followed by a cold night, a warm day and a warm night. The resolution of the MUPUS flight electronics is lower (ca. 13.5 significant bit) compared to that of the data logger (15 significant bit). However, in the figure only 8 bit information (i.e., 256 colours) is used, which is even less than that of the MUPUS flight model. The main feature in the data is the propagation of the diurnal heat wave, that is damped and phase shifted with depth. Close to the tip of the penetrator diurnal temperature variations are almost invisible. The fastest and largest temperature variations occur, as one would expect, near the surface. This characteristics is the main argument for the geometry of the sensors, which are short at the top and long at the bottom. Another feature is clearly visible in the field tests, but will be absent on the comet: very rapid temperature changes observed in all layers almost instantly and creating a vertical pattern in the data structure. This pattern is much weaker, or even absent, during night time. Obviously these temperature changes are triggered by rapid insulation variations at the surface, and the heat is transported via air convection inside the penetrator tube. A second test at the same location, performed immediately after the one shown here – but with sand filled into the tube – shows remarkably smoother temperature data. On the comet the tube will be "filled" with vacuum and therefore this problem should not occur.

4.2. Thermal diffusivity

In this instrument mode, the 16 temperature sensors are used as heaters (active mode). A typical result is shown in Figure 3b. This data set was acquired during the second thermal vacuum acceptance test for the MUPUS flight model, where the environmental temperature in the vacuum chamber was around 22 °C. The penetrator was mounted on a special rack, and the tube was not surrounded by a

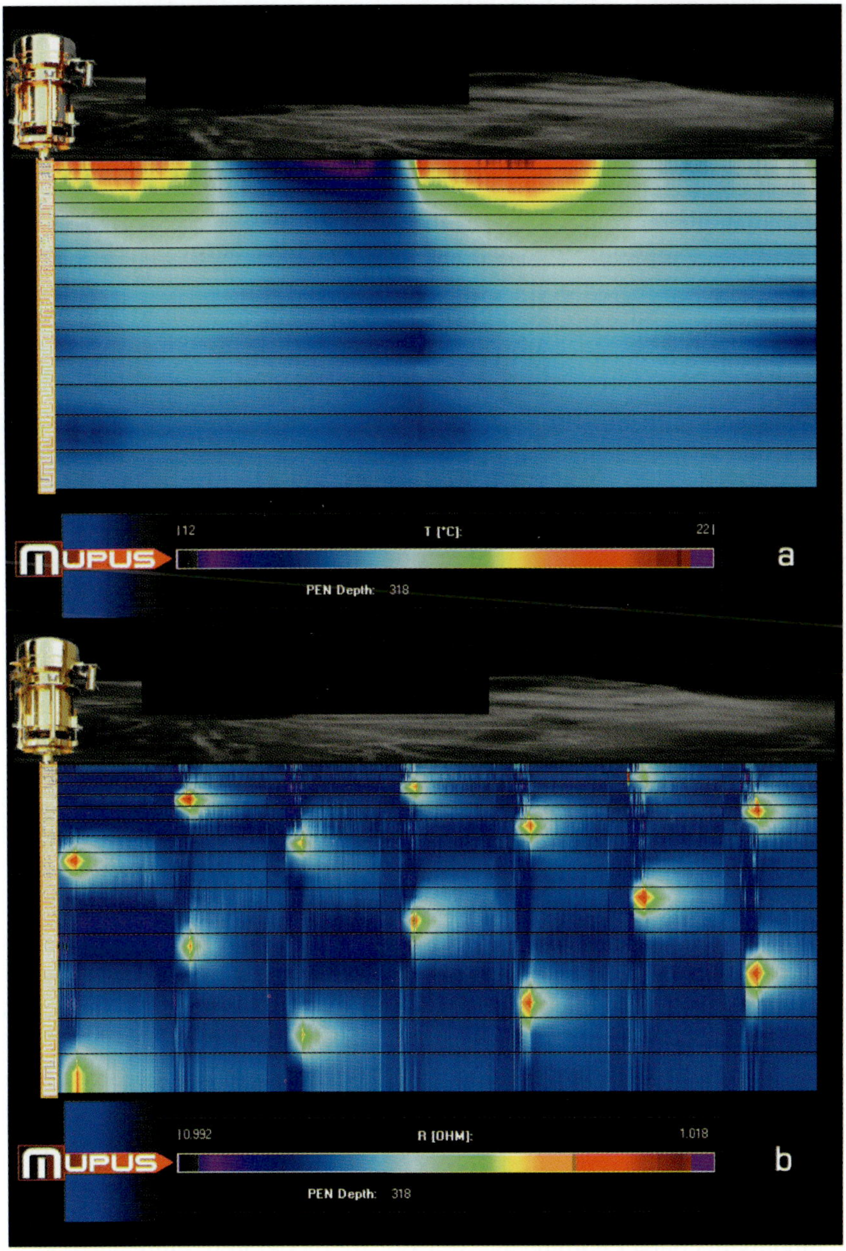

Figure 3. Laboratory and field tests with the temperature sensors implemented in MUPUS-PEN. In both panels, temperatures (or sensor resistances, respectively) are color coded and shown as a function of time (horizontal) and depth (vertical). The scale is shown below each data set. On the left hand side the penetrator is shown for comparison. Black horizontal lines mark sensor boundaries. (a) Measured temperature evolution in natural soil as a function of time and depth; total duration: 46 h; sampling interval: 60 sec. (b) Uncalibrated raw data obtained during the second flight model thermal vacuum acceptance test. The "temperature" is shown in terms of normalized sensor resistance values.

sample, i.e., heat exchange between the sensors in the penetrator and the environment could only take place by radiation and via the penetrator structure itself. The sensors were heated with the default settings, but with two sensors heated simultaneously. Therefore, one full cycle through all 16 sensors would take 8 steps, each one consisting of three basic phases: (1) Heating of the selected sensors, sampling interval "fast"; (2) cooling of the sensors, sampling interval "fast"; and finally (3) passive temperature measurements, as described in the previous section, for some time, with "long" sampling intervals. The typical duration of phases 1 and 2 was $\leq$ 10 min, whereas phase 3 could be anything $\geq$ 10 min.

The data set in Figure 3b shows one almost complete cycle (7 heating phases corresponding to 14 heated sensors). The effect of sensor heating can be easily seen. The figure also gives a nice impression of how the total energy input per depth region and time is optimized by the interlacing heating order. The data is shown in terms of sensor resistance normalized to the average sensor resistance, because the final calibration of the sensors is not yet completed. The maximum observed temperature increase induced by sensor heating was about 2%, which corresponds to approximately 5 K. For operations on the comet we must find a good compromise between two conflicting requirements: On the one hand, a large induced temperature increase allows us to derive the thermal diffusivity with a better accuracy than a small "peak", but – on the other hand – the total energy input into the comet should be kept reasonably low to disturb the natural temperature profile as little as possible. Since the heating power can be selected for each sensor individually, we are able to adjust the resulting temperature increase to about 5 K in each layer, even when the thermal conductivity of the medium differs by orders of magnitude. However, before useful measurements can be made and evaluated, a "first guess" for the settings is required. We plan to use several sources of information: remote sensing data provided by the orbiter together with thermophysical modelling, model fits of the temperature profiles measured with the MUPUS penetrator, and finally mechanical strength data provided by the anchor accelerometers together with empirical relationships between material strength and thermal conductivity.

Finally, Figure 3b demonstrates how well heat flow along the probe structure is suppressed: The sensors immediately above or below those which are actively heated "feel" the temperature increase in their vicinity quite tolerably; the observed temperature increase in these neighbouring sensors is typically 1/4 of the increase of the heated sensor. Since there are no gaps between sensors, it seems natural that the neighbours in contact with heated sensors are moderately influenced. However, sensors at a larger distance do not "feel" a significant effect. Taking into account that in this case the probe was not surrounded by a sample or medium that could conduct the heat effectively radially away, one can expect that on the comet the situation will be even better.

The remaining influence of the heat flow along the tube can be corrected by a sophisticated evaluation method solving an inverse heat conduction problem

(IHCP). Both – the hardware design optimization and the IHCP solution – are discussed in detail by Seiferlin et al., 2001), and the references given therein. We have studied temperature distributions on the comet using a one-dimensional numerical model, including heat transport via the vapour phase. The results of these numerical calculations were processed to simulate measurements with the MUPUS probe by averaging temperatures over the vertical length of the sensors, adding noise and reducing the resolution to 13.5 bit (see Seiferlin et al., 2002). For thermal parameters in the expected range, the results look similar to the field test data shown here.

4.3. Dynamic Penetrometry Tests

Figure 4 is a composite showing results of an anchor test performed with a prototype of the *Rosetta Lander* anchor. The projectile used had a diameter of 16 mm and a mass of 90 grams. It was shot into a sample of powdered CO_2-ice exhibiting a harder surface crust, which had formed by grain sintering in the uppermost 5–10 cm.

Figure 4a shows a part of the sample surface and two "shot holes" originating from anchor firing. At the impact sites small craters with diameters larger than the penetration channel are visible. In Figure 4b the deceleration profile measured by the shock accelerometer in the anchor is displayed, together with a fit based on a model calculation. In order to calibrate these data in terms of strength/cohesion, a quasi-static strength measurement was performed at a surface position close to the impact site. This measurement consisted of a slow (4 cm/s) penetration of a cylindrical rod with a 60° tip into the sample, whereby the resistance force was measured by a conventional force cell mounted at the top of the rod. The result of the quasistatic measurement is shown in Figure 4c, together with a simple approximation of the assumed strength profile. Phase shifts of the near surface strength and deceleration peaks are caused by differences in the tip geometry. By and large the main vertical structure of the layer seen in the quasistatic test, namely a strong surface crust and a slightly consolidated part below, is also well represented in the deceleration data. This demonstrates that the dynamic penetrometry method, together with an appropriate model of the tip geometry, should be able to detect at least variations and discontinuities in the vertical strength profile, when the anchors are fired into the cometary surface layers. A more detailed presentation and discussion on the penetrometry tests can found in recent publications by Kargl et al. (2001) and Kömle et al. (2001).

5. Summary

The MUPUS experiment described in this paper is a first, but important step towards the development of space proven, low weight equipment for measuring

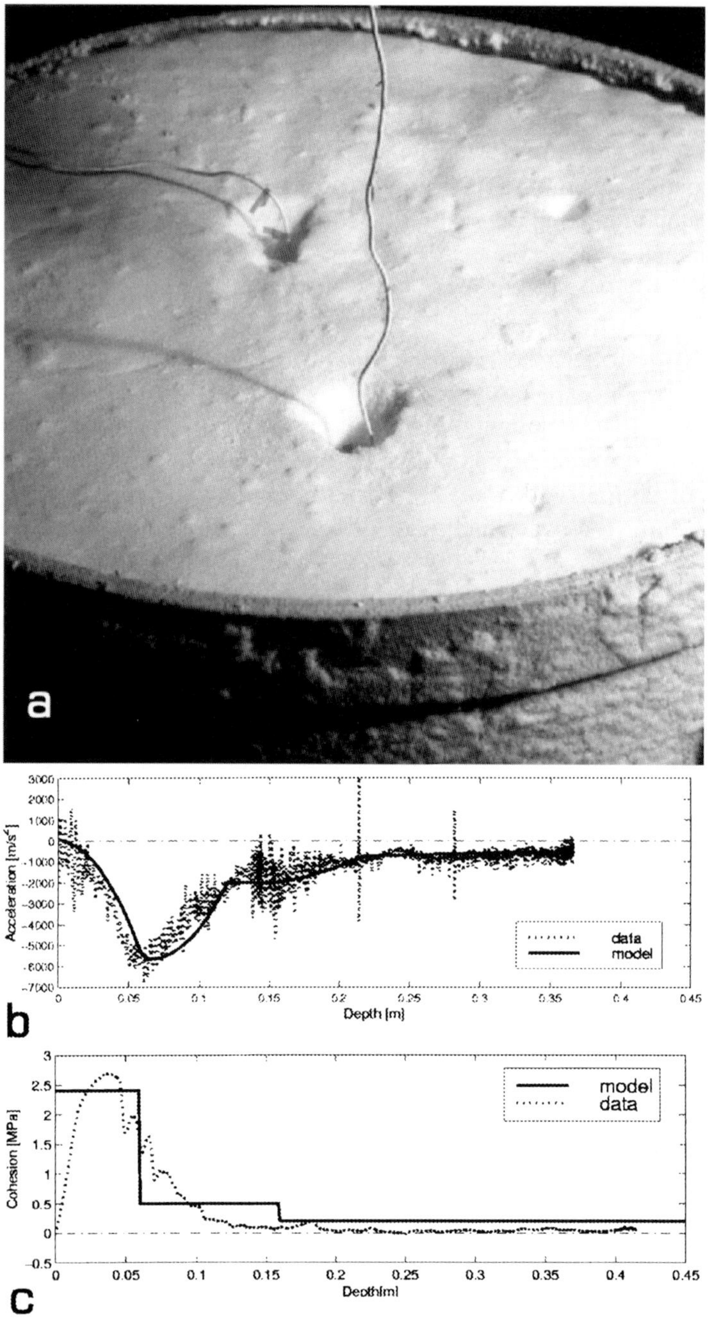

Figure 4. Example for anchor test shots performed at MPE Garching: (a) Surface of the sample used as a target, a sintered CO_2-crust with softer material below; the shot hole and the anchor cable are also visible. (b) The accelerometer signal recorded from the impact positioned close to the centre of the target. (c) Strength profile obtained by an independent quasi-static measurement with a conventional force cell.

thermo-mechanical properties of any planetary surface layer *in situ*. Apart from comets, it could be applied for lander missions to asteroids, the Moon, Mercury, Mars, or any of the icy moons of the outer solar system.

The field and laboratory test measurements demonstrate that the developed equipment is robust enough to survive both the harsh temperature conditions on the surface of a comet and to allow measuring thermal conductivity and the evolution of temperature profiles over a longer period. Moreover, tests with development models of the *Rosetta Lander* anchor have shown that it should be possible to detect at least major variations in material strength over the penetration path of the projectiles by using a shock accelerometer.

This possibility is particularly important for cometary nuclei, because we expect from theoretical modeling as well as from laboratory tests, that near–surface layers are strongly differentiated due to sublimation of volatile components. The associated sintering of icy particles leads to an increase of the material strength by orders of magnitude, and thus to the formation of strong local ice crusts. Especially ices more volatile than water ice (e.g., CO, CO_2) tend to form very hard crusts.

The results of the MUPUS experiment on the comet will for the first time provide *in situ* measurements on temperature evolution, thermal conductivity and material strength within those layers which control the "activity" of the nucleus. Together with the chemical analysis to be performed by the experiments COSAC and PTOLOMY and the acoustic sounding to be done by SESAME on the *Rosetta Lander* a quite complete characterisation of the landing site may be achieved (Biele, 2002). Nevertheless it must also be noted that even in the case of a full success we still have only a point measurement. Therefore for future lander missions to comets a more flexible concept (allowing similar measurements at many places of a comet nucleus) should be taken into consideration. For comets with larger mass, where anchoring to the surface is not such a severe requirement as in the case of 46P/Wirtanen, concepts like "surface hopping" might be re-considered. In addition to applications in future space missions, a MUPUS–type thermal probe is an ideal and until today unavailable tool for applications in snow and avalanche research and in geosciences. We are presently developing a system consisting of a thermal probe, field electronics and software (EXTASE: *EXperimental Thermal Probe for Applications in Snow Research and Earth Sciences*), designed for such terrestrial applications.

Acknowledgement

Apart from the coauthors T. Spohn and K. Seiferlin, who act as principal investigator and technical manager, respectively, it should also be mentioned that the MUPUS project is actually a joint effort of many people coming from various institutions. The authors would like to acknowledge their indirect contribution to the contents of this paper: Andrew J. Ball, Marek Banaszkiewicz, Johannes Benkhoff, Stanislaw Gadomski, Jerzy Grygorczuk, Marek Hlond, Günter Kargl, Jörg Knollenberg, Norbert I. Kömle, Konrad Kossacki, Jacek Krasowski, Ekkehardt Kührt, Mark R. Leese, Jacek Leliwa–Kopystynski, Wojciech Marczweski, Jerzy Roman, John C. Zarnecki. K. Seiferlin is supported by DLR grant RD-JX/50 QP 9709 4–ZA; N.I. Kömle and G. Kargl acknowledge support by the Austrian *Fonds zur Förderung der wissenschaftlichen Forschung* under project P15470.

References

Biele J.: 2002, these proceedings.

Capria, M. T., Coradini, A., and De Sanctis, M. C.: 2002, these proceedings.

Espinasse, S. J., Klinger, J., and Schmitt, B.: 1991, *Icarus* **92**, 350–365.

Gregorczyk, W., Jancewicz, B., and Marczewski, W.: 1999, in *Proceedings of the 23rd IMAPS Conference*, pp. 271–276.

Frew, D. J., Forrestal, M. J., and Hanchak, S. J.: 2001, in N. I. Kömle, G. Kargl, A. J. Ball, and R. D. Lorenz (eds.), *Penetrometry in the Solar System*, pp. 87–98.

Grün, E., Kochan, H., and Seidensticker, K. J.: 1991, *Geophys. Res. Lett.* **18**, 245–248.

Johnson, J. B.: 2001, in N. I. Kömle, G. Kargl, A. J. Ball, and R. D. Lorenz (eds.), *Penetrometry in the Solar System*, pp. 73–85.

Kargl, G., Macher, W., Kömle, N. I., Thiel, M., Rohe, C., and Ball, A. J.: 2001, *Planet. Space Sci.* **49**, 425–435.

Kömle, N. I. and Steiner, G.: 1992, *Icarus* **96**, 204–212.

Kömle, N. I., Kargl, G., Thiel, K., and Seiferlin, K.: 1996, *Planet. Space Sci.* **44**, 675–689.

Kömle, N. I., Ball, A. J., Kargl, G., Keller, T., Macher, W., Thiel, M., Stöcker, J., and Rohe, C.: 2001, *Planet. Space Sci.* **49**, 575–598.

Kossacki, K. J., Kömle, N. I., Leliwa-Kopystynski, J., and Kargl, G.: 1997, *Icarus* **128**, 127–144.

Kührt, E. and Keller, H. U.: 1994, *Icarus* **109**, 121–132.

Mekler Y. and Prialnik, D.: 1991, *Astrophys. J.* **366**, 318–323.

Seiferlin, K., Spohn, T., and Benkhoff, J.: 1995, *Adv. Space Res.* **15**(10), 35–38.

Seiferlin, K., Kömle, N. I., Kargl, G., and Spohn, T.: 1996, *Planet. Space Sci.* **44**, 691–704.

Seiferlin, K., Spohn, T., and the MUPUS Team: 2002, *ROSETTA ESA SP-1165*, in press.

Steiner, G. and Kömle, N. I.: 1992, *Icarus* **96**, 204–212.

IMPACT BREAK-UP OF COMETARY NUCLEI – CONCLUSIONS FROM IMPACT EXPERIMENTS

JACEK LELIWA-KOPYSTYŃSKI
Institute of Geophysics, Warsaw University, ul. Pasteura 7, 02-093 Warsaw, Poland
Space Research Center, Polish Academy of Sciences, ul. Bartycka 18A, 00-716 Warsaw, Poland
E-mail: jkopyst@mimuw.edu.pl

(Received 1 March 2002; Accepted 5 July 20002)

Abstract. Experiments of impact-generated break-up of icy and icy/mineral targets were performed. Formulae for the velocity of ejecta and for energy of disruption were fitted to the experimental data. An assumption that these formulae can be extrapolated for kilometer-size bodies enabled us to discuss the consequences of impacts on cometary nuclei and on planetesimals. It was found that the porosity of the targets as well as their composition (mineral to total mass ratio), are the crucial parameters.

Keywords: Cometary nuclei, ices, impacts

1. Experimental Background

We report here some planetological interpretations of the results of several impact experiments that have been performed in the Institute of Low Temperature Science of Hokkaido University (Arakawa et al., 2002). We studied the impacts on the porous targets made (a) of pure water ice, and (b) of water ice mixed with pyrophylite. The targets, with mass in the range 0.23–0.70 kg were prepared from the fine-grain sintered material. The row material was put into a cylindrical container, and it was pressurized until the required volume (therefore the required porosity) was obtained. Next, for temperature-driven sintering, the cylindrical targets were stored for several weeks in a cold room at -40 °C and atmospheric pressure. A target was impacted in the cold room at -10 °C by an impactor with mass 1.6 g. An impactor was accelerated in a vertical gas gun. The impact velocity v_i was in an interval 150–670 m s^{-1}. An experimental output for each shot consisted of high speed photo-camera registrations of the impact process. The post-impact mass distribution of the fragments was studied as well: We counted and weighted the recovered fragments. The recovered total mass was almost equal to the sum of initial mass of the target and the impactor ($\sim$99%).

The high speed records, 12 for each shot, allowed us to study the pattern of ejecta. The series of photos covered a time interval from 0 μs (time of impactor/target contact) to about 300 μs. We found that the maximum velocity of

Earth, Moon and Planets **90:** 283–291, 2002.

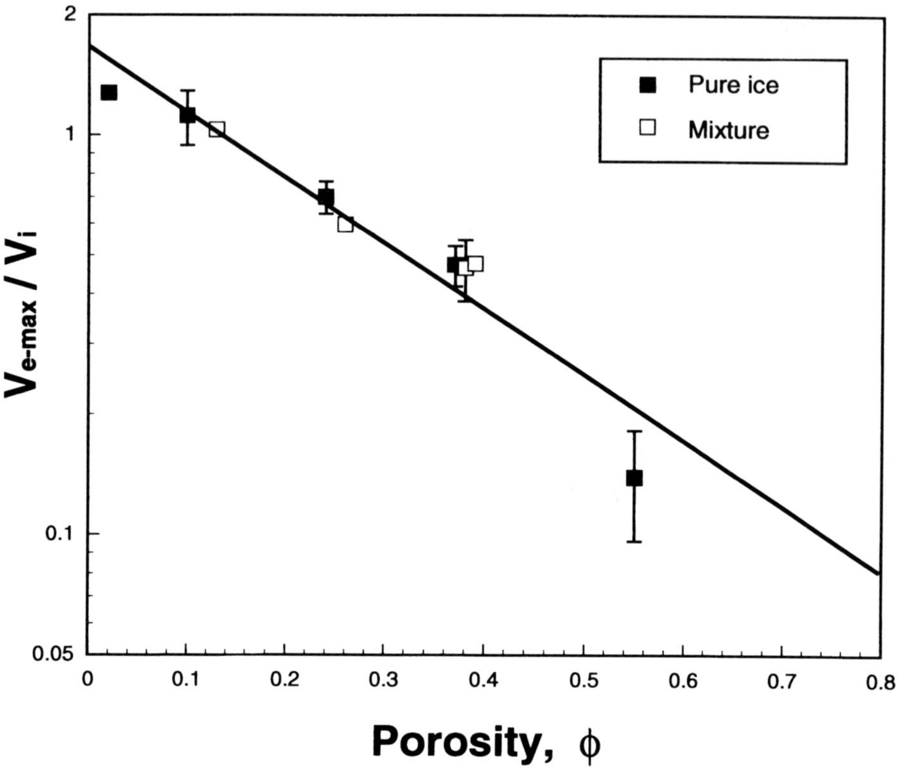

Figure 1. The porosity dependence of the maximum velocity of ejecta $v_{e-\max}$ normalized by the impact velocity v_i in pure ice and in the mixture target with mass ratio water ice to pyrophylite equal to one. Since there is no apparent difference between pure ice and mixture data, so the least-squares fit was applied to all data.

ejecta $v_{e-\max}$ versus impact velocity v_i and versus porosity ϕ of the target can be fitted as follows (see Figure 1):

$$v_{e-\max} = v_i \alpha 10^{-\beta\phi}, \tag{1}$$

where $\alpha = 1.66$ and $\beta = 1.65$. Formula (1) was fitted to the whole ensemble of data for pure icy targets as well as for the icy-pyrophylite mixtures with porosity 0–0.55. We note, that for low-porous targets ($\phi < 0.13$) some of the ejecta escape the target even faster than the impact velocity.

Let us define the specific energy Q (impact strength) required for impact generated destruction as the energy corresponding to such break-up of the target that the largest fragment has a mass equal to half of the initial mass of the target (Hartmann, 1979). For a comprehensive discussion of impact strength see Mizutani et al. (1990). Among other results derived from our experiments we have found that the specific energy Q depends on the target porosity ϕ:

$$Q = a(1 - \phi)^b. \tag{2}$$

This energy is considerably different for the icy targets and for the mixture targets. Indeed, the coefficients are: $a = 20.7$ J kg^{-1}, $b = -1.6$ for ice; and $a = 108$ J kg^{-1}, $b = 2.3$ for mixture. In our experiments, the mixture target denotes the composition ice/pyrophylite equal to 1 by mass or, equivalently, mineral mass to total mass is $C = 0.5$. The values of the coefficients b show that for the icy targets the impact strength increases with increasing porosity and for the mixture targets it decreases with increasing porosity.

Equation (2) can be generalized assuming that the coefficients a and b depend on C. Since we have only two points $C = 0$ and $C = 0.5$, we therefore fit the linear formulae

$$a = 20.7 + 174.6C \ \ [J\ kg^{-1}], \qquad b = -1.6 + 7.8C. \tag{3}$$

The ϕ-dependence of Q and C-dependence of b can be compared with the results of Davis and Ryan (1990). They studied the collisional disruption of the strong (low-porous) and weak (granular, high-porous) mortar targets. They concluded that the granular structure dissipates and absorbs a large fraction of collisional energy in the vicinity of the impact site rather than propagating it into the volume of the target.

In the following, the experimentally verified formulae (1)–(3) are applied to the discussion of the impact accretion and the impact destruction.

2. Accretion of a Projectile on the Target

Accretion of a projectile onto the target corresponds to the situation when maximum velocity of ejecta does not exceed the escape velocity from the target surface, $v_{e-\max} \leq v_{\rm esc}$. The latter $v_{\rm esc} = (GM/R)^{1/2}$, corresponds to escaping of ejecta to a circum-target orbit. The ejecta with lower velocity fall back onto the target. Since $v_{e-\max}$ is given by Equation (1) we have

$$v_{i-\rm acc}\alpha 10^{-\beta\phi} = (GM/R)^{1/2}$$

and thus the maximum velocity for accretion of an impactor on the target is

$$v_{i-\rm acc} = \alpha^{-1} 10^{-\beta\phi}(4\pi G\rho/3)^{1/2} R. \tag{4}$$

Here M, R, ρ are the mass, the radius, and the mean density of the target, respectively, and $G = 6.672 \times 10^{-11}$ m^3 kg^{-1} s^{-2}. The mean density ρ is related to the water ice density $\rho_{\rm ice}$, to the mineral density $\rho_{\rm rock}$, to the mineral concentration by mass C, and to the target porosity ϕ as follows

$$(1-\phi)/\rho = (1-C)/\rho_{\rm ice} + C/\rho_{\rm rock}. \tag{5}$$

The combination of Equations (4) and (5) gives the multiplicative formula

$$v_{i-\mathrm{acc}} = [\alpha^{-1}(4\pi/3G\rho_{\mathrm{ice}})^{1/2}] \times [1-(1-\rho_{\mathrm{ice}}/\rho_{\mathrm{rock}})C]^{-1/2} \times [(1-\phi)^{1/2}10^{\beta\phi}] \times R. \tag{6}$$

The successive factors are the constant value, the C-dependent factor, the ϕ-dependent factor, and the target radius R. We assume hereafter that $\rho_{\mathrm{ice}} = 940$ kg m^{-3}, and $\rho_{\mathrm{rock}} = 3000$ kg m^{-3}, and therefore

$$v_{i-\mathrm{acc}} = 3.09 \times 10^{-4} \times (1-0.6867C)^{-1/2} \times [(1-\phi)^{1/2}10^{1.65\phi}] \times R, \quad \mathrm{m\,s^{-1}}. \tag{7}$$

The C-dependent factor in Equation (7) only slightly increases, from 1 for $C = 0$ to 1.23 for $C = 0.5$ (typical composition of the cometary nuclei and our targets). On the other hand the ϕ-dependent factor increases considerably: from 1 for $\phi = 0$ (our targets) through 5.4 for $\phi = 0.55$ (our targets), to 9.3 for $\phi = 0.8$. Equation (1) corresponds to the range of experimental data $0 < \phi < 0.55$, therefore its extrapolation for very high porosity, e.g., for $\phi > 0.8$ or so, seems to be risky. The particular cases of Equation (7) are discussed in the following, simultaneously with the discussion of Equation (11) relevant to the impact-due disruption.

3. Impact-Due Disruption of the Target

Impact-due disruption of the target corresponds to that situation when the kinetic energy of the projectile with velocity v_{dis} (impact disruption velocity) is sufficient to break-up the target and, possibly, to disperse its fragments that are held together by mutual gravity:

$$0.5mv_{i-\mathrm{dis}}^2 = QM + 0.6fGM^2/R, \quad 0 \le f \le 1. \tag{8}$$

The right hand side of Equation (8) represents a sum of the disruption energy QM and the energy required for dispersion against gravity. The coefficient f depends on the mass distribution of the fragments:

$$f = 1 - \Sigma(m_i/M)^2(R/r_i) = 1 - \Sigma(m_i/M)^{5/3}. \tag{9}$$

The index i relates to the independent (separated) fragments. The case $f = 0$ corresponds to the situation when the fragments of the impacted and broken body are not dispersed at all. They remain together forming a rubble pile. On the contrary, the case $f = 1$ corresponds to gravitational energy of the infinitesimally small fragments that are dispersed away to infinity. In this case the gravity-dependent term in Equation (8) is equal to $0.6GM^2/R$. Any other value of the coefficient f is related to a particular mass distribution of the target fragments. For example, if as a result of an impact the largest fragment has mass equal to half of the target mass $M/2$ (note that the strength energy Q is defined for this mass of the largest fragment) and if the remaining fragments are infinitesimally small, then $f = 0.685$. For larger remaining fragments the coefficient $f < 0.685$.

From Equation (8) we can find the impact disruption velocity v_{dis}. Since $m = 4/3\, \pi r^3 \rho_{\mathrm{imp}}$ is the mass of an impactor assumed to be spherical (target is assumed spherical as well), we have:

$$v_{i-\mathrm{dis}}^2 = 2Q(\rho/\rho_{\mathrm{imp}})(R/r)^3 \left[1 + (0.8\pi f G\rho) R^2 f\right]. \tag{10}$$

The gravity-dependent term in Equation (10) increases as R^2. So, the increasing of the target size R leads to prevalence of the target self-gravity over its impact strength (see e.g., Holsapple, 1994; Melosh and Ryan, 1997). Since the impact strength Q is given by Equations (2) and (3) and the density of the target ρ is given by Equation (5), we can estimate the coefficient appearing in the self-gravity term:

$$0.8\pi G\rho/Q = 7.615 \times 10^{-9}(1 + 7.748C - 5.592C^2)^{-1}(1 - \phi)^{2.6-7.8C}. \tag{11}$$

It depends on the target composition C and on the target porosity ϕ, so we calculate it for those four pairs of (C, ϕ) parameters that correspond to the limits of our experimental data. The results allow us to calculate the values of the target critical radius R_{crit} that corresponds to the situation when self-gravity balances the impact strength:

$$(0.8\pi G\rho/Q) R_{\mathrm{crit}}^2 f = 1. \tag{12}$$

Following the discussion of Equation (9), we adopt $f = 0.5$ as the representative value of this coefficient. So, the particular values of R_{crit} are as follows.

For icy ($C = 0$) targets R_{crit} is 16.2 km ($\phi = 0$), and 45.8 km ($\phi = 0.55$), i.e., R_{crit} increases when porosity increases.

For mixture ($C = 0.5$) targets R_{crit} is 30.2 km ($\phi = 0$), and 18.0 km ($\phi = 0.55$), i.e., R_{crit} decreases when porosity increases.

Let us calculate the *minimum value of impact disruption velocity,* $v_{i-\mathrm{dis}}$. It is equal to the minimum velocity in the case of rubble pile formation, and therefore it corresponds to $f = 0$ in Equation (10) for $v_{i-\mathrm{dis}}$. (See Wilkinson et al., 2002 for the discussion of the rubble pile asteroids.) This equation with $f = 0$ and with Q given by Equations (2) and (3) reads as

$$v_{i-\mathrm{dis}}^2 = 2(20.7 + 174.6C)(1 - \phi)^{-1.6+7.8C}(\rho/\rho_{\mathrm{imp}})(R/r)^3. \tag{13}$$

Equation (13) shows that composition and porosity of the target are the crucial parameters for $v_{i-\mathrm{dis}}$, because of the factor $(1 - \phi)^{-1.6+7.8\,C}$.

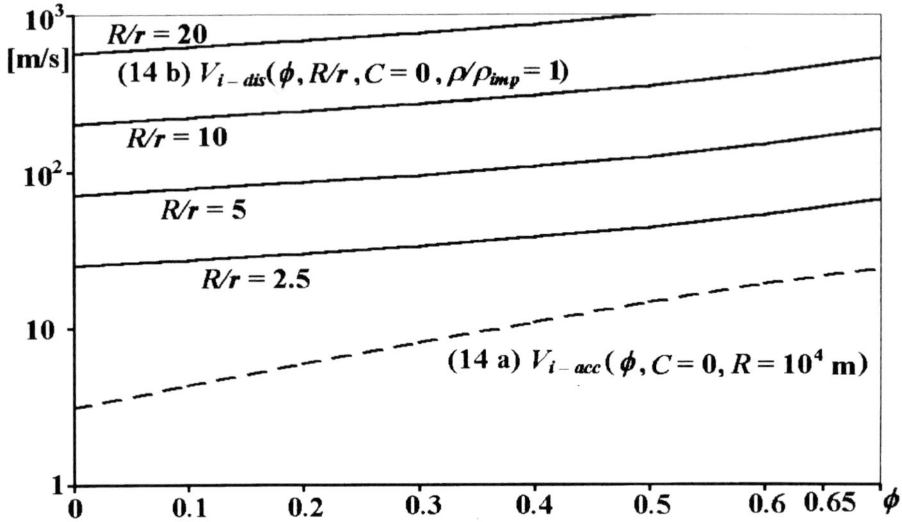

Figure 2. Plots of $v_{i-\mathrm{acc}}$ (dashed line) and $v_{i-\mathrm{dis}}$ (continuous lines) versus porosity ϕ for the icy targets, $C = 0$, according to Equations (14a) and (14b).

4. Discussion of the Results

An extrapolation of the results of the laboratory experiments onto the planetary scale of events is common in planetary sciences; nevertheless it bears a certain risk. An extrapolation of the size of the target from the laboratory scale to the cometary nucleus scale is spread out over at least three orders of magnitude. An extrapolation related to the impact energy scale is even much larger.

In our case the extrapolation concerns the ranges of validity of Equations (1)–(3). These equations were fitted to the laboratory experimental data and then they were applied to kilometer-size bodies to express the $v_{i-\mathrm{acc}}$, R_{crit}, and $v_{i-\mathrm{dis}}$ by means of Equations (7), (12), and (13), respectively. In the following we jointly apply formula (7) for impact accretion velocity $v_{i-\mathrm{acc}}$ and formula (13) for impact velocity $v_{i-\mathrm{dis}}$ of rubble pile formation. The velocities are in meters per second. Four cases of the target composition C are considered. Porosity ϕ is a parameter.

(a) A pure icy target, $C = 0$. We performed experiments for such a value of C. Application: For the *icy planetesimals.* Equations (7) and (13), illustrated in Figure 2, read as follows:

$$v_{i-\mathrm{acc}} = 3.09 \times 10^{-4}(1-\phi)^{1/2}10^{1.65\phi}R, \tag{14a}$$

$$v_{i-\mathrm{dis}} = 6.38(1-\phi)^{-0.8}(\rho/\rho_{\mathrm{imp}})^{1/2}(R/r)^{3/2}. \tag{14b}$$

(b) A mixture, $C = 0.205$ that corresponds to the particular case of $v_{i-\mathrm{dis}}$ being porosity-independent. This C is between our experimental values of $C = 0$ and $C = 0.5$. Application: For the *cometary nuclei with low contents of minerals.*

$$v_{i-\mathrm{acc}} = 3.39 \times 10^{-4}(1-\phi)^{1/2}10^{1.65\phi}R, \tag{15a}$$

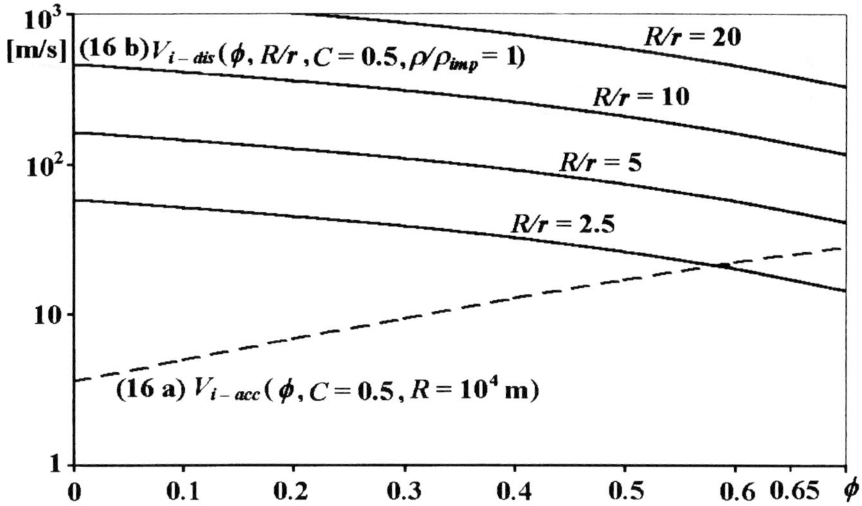

Figure 3. Plots of $v_{i-\text{acc}}$ and $v_{i-\text{dis}}$ versus porosity ϕ for the mixture targets, $C = 0.5$, according to Equations (16a) and (16b).

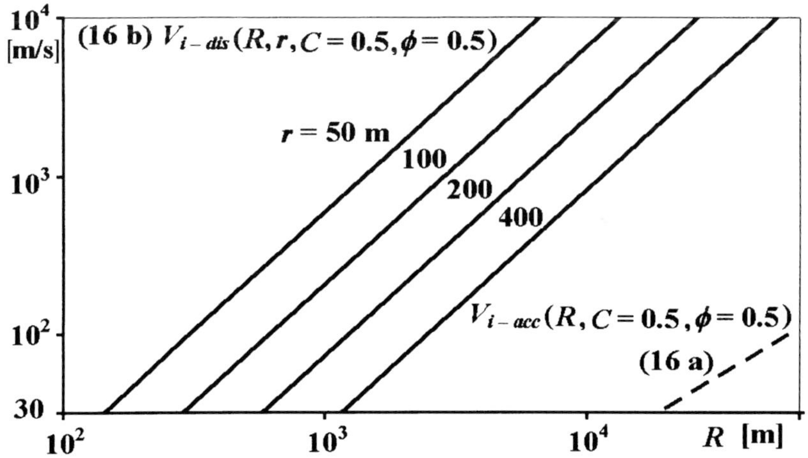

Figure 4. Plots of $v_{i-\text{acc}}$ and $v_{i-\text{dis}}$ versus target radius R according to Equations (16a) and (16b). The impactor radius r is a parameter. Mixture targets, $C = 0.5$, with porosity $\phi = 0.5$.

$$v_{i-\text{dis}} = 9.89(\rho/\rho_{\text{imp}})^{1/2}(R/r)^{3/2} \tag{15b}$$

Note, that "typical" density of the nuclei is about 500 kg m^{-3}. Equation (5) with ρ = 500 kg m^{-3} and $C = 0.5$ gives the porosity of the nucleus $\phi = 0.65$. See Figures 3 and 4 for illustration of Equations (16a) and (16b).

(c) A mixture, $C = 0.5$. We have experimental data for such a value of C. Application: For the *typical cometary nuclei.*

$$v_{i-\text{acc}} = 3.61 \times 10^{-4}(1 - \phi)^{1/2}10^{1.65\phi}R, \tag{16a}$$

$$v_{i-\mathrm{dis}} = 14.7(1-\phi)^{1.15}(\rho/\rho_{\mathrm{imp}})^{1/2}(R/r)^{3/2}. \tag{16b}$$

(d) A mineral target, $C = 1$. We have not our own experimental data for this C. So, the formula below holds provided Equation (3) is valid beyond $C = 0.55$. Possible application for two different classes of target bodies: For the *asteroids* (ϕ near 0) or for the *extinct cometary nuclei* (high ϕ).

$$v_{i-\mathrm{acc}} = 5.52 \times 10^{-4}(1-\phi)^{1/2}10^{1.65\phi}R, \tag{17a}$$

$$v_{i-\mathrm{dis}} = 19.8(1-\phi)^{3.1}(\rho/\rho_{\mathrm{imp}})^{1/2}(R/r)^{3/2}. \tag{17b}$$

The target-to-impactor density ratio ρ/ρ_{imp} is involved in all formulae for $v_{i-\mathrm{dis}}$. We can find this ratio from Equation (5) applied both to the target and to the impactor:

$$\begin{aligned}\rho/\rho_{\mathrm{imp}} &= \left[1-(1-\rho_{ice}/\rho_{rock})C_{\mathrm{imp}}\right] \,/\, \left[1-(1-\rho_{ice}/\rho_{rock})C\right] \\ &\quad \times(1-\phi)\,/\,(1-\phi_{\mathrm{imp}}).\end{aligned} \tag{18}$$

However, this procedure introduces two new parameters C_{imp} and ϕ_{imp} instead of the only one, that is, the density ratio ρ/ρ_{imp} itself. These new parameters are virtually unknown for any impact events that happen in the Solar System. Therefore Equation (18) is not used in the following. Instead, we simply assume that in the formulae (14b, 15b, 16b, and 17b) $\rho/\rho_{\mathrm{imp}} = 1$. It means that we consider such a class of impact events that the target and the projectile are built off similar material.

In Figures 2 and 3 we present plots of $v_{i-\mathrm{acc}}$ (dashed line) and $v_{i-\mathrm{dis}}$ (continuous lines) versus target porosity ϕ. Three (Figure 2) or four (Figure 3) regimes can be distinguished depending on whether the curves $v_{i-\mathrm{acc}}$ and $v_{i-\mathrm{dis}}$ intersect. These regimes are:

(i) The area below the curve $v_{i-\mathrm{acc}}$ as well as below the curve $v_{i-\mathrm{dis}}$. This is the cratering and regolith formation regime, without escape of the fragments (an accretion area).

(ii) The area above the curve $v_{i-\mathrm{acc}}$ and below the curve $v_{i-\mathrm{dis}}$. It is the cratering and regolith formation regime, with escape of some ejecta.

(iii) The area above the curve $v_{i-\mathrm{dis}}$ and below the curve $v_{i-\mathrm{acc}}$. This is the rubble pile formation regime.

(iv) The area above both the curves $v_{i-\mathrm{acc}}$ and $v_{i-\mathrm{dis}}$. This is the catastrophic mass loss regime.

Figure 4 shows that an impactor of the size of a few hundred meters with velocity of the order of several kilometers per second breaks up a nucleus of the size of a few kilometers. If porosity is higher than 0.5 the catastrophic disruption happens with less energy, i.e., for lower impact velocity $v_{i-\mathrm{dis}}$ (straight line for fixed r shifts to the right).

Acknowledgements

The author warmly acknowledges the Institute of Low Temperature Science, Hokkaido University at Sapporo for making the impact experiments possible that comprise the experimental base of this work. This paper was partially supported by the Polish Committee of Scientific Research, Grant No. 2.P03D.002.18.

References

Arakawa, M., Leliwa-Kopystynski, J., and Maeno, N.: 2002, *Icarus* **158**, 516–531.

Davis, D. R. and Ryan, E. V.: 1990, *Icarus* **83**, 156–182.

Hartmann, W. K.: 1979, *Proc. Lunar Planet. Sci.* **10**, 1887–1916.

Holsapple, K. A.: 1994, *Planet. Space Sci.* **42**, 1067–1078.

Melosh, H. J. and Ryan E. V.: 1997, *Icarus* **129**, 562–564.

Mizutani, H., Takagi, Y., and Kawakami, S.: 1990, *Icarus* **87**, 307–329.

Wilkinson, S. L., Robinson, M. S., Thomas, P. C., Veverka, J., McCoy, T. J., Murchie, S. L., Prockter, L. M., and Yeomans, D. K.: 2002, *Icarus* **155**, 94–103.

THERMOPHYSICAL MODELLING OF COMET P/BORRELLY EFFECTS OF VOLUME ENERGY ABSORPTION AND VOLUME SUBLIMATION

YURI V. SKOROV* and HORST U. KELLER
Max-Planck-Institut für Aeronomie, Max-Planck-Str. 2, D-37191 K.-Lindau, Germany

LAURENT JORDA
Laboratoire d'Astrophysique de Marseille, BP 12, 13376 Marseille Cedex 12, France

BJØRN J. R. DAVIDSSON
Department of Astronomy and Space Physics, Box 515, SE–751 20 Uppsala, Sweden

(Received 23 March 2002; Accepted 5 July 2002)

Abstract. In this work, we continue revising the theoretical basis of numerical models describing the transport of matter and energy inside a porous dust-ice mixture at low temperature. A model of a light-absorbing near-surface layer of a comet nucleus is investigated. Gas transport is considered simultaneously with the solution of the general heat transfer equation. The quasi-stationary temperature distribution and the H_2O mass flux and sublimation rate are computed for a nucleus model of comet 19P/Borrelly at the Deep Space 1 (DS1) encounter. The energy is deposited in a layer of about 20 particle radii: This corresponds to a solid-state greenhouse effect. The surface temperature of the layer-absorbing model as well as the gas production rate are significantly smaller than the ones in the surface-absorbing model. An active fraction of 40–50% would be required to explain the observed water production rate of P/Borrelly with our layer-absorption model at the time of the DS1 encounter.

Keywords: Cometary nuclei, P/Borrelly, surfaces, thermal models

1. Introduction

Usually, in modelling of cometary nuclei the reflectance and absorption of solar light are treated assuming that the material is totally opaque even at optical wavelength, thus all the solar energy is absorbed in an infinitely thin surface layer. In reality, the high porosity and the fraction of transparent ice allow the energy to penetrate into a subsurface layer at depths which depend on the physical properties of the subsurface layer. Recently, a first numerical model of the light–absorption surface layer of a cometary nucleus was presented and the optical characteristics of the media were evaluated for a large number of different regolith types by using a combination of Mie theory, the discrete dipole approximation, Hapke theory, and numerical solutions to the equation of radiative transfer (Davidsson and Skorov, 2002a). It was shown that, at least for highly porous regolith consisting of larger (100 μm–1 cm) solid grains of ice and dust, the commonly used surface energy

* Permanent affiliation: Keldysh Institute of Applied Mathematics, Russia

Earth, Moon and Planets **90:** 293–303, 2002.

absorption approximation is generally not valid. Solar radiation can penetrate to substantial depths, and this phenomenon can lead us to revise our physical approach of cometary nucleus thermophysical modelling. This last question was systematically investigated in our next paper of this series (Davidsson and Skorov, 2002b). We presented a first comparison between a surface energy absorption model (SEAM) and a layer energy absorption model (LEAM) for spherical rotating cometary nuclei. It was found that there were substantial differences between the models in terms of near surface temperature distribution and, as a result, in terms of gas production rate.

The present paper is the logical next step towards the development of a fully consistent description of mass and heat transfer on a cometary surface. We use the developed numerical thermophysical model with layer absorption of solar radiation in order to:

(i) Analyze distinctions between SEAM and LEAM (mainly in the context of effective gas production) for the P/Borrelly Deep Space 1 (hereafter DS1) flyby.

(ii) Apply the models to analyze the results obtained from observations and reconsider them in the light of the new model, which has both a layer absorption of radiation and volume character of sublimation.

2. Model: Equations and Parameters

We study the porous media formed by grains which are identical to the $\mathcal{F}$-grains in our previous article (Davidsson and Skorov, 2002a). The medium consists of large (radius of particle r_g = 1 cm or 1 mm) irregular and compact grains of clear ice contaminated by spherical core–mantle grains (23% in volume). The latter particles consist of a 0.07 μm silicate core (70% amorphous pyroxene, $MgSiO_3$, and 30% amorphous olivine, $MgFeSiO_4$, by mass), covered by a mantle of organic refractories. In this work we consistently use $r_p = r_g$, where r_p is pore radius.

A one-dimensional model describing simultaneously the energy and mass transport in a porous partially volatile medium (in particular dusty water ice) which was originally developed by Steiner et al. (1991) and has been recently extended by Skorov et al. (2001), is applied. We solve the coupled system of non-stationary heat transfer and gas diffusion equations together with appropriate boundary conditions. The complete model description is given in Davidsson and Skorov (2002b). Hereafter we show and discuss basic equations only. The thermal conduction equation with intrinsic energy sources due to both layer radiation absorption and gas sublimation-recondensation inside the media with volume porosity ψ can then be written as (Steiner et al., 1991):

$$(1-\psi)(\rho_i f_i c_i(T) + \rho_d f_d c_d(T))\frac{\partial T(x,t)}{\partial t} = \frac{\partial}{\partial x}\left(\kappa(T)\frac{\partial T(x,t)}{\partial x}\right)$$
$$-c_g\phi_g(p,T)\frac{\partial T(x,t)}{\partial x} - q(p,T)L - K\frac{S_\odot(1-A_{\mathrm{dh}})}{r_h^2}\frac{\partial F(x,t)}{\partial x}, \qquad (1)$$

where T is the temperature of the dusty ice, ρ_i and ρ_d are the densities of the solid ice and dust respectively, f_i and f_d are the volume fractions of ice and of dust in a solid medium, ϕ_g is the mass flux rate, calculated by Clausing's formula (Skorov and Rickman, 1995), and the volumic mass production rate q (kg m^{-3} s^{-1}) is given in Mekler et al. (1990).

The last term on the right–hand side of Equation (1) is a source term for solar energy absorbed by a unit volume. Here, $S_\odot$ is the solar constant, A_{dh} is the wavelength–integrated directional–hemispherical albedo, r_h is the heliocentric distance, $F(x,t)$ is the net downward directed energy flux (normalised to unity at the surface), and K is a parameter used to switch between a SEAM ($K=0$) and a LEAM ($K=1$). The flux profile $F(x,t)$ can be approximately described by:

$$F(x,t) = e^{-x/(\zeta\cos\gamma)}, \qquad (2)$$

where ζ is the e-holding scale, determined in the frame of our numerical model of the light–absorbing surface layer (Davidsson and Skorov, 2002a), and γ is the local incident angle.

In order to complete the model, we use a mass conservation equation for gas diffusion (Steiner et al., 1991) that connects pressure, mass flux rate, and mass production rate via the ideal gas law.

The boundary conditions on the surface for both the temperature and the pressure reflect the balance of either the energy or the mass flux. A detailed discussion of this point is presented by Davidsson and Skorov (2002b). Basic parameters used in the model are collected in Table I. The specific heat capacities of ice c_i, dust c_d and gas c_g are given by Enzian et al. (1997). Since the contribution of the internal energy from the gas is very small in comparison to that of the solid matrix, it is ignored in the present work. The conductivity of a solid mixture of ice and dust is given by Orosei et al. (1995):

$$\kappa(T) = h\left(\frac{\kappa_i(T)f_i + \kappa_d f_d}{f_i + f_d}\right), \qquad (3)$$

and is corrected for porosity via the Hertz factor h (Steiner et al., 1991), which is treated as a free parameter in the present work. Typical values of h for porous ice are in the range 10^{-1}–10^{-3} (Seiferlin et al., 1996). The conductivity of pure solid

TABLE I
Model parameters

Parameter	Symbol	Value
Solid density of ice	ρ_i	933 kg m^{-3}
Solid density of dust	ρ_d	2078 kg m^{-3}
Volume fraction of ice (solid mixture)	f_i	0.731
Volume fraction of dust (solid mixture)	f_d	0.269
Dust-to-ice mass ratio	μ	0.82
Porosity of the medium	ψ	0.7
e-folding scale	ζ	$3.9 \cdot r_g$ m
Albedo	A_{dh}	0.032

ice $\kappa_i(T)$ is given by Klinger (1981), $\kappa_i(T) = 567/T$ W m^{-1} K^{-1} and for the dust we use $\kappa_d = 3.1$ W m^{-1} K^{-1}, which is an average of the values given by Orosei et al. (1995) and Enzian et al. (1997). Hereby the saturation pressure of water vapor is given by Fanale and Salvail (1984). For the latent heat of ice sublimation the constant value $L = 2.778 \times 10^6$ J kg^{-1} is used in our calculations.

3. Observations of Comet P/Borrelly

Comet 19P/Borrelly was discovered in 1904 by A. Borrelly from Marseilles observatory. It has been observed visually in 1905–1932 by Van Biesbroek and several other astronomers (Sekanina, 1979 and references therein), who noted the elongated shape of its dust coma.

Observations of comet 19P/Borrelly in 1981 by the IUE permitted to measure a water production rate of 5.9×10^{28} s^{-1}. More recently, the comet was observed at the Nancay radiotelescope, which gives a good idea of the variations of the activity with heliocentric distance close to perihelion (Bockelée-Morvan et al., 2002; see Figure 1). We also have a very good idea of the water production rate at the Deep Space 1 encounter from measurements of UV and radio lines of OH and H_2O with the HST and ODIN satellites (Weaver et al., 2001; Lecacheux et al., private communication) almost during the flyby. All measurements are summarized in Fig. 1, where we show the water production rate vs. heliocentric distance in a log–log plot. We can fit a curve (dashed curve on Figure 1), that corresponds to a water production rate (in s^{-1}) of:

$$\log(Q[H_2O]) = 3.3 \times 10^{28} \left(\frac{q_h}{r_h}\right)^{10.8}, \tag{4}$$

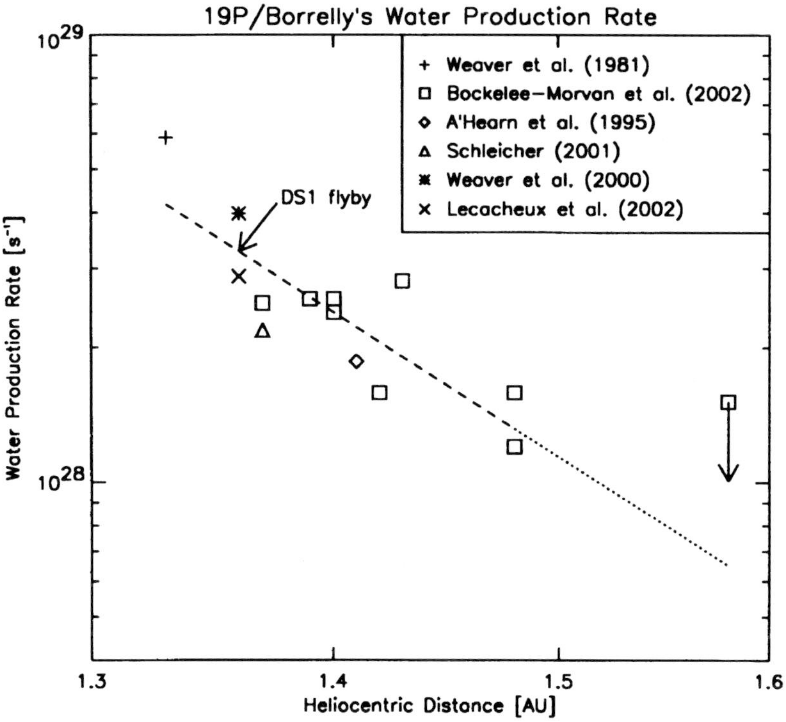

Figure 1. Water production rate measurements of comet 19P/Borrelly estimated from OH or H_2O UV or radio lines. The dashed line is a linear fit of the data which corresponds to the law $\log Q[H_2O] = 29.96 - 10.78 \log r_h$, from which a production rate of $\sim 3 \times 10^{28}$ s^{-1} is estimated at the time of the Deep Space 1 encounter. The residuals correspond to an uncertainty of ~25%.

where q_h is the perihelion distance of the comet. The residuals correspond to an uncertainty of ~25%. This relationship gives a production rate of $\sim 3 \times 10^{28}$ s^{-1} at the time of the DS1 encounter.

The size and shape of the comet nucleus have been estimated by Lamy et al. (1998) from a photometric partial lightcurve of the nucleus obtained after subtraction of the coma contribution in the central pixel of high-resolution HST images. The value of 1.8×4.4 km has been confirmed by the analysis of the images obtained by the MICAS camera onboard the DS1 spacecraft (Boice et al., 2002). The resulting active fraction calculated from the above water production rate is ~10% near perihelion (e.g., Lamy et al., 1998).

The orientation of the pole for the 2001 perihelion passage can be estimated from the dust fan observed on CCD images obtained in late 2001. The average direction calculated by several authors is $(\alpha, \delta) = (225 \pm 15°, -10 \pm 10°)$ in the equatorial reference system (Samarasinha and Mueller, 2002; Schleicher et al., presentation at IAU Coll. 186, Tenerife, January 2002; Thomas et al., 2001). This corresponds to an obliquity with respect to the orbital plane of ~100°. Note that this is one of the few comets for which there seems to be a general agreement on the orientation of the spin axis. According to this orientation, the spin axis was only $\sim 3^{\circ}$ away from the comet-Sun vector at the time of the DS1 encounter.

4. Application to Comet P/Borrelly and Comparison with Observations

Temperatures as functions of depth are shown in Figures 2–4 for the LEAM and SEAM. In Figure 2, we present the quasi-stationary distributions in the uppermost porous layer for three surface elements with solar zenith angles of 10°, 45° and 90°. The model comet is located at the heliocentric distance corresponding to the DS1 flyby of comet P/Borrelly. We further assumed a Hertz factor $h = 0.001$ and a pore and grain radius of 1 cm. Figure 3 shows the LEAM and SEAM temperature distributions for the same medium as in Figure 2, assuming millimeter grains, and hence, the *e*-folding factor is ten times smaller. We cannot accurately describe the effective heat conductivity of the modelled medium. Therefore we simply use the Hertz factor as a free model parameter in order to evaluate how sensitive the results are with respect to the conductivity process. In Figure 4, modelling results for a medium with millimetre grains and with a Hertz factor ten times larger ($h = 0.01$) are presented. A detailed discussion of specific features, characterizing temperature profiles obtained via SEAM can be found in our previous papers (Skorov et al., 2001; Davidsson and Skorov, 2002b). In the following, we will only emphasize the most prominent results.

Comparing the curves plotted in Figures 2–3, we see considerable qualitative differences between the calculated profiles for the LEAM and SEAM. For the LEAM, the solar energy is deposited in a layer of ~ 20 grain radii depth. The typical depth is independent of the incident angle and the value of the grain radius assumed in the model and on the incident angle. The deposited energy can not reradiate back because the material is opaque in the infrared. This corresponds to a visible solid-state greenhouse effect. If the conductivity is increased by increasing the Hertz factor (Figure 4), the size of the heated region enlarges too, and all gradients are weakened.

A very sharp temperature increase in the uppermost region and then generally convex monotonic curves are typical attributes of the profiles obtained in the LEAM. All spatial scales decrease with decreasing grain size and at the same time temperature gradients become sharper. For the LEAM, the peak temperature is reached at a depth of a few grain radii in all cases and not at the surface as in the case of the SEAM. The peak temperature in the LEAM is only a few degrees lower than in the SEAM. The surface temperature at the subsolar point is about 50 K lower in the LEAM than in the SEAM for grains of 1 cm, and only ~ 30 K lower for grains of 1 mm assuming a Hertz factor of $h = 0.001$. This temperature difference decreases with increasing Hertz factor between $h = 0.001$ (Figure 3) and $h = 0.01$ (Figure 4).

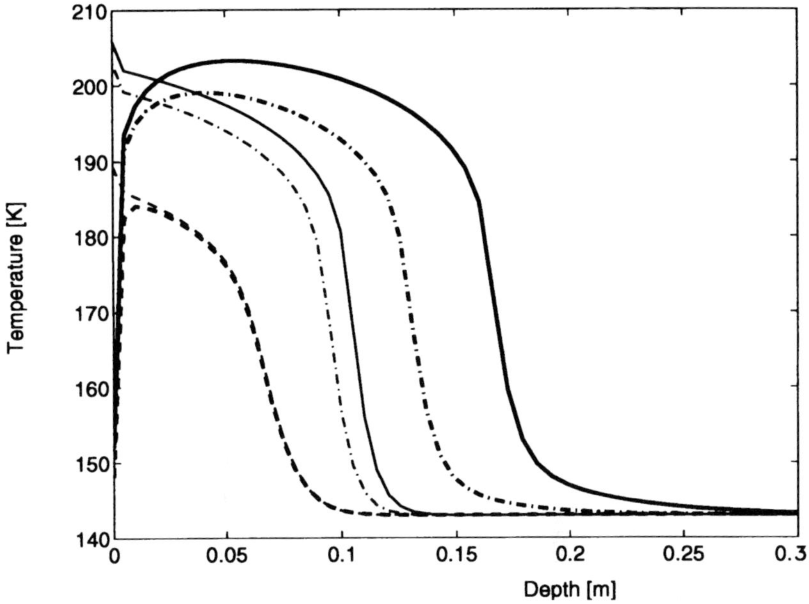

Figure 2. Temperature as function of depth for the SEAM (refer to normal curves) and LEAM (refer to bold curves). The incident angle γ is 0° (solid curves), 45° (dash–dotted curves), and 80° (dotted curves). Hertz factor is 0.001, pore radius is 1 cm.

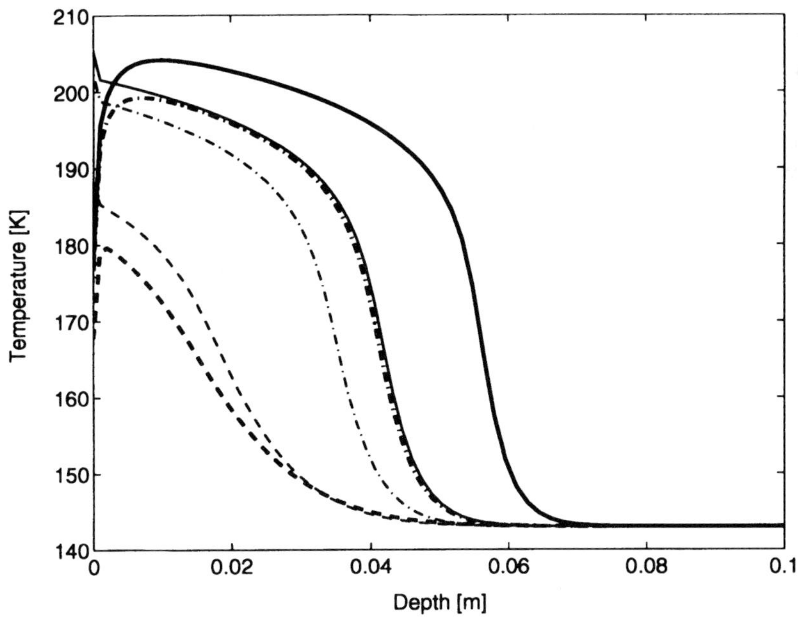

Figure 3. Temperature as function of depth for the SEAM (refer normal curves) and LEAM (refer to bold curves). The incident angle γ is 0° (solid curves), 45° (dash–dotted curves), and 80° (dotted curves). Hertz factor is 0.001, pore radius is 1 mm.

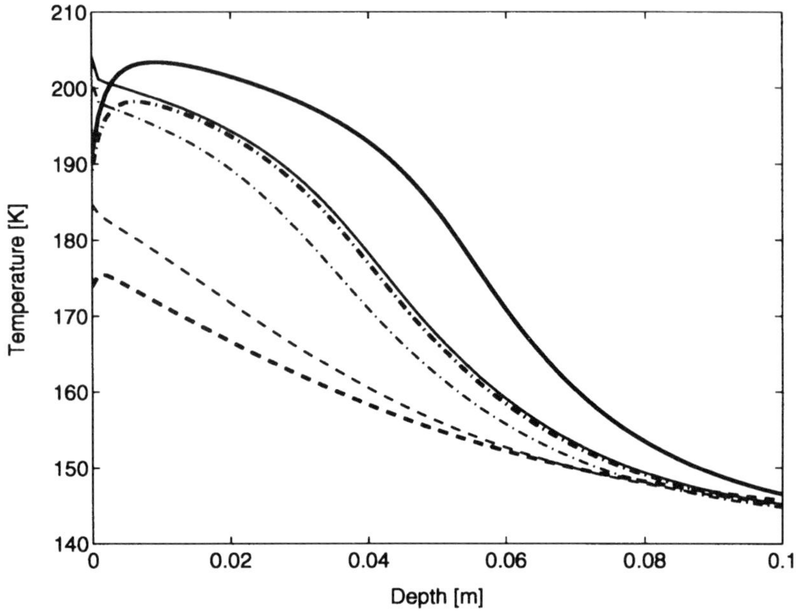

Figure 4. Temperature as function of depth for the SEAM (refer to normal curves) and LEAM (refer to bold curves). The incident angle γ is 0° (solid curves), 45° (dash–dotted curves), and 80° (dotted curves). Hertz factor is 0.01, pore radius is 1 mm.

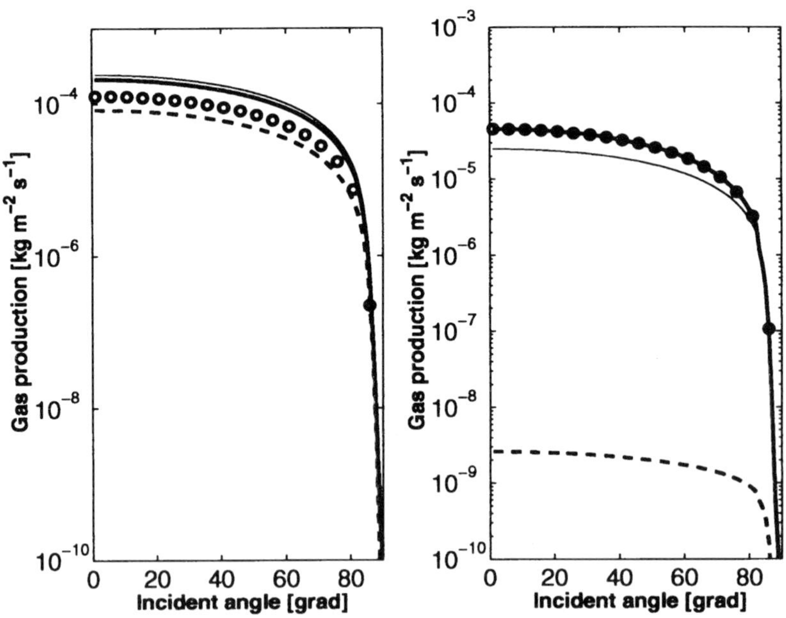

Figure 5. Production rates (contributions from pores – circles, "solid surface" – dotted curves, and their sum – solid curves) as function of latitude are plotted for the SEAM (left panel) and LEAM (right panel). Hertz factor is 0.001, pore radius is 1 cm. Thin solid curves show sum production for $h = 0.01$.

TABLE II
Total gas production for comet P/Borrelly (LEAM vs. SEAM)

Shape model	Gas production (10^{28} s^{-1})	Active fraction
$r_g = 0.01$ m $h = 0.001$		
LEAM	3.32	0.49
SEAM	15.31	0.11
$r_g = 0.001$ m $h = 0.01$		
LEAM	3.84	0.41
SEAM	13.83	0.12

Figure 5 shows the gas production rate as a function of local incident angle for the LEAM and for the SEAM. We present contributions from pores, the surface itself, and their sum. Because for the SEAM temperature is monotonically decreasing with depth, the gas flux from the surface is higher than the sublimation from pores (even for $\psi = 0.7$), although their magnitude is comparable. For the LEAM, the pore outgassing plays a dominant role for all considered cases and the sublimation from the "solid parts" of the cometary surface is negligible. Because a larger fraction of the energy is used to heat the subsurface layer in the LEAM, the calculated sublimation rates are lower by a factor of about 4–5 compared to that predicted by the SEAM. This factor varies very little with grain size. The decrease of pore size leads to relative suppression of processes determined by gas diffusion. As result for the fine grained medium ($r_g = 1$ mm) the total gas production is significantly reduced in the LEAM, but the opposite smaller effect is observed in the SEAM (thin curves).

We can calculate the expected gas production rates for the LEAM and SEAM and evaluate the active surface fraction. The results are presented in Table II. We list the active surface fraction for the LEAM and SEAM calculated assuming a spheroidal body with the size 4.4×1.8 km and its spin axis directed toward the Sun. In the SEAM, only a minor part of the surface of P/Borrelly should be active at the Deep Space 1 encounter. The calculated values ($\sim$10%) are in a good agreement with previous evaluations (Lamy et al., 1998). In the LEAM, the active fraction of the surface required to fit the measured water production rate of Figure 1 at the time of the DS1 encounter increases to 40–50%. Note that we neglect here possible contributions from sublimating icy grains in the coma and from outgassing under a non-volatile porous crust (Skorov et al., 1999, 2001).

Unfortunately, the effective observed temperature of the surface does not provide a selection criteria because of its uncertainty. Both for the LEAM (with 40–50% of active icy area) and for the SEAM (where the active area is only about

8–12%) we can achieve satisfactory agreement with the observational data. It is worth to note that if for the SEAM we can cover all physically reasonable temperature range, for the LEAM we can not get effective surface temperature higher than 300–310 K, keeping in mind that a significant part of the surface is active and therefore has lower temperature.

5. Conclusion

The main features of LEAM can be summarized as follows: (1) The surface temperature is substantially lower; (2) The total gas production rate is significantly lower on the irradiated side; (3) The source of most coma molecules is the near-surface interior (at a depth of $\sim$20 grain sizes) rather than the "surface" itself.

The analysis of the differences between the surface absorption model (SEAM) and a more realistic model in which volume absorption of energy occurs (LEAM) shows that in all investigated cases there are major differences between the two approaches, and therefore volume absorption of the energy should be considered in thermophysical models of comets.

For comet P/Borrelly, we find that for both cases, LEAM and SEAM, a significant part (50–90%) of the total cometary surface should be inactive.

References

Bockelée-Morvan, D., Biver, N., Colom, P., Crovisier, J., Festou, M. C., Henry, F., Lecacheux, A., and Weaver, H. A.: 2002, these proceedings.

Boice, D. C., Soderblom, L. A., Britt, D. T., Brown, R. H., Sandel, B. R., Yelle, R. V., Buratti, B. J., Hicks, M. D., Nelson, R. M., Rayman, M. D., Oberst, J., and Thomas, N.: 2002, these proceedings.

Davidsson, B. J. R. and Skorov, Yu. V.: 2002a, *Icarus* **156**, 223–248.

Davidsson, B. J. R. and Skorov, Yu. V.: 2002b, *Icarus* **159**, 239–258.

Enzian, A., Cabot, H., and Klinger, J.: 1997, *A&A* **319**, 995–1006.

Fanale, F. P. and Salvail, J. R.: 1984, *Icarus* **60**, 476–511.

Klinger, J.: 1981, *Icarus* **47**, 320–324.

Lamy, P. L., Toth, I., and Weaver, H. A.: 1998, *A&A* **337**, 945–954.

Mekler, Y., Prialnik, D., and Podolak, M.: 1990, *Astrophys. J.* **356**, 682–686.

Orosei, R., Capaccioni, F., Capria, M. T., Coradini, A., Espinasse, S., Federico, C., Salomone, M., and Schwehm, G. H.: 1995, *A&A* **301**, 613–627.

Samarasinha, N. H. and Mueller, B. E. A.: 2002, these proceedings.

Seiferlin, K. Spohn, T., and Benkhoff, J.: 1995, *Adv. Space Res.* **15**(10), 35–38.

Sekanina, Z.: 1979, *Icarus* **37**, 420–442.

Skorov, Yu. V. and Rickman, H.: 1995, *Planet. Space. Sci.* **43**, 1587–1594.

Skorov, Yu. V., Kömle, N. I., Keller, H. U., Kargl, G., and Markiewicz, W. J.: 2001, *Icarus* **153**, 180–196.

Skorov, Yu. V., Kömle, N. I., Markiewicz, W. J., and Keller, H. U.: 1999, *Icarus* **140**, 173–188.

Steiner, G., Kömle, N. I., and Kührt, E.: 1991, in N. I. Kömle, S. J. Bauer, and T. Spohn (eds.), *Theoretical Modeling of Comet Simulation Experiments*, pp. 11–29.

Thomas, N., A'Hearn, M. F., Boice, D. C., Britt, D. T., Meech, K. J., Sandel, B. R., Soderblom, L. A., and Yelle, R. V.: 2001, *Amer. Astron. Soc.* DPS meeting #33.
Weaver, H. A., Stern, S. A., Parker, J. W., and the HST-DS1 Comet Borrelly UV Team: 2001, *Amer. Astron. Soc.* DPS meeting #33, #58.06.

FROM A WEAK TO A STRONG COMET – 3D GLOBAL HYBRID SIMULATION STUDIES

THORSTEN BAGDONAT and UWE MOTSCHMANN
Institute for Theoretical Physics, Technical University of Braunschweig, Germany

(Received 21 February 2002; Accepted 12 June 2002)

Abstract. Plasma structures resulting from the solar wind interaction with weak comets are discussed. Numerical simulations using a newly developed hybrid code are presented. The simulations are primarily applied to quantitative data for comet Wirtanen, which will be the target of the Rosetta mission. It is expected that Wirtanen is very weak during the first encounter. The main purpose is the discussion of the different features of the plasma environment, such as the structured cycloidal plasma tail and non-linear Mach cones typical for weak comets and their relation to structures like shocklets, bow shock, diamagnetic cavity and the "classical" magnetotail found at stronger comets. Furthermore, the sensitivity of these various features in dependence on the plasma parameters is investigated.

Keywords: Comet, hybrid, simulation

1. Introduction

The Rosetta mission will reach comet Wirtanen at a distance of about 3.5 AU (Schwehm and Schulz, 1999b). At this far distance, the gas production of Wirtanen will be very small, if not vanish at all. This can be seen from observational data (Schwehm and Schulz, 1999a) and models which extrapolate these data to larger distances (Kührt, 1999). Even during the lander mission somewhere around 3 AU this situation does not change significantly. Therefore, Wirtanen will be the first weak comet observed *in situ*. The prediction of the plasma environment is interesting both from the view of fundamental plasma physics but also for experimental and technical reasons.

The main parameters for the qualitative behaviour of a weak comet are the outgassing rate and the Alfvénic Mach number of the superalfvénic and supersonic solar wind flow. Bogdanov et al. (1996) have divided this behaviour in different categories depending on these parameters. According to this Wirtanen will cover a huge parameter space on its way to the Sun.

The earliest numerical investigations done by Biermann et al. (1967) investigated fundamental plasma structures like the cometary bow shock, which were verified during flybys at Giacobini–Zinner, Halley and Grigg–Skjellerup. However, the plasma environment of weak comets look entirely different, as analytical results (Mendis and Flammer, 1984; Flammer et al., 1991), MHD (Gombosi et al., 1996),

Earth, Moon and Planets **90:** 305–321, 2002.

multi-fluid (Bogdanov et al., 1996; Sauer and Dubinin, 1998; Fischer, 1999) and hybrid simulations (Lipatov et al., 1997, 2002) show. The reason is that for weak comets the gyroradius of the cometary ions is larger than the typical scales of the coma.

Several different features have been discussed in the above papers. Here, we focus on the connection between these structures and their evolution for varying parameters. Therefore several simulations using the hybrid model with quantitative data for Wirtanen and some qualitative, more general, investigations were done. The results and the fundamental physical processes are discussed.

2. Model

2.1. Numerical Scheme

A new hybrid simulation code described by Bagdonat and Motschmann (2002) was used for this work. This code applies the explicit time step algorithm introduced by Matthews (1994) and a space discretisation scheme based on tensor calculus similar to the method used by Eastwood et al. (1995). The electrons are assumed to be a massless charge-neutralising fluid ($m_e = 0$), whereas the ions are treated as particles. The code solves the equations of motion for the ions, and a momentum conservation equation for the electron fluid which yields the equation for the electric field in the form

$$\mathbf{E} = -\frac{\mathbf{J}_{\text{ion}} \times \mathbf{B}}{\rho_c} + \frac{(\nabla \times \mathbf{B}) \times \mathbf{B}}{\mu_0 \rho_c} + \frac{\nabla p_e}{\rho_c} + \eta \frac{\nabla \times \mathbf{B}}{\rho_c}, \tag{1}$$

where ρ_c is the ion charge density, $\mathbf{J}_{\text{ion}}$ the ionic current density, p_e the electron pressure and η a constant taking account for anomalous resistivity effects.

From Faraday's law

$$\frac{\partial \mathbf{B}}{\partial t} = -\nabla \times \mathbf{E} \tag{2}$$

a time evolution equation for the magnetic field is obtained.

These equations are solved in three spatial dimensions using an arbitrary curvilinear grid. For the kinetic pressure of the electrons adiabatic behaviour is assumed, i.e.,

$$p_e = p_{e0} \left(\frac{\rho_e}{\rho_{e0}} \right)^{\kappa}, \tag{3}$$

where p_{e0} is given by the initial electron temperature by means of the electronic plasma beta β_e.

2.2. Units

The results are given mostly in dimensionless units as used in the code. All physical quantities A are normalised by

$$A^* = \frac{A}{A_0},$$

where A^* is the dimensionless quantity and A_0 some normalisation value. These normalisation values are chosen to be

B_0	background magnetic field
n_0	background ion density
$t_0 = \Omega_i^{-1}$	Ω_i background ion gyrofrequency
$x_0 = c/\omega_{p,i}$	$\omega_{p,i}$ background ion plasma frequency
$m_0 = m_p$	m_p proton mass
$v_0 = v_A$	v_A background Alfvén velocity
$J_0 = ev_A n_0$	e electron charge ($e > 0$)
$p_0 = B_0^2/2\mu_0$	

B_0 and n_0 are chosen according to the undisturbed solar wind values for each simulation run.

2.3. Boundary conditions

Because the solar wind flow is supersonic, especially for larger distances from the sun, where comets generally are weak, one does not expect to get much activity of waves and noise travelling upstream. Therefore very simple boundary conditions were chosen. The left side of the simulation box is an inflow boundary. Particles with a thermal distribution are newly inserted at each time step and all field quantities are kept constant. The right side is an outflow boundary. All particles which cross this boundary (and therefore leave the box) are deleted. All field quantities are linearly extrapolated from the upstream side. All other boundaries are either periodic in the case of a rectangular simulation box or also inflow-/outflow-boundaries in the case of non-orthogonal grids. Although this scheme is simple it did serve well in all presented simulations.

2.4. Modelling the comet

In the view of plasma physics a comet is a source of cometary ions which are picked up by the solar wind. We assume the comet to be spherical symmetric and to have a constant neutral gas production rate G. Only the most prominent gas component, which is H_2O is taken into account. This is well satisfied for most cases although for comets far away from the sun, where the surface temperature is

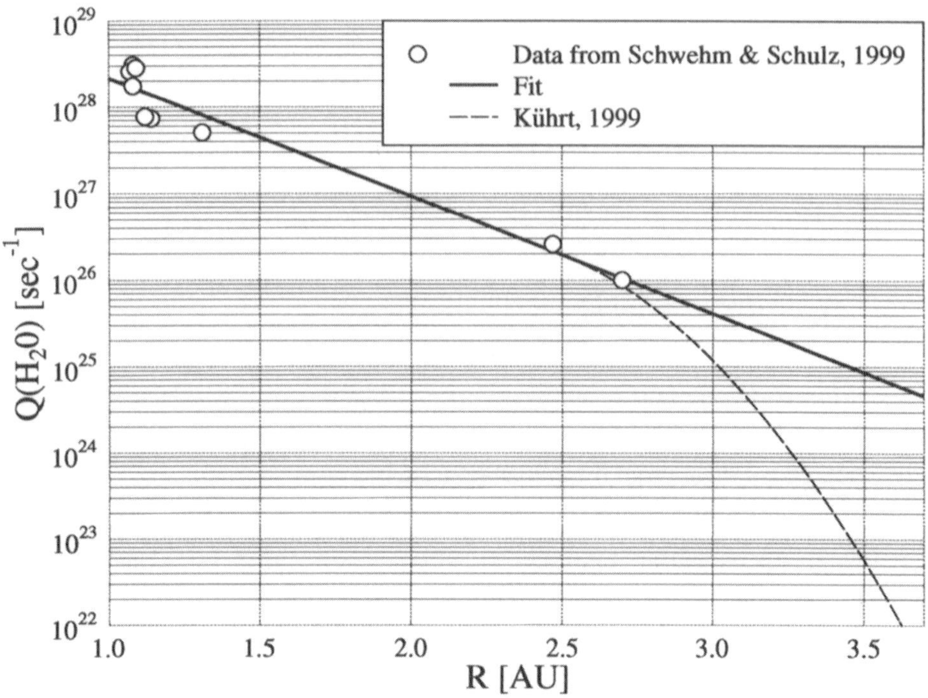

Figure 1. Water production rate of comet Wirtanen. Circles: Observed data. Solid line: Interpolation. Dashed line: Estimated behaviour for larger distances according to different models.

below the water sublimation temperature, CO and other components may play the dominant role (Kührt, 1999).

The neutral gas coma expands with an almost constant velocity v_0. It is ionised mainly by solar UV radiation with an ionisation rate ν. Other ionisation processes and charge exchange processes are not taken into account. With these assumptions the charge density growth rate of the produced O^+ ions is given by

$$\frac{d\varrho_h(r)}{dt} = \frac{e\nu G}{4\pi r^2 v_0} \exp\left(-\frac{\nu}{v_0}(r-a)\right), \tag{4}$$

where r is the distance from the cometary nucleus and a the radius of the nucleus. Because ν/v_0 is very small compared to the length scales under consideration, the exponential is almost constant and can be approximated to be 1 throughout the whole simulation area.

The comet is incorporated into the simulation by generating a certain number of cometary ions each time step according to the distribution given by Equation (4). Usually the cometary ions have a mass of $m_h = 16$ (O^+), except where noted.

The outgassing rates G were chosen arbitrarily for the more general simulations and by an interpolation of the data given by Schwehm and Schulz (1999a) for comet Wirtanen. Figure 1 shows the water production rate for comet Wirtanen. For large distances, where the surface temperature is in the range of the ice sublimation temperature one expects a deviation from the fit of the observed values (Kührt, 1999). However, a quantitative prediction at these distances depends on many parameters, which are not exactly known, such as thermal conductivity and rotation period of the nucleus. Therefore the values used here are a rough estimation of the expectations.

For the outgassing velocity v_0 a value of $v_0 = 1\text{km/s}$ is assumed, which corresponds to the thermal velocity of the outgassing atoms at a temperature of about 300K. The UV ionisation rate is assumed to be $\nu = 10^{-6}\text{s}^{-1}$ (Mendis et al., 1985; Gombosi et al., 1996). The main problem is the value of a, which is usually very small compared to the characteristic lengths of the plasma surrounding, and therefore also small compared to the grid size in a simulation run. This implies that by setting up a distribution as in Equation (4), one produces a very large amount of particles in almost one simulation cell, overestimating the density gradients at this cell, which leads to unwanted results. Therefore one has to choose an effective radius a', which should be in the range of 10% of the grid cell size. For stronger comets, where the solar wind does not reach the inner coma this does not influence the outcome of the simulation, but for weaker comets, where the solar wind penetrates the coma completely the overall outgassing rate is modelled too low compared to reality. Generally, this does not much influence the qualitative behaviour of the results, but may lead to some small errors in quantitative values.

The cometary ions are very cold compared to the solar wind ions, whereas the cometary electrons have a thermal energy of about 1 eV (Mendis et al., 1985) corresponding to $\beta_e = 0.2$, since they are produced by an ionisation process. However, this temperature is still different from the solar wind electron temperature. Unfortunately the hybrid description does not allow for different electron species with different temperatures, because β_e entering in Equations (3) and (1) is a global parameter. As will be described below, this parameter may play an important role for the physics near the nucleus and therefore it was decided to choose the β_e according to the temperature of the cometary electrons. Moreover, the above value of $\beta_e = 0.2$ is not very different from the typical solar wind value of $\beta_e = 0.4$ for larger distances from the Sun. The different ion temperatures, of course, can be taken into account separately due to the particle description of the ions.

2.5. SOLAR WIND PARAMETERS

For the more general simulations carried out, typical solar wind parameters at 1 AU were chosen, i.e., $B_0 = 5\text{nT}$, $n_0 = 5\text{cm}^{-3}$, $v_0 = 400\text{km/s}$ and $\beta_e = \beta_i = 0.4$. This yields an Alfvènic Mach number of $M_A \approx 10$.

For the more quantitative results concerning comet Wirtanen the values measured by IMP-8 and Voyager 2 (Richardson et al., 1995, 1996) were fitted using the Parker model (Parker, 1958).

The azimuthal angle between the magnetic field and the flow velocity is about 45 degrees at 1 AU and approaches almost 90 degrees around 3 AU. It is reasonable to use a perpendicular magnetic field at these large heliocentric distances to simplify the results and make the qualitative effects more clear.

The most complicated parameter is the resistivity constant η in Equation (1). The temperature dependent resistivity due to collisions can be neglected for the length and time scales under consideration. However, on length scales comparable to the ion gyroradius an anomalous resistivity may arise from microscopic instabilities which also depend on the local current density. Since the exact determination is difficult, it is convenient to choose η to yield a certain diffusion length for the magnetic field, which is small compared to the grid cell size (normally in the order of 10%) (Winske and Omidi, 1993).

3. Results

3.1. Test Particle Regime

In this section some well-known simple results should be shortly reviewed, which form the basics for the more complicated situations.

Figure 2 shows the density of cometary ions with mass $m_h = 16$ in a solar wind flow with $M_A = 5$. These ions enter the simulation box at position $x = -500$ and velocity $v = 0$. The cometary ions act as test particles, i.e., their density is so small that neither the magnetic field nor the solar wind flow are disturbed in any way.

In the solar wind rest frame the injected test particles have an relative motion with velocity v_h which is just M_A in the chosen units. According to a Lorentz transformation in this frame the magnetic field frozen-in in the solar wind has still the value $B_0 = 1$, whereas the electric field vanishes. Therefore the cometary ions start a gyration motion with gyroradius $r_h = m_h v_h / e B_0$ or $r_h = M_A m_h$ in dimensionless units. In the laboratory rest frame this gyromotion appears as a cycloid with height $M_A m_h$ and width $2\pi M_A m_h$ as can be verified in Figure 2. In this test particle case all cometary ions start with the same gyrophase and no gyrophase diffusion occurs.

Figure 3 shows the magnetic field produced by an immobile small ion cloud. Again, the field is oriented perpendicular to the simulation plane pointing inwards. In a supersonic flow a small local increase in the plasma density implies also an equivalent increase in the magnetic field. Then the incoming solar wind protons get deflected upwards. Therefore an area with increased density and field (above) and decreased density and field (below) is formed. The disturbance travels with the linear fast mode velocity given by

$$v_f^2 = (v_A^2 + v_S^2) + \sqrt{\frac{1}{4}(v_A^2 + v_S^2)^2 - v_A^2 v_S^2 \cos^2\theta}\,, \tag{5}$$

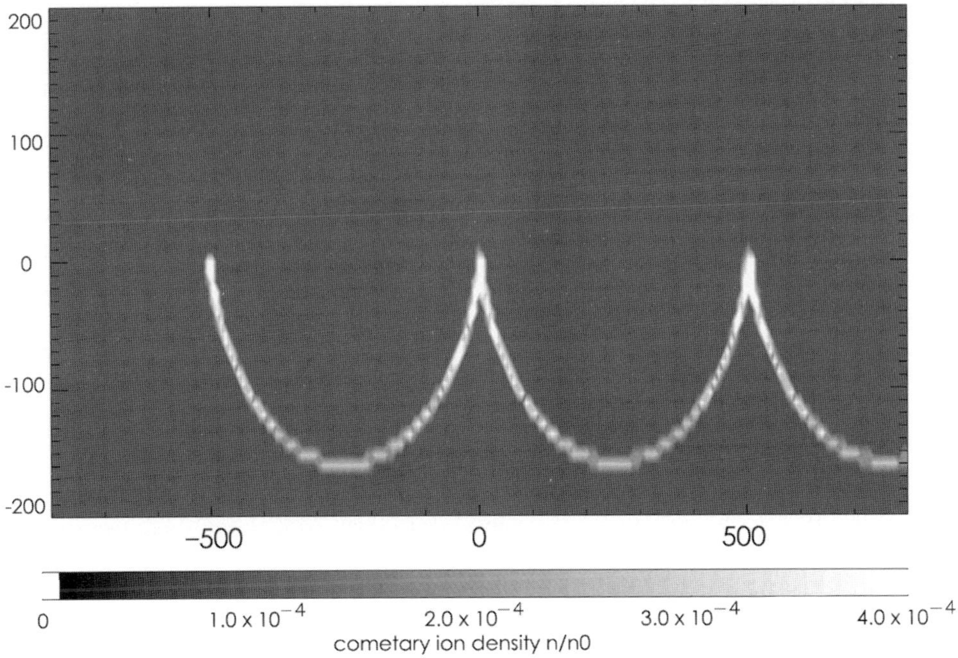

Figure 2. Cycloidal tail of a very weak comet. 2D simulation with $M_A = 5$, $m_h = 16$, $B_0 = 1$. Magnetic field is perpendicular to the shown plane and directed inwards. The solar wind flows from left to right.

where v_A, v_S, and θ are the Alfvèn speed, sound velocity, and the angle between the magnetic field and the wave vector, respectively. In Figure 3 $\theta = 90°$ is chosen. In dimensionless units v_S is simply given by

$$v_S = \sqrt{\frac{\beta_i + \beta_e}{2}} . \tag{6}$$

The wave front forms a Mach cone with an opening angle α given by $\sin\alpha = v_f/M_A$. Due to the deflection of the solar wind flow direction at the obstacle the cone is moved slightly upwards from the horizontal direction.

3.2. Linear Transition Region

Combining the two effects from the previous region for a very weak comet, one expects a cycloidal tail with an attached Mach cone. Such a situation is shown in Figure 4 at a much smaller scale compared to Figure 2. The results of Figure 4 are obtained from a 3D simulation for comet Wirtanen at 3.5 AU.

The main part of the tail forms the beginning of the cycloid moving downwards with a small cometary ion density. As expected, this part of the tail has an increased magnetic field. Furthermore one can see the weak suppression and

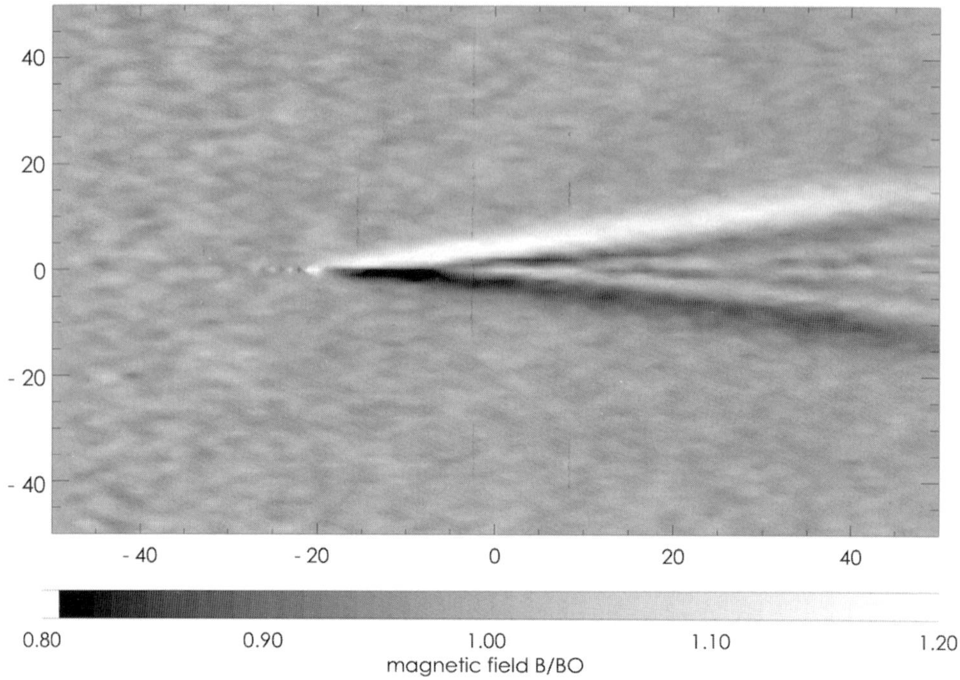

Figure 3. Linear fast mode Mach cone produced by an immobile obstacle.

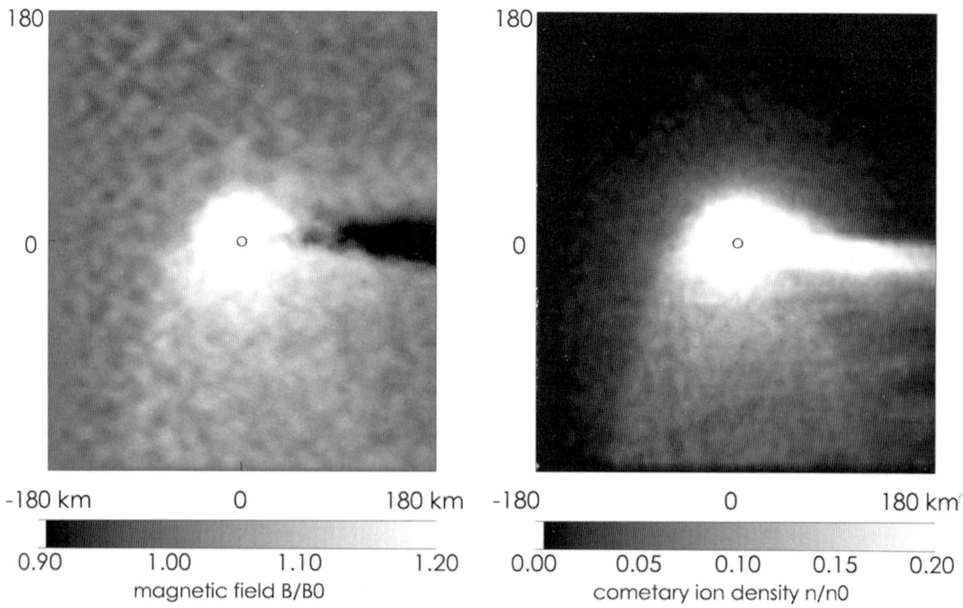

Figure 4. 3D Simulation for Comet Wirtanen at 3.5 AU. $M_A = 15$, $G = 5 \cdot 10^{23}$ s^{-1}, $B_0 = 0.4$ nT, $n_0 = 0.4$ cm^{-3}. The shown plane is a cut perpendicular to the magnetic field (pointing inwards) and parallel to the solar wind flow (from left to right). The circle marks the position of the nucleus.

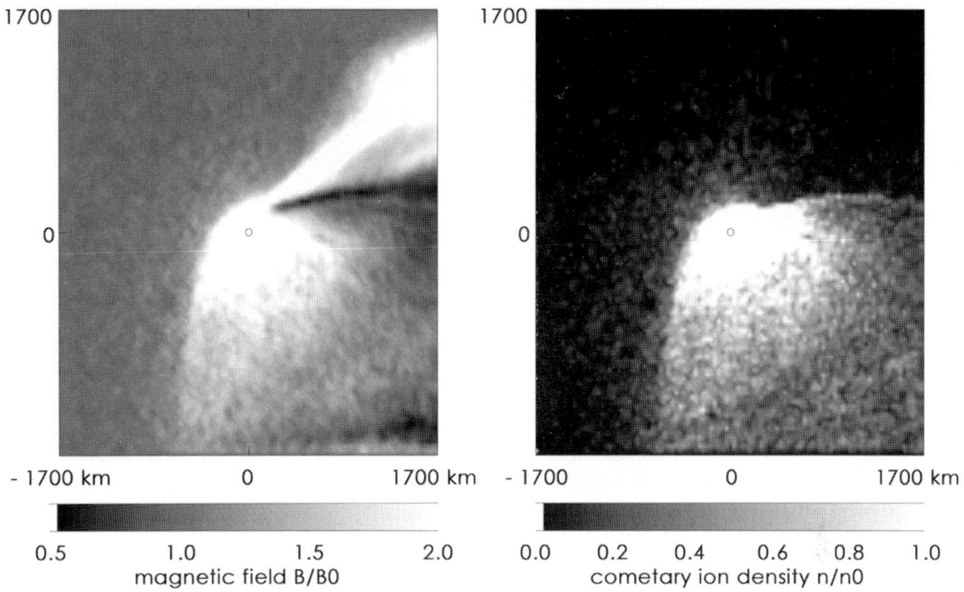

Figure 5. Comet Wirtanen at 3.25 AU. $M_A = 10$, $G = 1.8 \cdot 10^{25}\mathrm{s}^{-1}$, $B_0 = 1.13\mathrm{nT}$, $n_0 = 0.66\mathrm{cm}^{-3}$.

depression Mach cone in the magnetic field due to the deflection of the incoming flow. The most interesting feature, however, is the sharp part of the plasma tail with high cometary ion density moving along the area with decreased magnetic field. Whereas the cycloidal, broad tail is formed by a pickup process, this part of the tail has a completely different physics.

Due to the relatively strong gradient in the magnetic field strength between the inner coma region and the depression cone an electric field arises which drags the cometary ions into this region. Additionally the thermal pressure pushes the cometary ions into the region with lower solar wind density.

Figures 5 and 6 show results for Wirtanen at 3.25 AU and 3 AU, respectively. The overall picture (cycloidal tail with attached Mach cone) remains the same, however, the scales become larger when approaching the Sun. There is still a weak remainder of the sharp plasma tail from Figure 4 visible which runs along the area with decreased magnetic field.

The absolute value of the magnetic field in the inner coma becomes larger due to the overall increase of the cometary ion density. Therefore also the deflection of the solar wind becomes stronger, the suppression cone rises up in Figure 5 and the magnetic field strength increases. The gyroradius of the cometary ions becomes smaller so that the cycloidal tail is pushed more and more to the horizontal direction. But the gyroradius is still much larger than the size of the coma.

In Figure 6 the density of the suppression cone reaches a critical value and shows the onset of non-linear behaviour described in the next section. There are

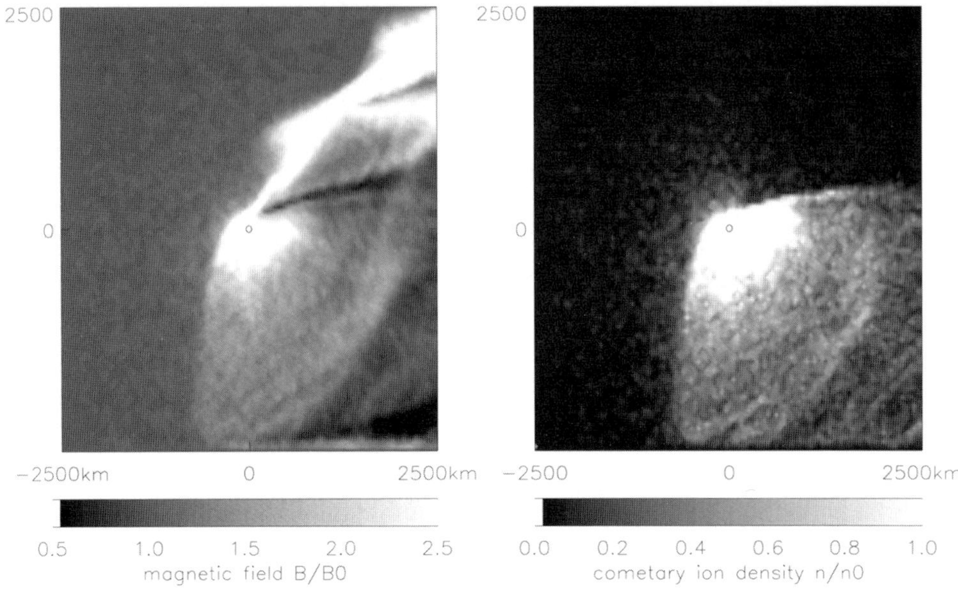

Figure 6. Comet Wirtanen at 3 AU. $M_A = 10$, $G = 3 \cdot 10^{25}\text{s}^{-1}$, $B_0 = 1.2\text{nT}$, $n_0 = 0.8\text{cm}^{-3}$.

also some transversal structures visible in the cycloidal tail, which are also described below.

Unfortunately, by that time the Rosetta spacecraft will be completely inside the homogenous region of the coma so that these interesting structures may not be accessible to observation.

Flammer et al. (1991) have carried out some analytical calculations for comet Kopff at large heliocentric distances. Figure 7 (which represents cuts along the sun-comet line of Figures 4–6 and 10) should be compared to their results. They obtain a monotonic increase in the magnetic field up to the nucleus itself similar to that in Figure 7 up to 3 AU. However, our simulations yield a maximum in the field strength somewhat before the nucleus' surface and the relative increase is stronger. Flammer et al. (1991) use a simplified one-dimensional model which is only valid if the scale of the coma is large compared to the gyroradius of the cometary ions. This is not the case for Wirtanen, which is weaker than comet Kopff.

3.3. Non-linear Regime

Figure 8 shows a global picture of the plasma structures similar to the situation of Figure 6. The outgassing rate is slightly larger but the simulation uses values for n_0 and B_0 typical for 1 AU. Two features are visible.

The tail is still of cycloidal shape but it shows a pronounced transversal structuring. This was already found in bi-ion fluid simulations (Bogdanov et al., 1996; Sauer and Dubinin, 1998) and explained by means of bi-ion acoustic waves gener-

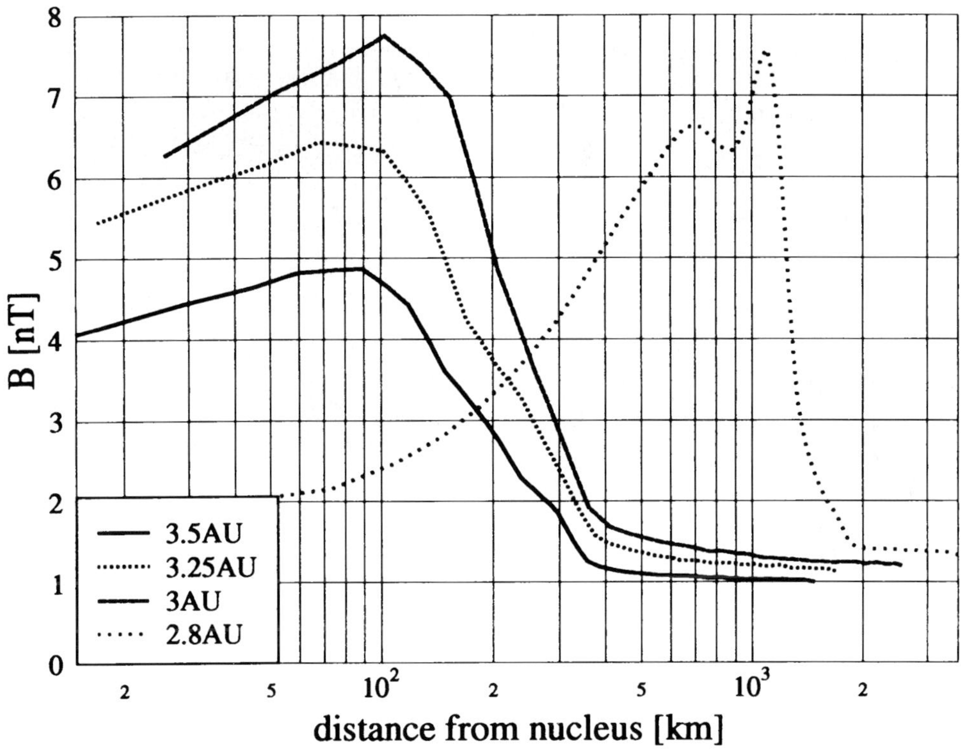

Figure 7. Magnetic field strength upstream of the nucleus along the sun-comet line for different heliocentric distances.

ated in the massloading area of the tail. There are several observations of such node-like structures moving along the plasma tail (cf. Fischer, 1999), but it is presently unclear, if the observed structures are of the same nature as the ones obtained in the simulations.

The other feature is the formation of multiple Mach cones. This was also found in hybrid simulations by Lipatov et al. (1997) and Sauer and Dubinin (1998) but no discussion was performed.

First of all, these wave fronts show no shock-like behaviour. The solar wind which crosses the fronts behaves adiabatically. This can be seen from Figure 9 where the adiabatic exponent was estimated by calculating the solar wind pressure directly from the particle distribution function at various points along the front. The double logarithmic plot verifies the adiabatic behaviour with an adiabatic exponent of $\kappa \approx 2$ obtained by the linear fit. The value of 2 instead of $5/3$ is due to the fact that thermodynamic coupling of the motion is only effective perpendicular to the magnetic field in a collisionless plasma. Therefore the thermodynamics of the system is quasi two-dimensional implying $\kappa = 2$. If these wave fronts would be shockwaves of some kind, a higher adiabatic exponent is to be expected (Mandt and Kan, 1990).

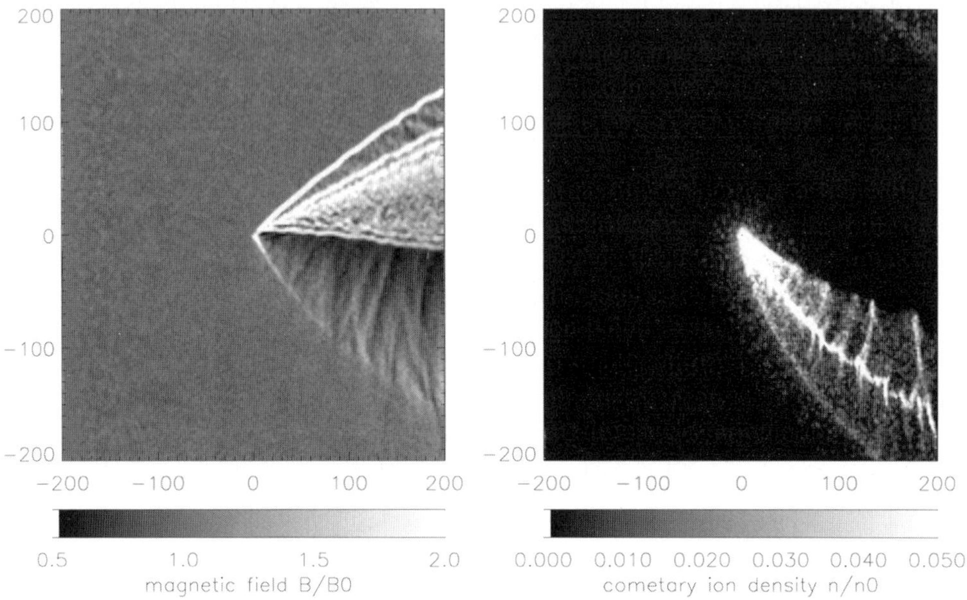

Figure 8. Multiple Mach cones and tail structures for a weak comet at 1 AU. $M_A = 10$, $G = 10^{26}\text{s}^{-1}$, $B_0 = 5\text{nT}$, $n_0 = 5\text{cm}^{-3}$. The length scale here is $c/\omega_{pi} \approx 100\text{km}$. The simulation box has a size of about 40,000 km.

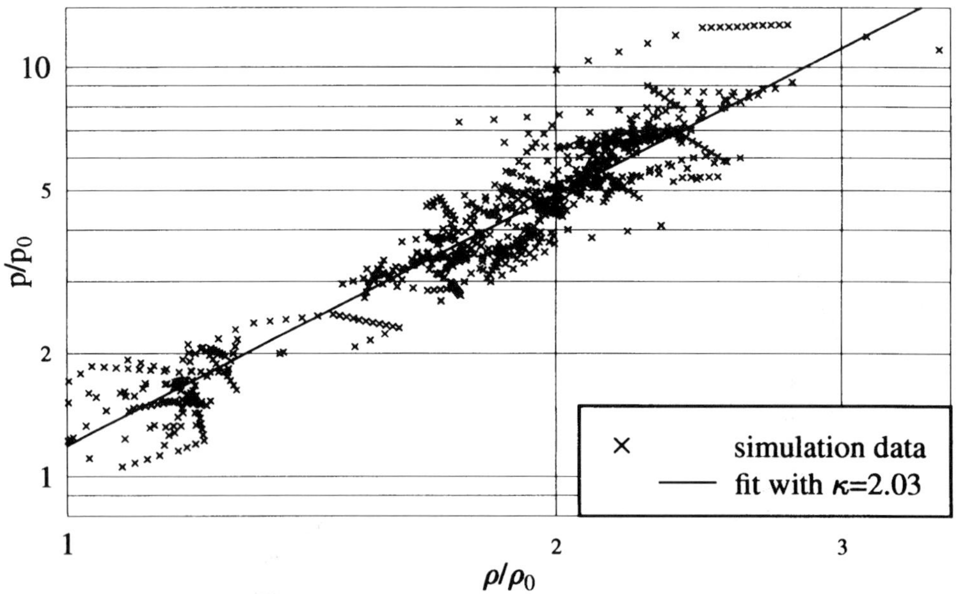

Figure 9. Estimation of the adiabatic exponent for the non-linear Mach cones.

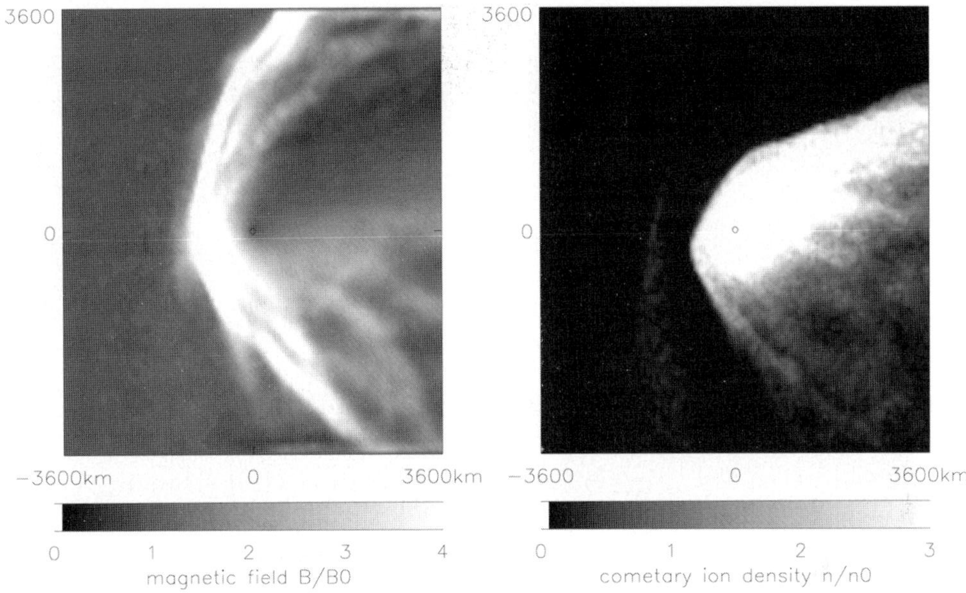

Figure 10. Wirtanen at 2.8 AU. $M_A = 8$, $G = 7.5 \cdot 10^{25}$ s^{-1}, $B_0 = 1.3$ nT, $n_0 = 0.9$ cm^{-3}.

The magnetic field amplitude of these multiple Mach cones is in the non-linear regime. In this case the wave velocity depends on the amplitude (Mann, 1995; Cap, 1994). High amplitude structures run faster than those with low amplitudes. Therefore the cones in Figure 8 are probably solitons. Each of them has a different magnetic field amplitude and forms a different angle with the solar wind flow.

3.4. Transition to Shock Regime

Going back to Wirtanen, Figure 10 shows a simulation result for a distance of 2.8 AU. The qualitative behaviour has changed dramatically compared to the situation at 3 AU in Figure 6. First of all, now the tail moves no longer cycloidal but horizontally and slightly upwards. The situation becomes more symmetric, and the Mach cone extends also below the tail. It should be noted that at this point the transition region with increased magnetic field seems to be oscillating. Moreover, the magnetic field in the inner coma is decreased to a value of nearly zero. This is the onset of a diamagnetic cavity formation.

The underlying physical process is governed by different competing effects. The solar wind flow transports magnetic flux into the coma. Furthermore magnetic field diffusion also tends to transport flux inbound. On the other hand the outgassing velocity of the cometary ions forms a flow which transports flux outbound. However, the relatively hot electrons in the inner coma are much more important. The high gradient of the electron pressure establishes a radial electric field according to Equation (1). The cometary ions are accelerated by this radial field. This process

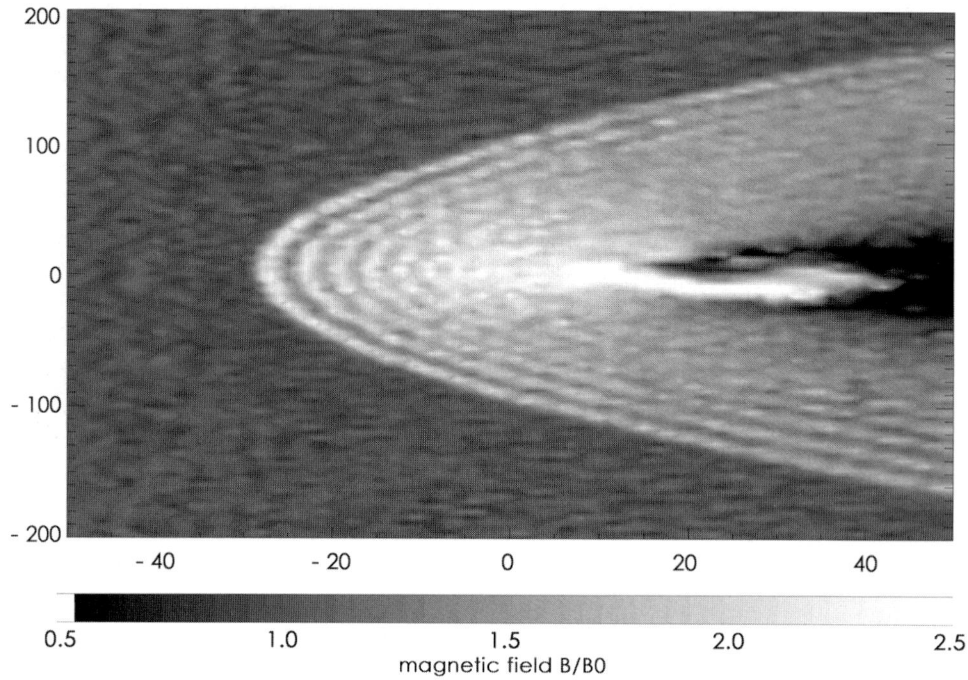

Figure 11. Illustration of multiple shocklets travelling upstream in front of an immobile ion cloud with high density.

is accompanied by an increase of the radial flow density and by the transport of magnetic field out of the coma. This establishes a boundary just at the position where this outward flow balances the inward flow (Neubauer, 1991). Note, that this mechanism of forming a diamagnetic cavity works without any currents arising from a collision-induced resistivity.

When the accelerated cometary ions reach this boundary and enter the area with high magnetic field, they immediately start a gyromotion with rather small gyroradius. Then they are deflected to the downstream side, where the magnetic field is reduced strongly. In this area with reduced magnetic field they flow downstream. Therefore the tail is now directed horizontally.

Due to the horizontal orientation of the tail the magnetic suppression Mach cones can now extend around the whole coma and the situation becomes more and more symmetrically. When the cometary ion density is further increased a situation as shown in Figure 11 arises. The “Mach cones” become completely symmetrically. Their amplitude is large enough so that their velocity becomes high enough to travel upstream. They evolve into a set of soliton-like wavefronts, which are called “shocklets” (Omidi and Winske, 1990). These shocklets travel upstream until their amplitude decreases to reach a stationary state. The solar wind flow speed is decreased slightly by each of these shocklets decelerating it to subsonic speed. This situation is described in some detail by Omidi and Winske (1990). However, this

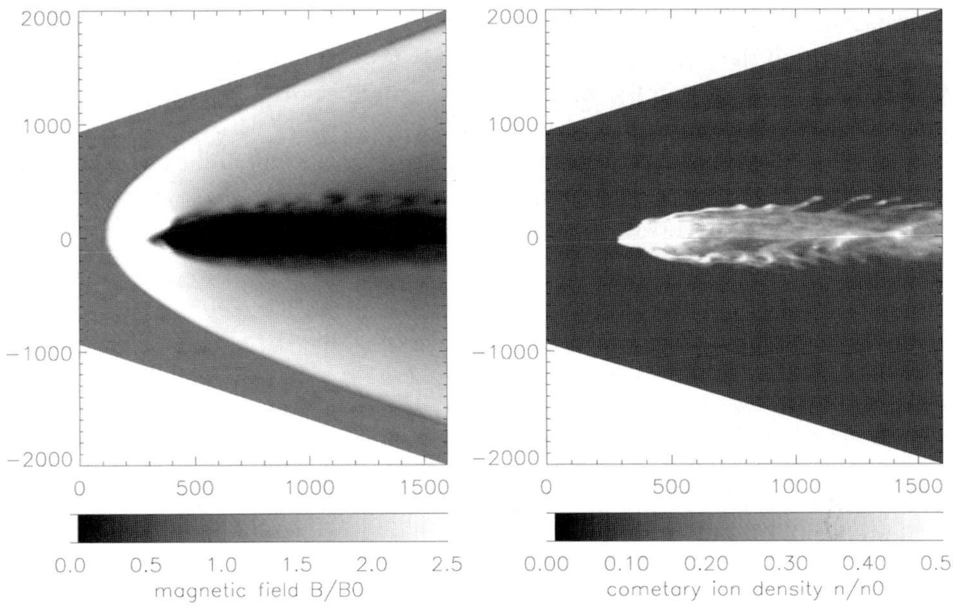

Figure 12. Parabolic shock for a weak comet below 1 AU. $M_A = 6$, $G = 5 \cdot 10^{27}$ s^{-1}, $B_0 = 18$ nT, $n_0 = 17$ cm^{-3}.

mechanism can only work up to a certain critical Alfvèn Mach number. Above this critical Mach number the well-known parabolic shock is formed as shown in Figure 12. Figure 12 represents a weak comet at some distance below 1 AU, and should be compared to similar results of 2D hybrid simulations done by Hopcroft and Chapman (2001). Note, that in Figure 12 a non-orthogonal cone-shaped grid was used.

4. Summary

The emphasis of this paper is the trace of typical plasma structures in the environment of weak comets on their way towards the Sun. We choose Wirtanen as our reference comet as it is target of the Rosetta mission. Two- and three-dimensional simulations have been carried out using a newly developed hybrid code described by Bagdonat and Motschmann (2002).

We start our survey far away from the Sun when the comet is extremely weak. Then the density of the cometary ions is much lower than the solar wind density. The cometary ions may be regarded as test particles. In this regime the plasma tail forms a cycloid.

Approach of the comet to the sun increases the production rate. The first reaction to the solar wind is expressed by a weak linear Mach cone attached to the nucleus. The magnetic field increases in the whole coma.

Further increasing production rate provides a steepening of the Mach cone due to the increase of the non-linear fast mode velocity. The tail remains cycloidal. A secondary sharp tail along the magnetic field gradient of the depression cone is formed. This situation is highly asymmetric.

In a next evolution step the Mach cone splits up in several adiabatic non-linear fast mode solitons. The situation remains asymmetric with a cycloidal tail which shows a node-like transversal structuring probably due to bi-ion acoustic waves.

A further step is the onset of a diamagnetic cavity formation. When the cometary ion density is high enough, the outward flow, mainly established by the cometary electron pressure, may balance or deflect the inward flow. The magnetic pile-up becomes strong enough to deflect all cometary ions towards the horizontal flow direction. The cycloidal tail no longer exists. The Mach cones can extend around the whole coma and the situation becomes symmetric.

When the amplitude of the non-linear Mach cones becomes strong enough they may travel upstream forming a set of shocklets, which was already observed at comet Giacobini-Zinner (Omidi and Winske, 1990). For even larger obstacles or higher Mach numbers the well-known parabolic shock is formed.

Acknowledgements

This work is supported by the Deutsche Forschungsgemeinschaft through the grant MO 539/10-1.

References

Bagdonat, T. and Motschmann, U.: 2002, '3D Hybrid Simulation Code Using Curvilinear Coordinates', *J. Comp. Phys.*, in press.

Biermann, L., Brosowski, B., and Schmidt, H. U.: 1967, 'The Interaction of the Solar Wind with a Comet', *Solar Phys.* **1**, 254–283.

Bogdanov, A., Sauer, K. Baumgärtel, K., and Srivastava, K.: 1996, 'Plasma Structures at Weakly Outgassing Comets – Results from Bi-Ion Fluid Analysis', *Planet. Space Sci.* **44**, 519–528.

Cap, F.: 1994, *Lehrbuch der Plasmaphysik und Magnetohydrodynamik*, Springer, Wien, New York.

Eastwood, J. W., Arter, W., Brealey, N. J., and Hockney, R. W.: 1995, 'Body-Fitted Electromagnetic PIC Software for Use on Parallel Computers', *Comp. Phys. Commun.* **87**, 155–178.

Fischer, C. T.: 1999, 'Beobachtung und Simulation von Strukturen im Plasmaschweif eines Kometen', Ph.D. Thesis, Technical University of Braunschweig.

Flammer, K. R., Mendis, D. A., Northrop, T. G., and Whipple, E. C.: 1991, 'A Self-Consistent Model for the Particles and Fields in the Environment of a Weakly Outgassing Comet', *J. Geophys. Res.* **96**, 15907–15916.

Gombosi, T. I., DeZeeuw, D. L., and Häberli, R. M.: 1996, 'Three-Dimensional Multiscale MHD Model of Cometary Plasma Environments', *J. Geophys. Res.* **101**(A7), 15233–15253.

Hopcroft, M. W. and Chapman, S. C.: 2001, '2D Hybrid Simulations of the Solar Wind Interaction with a Samll Scale Comet in High Mach Number Flows', *Geophys. Res. Lett.* **28**, 1115–1118.

Kührt, E.: 1999, 'H_2O-Activity of Comet Hale–Bopp', *Space Sci. Rev.* **90**, 75–82.

Lipatov, A. S., Motschmann, U., and Bagdonat, T.: 2002, '3D Hybrid Simulation of the Solar Wind-Weak Comet Interaction', *Planet. Space Sci.* **50**, 403–411.
Lipatov, A. S., Sauer, K., and Baumgärtel, K.: 1997, '2.5D Hybrid Code Simulation of the Solar Wind Interaction with Weak Comets and Related Objects', *Adv. Space Res.* **20**, 279–282.
Mandt, M. E. and Kan, J. R.: 1990, 'Dispersive and Viscous Scale Lengths in the Two-Stage Ion Heating at Quasi-Parallel Collisionless Shocks', *J. Geophys. Res.* **95**(A5), 6353–6362.
Mann, G.: 1995, 'Simple Magnetohydrodynamic Waves', *J. Plasma Phys.* **53**, 109–125.
Matthews, A. P.: 1994, 'Current Advance Method and Cyclic Leapfrog for 2D Multispecies Hybrid Plasma Simulations', *J. Comput. Phys.* **112**, 102–116.
Mendis, D. A. and Flammer, K. R.: 1984, 'The Multiple Modes of Interaction of the Solar Wind with a Comet as it Approaches the Sun', *Earth Moon Planets* **31**, 301.
Mendis, D. A., Houpis, H. L. F., and Marconi, M. L.: 1985, 'The Physics of Comets', *Fund. Cosmic Phys.* **10**, 1–380.
Neubauer, F. M.: 1991, 'The Magnetic Field Structure of the Cometary Plasma Environment', in R. L. Newburn, Jr. (ed.), *Comets in the Post-Halley Era*, Vol. 2. Kluwer Academic Publishers, Dordrecht, pp. 1107–1124.
Omidi, N. and Winske, D.: 1990, 'Steepening of Kinetic Magnetosonic Waves into Shocklets: Simulations and Consequences for Planetary Shocks and Comets', *J. Geophys. Res.* **95**, 2281–2300.
Parker, E. N.: 1958, 'Dynamics of the Interplanetary Gas and Magnetic Fields', *Astrophys. J.* **128**, 664–676.
Richardson, J. D., Belcherand, J. W., Lazarus, A. J., Paularena, K., and Gazis, P. R.: 1996, 'Statistical Properties of the Solar Wind', in D. Winterhalter, J. Gosling, S. Habbal, W. Kurth, and M. Neugebauer (eds.), *AIP Conference Proceedings*, Vol. 382, pp. 483–486.
Richardson, J. D., Paularena, K. I., Lazarus, A. J., and Belcher, J.: 1995, 'Radial Evoultion of the Solar Wind from IMP 8 to Voyager 2', *Geophys. Res. Lett.* **22**, 1469–1472.
Sauer, K. and Dubinin, E.: 1998, 'Bi-Ion Fluid Simulations of Solar Wind Massloading', MPAE Report, MPAE-W-100-99-02, Max-Planck-Institut für Aeronomie, Katlenburg-Lindau.
Schwehm, G. and Schulz, R.: 1999a, 'Coma Composition and Evolution of Rosetta Target Comet Wirtanen', *Space Sci. Rev.* **90**, 321–328.
Schwehm, G. and Schulz, R.: 1999b, 'Roestta goes to Comet Wirtanen', *Space Sci. Rev.* **90**, 313–319.
Winske, D. and Omidi, N.: 1993, *Hybrid Codes: Methods and Applications*, Chap. 5, Terra, Tokyo, p. 113.

CHEMICAL COMPOSITION DIVERSITY AMONG 24 COMETS OBSERVED AT RADIO WAVELENGTHS

NICOLAS BIVER
ESA/ESTEC, Sci-Sb, Keplerlaan 1, 2200-AG Noordwijk, The Netherlands and LESIA, Observatoire de Paris-Meudon, 5 pl. J. Janssen, F-92195 Meudon, France
E-mail: Nicolas.Biver@obspm.fr

DOMINIQUE BOCKELÉE-MORVAN, JACQUES CROVISIER, PIERRE COLOM and FLORENCE HENRY
LESIA, Observatoire de Paris-Meudon, 5 pl. J. Janssen, F-92195 Meudon, France

RAPHAËL MORENO
IRAM, 300, rue de la Piscine, F-38406 Saint Martin d'Hères, France

GABRIEL PAUBERT
IRAM, Avd. Divina Pastora, 7, 18012 Granada, Spain

DIDIER DESPOIS
Observatoire de Bordeaux, BP 89, F-33270 Floirac, France

DARIUSZ C. LIS
Caltech, MS 320-47, Pasadena, CA 91125, USA

(Received 21 March 2002; Accepted 19 May 2002)

Abstract. We present a comparative study on molecular abundances in comets based on millimetre/submillimetre observations made with the IRAM 30-m, JCMT, CSO and SEST telescopes. This study concerns a sample of 24 comets (6 Jupiter-family, 3 Halley-family, 15 long-period) observed from 1986 to 2001 and 8 molecular species (HCN, HNC, CH_3CN, CH_3OH, H_2CO, CO, CS, H_2S). HCN was detected in all comets, while at least 2 molecules were detected in 19 comets.

From the sub-sample of comets for which contemporary H_2O production rates are available, we infer that the HCN abundance relative to water varies from 0.08% to 0.25%. With respect to other species, HCN is the molecule which exhibits the lowest abundance variation from comet to comet. Therefore, production rates relative to that of HCN can be used for a comparative study of molecular abundances in the 19 comets. It is found that: CH_3OH/HCN varies from ≤ 9 to 64; CO/HCN varies from ≤ 24 to 180; H_2CO/HCN varies between 1.6 and 10; and H_2S/HCN varies between 1.5 and 7.6.

This study does not show any clear correlation between the relative abundances and the dynamical origins of the comets, or their dust-to-gas ratios.

Keywords: Comets, molecules, radio observations

1. Introduction

The last decade has proven the efficiency of microwave observations in investigating the chemical composition of cometary atmospheres. The composition of

Earth, Moon and Planets **90:** 323–333, 2002.

cometary nuclei is of strong interest for understanding their origin. As these objects should not have evolved so much since their formation, their composition provides clues to the composition in the outer regions of the solar nebula where they formed. The ultimate goal of this chemical composition survey of comets is to find out eventually comet families according to their chemical content and link them to their origin (e.g., Jupiter-family comets or Oort–Cloud comets) and likely formation region (giant planets region to Kuiper Belt). This paper provides a first synthesis of a 15-year investigation at radio wavelengths.

Up to now, chemical diversity among comets has been investigated from narrowband photometry of radicals in the visible (A'Hearn et al., 1995). A class of *carbon-depleted* comets, tentatively associated with Jupiter-family comets, was noted. One of the goals of the present study is to investigate whether such a diversity can be confirmed by the direct observation of the sublimation products of nucleus ices.

2. Observations

The radio lines of 8 molecules (HCN, HNC, CH_3CN, CH_3OH, H_2CO, CO, CS, H_2S) were searched for with the Institut de Radio Astronomie Millimétrique (IRAM) 30-m or $4 \times -5 \times 15$-m Plateau-de-Bure interferometer, the James Clerk Maxwell Telescope (JCMT) 15-m, the Caltech Submillimeter Observatory (CSO) 10.4-m or the Swedish-ESO Submillimetre Telescope (SEST) 15-m radio telescopes. A list of the 24 comets detected in at least one line, with particularities and references, is given in Table I.* From a statistical point of view, at least 5 of these molecules were searched for in 11 of the comets and at least two molecules were detected in 19 comets out of the 24.

3. Data Reduction and Production Rates

In most cases the production rates were derived following the methodology and parameters described in Biver et al. (1999a). To model the gas density in the coma, an isotropic steady-state outflow described by a Haser density radial profile was assumed, unless stated otherwise in the references given in Table I. The line shapes have been used to derive an estimate of the gas outflow velocity (v_{exp}). When available, rotational temperatures of methanol or other species were used to derive the gas temperature. Otherwise, the gas temperature was assumed on the basis of observations in comets of similar activity at similar heliocentric distances.

* We have excluded from this study comets C/1997 J2 (Meunier–Dupouy), observed at 3 AU in which only CO and CH_3OH were detected, and 29P/Schwassmann–Wachmann 1, observed at 6 AU in which only CO was detected. The upper limits obtained on HCN production rates in both comets do not provide any significant constraint.

TABLE I
Observed comets and references

Comet	N[a]	Origin[b]	Particularities	References[c]
2P/Encke	–	JF		
10P/Tempel 2	20	JF	Strong jet	
19P/Borrelly	19	JF	Strong jet	[1]
21P/Giacobini–Zinner	21	JF	No or low CO	
22P/Kopff	22	JF	Low CO	[2]
45P/Honda-M.-P.	–	JF		[2]
1P/Halley	–	HF		[3]
23P/Brorsen–Metcalf	–	HF		[2, 4]
109P/Swift–Tuttle	25	HF	Strong jet	[2, 5]
			HCN, CH_3OH rich	[6]
C/1989 X1 (Austin)	1	LPC, DN	CH_3OH rich	[2, 4, 7, 8]
C/1990 K1 (Levy)	2	LPC		[2, 4, 7, 8]
C/1995 O1 (Hale–Bopp)	3	LPC, old	HCN, H_2CO, H_2S, CO, CS rich	[9, 10]
C/1996 B1 (Szczepanski)	4	LPC, old	CH_3OH rich	[2]
C/1996 B2 (Hyakutake)	5	LPC	CO rich	[11]
C/1996 Q1 (Tabur)	6	LPC, old	CH_3OH poor disintegrated	[2]
C/1998 M5 (LINEAR)	–	LPC		
C/1998 P1 (Williams)	–	LPC	CO rich	
C/1998 T1 (LINEAR)	10	LPC		
C/1998 U5 (LINEAR)	–	LPC, old		
C/1999 H1 (Lee)	12	LPC	CH_3OH rich	[12]
C/1999 S4 (LINEAR)	13	LPC, DN	CH_3OH, CO poor disintegrated	[13]
C/1999 T1 (McNaught-H.)	14	LPC	CO rich	[14]
C/2000 WM_1 (LINEAR)	15	LPC	CO, H_2S poor	
C/2001 A2 (LINEAR)	16	LPC	CO poor, H_2S rich	[14]

[a] Reference number in Figures 2–4.
[b] JF: Jupiter family; HF: Halley family; LPC: long-period comet; DN: dynamically new (semi-major axis $a > 10\,000$ AU); old: orbital period $< 10\,000$ years.
[c] [1]: Bockelée-Morvan et al. (2002); [2]: Biver (1997); [3]: Despois et al. (1986); [4]: Colom et al. (1992); [5]: Bockelée-Morvan et al. (1994b); [6]: Despois et al. (1996); [7]: Bockelée-Morvan et al. (1994); [8]: Crovisier et al. (1991); [9]: Biver et al. (2002); [10]: Bockelée-Morvan et al. (2000); [11]: Biver et al. (1999a); [12]: Biver et al. (2000); [13]: Bockelée-Morvan et al. (2001); [14]: Crovisier et al. (2001).

TABLE II
Molecular abundances in 24 comets

Molecule	Comets observed		Q/Q_{water}		Q/Q_{HCN}	
	Detected	Upper limit	Mini	Maxi	Mini	Maxi
HCN	24	0	0.08%	0.25%	1	1
HNC	5	2	<0.003%	0.035%	<0.03	0.17
CH_3CN	4	0	0.013%	0.035%	0.08	0.23
CH_3OH	15	2	<0.9%	6.2%	<9	64
CO	5	4	<1.7%	23%	<19	180
H_2CO	13	2	0.13%	1.3%	1.6	10
H_2S	11	3	0.12%	1.5%	1.5	7.6
CS	9	0	0.05%	0.17%	0.5	1.2

Water outgassing rates were taken from the current literature, namely Crovisier et al. (2002a, b), when available, or extrapolated on the basis of the magnitude–water production rates correlation formula $\log Q_{H_2O} = 30.74 - 0.24(m_1 - 5 \log \Delta)$, from Jorda et al. (1992).

4. Comparison of Chemical Abundances

Most relative abundances were derived from the ratio of production rates to water outgassing rates. HCN was also used as a reference as it is the easiest molecule to detect close to the Sun and was detected in the 24 comets observed within 1.5 AU from the Sun. Since HCN was often observed quasi simultaneously and with a similar beam size as the other molecules studied here, the abundances relative to HCN are more reliable. They have been used in Figures 3–4.

The sample (Table I) comprises 4 Jupiter-family comets, 2 Halley-family comets and 13 long-period comets (2 dynamically new, 3 "old" with period of less than 10 000 years), in which HCN plus at least one other molecule were detected. Figure 1 shows the current status of the histogram analysis of the abundances relative to water (HNC/HCN for HNC) in those comets, in which some of the 8 molecular species abundances could be measured. Table II summarizes the abundance ranges observed.

No significant difference could be found between the different comet classes, but it must be emphasized that our analysis is still preliminary, and that the sizes of the sub-samples are limited. We also investigated the possibility of a correlation between relative molecular abundances with the dust-to-gas ratio of the comets, traced by the $Af\rho/Q[\text{CN}]$ ratio deduced from visible spectrophotometry. No trend could be found.

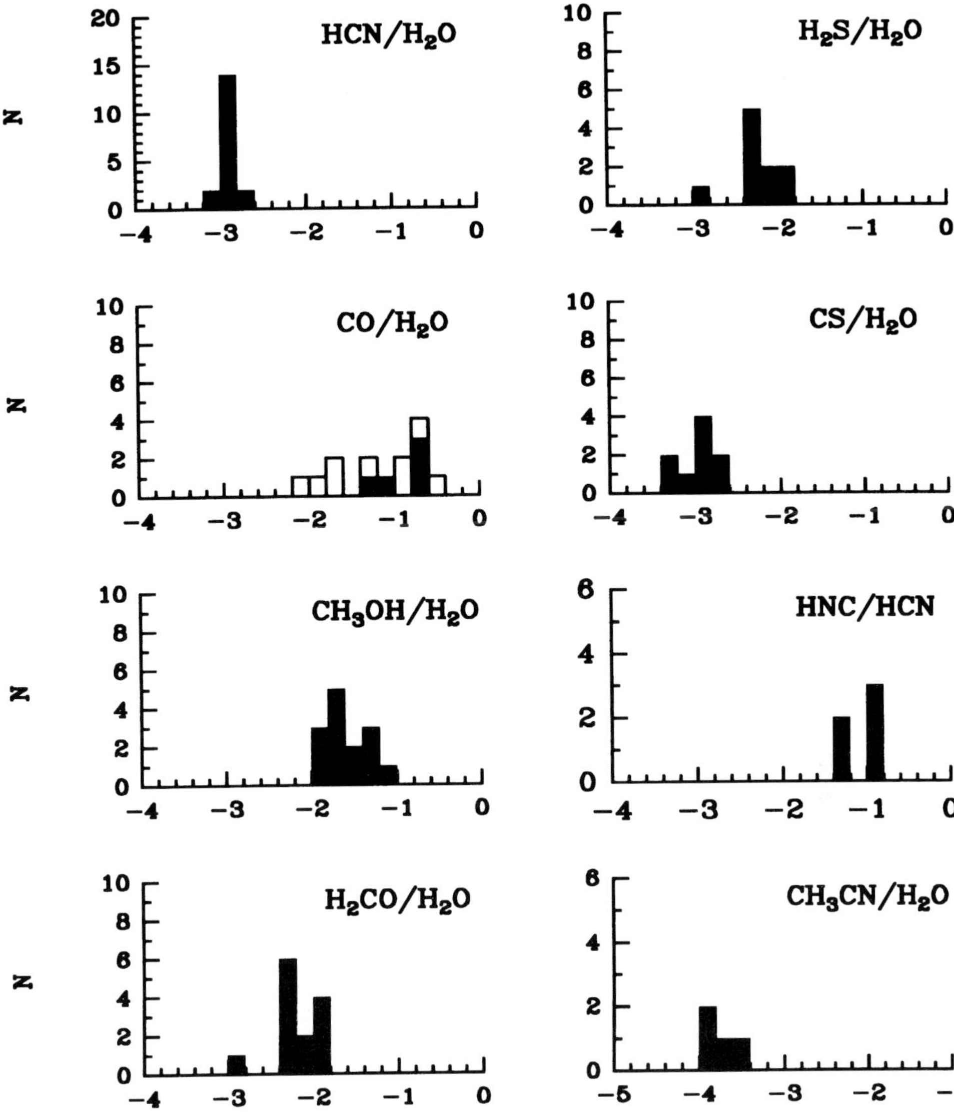

Figure 1. Histogram analysis of the abundances relative to water (excepted for HNC, for which the HNC/HCN abundance ratio is of higher significance) of all eight molecules studied here. Comets are grouped by bins of 0.2 in log of the molecular abundance ratios. In the case of CO, unfilled bars are additional information provided by the more extensive study carried out in the infrared (DiSanti et al., 2000 and previous observations) or UV (Feldman et al., 1997). See also Huebner (2002) and Bockelée-Morvan and Crovisier (2002) for more extensive studies of the chemical content of comets.

5. Results

5.1. CN-BEARING MOLECULES

HCN was the first molecule firmly detected at millimetre wavelengths in a comet, in 1P/Halley in 1985. It has subsequently been detected in 23 comets. Its measured abundance relative to water ranges between 0.08% and 0.25% – when a water outgassing rate is available – and is one of the least variable from comet to comet (Figures 1 and 2). For this reason, HCN is also often taken as a reference.

CH_3CN has been detected in only 4 comets (including one marginal detection) since 1996, with an abundance relative to HCN between 0.08 and 0.23.

HNC has been detected in 5 comets since 1996, but its origin is puzzling. While production through chemical reaction in the coma of Hale–Bopp was invoked to explain its increasing abundance closer to the Sun, it cannot explain abundances relative to HCN in the range 6–17% observed in other comets. It also seems that there is a general trend of increasing HNC/HCN ratio for decreasing r_h in all comets.

5.2. CO-BEARING MOLECULES

Methanol (CH_3OH) has been detected in 17 comets since C/1989 X1 (Austin) in 1990. Several lines have often been simultaneously detected, enabling temperature measurements. Its abundance relative to water (Table II, Figures 1–2) varies at least twice as much as for HCN: <0.9% to 6.2%, with most comets (including all Jupiter-family comets) being in the 1.5–2.8% range.

Formaldehyde (H_2CO): This molecule has been detected in 13 comets since 1989. We have assumed a distributed source with a Haser equivalent density distribution and a parent lifetime of about 10 000 s at $r_h = 1$ AU, as suggested by many observations (Colom et al., 1992; Biver et al., 1999a, 2002) – this is equivalent to a parent scale-length of 8000 $r_h^{1.5}$ for an average expansion velocity of 0.8 $r_h^{-0.5}$ km s^{-1} (Biver et al., 1999a, 2000). Its abundance relative to water varies by one order of magnitude and a factor of 6 relative to HCN. Curiously, the comets that are abundant in formaldehyde are also abundant in methanol (Figure 3), but owing to the long lifetime of CH_3OH (about 77 000 s at 1 AU), it is not a likely parent molecule for H_2CO. Large polymers thermally degraded on grains are a more likely parent source for formaldehyde, as suggested by Cottin et al. (2001).

CO: other techniques (infrared and UV) are usually more sensitive to measure the CO abundance in comets close to the Sun. But this molecule can have a high abundance (CO/HCN $>$ 85 in 4 long-period comets) above $\approx$ 10% relative to water close to the Sun and much higher further away. Far from the Sun, CO is the main driver of cometary activity and can be detected in the radio out to more than 14 AU (Biver et al., 2002; Biver, 2001). In 4 other comets (including short-period comets) we found CO/HCN below 40. Hence CO might not be the second most abundant species in these comets.

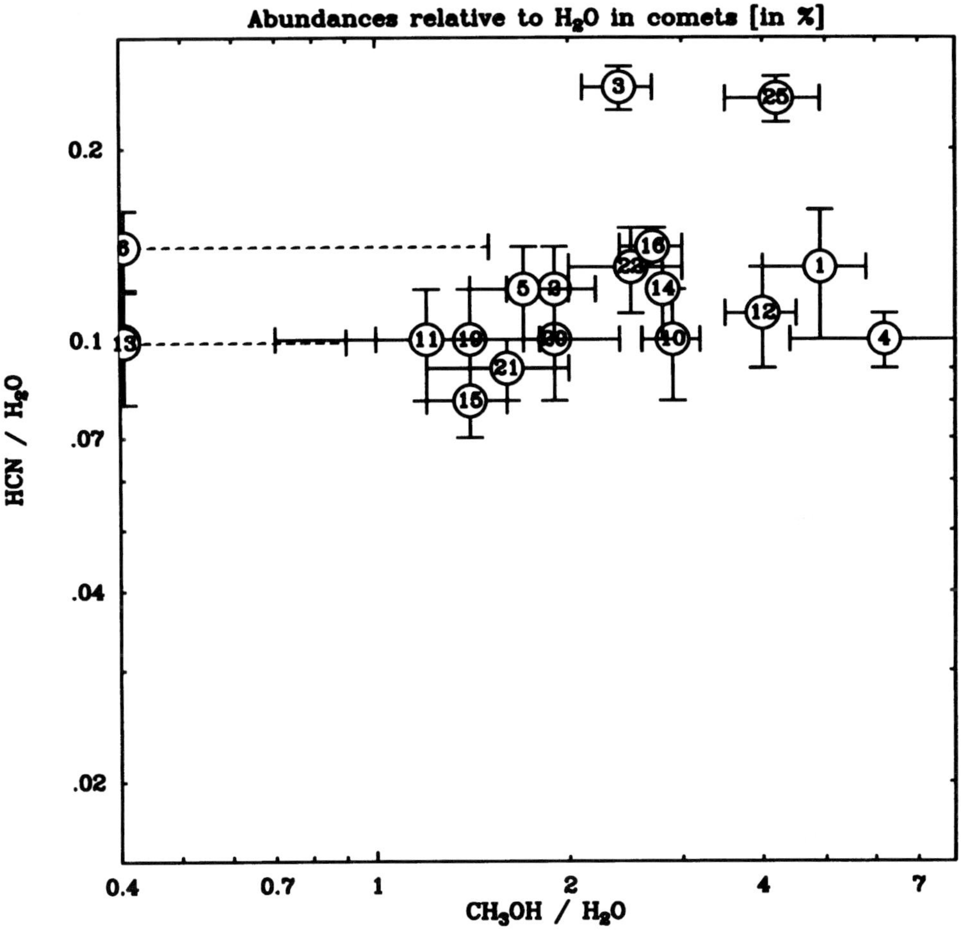

Figure 2. HCN abundance versus CH_3OH abundance (relative to water) in comets. The numbers inside the circles refer to comets as listed in Table I. Dotted line error bars with comet number on one of the axis are 3-σ upper limits for non-detections.

5.3. SULPHUR-BEARING MOLECULES

Hydrogen sulphide (H_2S) was first detected in comets in 1990 (Crovisier et al., 1991) and has been detected in 11 comets since. Its abundance relative to HCN varies by a factor 5 – and even by one order of magnitude with respect to water – but most comets have a ratio H_2S/HCN between 4 and 6 (Figure 4). Variations in H_2S abundances seem completely uncorrelated with dynamical origin.

CS has been recently observed in the radio but was observed in comets in the UV well before 1996. So far we have assumed (in the 9 comets in which we detected CS) that it is the photo-dissociation product of CS_2 (Biver et al., 1999a). However, it was found that the CS/HCN ratio decreased with heliocentric distance as $r_h^{-0.7}$ (Biver et al., 1999a, 2002). This trend is observed in some other comets

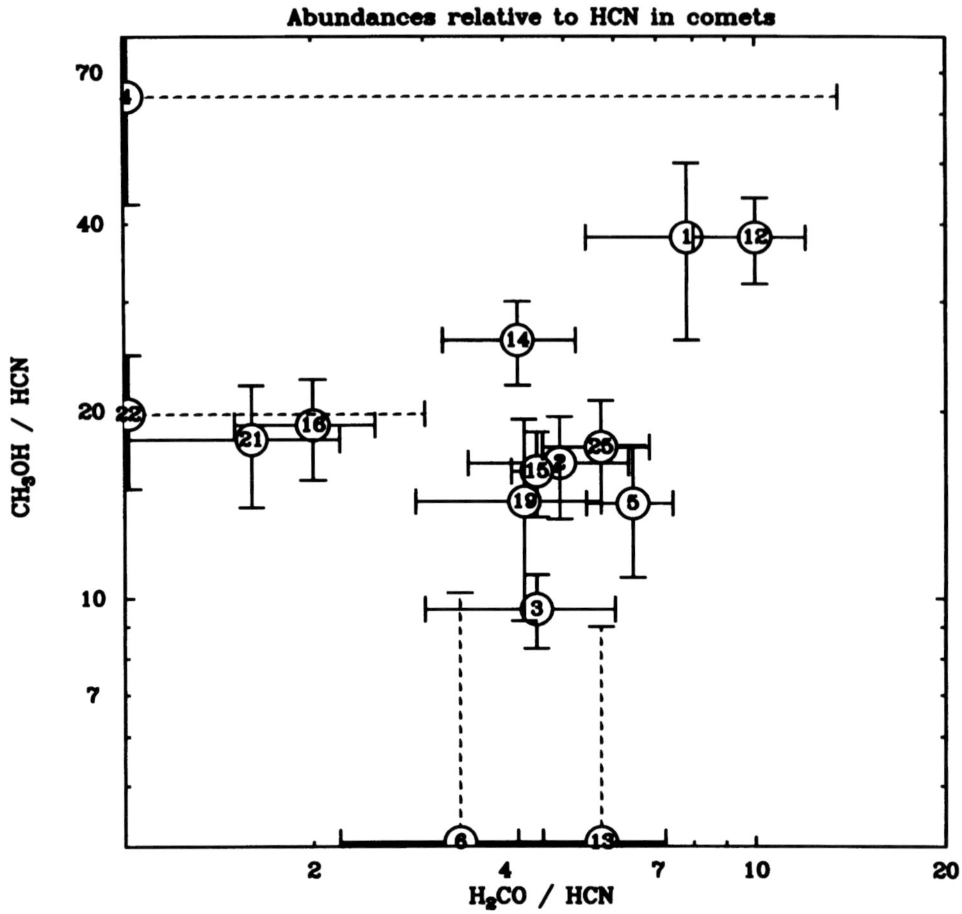

Figure 3. CH_3OH abundance versus H_2CO abundance (relative to HCN) in comets. The comets that are rich in methanol seem also to be abundant in formaldehyde.

and was previously observed in comet 1P/Halley by Meier and A'Hearn (1997). This is puzzling as CS_2 is a very volatile molecule and is not expected to sublimate less easily than H_2O or HCN further away from the Sun. Taking out this trend with r_h, the CS/HCN ratio is close to 0.8 at $r_h \approx 1$ AU in most comets.

6. Conclusion

This work is still in an early phase: A larger sample of comets and a multi-dimensional analysis of the grouping of the comets according to their chemical composition are obviously needed. Heliocentric trends (for HNC, CS and formaldehyde) should be further understood before any definitive conclusions can be drawn. They possibly indicate that the chemical composition of cometary comae is

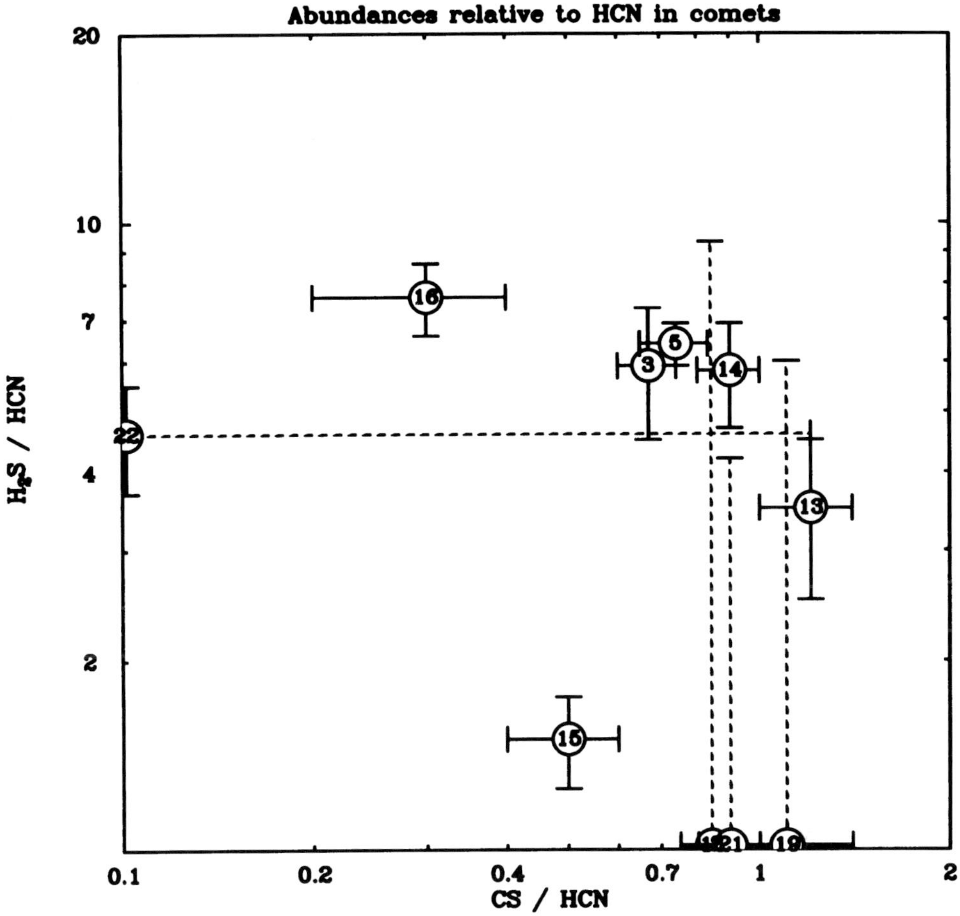

Figure 4. H_2S abundance versus CS abundance (relative to HCN) in comets.

probably not fully representative of nuclei composition. In the current state, comparison for similar heliocentric distances only reveals that Jupiter-family comets, likely formed in the Kuiper Belt, do not show distinct compositions from the others, presumably coming from the Oort Cloud. But one might also expect a larger scatter among long-period comets if they were formed over a wider range of heliocentric distances (Jupiter to Neptune region) in the proto-planetary nebula.

Acknowledgements

We are grateful to the IRAM, JCMT, CSO and SEST staff and other observers for their assistance during the observations. IRAM is an international institute cofunded by the CNRS, the Max Planck Gesellschaft and the Instituto Geografico Nacional, Spain. The JCMT is operated in behalf of the Particle Physics and Astronomy Research Council of the United Kingdom, the Netherlands Organization for Scientific Research, and the National Research Council of Canada. The CSO is supported by National Science Foundation grant AST 99-80846. The SEST is operated jointly by the Swedish National Facility for Radio Astronomy, and by the European Southern Observatory.

References

A'Hearn, M. F., Millis, R. L., Schleicher, D. G., Osip, D. J., and Birch, P. V.: 1995, 'The Ensemble Properties of Comets: Results from Narrowband Photometry of 85 Comets, 1976–1992', *Icarus* **118**, 223–270.

Biver, N: 1997, 'Molécules Mères Cométaires: Observations et Modélisations', Ph.D. Thesis, Paris-7 University.

Biver, N.: 2001, 'Correlation between Visual Magnitudes and Outgassing Rate of CO in Comets Beyond 3 AU', *Int. Comet Quart.* **23**(3), 85–93.

Biver, N., Bockelée-Morvan, D., Colom, P. et al.: 1999, 'Long-term Evolution of the Outgassing of Comet Hale–Bopp from Radio Observations', *Earth Moon Planets* **78**, 5–11.

Biver, N., Bockelée-Morvan, D., Colom, P. et al.: 2002, 'The 1995–2002 Long-Term Monitoring of Comet C/1995 O1 (Hale–Bopp) at Radio Wavelength', these proceedings.

Biver, N., Bockelée-Morvan, D., Crovisier, J. et al.: 1999, 'Spectroscopic Monitoring of Comet C/1996 B2 (Hyakutake) with the JCMT and IRAM Radio Telescopes', *Astron. J.* **118**, 1850–1872.

Biver, N., Bockelée-Morvan, D., Crovisier, J. et al.: 2000, 'Spectroscopic Observations of Comet C/1999 H1 (Lee) with the SEST, JCMT, CSO, IRAM, and Nançay Radio Telescopes', *Astron. J.* **120**, 1554–1570.

Bockelée-Morvan, D. and Crovisier, J.: 2002, 'Lessons of Comet Hale–Bopp for Coma Chemistry: Observations and Theory', these proceedings.

Bockelée-Morvan, D., Biver, N., Colom, P., Crovisier, J., Festou, M. C., Henry, F., Lecacheux, A., and Weaver H. A.: 2002, 'Radio Investigation of 19P/Borrelly in Support to the Deep Space 1 Flyby', these proceedings.

Bockelée-Morvan D., Biver, N., Moreno, R. et al.: 2001, 'Outgassing Behavior and Composition of Comet C/1999 S4 (LINEAR) during its Disruption', *Science* **292**, 1339–1343.

Bockelée-Morvan, D., Crovisier, J., Colom, P., and Despois, D.: 1994a, 'The Rotational Lines of Methanol in Comets Austin 1990 V and Levy 1990 XX', *Astron. Astrophys.* **287**, 647–665.

Bockelee-Morvan, D., Padman, R., Davies, J. K., and Crovisier, J.: 1994b, 'Observations of Submillimetre Lines of CH_3OH, HCN, and H_2CO in Comet P/Swift–Tuttle with the James Clerk Maxwell Telescope', *Planet. Space Sci.* **42**, 193–198.

Bockelée-Morvan, D., Lis, D. C., Wink, J. E. et al.: 2000, 'New Molecules Found in Comet C/1995 O1 (Hale–Bopp). Investigating the Link between Cometary and Interstellar Material', *Astron. Astrophys.* **353**, 1101–1114.

Colom, P., Crovisier, J., Bockelée-Morvan, D., Despois, D., and Paubert, G.: 1992, 'Formaldehyde in Comets I. Microwave Observations of P/Brorsen-Metcalf (1989 X), Austin (1990 V) and Levy (1990 XX)', *Astron. Astrophys.* **264**, 270–281.

Cottin, H., Gazeau, M. C., Benilan, Y., and Raulin, F.: 2001, 'Polyoxymethylene as Parent Molecule for the Formaldehyde Extended Source in Comet Halley', *Astrophys. J.* **556**, 417–420.

Crovisier, J., Biver, N., Moreno, R. et al.: 2001, 'Spectroscopic Investigation of Comets C/1999 T1 (McNaught–Hartley) and C/2001 A2 (LINEAR) at Radio Wavelengths: Evidence for Chemical Diversity', *B.A.A.S.* **33**, 1121.

Crovisier, J., Bockelée-Morvan, D., Colom, P., Despois, D., and Paubert, G.: 1991, 'Microwave Observations of Hydrogen Sulfide and Searches for Other Sulfur Compounds in Comets Austin (1989c1) and Levy (1990c)', *Icarus* **93**, 246–258.

Crovisier, J., Colom, P., Gérard, E., and Bockelée-Morvan, D.: 2002a, 'Observation of the OH Radio Lines of Comets at Nançay – Recent Developments', these proceedings.

Crovisier, J., Colom, P., Gérard, E., Bockelée-Morvan, D., and Bourgois, G.: 2002b, 'Observations of the OH 18-cm Lines in Comets at Nançay: The Data Base. Observations made from 1982 to 1999', *Astron. Astrophys.* **393**, 1053–1064.

Despois, D., Biver, N., Bockelée-Morvan, D., Colom, P., Crovisier, J., and Paubert, G.: 1996, 'Radio Line Observations of Comet 109P/Swift-Tuttle at IRAM', *Planet. Space Sci.* **44**, 529–539.

Despois, D., Crovisier, J., Bockelée-Morvan, D., Gérard, E., and Schraml, J.: 1986, 'Observations of Hydrogen Cyanide in Comet Halley', *Astron. Astrophys.* **160**, L11–L12.

DiSanti, M. A., Mumma, M. J., Dello Russo, N., Magee-Sauer, K., Novak, R., and Rettig, T. W.: 2000, 'A Comparison of CO Abundances from Infrared Observations of Recent Comets', *B.A.A.S.* **32**, 1071.

Feldman, P. D., Festou, M. C., Tozzi, G. P., and Weaver, H. A.: 1997, 'The CO_2/CO Abundance Ratio in 1P/Halley and Several Other Comets Observed by IUE and HST', *Astrophys. J.* **475**, 829–834.

Huebner, W. F.: 2002, 'Composition of Comets: Observations and Models', these proceedings.

Jorda, L., Crovisier, J., and Green, D. W. E.: 1992, 'The Correlation between Water Production Rates and Visual Magnitudes in Comets', in *Asteroids, Comets, Meteors 1991, Lunar and Planetary Inst.*, pp. 285–288.

Meier, R. and A'Hearn, M. F.: 1997, 'Atomic Sulfur in Cometary Comae Based on UV Spectra of the S I Triplet near 1814 Å', *Icarus* **125**, 164–194.

Meier, R., Eberhardt, P., Krankowsky, D., and Hodges, R. R.: 1993, 'The Extended Formaldehyde Source in Comet Halley', *Astron. Astrophys.* **277**, 677–690.

Rodgers, S. D. and Charnley, S. B.: 1998, 'HNC and HCN in Comets', *Astrophys. J. Lett.* **501**, L227–L230.

Wink, J., Bockelée-Morvan D., Despois, D. et al.: 1999, 'Evidences for Extended Sources and Temporal Modulations in Molecular Observations of C/1995 O1 (Hale–Bopp) at IRAM Interferometer', *Earth Moon Planets* **78**, 63.

NITROGEN SULFIDE IN COMETS HYAKUTAKE (C/1996 B2) AND HALE–BOPP (C/1995 O1)

MARCUS V. CANAVES* and AMAURY A. DE ALMEIDA**
Department of Astronomy, Institute of Astronomy, Geophysics and Atmospheric Sciences, University of São Paulo, Main Campus, Rua do Matão 1226, CEP 05508-0900, São Paulo, São Paulo, Brazil

DANIEL C. BOICE***
Southwest Research Institute, Department of Space Science, Instrumentation and Space Research Division, P.O. Drawer 28510, San Antonio, Texas 78228-0510, USA

GILBERTO C. SANZOVO‡
Laboratory of Molecular Astrophysics, Department of Physics, State University of Londrina, Perobal, CEP 86051-970, Londrina, Paraná, Brazil

(Received 4 March 2002; Accepted 20 June 2002)

Abstract. The chemistry of both nitrogen and sulfur presents interesting problems in comets. In this paper, we use a model of cometary comae with gas-phase chemical kinetics and gas dynamics to predict molecular abundances in the inner coma region for two of the brightest comets in the past 20 years, Hyakutake (C/1996 B2) and Hale–Bopp (C/1995 O1). In this progress report we concentrate on the gas-phase chemistry of the nitrogen sulfide (NS) radical at a heliocentric distance of 1 AU to study the abundance of NS using a detailed photo and chemical reaction network with over 100 species and about 1000 reactions. The results are compared with recent observations of Comet Hale–Bopp and reveal that conventional gas-phase reactions schemes do not produce NS in sufficient quantities to explain the observations. We plan to continue the refinement of the model to improve agreement with observational constraints.

Keywords: Abundance, comet C/1995 O1 Hale–Bopp, comet C/1996 B2 Hyakutake, cometary comae, nitrogen sulfide

1. Introduction

Nitrogen sulfide (NS) was first identified in the giant molecular cloud, Sgr B2 (Gottlieb et al., 1975) at 115 GHz (transition $J = 5 \rightarrow 3$), and is now known to be common in dense interstellar clouds in regions of massive star formation (McGonagle et al., 1992). It has been detected also in cold dark clouds (McGonagle et al., 1994). Recently, Irvine et al. (2000) reported the first detection of the NS radical in the coma of a comet using the James Clerk Maxwell Telescope (JCMT) on March 22, 1997, to observe Hale–Bopp (C/1995 O1) at a heliocentric distance of

* E-mail: mcanaves@astro.iag.usp.br
** E-mail: amaury@astro.iag.usp.br
*** E-mail: dboice@swri.edu
‡ E-mail: gsanzovo@uel.br

0.932 AU and a geocentric distance of 1.315 AU. This constitutes the first cometary molecular species to contain both nitrogen and sulfur elements observed within the collision-dominated portion of the coma. In the present study, we show the results of a cometary coma model with reactive gas dynamics to study the chemical evolution of sulfur-bearing species for the inner coma with conditions appropriate for Hyakutake (C/1996 B2) and Hale–Bopp in order to estimate the abundance of the NS radical in the coma. We seek also to study the sensitivity of the results on the free parameters in the model, such as the abundances of the minor species on the nucleus surface.

2. Model Summary

In this work, we obtain a self-consistent model of coma chemistry combined with fluid dynamics appropriate for a steady, spherically symmetric outflow and neglecting solar wind interaction. In the collision-dominated region of the coma, we represent the gas as four fluids, treating the electrons, fast atomic hydrogen, and fast molecular hydrogen separately from the remaining neutrals and ions. The system of coupled ordinary differential equations is solved using the Gear method (Gear, 1971; Kahaner, 1989) for stiff differential equations. A full description of the model is given by Benkhoff and Boice (1996), Boice et al. (1990), Huebner et al. (1991) and references therein. The model includes 152 species linked by about 1000 reactions with photorates taken from Huebner et al. (1992), Schmidt et al. (1988), and the UMIST reaction rate database (Le Teuff et al., 2000; Millar et al., 1997). The effects of dust in the coma are ignored in this study which limits our preliminary results to identify important gas-phase processes.

The starting conditions of our coma calculations are determined by an assumed chemical composition of the nucleus where the goal is to adjust the composition until the calculation is consistent with the coma observations. From energy balance and the equation of state for an ideal gas, we can calculate pressure p_0, number density N_0, specific molecular production rate Z_0, sound speed v_s of the gas, and surface gas temperature T_0. The visual albedo A_v and the infrared emissivity ε_{ir} are assumed parameters in such calculations as well as the mean latent heat $L(T)$, the mean molecular weight M and the adiabatic exponent γ (ratio of specific heats) which are determined from the assumed constituent frozen gases in the nucleus.

3. The Chemistry of Nitrogen Sulfide

The chemistry of sulfur in cometary coma has been a matter of interest since the pioneering spectrographic study of Smith et al. (1976) and Smith et al. (1980) on comet West (1976 VI = C/1975 V1-A) which revealed for the first time ultraviolet emissions from sulfur atoms and carbon monosulfide (CS) molecules. This was strengthened by the discovery of S_2 in the ultraviolet spectra of comet IRAS-Araki-Alcock (I-A-A) (1983 VII = C/1983 H1) by A'Hearn et al. (1983) using the *International Ultraviolet Explorer* (IUE) satellite. On May 11.5, 1983, comet I-A-A had its closest approach to Earth at a distance of 0.0312 AU, the closest in about 200 years since comet Lexell (D/1770 L1) on Jul. 1.7, 1770 (0.0151 AU) and this was the first detection of a molecular sulfur species in any astrophysical object.

After the advent of comets Hyakutake and Hale–Bopp, our knowledge about sulfur-bearing molecules increased considerably. Building on this knowledge, we have managed to identify a chemical network mechanism to compute the spatial distribution of the number densities of NS at various cometocentric distances in the inner coma region from the nucleus of these comets.

The main process, to form the NS radical is electron dissociative recombination of the HNS^+ ion,

$$\mathrm{HNS}^+ + e^- \rightarrow \mathrm{NS} + \mathrm{H} \tag{1}$$

while another reaction to form NS is the neutral rearrangement,

$$\mathrm{HS} + \mathrm{N} \rightarrow \mathrm{NS} + \mathrm{H} \tag{2}$$

with rate coefficients $k_1 = 3.0 \times 10^{-7}(T/300)^{-0.5}$ cm^3 s^{-1} and $k_2 = 1.73 \times 10^{-11}(T/300)^{0.5}$ cm^3 s^{-1}, respectively. A network of six reactions for destruction is used in our model with two reactions of positive ion charge transfer,

$$\mathrm{H}^+ + \mathrm{NS} \rightarrow \mathrm{NS}^+ + \mathrm{H} \tag{3}$$

and

$$\mathrm{C}^+ + \mathrm{NS} \rightarrow \mathrm{NS}^+ + \mathrm{C} \tag{4}$$

with rate coefficients $k_3 = 4.70 \times 10^{-9}$ cm^3 s^{-1} and $k_4 = 7.60 \times 10^{-10}$ cm^3 s^{-1}, respectively. We also have three reactions of positive ion-atom interchange,

$$\mathrm{C}^+ + \mathrm{NS} \rightarrow \mathrm{CS}^+ + \mathrm{N} \tag{5}$$

$$\mathrm{H}_3^+ + \mathrm{NS} \rightarrow \mathrm{HNS}^+ + \mathrm{H}_2 \tag{6}$$

and

$$\mathrm{HCO}^+ + \mathrm{NS} \longrightarrow mbox\,HNS^+ + \mathrm{CO}, \tag{7}$$

where $k_5 = 7.60 \times 10^{-10}$ cm^3 s^{-1}, $k_6 = 2.80 \times 10^{-9}$ cm^3 s^{-1} and $k_7 = 1.10 \times 10^{-9}$ cm^3 s^{-1}, respectively. Rates for reactions (1) through (7) are taken from Le Teuff et al. (2000). To the best of our knowledge, the lifetime and the rate coefficient for

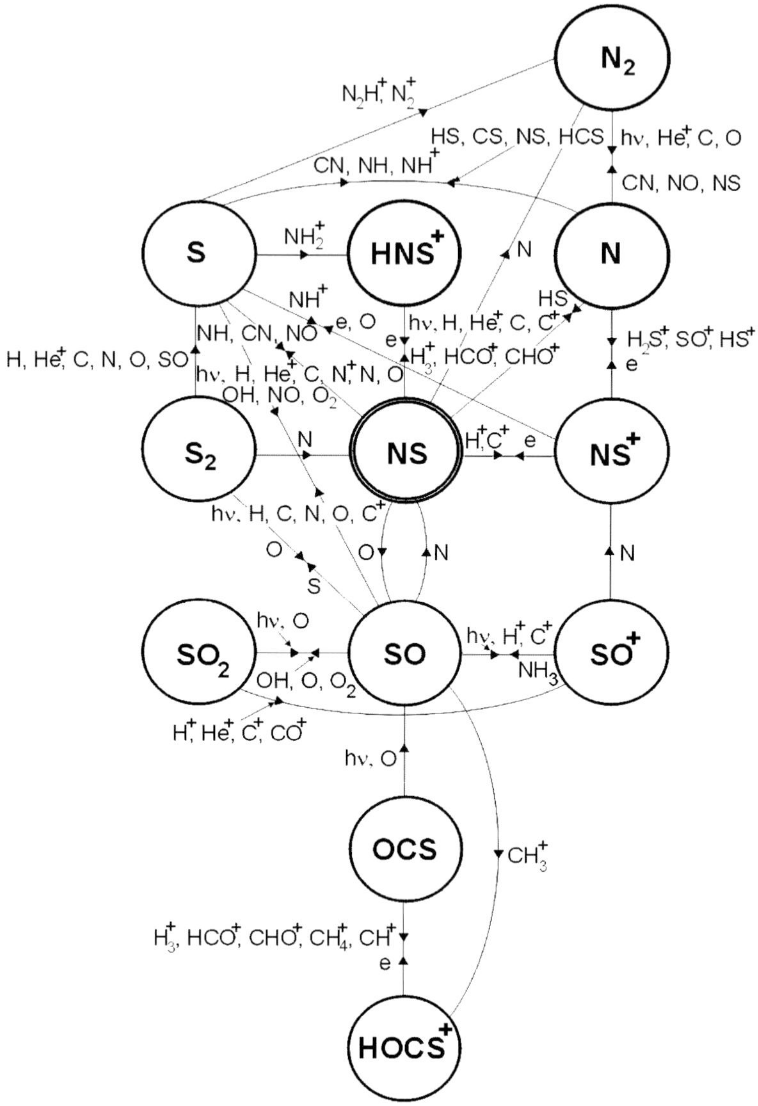

Figure 1. Schematic diagram showing the major sulfur reactions relevant to the inner coma for the NS radical. Symbols next to the arrows are the reactants; $h\nu$ indicates photodissociation.

photodissociation of NS has not been measured or calculated. Our estimate for the rate coefficient for the photodissociation of nitrogen sulfide

$$NS + h\nu \rightarrow N + S \tag{8}$$

is $k_8 \sim 1 \times 10^{-6}\ s^{-1}$. The major sulfur reactions for the NS radical are shown schematically in Figure 1.

TABLE I

Initial chemical abundances used in the model for comet Hyakutake at $r_h \sim 1$ AU

Molecular species	Molecular abundance	Reference
H_2O	100	–
CO_2	7	a, g
CO	30	d, e
CH_3OH	2	d
H_2CO	1	d
H_2CS	0.02	i
CH_4	0.7	e
NH_3	0.5	f
HCN	0.1	d
HNC	0.01	d, c
N_2	0.02	i
H_2S	0.6	a
CS_2	0.1	a
OCS	0.3	a
C_2H_2	0.5	b
C_2H_6	0.4	e
S_2	0.005	h

[a]Bockelée-Morvan (1997); [b]Brooke et al. (1996); [c]Irvine (1996); [d]Lis et al. (1997); [e]Mumma et al. (1996); [f]Palmer et al. (1996); [g]Weaver et al. (1996); [h]Woodney et al. (1997); [i]adopted here.

4. Model Results for Comet Hyakutake

In our model for comet Hyakutake, the fundamental parameters are: Heliocentric distance $r_h = 1$ AU, a nucleus with a radius $r_n = 2.24$ km (Lisse et al., 1999, Sarmecanic et al., 1997), a mean visual albedo $\overline{A}_v = 0.04$ and a mean infrared emissivity $\overline{\varepsilon}_{ir} = 0.96$. The initial chemical abundances in the nucleus are listed in Table I.

We assume that the ice is sublimating at the surface of the nucleus and that the vapor is in equilibrium with the surface. The vapor has a temperature that will be quite below the black body equilibrium temperature because the vaporization carries away much of the heat as latent heat of vaporization (A'Hearn and Festou, 1990). In this work, the sublimation (vaporization) will be controlled by H_2O with a temperature of 200 K at a heliocentric distance of 1 AU. The solution of the energy balance equations at the surface using these parameters yields a specific gas production rate $Z_0 = 9.7 \times 10^{17}$ molecules cm^{-2} s^{-1}, a total gas production rate $Q_0 = 6.1 \times 10^{29}$ molecules s^{-1}, and a sound speed $v_s = 0.32$ km s^{-1} at the nuclear surface.

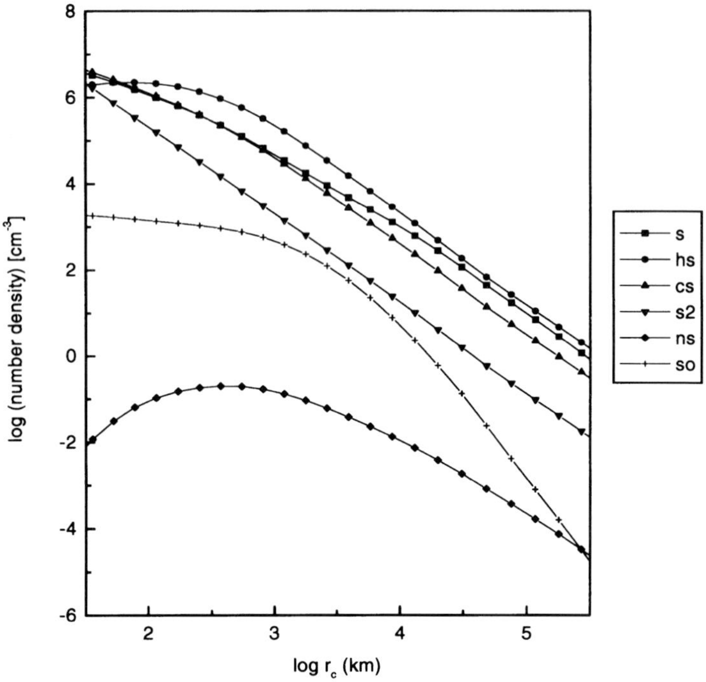

Figure 2. Number density of the main sulfur-bearing species as a function of cometocentric distance for comet Hyakutake.

Figure 2 shows the number density profiles of the sulfur-bearing species as a function of cometocentric distance for comet Hyakutake. At a distance of $\sim 5.80 \times 10^3$ km from the nucleus, the neutral gas consists of 58.4% H_2O while at a distance of 1.32×10^6 km, atomic oxygen is the most abundant species (50.4%).

In the inner coma region at $\sim 5.80 \times 10^3$ km, the plasma without solar wind interaction consists of 74% H_3O^+, 9.2% H_2O^+, 1.3% NH_4^+, 2% CO^+, 0.8% OH^+, 0.8% NH_3^+ and 0.3% O^+. In Figure 3 we show the profiles for thermodynamic variables: (a) Total velocity of the flow, (b) gas temperature, (c) electron temperature and (d) total pressure of the flow (sum of the partial pressures of the chemical species).

For the flow velocity at far distances, the solution exhibits an increase with radial distance due to the surplus energy of the dissociation processes. There is a rapid decrease of electron and gas temperatures due to adiabatic expansion cooling that dominates the photolytic heating close to the nucleus. For the gas temperature, the initial value of 200 K falls to about 11.5 K at a cometocentric distance of r_c = 171 km. If we assume equipartition, some fraction of the excess energy of dissociative and chemical heating will go into translation and rotation and cause progressive increase of the coma temperature until it reaches its peak with a value of 128 K near the outer fringes of the collision zone. In the inner coma the fluid approximation for the electron gas is appropriate because they thermalize rapidly

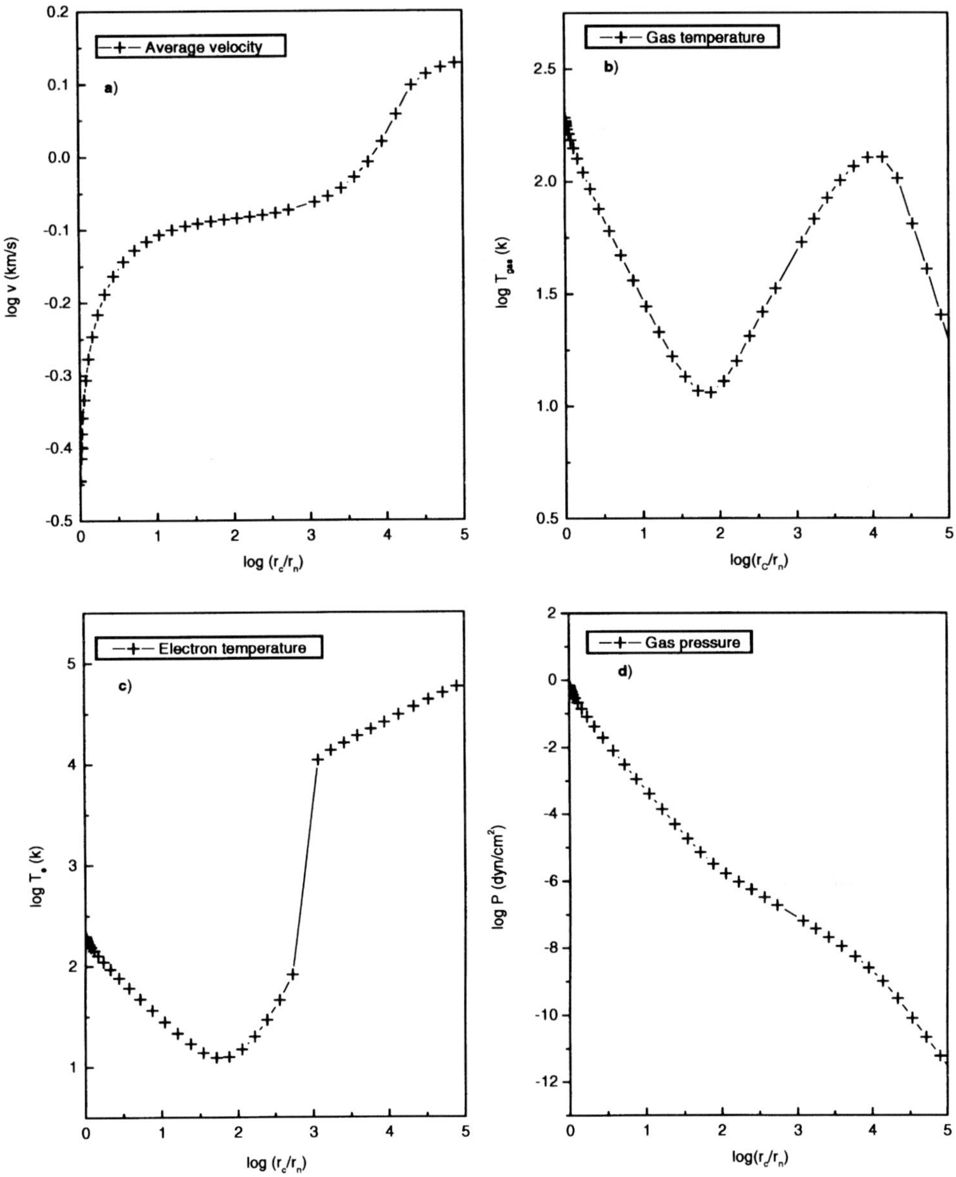

Figure 3. Thermodynamic variables as a function of cometocentric distance for comet Hyakutake.

through inelastic collisions with polar molecules, primarily water, but the increase of electron temperature in Figure 3c signifies the thermal decoupling from the coma gas.

In the model, the cross section radius for collisions between molecules of the fluid is $a = 4.4$ Å. The ratio between the mean free path of fluid molecules ($\Lambda_f = 1/\sqrt{2}\sigma n$ where $\sigma = \pi a^2$) and the nuclear radius (r_n) of the comet gives

TABLE II

Initial chemical abundances used in the model for comet Hale–Bopp at $r_h \sim 1$ AU

Molecular species	Molecular abundance	Reference
H_2O	100	–
CO_2	6	e
CO	23	c
CH_3OH	2.4	c
H_2CO	1	d
H_2CS	0.02	h
CH_4	0.6	g
NH_3	0.7	a
HCN	0.25	c
HNC	0.04	b, f
N_2	0.02	i
H_2S	1.5	c
CS_2	0.17	c
OCS	0.4	c
C_2H_2	0.1	g
C_2H_6	0.3	g
S_2	0.005	d

[a]Bird et al. (1997); [b]Biver et al. (1997); [c]Bockelée-Morvan et al. (2000); [d]Crovisier (1998); [e]Crovisier et al. (1997); [f]Irvine et al. (1998); [g]Weaver et al. (1997b); [h]Woodney et al. (1997); [i]adopted here.

us an estimate where the fluid flow goes over into free molecular flow. For comet Hyakutake we have $\Lambda_f/r_n \approx 1$ at a distance $\sim 3.72 \times 10^2$ km from the nucleus, that should be a first order estimate of the collision zone.

Our model gives a total column density of NS of $\sim 1.6 \times 10^7$ cm^{-2} and an estimate for the number density of NS in comet Hyakutake shows an upper limit of ~ 0.02 cm^{-3} at a distance $r_c \sim 3.72 \times 10^2$ km. However, there are no observational estimates of this species for comet Hyakutake.

5. Model Results for Comet Hale–Bopp

The model parameters which we use for comet Hale–Bopp are: Heliocentric distance $r_h = 1$ AU, a nucleus radius $r_n = 20$ km (Sekanina, 1997; Weaver et al., 1996, 1997a; Weaver and Lamy, 1997), a mean visual albedo $\overline{A}_v = 0.04$, and a mean infrared emissivity $\overline{\varepsilon}_{ir} = 0.96$. The initial chemical abundances in the nucleus are listed in Table II.

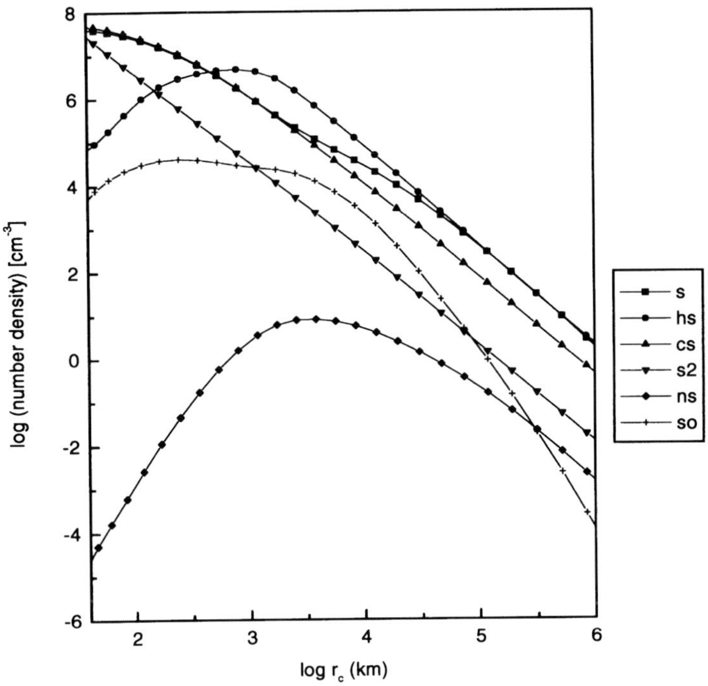

Figure 4. Number density of the main sulfur-bearing species as a function of cometocentric distance for comet Hale–Bopp.

We also assume in this model a sublimation controlled by H_2O with a temperature of 200 K at a heliocentric distance of 1 AU. The solution of the energy balance equations yields a specific gas production rate $Z_0 = 2 \times 10^{17}$ molecules cm^{-2} s^{-1}, a total gas production rate $Q_0 = 1.0 \times 10^{31}$ molecules s^{-1}, and a sound speed $v_s = 0.32$ km s^{-1} at the nuclear surface.

Figure 4 shows the number density profiles of the sulfur-bearing species for comet Hale–Bopp. At a distance of $\sim 8.50 \times 10^3$ km from the nucleus, the neutral gas consists of 58% H_2O and 17% CO, while at a distance of 2.13×10^6 km, atomic hydrogen is the most abundant species with 48%. In the inner coma, the plasma without solar wind interaction consists of 74% H_3O^+, 4.1% H_2O^+ and S^+ is the most abundant atomic ion with 0.9% at a distance of $\sim 8.50 \times 10^3$ km. In Figure 5, we show the profiles for thermodynamic variables. We note that the profiles of electron temperature, average velocity and gas temperature are higher than those of comet Hyakutake. For the gas temperature, the initial value of 200 K falls to about 22.2 K at a cometocentric distance of $r_c = 536$ km. Beyond that distance, the temperature increases gradually, reaching a value of 369 K at 1.91×10^5 km. In the case of comet Hale–Bopp, the ratio Λ_f/r_n becomes ≈ 1 at a distance $\sim 3.8 \times 10^3$ km.

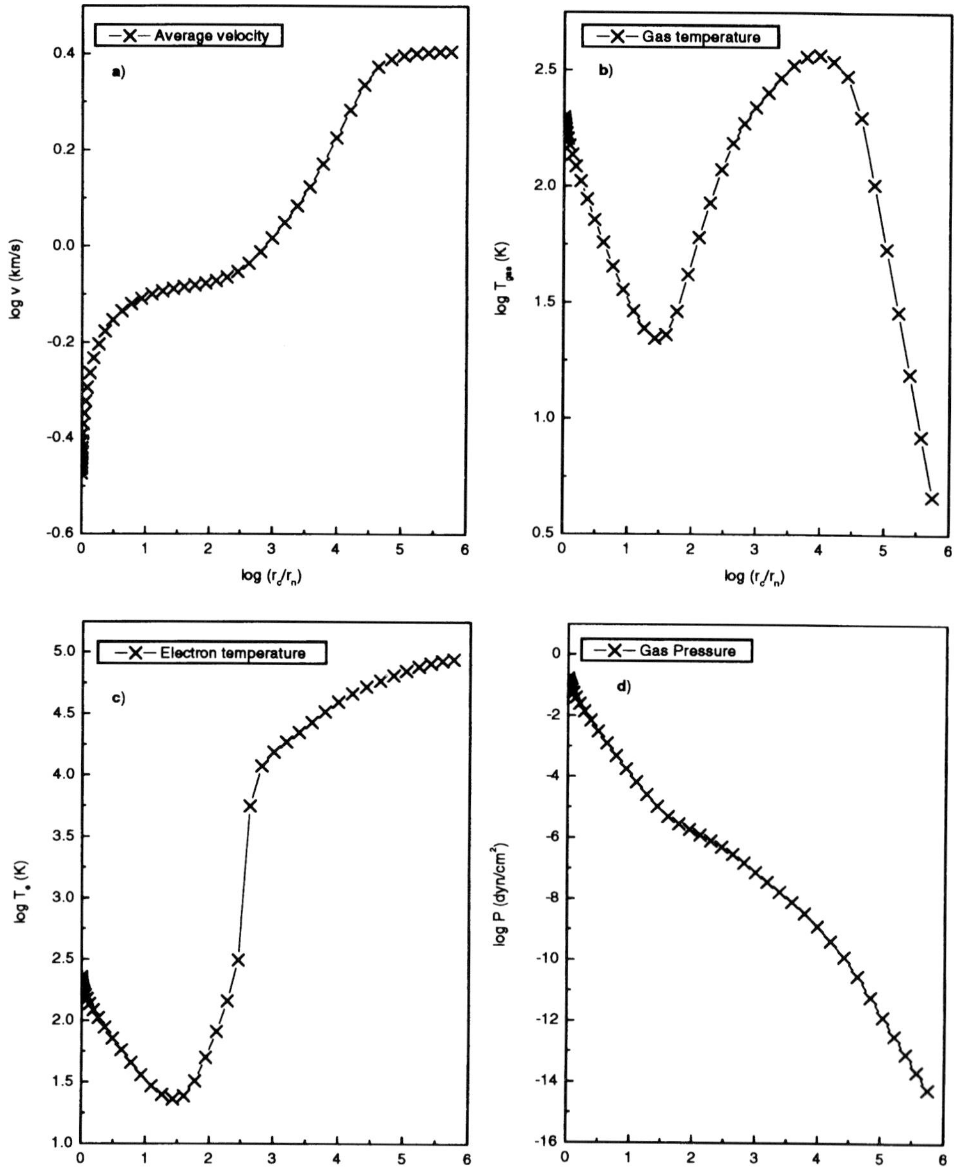

Figure 5. Thermodynamic variables as a function of cometocentric distance for comet Hale–Bopp.

Cometary nitrogen sulfide was first observed by Irvine et al. (2000) in comet Hale–Bopp with an abundance relative to water of at least a few hundredths of a percent. They reported a total column density of $\sim 6.8 \times 10^{12}$ cm^{-2} for nitrogen sulfide by observing the $J = 15/2 \rightarrow 13/2$ parity-e and parity-f transitions. In our model, the total column density of NS is $\sim 4.0 \times 10^{9}$ cm^{-2} which is smaller than the observational value of Irvine et al. (2000) by a factor of 1.7×10^3 and higher by a factor of ~ 250 for comet Hyakutake. The maximum number density was 1 cm^{-3} at a distance $r_c \sim 3.8 \times 10^3$ km.

6. Conclusions

There is a consensus that cometary nuclei are remnants of the planetesimal population of the outer parts of the Solar Nebula and that the composition of their volatiles should provide clues to the composition of the Solar Nebula. There is good agreement for some species, while many others remain uncertain.

In this paper, we show the abundance profiles for sulfur-bearing species, including nitrogen sulfide, as a function of cometocentric distance with conditions appropriate to comets Hyakutake and Hale–Bopp at 1 AU from the Sun. From the modelling we estimate for comet Hyakutake a column density of the radical NS of $\sim 1.6 \times 10^7$ cm^{-2} and a number density of 0.02 cm^{-3} for $r_c \sim 3.72 \times 10^2$ km. In the case of comet Hale–Bopp, our model predicts a column density of $\sim 4.0 \times 10^9$ cm^{-2}, while the number density has an upper limit of ~ 1 cm^{-3} at $r_c \sim 3.8 \times 10^3$ km. It appears that the abundances predicted here by gas-phase models for the radical NS are much smaller than those observed by Irvine et al. (2000). The chemical reactions shown in Section 3 and schematically in Figure 1 are a conventional network to study nitrogen sulfide in a comet, containing the main formation mechanisms for NS under dark cloud conditions, i.e., reactions (1) and (2). McGonagle et al. (1994) reported that the measured abundances of interstellar nitrogen sulfide were significantly higher than those predicted by some recent gas-phase, ion-molecule models. These authors concluded that the abundance of NS depends on a relative long series of reactions in gas-phase models. We believe that a similar series of reactions may also occur in a cometary coma. Clearly there is an important unknown parent molecular species (X-NS) which photodissociates to produce NS.

We continue the refinement of the model, through the inclusion of additional collisional and photolytic gas-phase reactions and other physical processes (dust) aimed at further improvement of the accuracy of the abundances with respect to observational constraints.

Acknowledgements

M. V. Canaves and A. A. de Almeida gratefully acknowledge support from FAPESP (grant no, 99/07372-1). D. C. Boice is thankful to FAPESP (grant no. 01/03268-7) and to the hospitality of IAG/USP during his extend stay in 2001. G. C. Sanzovo is grateful to UEL, Londrina, for a leave of absence. We appreciate the constructive comments of the anonymous referee.

References

A'Hearn, M. F. and Festou, M. C.: 1990, 'The Neutral Coma', in W. F Huebner (ed.), *Physics and Chemistry of Comets*, Springer-Verlag.

A'Hearn, M. F., Feldman, P. D., and Schleicher, D. G.: 1983, 'The Discovery of S_2 in Comet IRAS-Araki-Alcock 1983d', *ApJ (Lett.)* **274**, L99–L103.

Benkhoff, J. and Boice, D. C.: 1996, 'Modeling the Thermal Properties and the Gas Flux From a Porous, Ice-Dust Body in the Orbit of P/Wirtanen', *P&SS* **44**, 665–673.

Bird, M. K., Janardhan, P., Wilson, T. L., Huchtmeier, W. K., Gensheimer, P., and Lemme, C.: 1997, 'K-Band Radio Observations of Comet Hale–Bopp: Detections of Ammonia and (Possibly) Water', *Earth Moon Planets* **78**, 21–28.

Biver, N., Bockelée-Morvan, D., Colom, P., Crovisier, J., Germain, B., Lellouch, E., Davies, J. K., Dent, W. R. F., Moreno, R., Paubert, G., Wink, J., Despois, D., Lis, D. C., Mehringer, D., Benford, D., Gardner, M., Phillips, T. G., Gunnarson, M., Rickman, H., Winnberg, A., Bergman, P., Johansson, L. E. B., and Rauer, H.: 1997, 'Long-Term Evolution of the Outgassing of Comet Hale–Bopp from Radio Observations', *Earth Moon Planets* **78**, 5–11.

Bockelée-Morvan, D.: 1997, 'Cometary Volatiles: The Status after Comet C/1996 B2 Hyakutake', in E. F. van Dishoek (ed.), *Molecules in Astrophysics: Probes and Processes*, IAU, Symp. 178, pp. 219–235.

Bockelée-Morvan, D., Lis, D. C., Wink, J. E., Despois, D., Crovisier, J., Bachiller, R., Benford, D. J., Biver, N., Colom, P., Davies, J. K., Gérard, E., Germain, B., Houde, M., Mehringer, D., Moreno, R., Paubert, G., Phillips, T. G., and Rauer, H.: 2000, 'New Molecules Found in Comet C/1995 O1 (Hale–Bopp): Investigating the Link between Cometary and Interstellar Material', *A&A* **353**, 1101–1114.

Boice, D. C., Huebner, W. F., Sablik, M. J., and Konno, I.: 1990, 'Distributed Coma Sources and the CH_4/CO Ratio in Comet Halley', *Geophys. Res. Lett.* **17**, 1813–1816.

Brooke, T. Y., Tokunaga, A. T., Weaver, H. A., Crovisier, J., Bockelée-Morvan, D., and Crisp, D.: 1996, 'Detection of Acetylene in the Infrared Spectrum of Hyakutake', *Nature* **383**, 606–608.

Crovisier, J.: 1998, 'Physics and Chemistry of Comets: Recent Results from Comets Hyakutake and Hale–Bopp: Answers to Old Questions and New Enigmas', *Faraday Discuss.* **109**, 437–452.

Crovisier, J., Leech, K., Bockelée-Morvan, D., Brooke, T. Y., Hanner, M. S., Atieri, B., Keller, H. U., and Lellouch, E.: 1997, 'The Spectrum of Comet Hale–Bopp (C/1995 O1) Observed with the Infrared Space Observatory at 2.9 AU from the Sun', *Science* **275**, 1904–1907.

Gear, C. W.: 1971, *Numerical Initial Value Problems in Ordinary Differential Equations*, Prentice Hall, Englewood Cliffs, NJ.

Gottlieb, C. A., Ball, J. A., Gottlieb, E. W., Lada, C. J., and Penfield, H.: 1975, 'Detection of Interstellar Nitrogen Sulfide', *ApJ* **200**, L147–L149.

Huebner, W. F., Boice, D. C., Schmidt, H. U., and Wegmann, R.: 1991, 'Structure of the Coma: Chemistry and Solar Wind Interaction', in R. L. Newburn Jr., M. Neugebauer, and J. Rahe (eds.), *Comets in the Post-Halley Era*, Vol. 2, Kluwer Academic Publishers, Dordrecht, pp. 907–936.

Huebner, W. F., Keady, J. J., and Lyon, S. P.: 1992, *Solar Photo Rates for Planetary Atmospheres and Atmospheric Pollutants*, Kluwer Academic Publishers, Dordrecht.

Irvine, W. M.: 1996, 'Spectroscopic Evidence for Interstellar Ices in Comet Hyakutake', *Nature* **383**, 418–420.

Irvine, W. M., Bergin, E. A., Dickens, J. E., Jewitt, D., Lovell, A. J., Matthews, H. E., Schloerb, F. P., and Senay, M. C.: 1998, 'Chemical Processing in the Coma as the Source of Cometary HNC', *Nature* **393**, 547–550.

Irvine, W. M., Senay, M., Lovell, A. J., Matthews, H. E., McGonagle, D., and Meier, R.: 2000, 'Note: Detection of Nitrogen Sulfide in Comet Hale–Bopp', *Icarus* **143**, 412–414.

Kahaner, D.: 1989, *Numerical Methods and Software*, Prentice-Hall.

Le Teuff, Y. H., Millar, T. J., and Markwick, A. J.: 2000, 'The UMIST Database for Astrochemistry 1999', *A&AS* **146**, 157–168.

Lis, D. C., Keene, J., Young, T. G., Phillips, T. G., Bockelée-Morvan, D., Crovisier, J., Schilke, P., Goldsmith, P. F., and Bergin, E. A.: 1997, 'Spectroscopic Observations of Comet C/1996 B2 (Hyakutake) with the Caltech Submillimeter Observatory', *Icarus* **130**, 355–372.

Lisse, C. M., Fernández, Y. R., Kundu, A., A'Hearn, M. F., Dayal, A., Deutsch, L. K., Fazio, G. G., Hora, J. L., and Hoffmann, W. F.: 1999, 'The Nucleus of Comet Hyakutake (C/1996 B2)', *Icarus* **140**, 189–204.

McGonagle, D., Irvine, W. M., and Minh, Y. C.: 1992, 'Nitrogen Sulfide (NS) in Star Forming Regions', in P. D. Singh (ed.), *Astrochemistry of Cosmic Phenomena*, IAU Symp. 150, Kluwer Academic Publishers, Dordrecht, p. 227.

McGonagle, D., Irvine, W. M., and Ohishi, M.: 1994, 'Nitrogen Sulfide in Quiescent Dark Clouds', *ApJ* **422**, 621–625.

Millar, T. J., Farquhar, P. R. A., and Willacy, K.: 1997, 'The UMIST Database for Astrochemistry 1995', *A&A* **121**, 139–185.

Mumma, M. J., Disanti, M. A., Dello Russo, N., Fomenkova, M., Magee-Sauer, K., Kaminski, C. D., and Xie, D. X.: 1996, 'Detection of Abundant Ethane and Methane, along with Carbon Monoxide and Water in Comet C/1996 B2 Hyakutake: Evidence for Interstellar Origin', *Science* **272**, 1310–1314.

Palmer, P., Wootten, A., Butler, B., Bockelée-Morvan, D., Crovisier, J., Despois, D., and Yeomans, D. K.: 1996, 'Comet Hyakutake: First Secure Detection of Ammonia in a Comet', *BAAS* **188**, 62.12.

Sarmecanic, J., Fomenkova, M., Jones, B., and Lavessi, T.: 1997, 'Constraints on the Nucleus and Dust Properties from Mid-Infrared Imaging of Comet Hyakutake', *ApJ (Lett.)* **483**, L69.

Schmidt, H. U., Wegmann, R., Huebner, W. F., and Boice, D. C.: 1988, 'Cometary Gas and Plasma Flow with Detailed Chemistry', *Comp. Phys. Commun.* **49**, 17–59.

Sekanina, Z.: 1997, 'A Determination of the Nuclear Size of Comet Hale–Bopp (C/1995 O1)', *Earth Moon Planets* **77**, 147–153.

Smith, A. M., Bohlin, R. G., and Stecher, T. P.: 1976, 'Ultraviolet Objective Grating Imagery of Comet West', *BAAS* **8**, 313.

Smith, A. M., Stecher, T. P., and Casswell, L.: 1980, 'Production of Carbon, Sulfur, and CS in Comet West', *ApJ* **242**, 402–410.

Weaver, H. A. and Lamy, P. L.: 1997, 'Estimating the Size of Hale–Bopp's Nucleus', *Earth Moon Planets* **79**, 17–33.

Weaver, H. A., Feldman, P. D., A'Hearn, M. F., and Arpigny, C.: 1997a, 'The Activity and Size of the Nucleus of Comet Hale–Bopp (C/1995 O1)', *Science* **275**, 1900–1904.

Weaver, H. A., Brooke, T. Y., Chin, G., Kim, S. J., Bockelée-Morvan, D., and Davies, J. K.: 1997b, 'Infrared Spectroscopy of Comet Hale–Bopp', *Earth Moon Planets* **78**, 71–80.

Weaver, H. A., Feldman, P. D., McPhate, J. B., A'Hearn, M. F., Arpigny, C., Brandt, J. C., Randall, C. E., and Stern, S. A.: 1996, 'HST Observations of C/1995 O1 (Hale–Bopp)', *BAAS* **1094**, 09.31.

Woodney, L. M., A'Hearn, M. F., McMullin, J., and Samarasinha, N.: 1997a, 'Sulfur Chemistry at Millimeter Wavelenghts in C/Hale–Bopp', *Earth Moon Planets* **78**, 69–70.

Woodney, L. M., McMullin, J., and A'Hearn, M. F.: 1997b, 'Detection of OCS in Comet Hyakutake (C/1996 B2)', *P&SS* **45**, 717–719.

CHEMICAL PROCESSES IN COMETARY COMAE

S. B. CHARNLEY[1], S. D. RODGERS[1], H. M. BUTNER[2] and P. EHRENFREUND[3]
[1] *Space Science Division, NASA Ames Research Center, MS 245-3, Moffett Field, CA 94035, USA*
[2] *SMTO-Steward Observatory, University of Arizona, 933 N. Cherry Avenue, Tucson, AZ 85721, USA*
[3] *Leiden Observatory, P.O. Box 9513, 2300 RA Leiden, The Netherlands*

(Received 21 April 2002; Accepted 2 June 2002)

Abstract. Recent developments in the chemical modelling of cometary comae are described. We discuss the cyanide chemistry and present new HCN observations of the recent comet C/2002 C1 (Ikeya–Zhang). The connection between interstellar and cometary organic molecules is discussed from the perspective of recent theories of interstellar gas-grain chemistry.

Keywords: Comets, general-comets, individual, (Hale–Bopp)

1. Introduction

Comets are the most volatile-rich and pristine objects in our solar system. Although comets will have undergone some processing during their $\sim$4.5 Gyr lifetime, their chemical composition will fundamentally reflect the conditions in the outer solar nebula when they were formed (Mumma et al., 1993; Irvine et al., 2000). Determining molecular abundances in cometary ices therefore allows us to place important constraints on models of the protosolar nebula, and planetary formation in general. A key question, when considering exactly when, where and how comets formed, is to what extent comets consist of relatively pure interstellar material? Detailed studies of interstellar solid state absorption bands made with the *Infrared Space Observatory (ISO)*, theoretical and observational progress in the understanding of interstellar gas-grain chemistry, and the recent apparitions of bright comets, particularly Hale–Bopp, have all contributed to a deeper understanding of the chemical similarities and differences between interstellar and cometary material (Table 1; Ehrenfreund and Charnley, 2000). Until the *Rosetta* and *Deep Impact* space missions, we have no direct way to determine the composition of the nucleus and must rely on gas phase observations of coma molecules to estimate nuclear molecular abundances. However, a great deal of chemical processing can occur in the coma and so accurate chemical models are of importance when using coma observations to derive nuclear ice composition.

Earth, Moon and Planets **90:** 349–360, 2002.

TABLE I

Representative compositions of the gas in a cold molecular cloud (L134N), protostellar ices (NGC7538: IRS9), protostellar hot core gas (Sgr B2(N)), and in a cometary coma (Hale–Bopp)

Molecule	L134N	NGC7538: IRS9	Sgr B2(N)	Hale–Bopp
H_2O	<3	100	>100	100
CO	1000	16	1000	20
CO_2	–	20	–	6–20
H_2CO	0.25	5	>0.005	1
CH_3OH	0.04	5	2	2
NH_3	2.5	13	–	0.7–1.8
CH_4	–	2	–	0.6
C_2H_2	–	<10	–	0.1
C_2H_6	–	<0.4	–	0.3
HCOOH	0.004	3	>0.003	0.06
CH_2CO	<0.009	–	0.002	<0.03
CH_3CHO	0.008	–	0.002	0.02
c-C_2H_4O	–	–	0.001	–
CH_3CH_2OH	–	<1.2	0.01	<0.05
CH_3OCH_3	–	–	0.03	<0.45
$HCOOCH_3$	<0.02	–	0.02	0.06
CH_3COOH	–	–	0.0008	–
CH_2OHCHO	–	–	0.003	–
OCN^-	–	1	–	–
HNCO	–	–	0.006	0.06–0.1
NH_2CHO	<0.001	–	0.002	0.01
HCN	0.05	–	>0.05	0.25
HNC	0.08	–	>0.001	0.04
CH_3CN	<0.01	–	0.3	0.02
CH_3NC	–	–	0.015	–
HC_3N	0.002	–	0.05	0.02
C_2H_3CN	–	–	0.6	–
C_2H_5CN	–	–	0.006	–
H_2S	0.01	–	–	1.5
OCS	0.02	0.05	>0.02	0.5
H_2CS	0.008	–	0.2	0.02
SO	0.25	–	0.2	0.2–0.8
SO_2	0.005	–	0.3	0.1

Abundances for interstellar ices and volatiles in comet Hale–Bopp are normalized to H_2O. Gaseous abundances for the Sgr B2(N) hot core and L134N are normalized to CO. Adapted from Ehrenfreund and Charnley (2000) and Charnley et al. (2002).

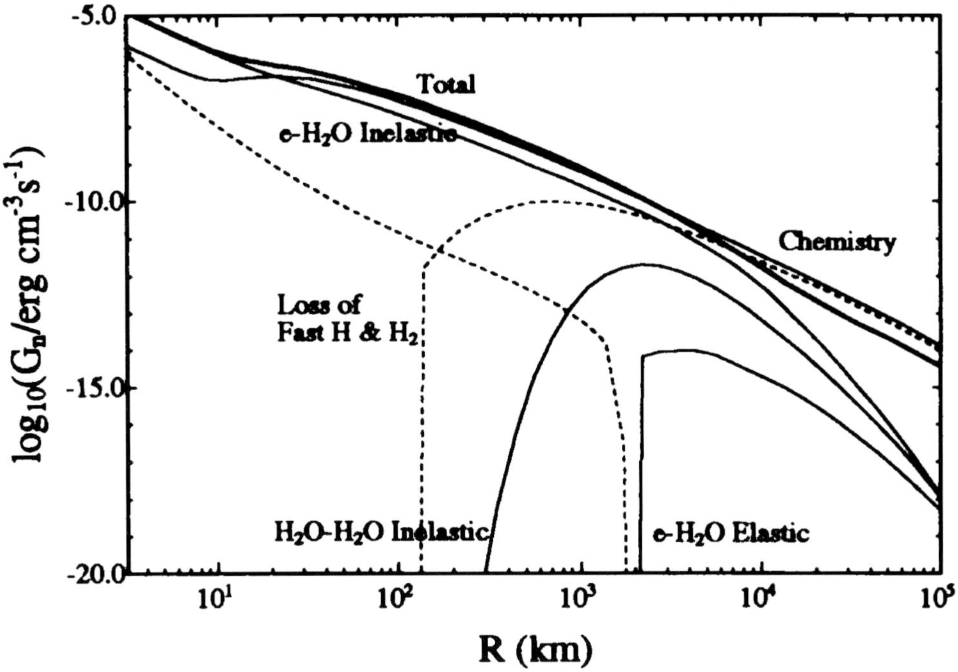

Figure 1. Major heating (full lines) and cooling (broken lines) mechanisms for coma neutrals (from Rodgers and Charnley, 2002a).

2. Modelling Coma Chemistry

Following sublimation from the nucleus, parent molecules are exposed to the solar UV field and they are photodissociated and ionized, creating distinct plasma and neutral fluid components, as well as populations of supra-thermal hydrogen atoms and molecules. The initial expansion of the gas is adiabatic and the temperature falls. Further out, photodissociation and ionization reactions heat the gas and the temperatures rise. A rich chemistry can ensue in the coma, involving processes such as electron impact ionization/dissociation, charge transfer, proton transfer, ion-atom interchange, and neutral rearrangement (see Schmidt et al., 1988). Also, sublimation from and/or photodestruction of dust injects new molecules into the coma.

Hydrodynamics and chemistry are intimately linked. Figure 1 shows that many processes affect the thermal balance of the neutral fluid. Chemistry can affect hydrodynamics: exothermic ($\sim$1 eV) proton transfer reactions are an important heat source in the inner coma, and cause the ion temperature to decouple from the neutral and electron temperatures (Körösmezey et al., 1987). Hydrodynamics can affect the chemistry: large variations in the electron temperature throughout the coma strongly affect the local rates of ion-electron recombination reactions.

We have developed a hydrodynamical-chemical model of the coma (Rodgers and Charnley, 2002a). A multifluid treatment is necessary and the model calculates the coupled chemical and dynamical evolution in the outflowing coma, computing in a self-consistent fashion molecular abundances, individual densities and temperatures for the plasma components and neutrals, as well as the inclusion of fast hydrogen atoms as a distinct component. It is similar to earlier models which also employed multifluid hydrodynamics (e.g., Huebner, 1985; Körösmezey et al., 1987) but, more pragmatically, we have the capability to study a wide range of different chemistries with relative ease. This model has been the basis of several recent chemical studies (Rodgers and Charnley, 1998, 2001b, c).

3. HCN and HNC in Comets

The HNC/HCN ratio varies greatly in different regions of the interstellar medium, with an inverse temperature dependence: In cold dark clouds the ratio can be higher than unity, whereas in hot cores it is small. At present we are unable to explain the excess of HNC in cold clouds (see Talbi et al., 1996). HCN has been seen in many comets, but HNC was detected for the first time in comet Hyakutake, where it was present with an abundance relative to HCN of 6% (Irvine et al., 1996). HNC was subsequently detected in comets Hale–Bopp and Lee, and the HNC/HCN ratio in Hale–Bopp showed a strong increase as the comet approached perihelion, from $<2\%$ at 2.5 AU to $\sim 16\%$ at 1 AU (Irvine et al., 1998). The Hyakutake results were originally interpreted as proof that cometary ices contain unprocessed material from the ISM, but the observations of Hale–Bopp proved conclusively that HNC must be a daughter species, and that its production is related to the solar photon flux and/or coma temperature.

Theoretical models can provide insight into possible chemical production routes for HNC in the coma (Rodgers and Charnley, 1998, 2001c). One must be careful to consider proton transfer reactions and to add or exclude all exoergic or endoergic reactions of this type (Hunter and Lias, 1998). This has the important consequence that ion-molecule chemistry is unable to synthesise the observed quantities of HNC since, for example, $HCNH^+$ will react with methanol to form HCN, but not HNC, thus quenching HNC production in methanol-rich comets (Rodgers and Charnley, 1998). However, endoergic isomerisation reactions of HCN, driven by suprathermal H atoms produced in the photodissociation of parent molecules, may be efficient in large, active comets such as Hale–Bopp.

The observed HNC/HCN ratio in Comet Lee, of about 12%, effectively excludes bimolecular gas-phase reactions as the source of HNC in comets less active than Hale–Bopp. Detailed theoretical modelling indicates that the HNC observed in comets Hyakutake and Lee must be produced via photodissociation of some unknown parent(s), either organic dust particles or large molecules (Rodgers and Charnley, 2001c). Most small molecular candidates can be ruled out (e.g., HNCO,

CH_2NH). Photo-fragmentation of large organic molecules/particles has been proposed as an explanation of extended coma distributions of several molecules and radicals, e.g., CO, C_2, C_3, H_2CO, CN and NH_2. It appears likely that this process is also a common source of HNC, as well as perhaps accounting for the extended sources of HCN and CN. Potential candidates for the unknown parent(s) include hexamethylenetetramine (Bernstein et al., 1995) and polyaminocyanomethylene (Rettig et al., 1992).

These analyses suggest there may be two viable sources of HNC in comets. Photo-degradation being the most common, but with isomerisation reactions driven by fast H atoms playing an important role in large, active comets. Clearly, one would wish to know the HNC/HCN ratio, and its heliocentric variation, in many more comets. Mapping the spatial distribution of HCN and HNC (e.g., Veal et al., 2000; Blake et al., 1999) may also shed light on their origin and relationship with other coma molecules.

3.1. HCN in Comet Ikeya–Zhang

Using the Kitt Peak 12m telescope and the Submillimeter Telescope Observatory (SMTO) we have mapped the HCN emission from the coma of the recent bright Comet C/2002 C1 (Ikeya–Zhang). The comet ephemerides used were based on IAU Circulars and were calculated and updated as new circulars were released. The millimeter observations at the 12 m had typical system temperatures of less than 800 K for 1 mm observations. The backends used were 100 kHz filterbanks and 250 kHz filterbanks in parallel configuration. Calibration was done with a vane observation every 6 to 10 min. Submillimetre observations at the SMTO used either of the single channel, double-sideband, receivers (1.3 mm or 0.87 mm). Backends include the 250 kHz filterbanks, which is used to display the results in this paper. System temperatures varied from 400 K to 1000 K depending on weather conditions. Pointing/focus was done on a nearby planet and is accurate to better than 2 arcsec. Calibration was done with a hot vane observation every 10 minutes.

Figure 2 shows a superposition of coma maps in the J = 3-2 and J = 4-3 lines of HCN. The data was taken on different days (March 31 2002 for the HCN(4-3) map and April 10 2002 for the HCN(3-2) map). Both maps have been integrated between -2 and 2 km/s of the line center, which represents the maximum line extent seen in either line. The HCN(4-3) data are in the form of T_A^* (which should be multiplied by 0.50 to convert to T_R^*) and the HCN(3-2) data are in the form of T_R^*. To convert the HCN(4-3) data to T_R^*, 1 sigma equals 1.02 K km/s. The beams were 21 arcsec with 20 arcsec spacing for J = 4-3, and 24 arcsec with 15 arcsec spacing for J = 3-2.

The difference between the centre of the map (0,0) and the HCN(4-3) peak is real – about 10 arcsec. We can rule out differences in rotational emission for the observed J levels since one would expect that the HCN(3-2) emission would be more extended as one moves away from the inner coma, where the gas density is

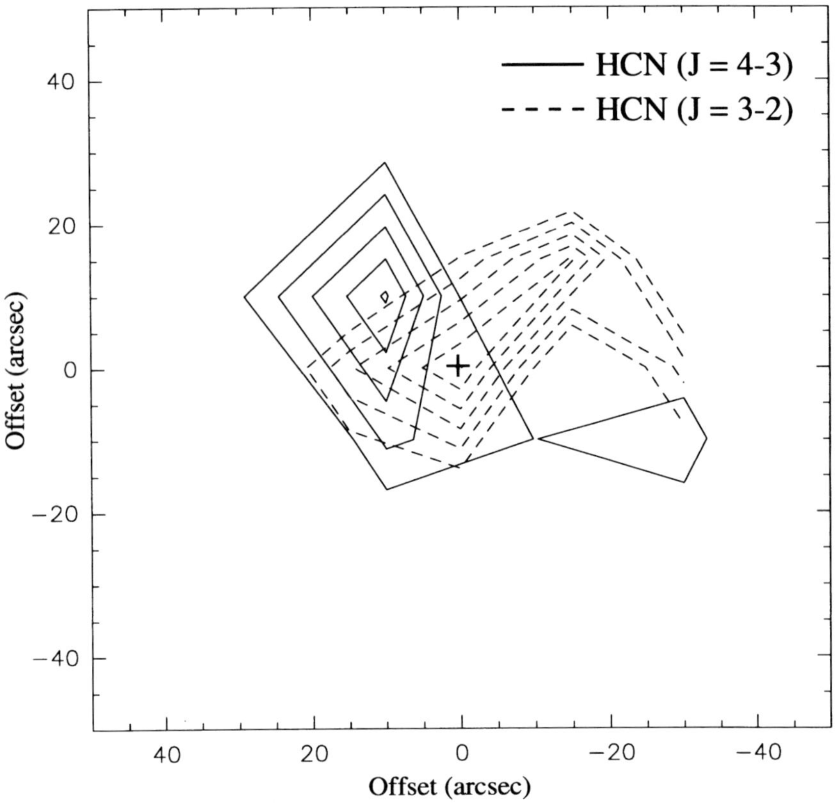

Figure 2. Maps of HCN emission in the coma of comet Ikeya–Zhang. Solid lines are HCN $J = 4\text{-}3$ and broken lines are HCN $J = 3\text{-}2$. The cross marks the nucleus position. For the $J = 4\text{-}3$ map, the lowest contour is 3 times the rms value and the contours are in steps of one sigma (i.e., 3, 4, 5, 6 and 7 sigma). The peak flux is 3.6 K km/s (7 sigma) and one sigma equals 0.51 K km/s. For the $J = 3\text{-}2$ map, the lowest contour is 20 times the rms value and the contours are in steps of 3 sigma each. The peak flux is 0.9 K km/s (35 sigma) and one sigma equals 0.026 K km/s. In both cases the sigma is the baseline rms per channel divided by the number of channels that are covered in the integration.

falling. The most likely explanation is that when the HCN(4-3) observations were made an extended source of HCN existed in the coma, whereas about 10 days later this had dissipated and the major contributor to HCN emission, as determined by the 3-2 line, was from molecules coming directly from the nucleus. This interpretation supports the idea that there are two sources of HCN in cometary comae. One source is pristine interstellar HCN from the nucleus, the other is photolytic decomposition of some organic parent molecule. This parent may be the same as that which is responsible for cometary HNC and hence, depending on the product branching ratios for photolysis of this unknown molecule, the HNC/HCN ratio in comets may actually reflect the nuclear ratio [organic parent]/HCN.

We have also measured the HCN production rates at several epochs and measured the HNC/HCN ratio. These results will be reported elsewhere (Butner et al.,

2002), however we note here that the HNC/HCN ratio in Ikeya–Zhang underwent a sharp increase between 0.96 AU and 0.73 AU, rising from $\sim$3% to around 20%.

4. Organic Interstellar Molecules in Comets

Table I shows that, although there are some discrepancies, there is a strong correspondence between cometary molecules and those detected in interstellar ice mantles or in warm gas containing these evaporated ices – so-called "hot cores" (e.g., Charnley et al., 1992). We can strengthen this connection both by modelling coma organic chemistry and by exploring the implications interstellar grain chemistry theories have for cometary composition.

The brightness of comet Hale–Bopp allowed many organic molecules to be positively detected (Bockelée-Morvan et al., 2000). Many of the organic species detected, such as $HCOOCH_3$, HC_3N, and CH_3CN, are also seen in hot cores where they are thought to be daughter species formed by gas-phase reactions (Rodgers and Charnley, 2001a). Thus, it is possible these molecules were actually created in the coma of Hale–Bopp. Coma modelling utilising the interstellar organic chemistry of these molecules indicates that chemical reactions *cannot* form sufficient quantities of HCOOH, $HCOOCH_3$, HC_3N and CH_3CN (Rodgers and Charnley, 2001b). This result confirms that these species originate in the nuclear ices and suggests that, prior to incorporation into cometesimals, primordial interstellar grain mantles ices were evaporated, perhaps within the protosolar nebula, underwent a hot core-like phase of chemical processing, and then recondensed.

Figure 3 depicts chemistry driven by single atomic addition reactions on grains, with the further constraint of *radical stability* imposed on the intermediate organic radicals (Charnley, 1997, 2001). This scheme accounts for the presence in hot core precursor ices, and presumably also cometary ices, of small organic molecules containing the carbonyl group. The basic organic chemistry is driven primarily by C and N additions. Further H additions lead to a saturated mantle rich in alcohols although energy barriers for H additions to double and triple bonds mean that complete saturation does not occur and some aldehydes can persist. Acetaldehyde has been tentatively observed in Hale–Bopp (Crovisier et al., 1999) and this theory suggests that, depending on the reduction state of the cometary ices, ketene and/or ethanol should also have been present. Comets should also contain the ring isomer of acetaldehyde, ethylene oxide ($c-C_2H_4O$). Ethylene oxide is formed by O atom addition to the C_2H_3 radical, which is formed as an intermediate in the reduction of acetylene to ethene and ethane; this sequence has been proposed as the origin of cometary ethane (Mumma et al., 1996). The HCOOH, HNCO and NH_2CHO observed in both hot cores and comets are also accounted for by the theory. Stronger support for a connection to cometary material comes from PUMA mass spectrometry data of comet Halley. Kissel and Krueger (1987) identified the

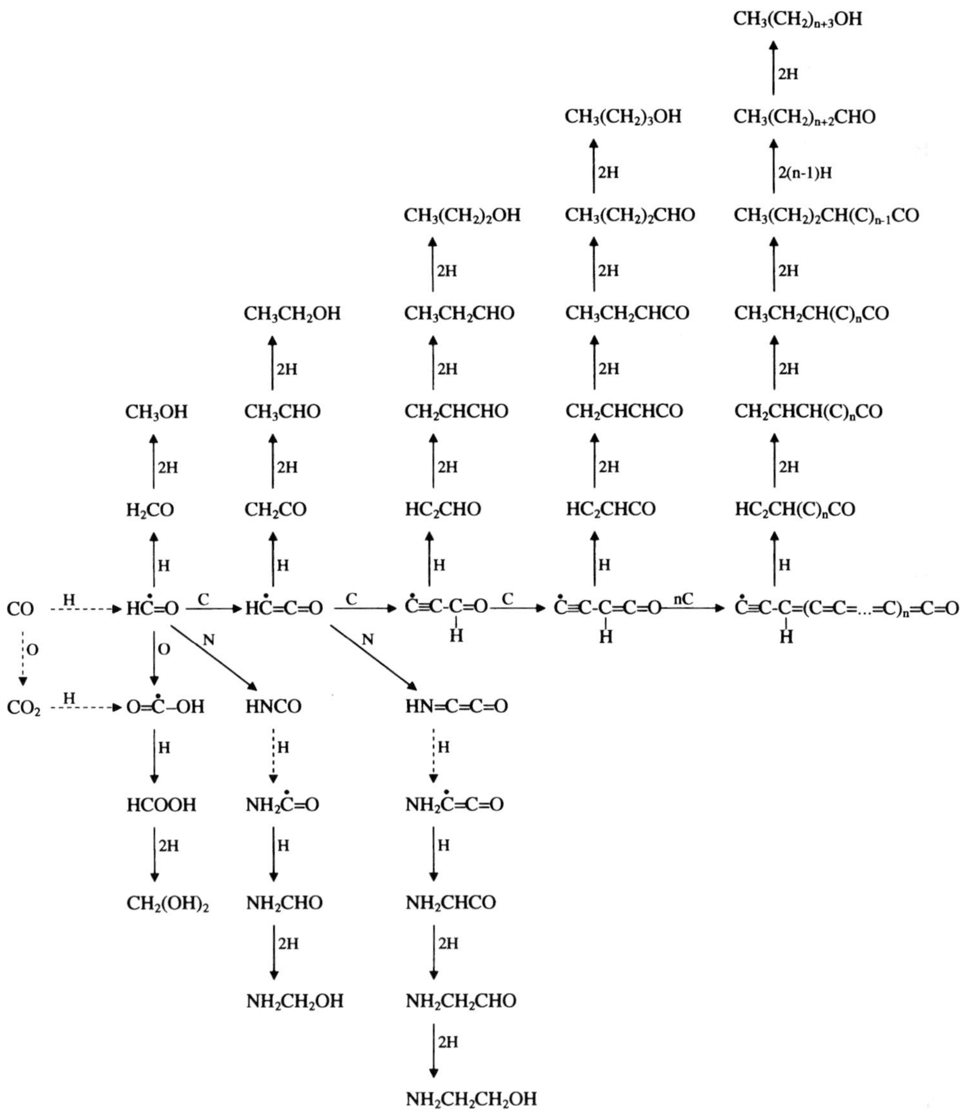

Figure 3. Interstellar grain surface chemistry (adapted from Charnley, 2001). Hydrogen atom addition to unsaturated molecules creates reactive radicals and a rich organic chemistry seeded by carbon monoxide ensues. Broken arrows indicate reactions with activation energy barriers; where 2H is shown, a barrier penetration reaction followed by an exothermic addition is implicitly indicated.

mass peaks at $m/Z = 44$, 46 and 48 with protonated HNCO (or HOCN), NH_2CHO and NH_2CH_2OH (aminomethanol).

A'Hearn et al. (1995) demonstrated that short-period comets are much more likely to be depleted in the carbon chains C_2 and C_3 than long-period comets. The current consensus is that Kuiper belt comets most probably formed near their current location, whereas Oort cloud comets originally formed somewhere in the giant planet region of the protosolar nebula. Understanding this differentiation may therefore provide important cosmogonic information. Recent work suggests that C_2 and C_3 could be produced in the coma from C_2H_2, C_2H_6 and C_3H_4 (Helbert et al., IAU Colloquium 186 presentation, Tenerife 2002). Alternatively, the chemistry of Figure 3 lends support to the idea that photolysis of nuclear propynal (HC_2CHO) could be the parent of C_3 and some C_2 (Krasnopolsky, 1991). Observational upper limits on the nuclear abundance of propynal are not very restrictive (Crovisier et al., 1993). One may speculate that C_2/C_3 differentiation amongst comet families, as well as other relative molecular depletions (CO, CH_3OH, CH_4, C_2H_2, C_2H_6; Mumma et al., 2000, 2001; Bockelée-Morvan et al., 2001), could be traced back to differences on the environment in which the chemistry of Figure 3 originally occurred.

Ion-molecule chemistry at the temperatures of molecular clouds ($\sim 10 - 30$K leads to significant isotopic fractionation of D, ^{13}C and ^{15}N in interstellar molecules (Tielens and Hagen, 1982; Langer et al., 1984; Millar et al., 2000; Terzieva and Herbst, 2001). Hence, measuring these ratios in cometary parent molecules can strengthen the link with an interstellar provenance, and also offers the possibility of extracting important information regarding the origin, pristinity and formation temperature of the nuclear ices. The only deuterated molecules observed in comets are HDO and DCN. The observed HDO/H_2O ratios lie in the range 5.7–6.6 $\times$ 10^{-4} (e.g., Meier et al., 1998a) and a DCN/HCN ratio of 0.002 was determined in Hale–Bopp (Meier et al., 1998b). However, rapid chemical reactions in the coma could in principle alter D/H ratios, and this would compromise any comparisons with interstellar chemistry. Comet modelling of deuterium chemistry in the coma has demonstrated that, as in hot cores, post-evaporation chemistry does not significantly alter the initial D/H ratios (Rodgers and Charnley 2002a). Therefore, for parent species we can use the observed coma ratios to infer the nuclear D/H ratio, and for daughter species we can use these ratios to constrain the nature of the parent and the chemical mechanism by which the species are formed. In particular, if the DNC/HNC ratio can be determined in a comet, we may be able to ascertain the origin of cometary HNC, which is currently not well understood (see above).

5. Conclusions

Modelling of coma chemistry can constrain nuclear composition and isotopic fractionation, as well as the identification of the most important formation processes for individual molecules. We have highlighted this approach in studies of the important HNC/HCN ratio. More measurements of this ratio as a function of heliocentric distance are needed, as are maps of the individual molecules. Reactions involving suprathermal hydrogen atoms is an efficient pathway for HNC production in some comets. Fast H atoms may play an important general role in inner coma chemistry, particularly so in the sulphur chemistry where the barrier for H abstraction from H_2S is the lowest amongst the most abundant parent molecules.

Existing measurements of the $^{15}N/^{14}N$ ratio in CN and HCN (Crovisier, 1999) show them to be consistent with bulk Solar System values. The measurement of an enhanced $^{15}NH_3/^{14}NH_3$ ratio in a comet would provide positive evidence for the interstellar fractionation scenario proposed by Charnley and Rodgers (2002) which also gives a possible explanation of the widespread depletion of N_2 found in comets.

The detection of even more deuterated species would greatly constrain the connection with interstellar molecules. In particular, recent observations of low-mass protostellar cores yield very large $HDCO/H_2CO$ and D_2CO/H_2CO ratios (Loinard et al., 2001) – measuring formaldehyde deuteration should be an important goal for the future. Determination of the DNC/HNC ratio would constrain some coma chemistry scenarios (Rodgers and Charnley, 2002a).

Anomalous water emission has recently been discovered in the inner envelope of the carbon-rich AGB star IRC+10216 (Melnick et al., 2001). This has been attributed to the sublimation of a population of comets, residing at a radius corresponding to that of the Kuiper Belt in the Solar System. The discovery of HDO molecules in IRC+10216 would be the definitive test of this scenario, as stars on the AGB have consumed their deuterium (Rodgers and Charnley, 2002b). As HDO molecules would also be present during the O-rich phase of evolution, it would also be an excellent general probe of extrasolar cometary material at the final stages of stellar evolution.

Finally, cometary ortho:para spin ratios have been measured for H_2O in a number of comets (Mumma et al., 1993; Irvine et al., 2000) and for H_2CO, CH_4, $NH_3(NH_2)$ and CH_3OH in Hale–Bopp (Crovisier, 1999; Kawakita et al., 2001). The observed ratios imply formation temperatures for cometary molecules of typically 25–35 K; this range has important cosmogonic implications. However, it is not certain that the observed values accurately reflect the ratios in the nucleus. The presence of chemically significant abundances of H^+ and protonated ions in the inner coma leads to the possibility that rapid ion-molecule spin-exchange (Dalgarno et al., 1973) and proton transfer reactions (Kahane et al., 1984) could alter these ratios from those originally present in the nucleus. Detailed models of the coma chemistry are necessary in order to ensure that coma observed ortho:para ratios accurately reflect the nuclear values.

Acknowledgements

This work was supported by NASA's Exobiology and Planetary Atmospheres Programs, through NASA Ames Interchange NCC2-1162, and by the Netherlands Research School for Astronomy (NOVA).

References

A'Hearn, M. F., Millis, R. L., Schleicher, D. G., Osip, D. J., and Birch, P. V.: 1995, *Icarus* **118**, 223–270.

Bernstein, M. P., Sandford, S. A., Allamandola, L. J., Chang, S., and Scharberg, M. A.: 1995, *ApJ* **454**, 327–1995.

Blake, G. A., Qi, C., Hogerheijde, M. R., Gurwell, M. A., and Muhleman, D. O.: 1999, *Nature* **398**, 213–216.

Bockelée-Morvan, D. et al.: 2000, *A&A* **353**, 1101–1114.

Bockelée-Morvan, D. et al.: 2001, *Science* **292**, 1339–1443.

Butner, H. M., Rodgers, S. D., Charnley, S. B., and Ehrenfreund, P.: 2002, *ApJ*, submitted.

Charnley, S. B.: 1997, in C. B. Cosmovici, S. Bowyer, and D. Werthimer (eds.), *Astronomical and Biochemical Origins and the Search for Life in the Universe*, Editrice Compositori, Bologna, p. 89.

Charnley, S. B.: 2001, in F. Giovannelli (ed.), *The Bridge between the Big Bang and Biology*, Consiglio Nazionale delle Ricerche, Italy, p. 139.

Charnley, S. B. and Rodgers, S. D.: 2002, *ApJ (Lett.)* **569**, L133–137.

Charnley, S. B., Rodgers, S. D., Kuan, Y.-J., and Huang, H.-C.: 2002, *Adv. Space Res.* in press.

Charnley, S. B., Tielens, A. G. G. M., and Millar, T. J.: 1992, *ApJ* **399**, L71–74.

Crovisier, J.: 1999, in J. M. Greenberg and A. Li (eds.), *Formation and Evolution of Solids in Space*, Kluwer Academic Publishers, Dordrecht, pp. 389–426.

Crovisier, J., Bockelée-Morvan, D., Colom, P., Despois D., and Paubert G.: 1993, *A&A* **269**, 527–540.

Crovisier, J. et al.: 1999, *AAS DPS* **31**, 32.02.

Dalgarno, A., Black, J. H., and Weisheit, J. C.: 1973, *Astrophys. Lett.* **14**, 77–79.

Ehrenfreund, P. and Charnley, S. B.: 2000, *ARA&A* **38**, 427–483.

Huebner, W. F.: 1985, in Diercksen et al. (eds.), *Molecular Astrophysics*, Reidel, Dordrecht, pp. 311–330.

Hunter, E. P. and Lias, S. G.: 1998, in W. G. Mallard and P. J. Linstrom (eds.), *NIST Chemistry WebBook, NIST Standard Reference Database No. 69*, NIST, Gaithersburg MD, (http://webbook.nist.gov).

Irvine, W. M. et al.: 1996, *Nature* **383**, 418–421.

Irvine, W. M. et al.: 1998, *Nature* **393**, 547–550.

Irvine, W. M., Schloerb, F. P., Crovisier, J., Fegley, B., and Mumma, M. J.: 2000, in V. Mannings, A. P. Boss, and S. S. Russell (eds.), *Protostars and Planets IV*, University of Arizona Press, Tucson, pp. 1159–1200.

Kahane, C., Lucas, R., Frerking, M. A., Langer, W. D., and Encrenaz P.: 1984, *A&A* **137**, 211–222.

Kawakita, H. et al.: 2001, *Science* **294**, 1089–1091.

Kissel J. and Krueger F. R.: 1987, *Nature* **326**, 755–760.

Körösmezey, A. et al.: 1987, *J. Geophys. Res.* **92**, 7331–7340.

Krasnopolsky, V. A.: 1991, *A&A* **245**, 310–315.

Langer, W. D., Graedel, T. E., Frerking, M. A., and Armentrout, P. B.: 1984, *ApJ* **277**, 581–604.

Loinard, L., Castets, A., Ceccarelli, C., Caux, E., and Tielens, A. G. G. M.: 2001, *ApJ* **552**, L163–166.

Meier, R. et al.: 1998a, *Science* **279**, 842–846.

Meier, R. et al.: 1998b, *Science* **279**, 1707–1710.

Melnick, G. J., Neufeld, D. A., Ford, K. E. S., Hollenbach, D. J., and Ashby, M. L. N.: 2001, *Nature* **412**, 160–163.

Millar, T. J., Roberts, H., Markwick, A. J., and Charnley, S. B.: 2000, *Phil. Trans. Roy. Soc. Lond. A.* **358**, 2535–2547.

Mumma, M. J., DiSanti, M. A., Dello Russo, N., Fomenkova, M., Magee-Sauer, K., Kaminski, C. D., and Xie, D. X.: 1996, *Science* **272**, 1310–1314.

Mumma, M. J., DiSanti, M. A., Dello Russo, N., Magee-Sauer, K., and Rettig, T. W.: 2000, *ApJ* **531**, L155–159.
Mumma, M. J., Weissman, P. R., and Stern, S. A.: 1993, in E. H. Levy and J. Lunine (eds), *Protostars and Planets III*, University of Arizona Press, Tucson, pp. 1177–1252.
Mumma, M. J. et al.: 2001, *Science* **292**, 1334–1339.
Rettig, T. W., Tegler, S. C., Pasto, D. J., and Mumma, M. J.: 1992, *ApJ* **398**, 293–298.
Rodgers, S. D. and Charnley, S. B.: 1998, *ApJ* **501**, L227–230.
Rodgers, S. D. and Charnley, S. B.: 2001a, *ApJ* **546**, 324–329.
Rodgers, S. D. and Charnley, S. B.: 2001b, *MNRAS* **320**, L61–64.
Rodgers, S. D. and Charnley, S. B.: 2001c, *MNRAS* **323**, 84–92.
Rodgers, S. D. and Charnley, S. B.: 2002a, *MNRAS* **330**, 660–674.
Rodgers, S. D. and Charnley, S. B.: 2002b, *Planet. Space Sci.*, in press.
Schmidt, H. U., Wegmann, R., Huebner, W. F., and Boice, D. C.: 1988, *Comp. Phys. Comm.* **49**, 17.
Talbi, D., Ellinger, Y., and Herbst, E.: 1996, *A&A* **314**, 688–692.
Terzieva, R. and Herbst, E.: 200, *MNRAS* **317**, 563–568.
Tielens, A. G. G. M. and Hagen, W.: 1982, *A&A* **114**, 245–260.
Veal, J. M. et al.: 2000, *AJ* **119**, 1498–1511.

OPTICAL SPECTROSCOPY OF COMET C/2000 WM1 (LINEAR) AT THE GUILLERMO HARO ASTROPHYSICAL OBSERVATORY IN MEXICO

KLIM I. CHURYUMOV[1], IGOR V. LUK'YANYK[1], ALEXEI A. BEREZHNOI[2,3], VAHRAM H. CHAVUSHYAN[2], LOURDES S. SANDOVAL[4] and ALEJANDRO A. PALMA[2,4]

[1]*Astronomical Observatory, Kyiv National Shevchenko University, Kyiv, Ukraine;* [2]*Instituto Nacional de Astrofisica, Optica y Electronica, Tonantzintla, Puebla, Mexico;* [3]*Sternberg Astronomical Institute, Moscow, Russia;* [4]*Benemerita Universidad Autonoma de Puebla, Puebla, Mexico*

(Received 27 February 2002; Accepted 29 June 2002)

Abstract. We present a preliminary analysis of medium resolution optical spectra of comet C/2000 WM1 (LINEAR) obtained on 22 November 2001. The emission lines of the molecules C_2, C_3, CN, NH_2, H_2O^+ and presumably CO (Asundi and triplet bands) and C_2^- were identified in these spectra. By analysing the brightness distributions of the C_2, C_3, CN emission lines along the slit of the spectrograph we determined some physical parameters of these neutrals, such as their lifetimes and expansion velocities in the coma. The Franck–Condon factors for the CO Asundi bands and C_2^- bands were calculated using a Morse potential model.

Keywords: C_2, C_3, CN, comet C/2000 WM1 (LINEAR), expansion velocity, Franck–Condon factors, negative ions, optical spectroscopy

1. Introduction

The spectra of comets contain many emission bands of simple molecules. The study of cometary spectra can give information about the chemical composition of comet nuclei. The negative ions play an important role in cometary chemistry. Negatively charged molecular radicals were identified in the inner coma of comet Halley (2300 km from the nucleus) on the basis of mass spectra obtained on board the Giotto spacecraft on 14 March 1986 (Chaizy et al., 1991). The concentration of negative ions with 22–65 amu was estimated to be 0.05 cm^{-3}. The C_2^- ion was detected in the optical spectrum of comet Scorichenko–George (1990, VI) (Churyumov et al., 1993) and its column density was estimated to be 10^{12} cm^{-2}.

Here we present medium-resolution optical spectra of comet C/2000 WM1 (LINEAR) obtained with a long slit spectrograph. Such spectra allow to calculate some physical parameters of the neutral cometary atmospheres (gas expansion velocities, lifetimes of species), to search for new cometary emission lines, to estimate parameters of the gas and dust production of the comet nucleus, to detect the cometary luminescence continuum of non-solar nature. We will also present Franck–Condon factor calculations for some of the identified molecules.

2. Observations and Data Reduction

Spectroscopic observations of comet C/2000 WM1 were obtained with the 2.12-m telescope of the Guillermo Haro Observatory in Cananea, Sonora, Mexico, operated by the National Institute of Astrophysics, Optics and Electronics. The Boller and Chivens spectrograph (long slit and CCD) was used with a diffraction grating of a reciprocal dispersion of 3.5 Å per pixel and a spectral resolution of 15 Å on 22 November, and a dispersion of 1.4 Å per pixel and a spectral resolution of 5.4 Å on 24 November 2001. The 2.5 arcsec wide slit was oriented along the comet tail. The length of the slit is 2.5 arcmin. Gratings with 150 l/mm and 300 l/mm were used, respectively. Four spectra (spectral range: 4000–7000 Å S/N $\sim$ 40) were obtained on 22 November, and five spectra (spectral range: 4000–5700 Å S/N $\sim$ 50) on 24 November 2001. The exposure time was 30 min for each spectrum. Helium-argon spectra were used for the wavelengths calibration. All spectra were processed with the help of ESO-MIDAS standard reduction context "LONG" and the Research System IDL computer programs allowing for reductions of the CCD bias level, cosmic ray particles, flat fielding, and night sky contribution.

3. Determination of Gas Velocity and Lifetime of Molecules

The production rates of C_2, CN and CO^+ were estimated from spectrophotometric observations of comet C/2000 WM1 obtained for instance at a heliocentric distance of $r = 2.79$ AU (Czabo et al., 2002). From our spectroscopic observations at $r = 1.35$ AU detailed information about physical parameters of the gaseous components of the neutral cometary atmosphere can be retrieved. In order to determine the gas component expansion and the lifetime, τ, of the particles we produced spatial profiles for the C_2, C_3 and CN emission lines (Figure 1). The obtained monochromatic profiles were analyzed according to Shulman's model (Shulman, 1970). From this model the surface brightness can be determined by the following formula:

$$I(\rho, \phi) = \frac{A}{\rho}(1 - \frac{2\rho}{L_s}\sin\Theta\cos\phi)\int_{\frac{\rho}{L_d}}^{\infty} K_0(y)dy - \frac{2A}{L_s}\cos\Theta E_i(-\frac{\rho}{L_d}),$$

where A is a proportional coefficient, Θ is the phase angle of the comet, $L_d = v\tau$ is the characteristic scale of the molecule path, τ is the molecular lifetime, $L_s = 2v^2/a$ is a characteristic scale of the region of the spherical symmetry, a is the radiation acceleration of molecules, E_i is the exponential integral function.

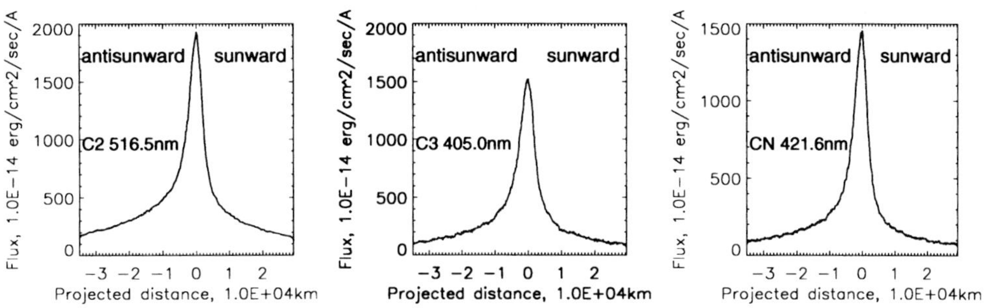

Figure 1. Distribution of surface brightness along the slit for three emission bands C_2, C_3 and CN.

TABLE I

Physical parameters of gaseous cometary components C_2, C_3 and CN

Species	Velocity, ms^{-1}	Lifetime, 10^6 s
C_2 (5165 Å)	482	1.45
C_3 (4050 Å)	240	0.2
CN (4200 Å)	88	1.0

The logarithm of the asymmetry of the line brightness profile is

$$\lg \frac{I(\rho, \phi + \pi)}{I(\rho, \phi)} = 1.72 \frac{\rho}{L_s} \sin \Theta_0 \cos \phi,$$

$$\lg I(\rho, \phi + \pi) I(\rho, \phi) = \text{const} + \lg \frac{L_d}{\rho} \int_{\frac{\rho}{L_d}}^{\infty} K_0(y) dy.$$

The solution of these equations gives values for v and τ (Table I). From this table we can see that estimated expansion velocities of the C_2, C_3 and CN molecules in the coma of comet C/2000 WM1 differ noticeably from the gas expansion velocities determined by Delsemme's formula (Delsemme, 1982) which gives values in the range 500–600 ms^{-1}.

4. Possible Detection of C_2^- and CO in Comet C/2000 WM1

The catalogs of the spectral lines in comet Brorsen–Metcalf (Brown et al., 1996) and comet Bradfield 1980 Y1 (Cosmovici et al., 1982) were used for identification of the emission lines in the spectra of comet C/2000 WM1 (LINEAR). Figure 2 shows the distribution of energy in the spectrum obtained in the coma region at the

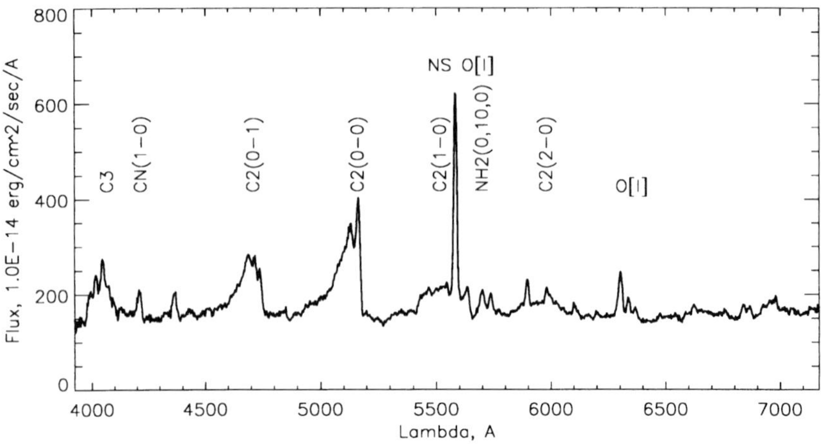

Figure 2. The spectrum of comet C/2000 WM1 (LINEAR) on 22 November 2001.

distance ~ 5000 km from the nucleus (position angle of the prolonged radius vector in the sunward direction). In four spectra we found emission lines of the following neutral molecules: 110 were of C_2, 55 of NH_2, 8 of CN, 4 of C_3, 10 of CO (Asundi and triplet bands), 3 of CH. We also observed the following ion species: H_2O^+ (5 emission lines) and C_2^- (presumably two lines). The comparison of the relative intensities of the identified emission lines of 6-0 (7229.12 Å), 7-0 (6726.47 Å) Asundi and 5-0 (5670.76 Å) triplet bands of the CO molecule (see Figure 3) with corresponding values of Franck–Condon factors shows a good agreement (Table III and Kuzmenko et al., 1984). However, for 11-1 (5761.05 Å), 8-0 (6244.56 Å, 6257.85 Å, 6267.80 Å), 8-1 (7007.06 Å) Asundi and 15-3 (4445.47 Å), 11-2 (4935.93 Å) triplet bands the Franck–Condon factors (see Section 5) are very low. Possibly, other molecules are responsible for these emission lines. Unfortunately, the atmospheric conditions during the observations were not sufficient to ensure reliable photometric measurements of the absolute emission line fluxes. So our identification of CO in the spectrum of comet C/2000 WM1 is tentative.

In the spectrum we also found two emission lines whose wavelengths are close to the theoretical wavelengths of the spectral lines of the C_2^- $B^2\Sigma_u^+ - X^2\Sigma_g^+$ bands (see Figure 3 for vibrational transition 1-0). This corresponds to wavelengths 4902.0 Å (λ_{teor}= 4902.10 Å, vibrational transition 1-0) and 5363.3 Å(λ_{teor}= 5363.37 Å, vibrational transition 1-1). The relative emission intensities (for 4902.0 Å- I_{rel}= 0.26 and for 5363.3 Å - I_{rel}= 0.3) are in good agreement with the Franck–Condon factor values for these transitions (see Table II and Kuzmenko et al., 1984). As there are no other C_2^- transitions with high Franck–Condon factors in the comet spectrum (for example λ_{teor}= 5415.87 Å, vibrational transition 0-0, λ_{teor}= 5912.69 Å, vibrational transition 1-2, λ_{teor}= 5987.82 Å, vibrational transition 0-1), the C_2^- detection is very tentative.

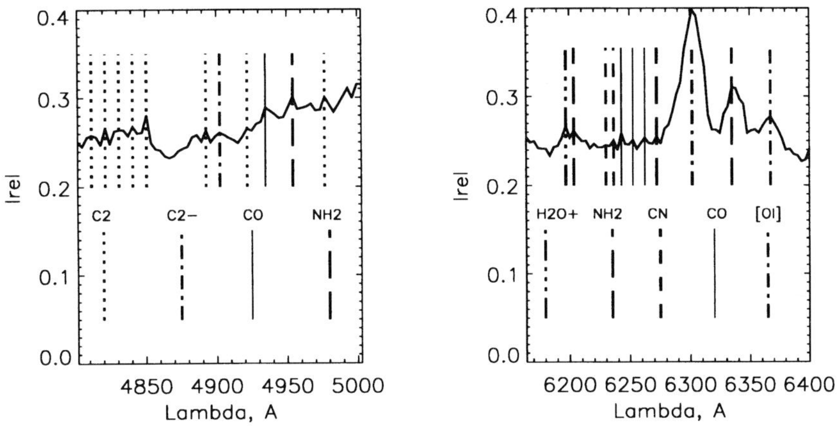

Figure 3. Two spectral regions of comet C/2000 WM1 spectrum obtained on 22 November 2001 (with spectral lines of the CO molecule for λ = 4935.93 Å, 6244.56 Å, 6257.85 Å, 6267.80 Åand with the spectral line of the C_2^- negative ion for λ = 4902.02 Å).

Let us consider a possible C_2^- detection in comet C/2000 WM1. The column density of negative ions with 22–65 amu in comet Halley is estimated to be 10^7 cm^{-2} (Chaizy et al., 1991). This value is about five orders of the magnitude smaller than that of the column density of C_2^- ions in comet Scorichenko–George. For comet Halley, theoretical calculations predict a negative ion abundance as low as 10^{-6} to 10^{-10} of electron densities in the inner coma (Wekhof, 1981). Such theoretical predictions agree with the experimental results of investigations on comet Halley and disagree with those of high C_2^- abundance in comet Scorichenko–George. Special conditions are required for high density of C_2^- in comets. First, we need regions with high electron densities because C_2^- formation most likely happens via electron attachment to C_2. Second, due to a short C_2^- photodissociation lifetime this ion can be destructed slowly only in the inner coma. Third, the rate constant for a C_2^- reaction with atomic hydrogen is much higher than that for a reaction with molecular hydrogen (Barckholtz et al., 2001). All these considerations led us to conclude that the C_2^- abundance is highest in the inner comae of comets. It is more likely to detect this ion in water-depleted comets because the atomic hydrogen concentration may be low in such comets. In the outer coma the C_2^- abundance may be higher than that of other negative ions such as CN^- and OH^-, because the C_2 scale length is about ten times higher than that of CN (Combi and Fink, 1997).

Another feature of the optical spectrum of comet WM1 indicates the presence of weak CO Asundi and triplet bands. The first detection of such bands in the optical spectrum was reported for comet Bradfield 1980 Y1 (Cosmovici et al., 1982). The authors considered dissociative recombination of CO_2^+ or HCO^+ as a possible mechanism for explaining the origin of Asundi and triplet bands. It was also noted that reactions of dissociative recombination require high electron densit-

TABLE II

Franck–Condon factors for C_2^- $B^2\Sigma_u^+ - X^2\Sigma_g^+$ transitions. Spectroscopic constants ω_e and $\omega_e x_e$, μ and r_e were taken from Mead et al. (1985)

$v_2 \backslash v_1$	0	1	2	3	4	5	8
0	0.7179	0.2494	0.0311	0.0015	$< 10^{-4}$	$< 10^{-4}$	$< 10^{-4}$
1	0.2265	0.3268	0.3648	0.0766	0.0052	0.0001	$< 10^{-4}$
2	0.0465	0.2943	0.1224	0.3994	0.1264	0.0109	0.0002
3	0.0078	0.1005	0.2804	0.0299	0.3881	0.1748	0.0183
4	0.0011	0.0236	0.1437	0.2309	0.0011	0.3531	0.2188
5	0.0002	0.0045	0.0446	0.1699	0.1721	0.0057	0.3084
6	$< 10^{-4}$	0.0007	0.0105	0.0671	0.1792	0.1176	0.0253
7	$< 10^{-4}$	0.0001	0.0021	0.0192	0.0879	0.1747	0.0732
8	$< 10^{-4}$	$< 10^{-4}$	0.0004	0.0044	0.0298	0.1048	0.1605

ies. Interestingly, C_2^- bands were observed simultaneously before our observations only in comet Scorichenko–George (Churyumov et al., 1993). For C_2^- formation high electron densities are also required.

5. C_2^- and CO Franck–Condon Factors Calculations

Franck–Condon factors (FCF) play a fundamental role for the determination of the column density of cometary diatomic molecules (Churyumov et al., 1993). These factors can be determined if the harmonic frequency ω_e, the anharmonic frequency $\omega_e x_e$, the reduced mass μ, and the internuclear distance r_e are known. This involves the calculation of an overlap between vibrational wave functions. Algebraic and numerical techniques have been devised to perform such calculations, one of the most accurate methods is the one based on the Morse potential and Rydberg–Klein–Rees models. For the present case we have chosen a Morse potential model based on a Simpson composite quadrature and compare its precision with the Rydberg–Klein–Rees calculations available in the literature.

This special numerical technique avoids overflows and underflows upon evaluation of the normalization constants and Laguerre polynomials. The numerical evaluation was done in the closed interval [0.4 Å; 2.5 Å] with a step size of h = 0.01 Å. Although there are many ways to calculate Franck–Condon factors using a Morse oscillator, a special software was written following the technique (Halmann and Laulicht, 1965). The spectroscopic data for this ion, obtained by photodetachment spectroscopy, were used. Jones et al. (1980) report the following spectroscopic constants for the C_2^- ion: ω_e = 1969.076 (0.165), $\omega_e x_e$ = 14.902 (0.098) for $B^2\Sigma_u^+$ state; ω_e = 1781.329 (0.066), $\omega_e x_e$ = 11.719 (0.018) for the

TABLE III

Morse Franck–Condon factors of CO for the transition $a'^3\Sigma^+ - a^3\Pi$. Spectroscopic data were taken from Huber and Herzberg (1976)

$v_2 \setminus v_1$	0	1	2	3	4	5	6
0	0.0413	0.1119	0.1666	0.1810	0.1606	0.1236	0.0858
1	0.1525	0.1945	0.0999	0.0141	0.0048	0.0429	0.0813
2	0.2563	0.0808	0.0033	0.0746	0.0957	0.0472	0.0051
3	0.2584	0.0012	0.1119	0.0667	0.0003	0.0347	0.0717
4	0.1738	0.1051	0.0794	0.0070	0.0812	0.0519	0.0015
5	0.0819	0.2056	0.0003	0.1048	0.0273	0.0147	0.0664
6	0.0277	0.1756	0.0981	0.0484	0.0329	0.0762	0.0064
7	0.0068	0.0887	0.1941	0.0100	0.0992	0.0003	0.0576

$X^2\Sigma_g^+$ electronic state. In this paper the Franck–Condon factors were calculated for $v_1 = 0$–7 and $v_2 = 0$–10 with low accuracy, i.e., with only one significant digit. Mead et al. (1985), by using ultrahigh resolution spectroscopy, report more accurate values: $\omega_e = 1969.542$ (0.084), $\omega_e x_e = 15.100$ (0.057), $\mu = 6.0$, $r_e = 1.2234$ Åfor $B^2\Sigma_u^+$ state; $\omega_e = 1781.202$ (0.020), $\omega_e x_e = 11.6716$ (0.0048), $\mu = 6.0$, $r_e = 1.2684$ Åfor the $X^2\Sigma_g^+$ electronic state. The standard deviation between the Franck–Condon factors based on the spectroscopic constants of Jones et al. (1980) and Mead et al. (1985) is 0.002. By taking into account the standard deviation of the harmonic and anharmonic frequencies with spectroscopic data of Mead et al. (1985), we obtain a standard deviation of Morse Franck–Condon factors of 5×10^{-4}. These results can be used for detecting the presence of C_2^- in comets, atmospheres of cold carbon stars, and in diffuse molecular clouds.

Additionally, the values of Franck–Condon factors were calculated for some vibrational transitions of the triplet and Asundi bands which we presumably identified in the spectrum of comet C/2000 WM1 (LINEAR): For the triplet bands – 5-0, FCF = 0.135; 11-2, FCF = 0.019; 15-3, FCF = 0.00012 and for the Asundi bands -8-0, FCF = 0.0012, 8-1, FCF = 0.0291; 11-1, FCF = 0.000092. For the calculation of the Asundi FCF values we used the following spectroscopic constants: $\omega_e = 1228.6$, $\omega_e x_e = 10.468$, $\mu = 6.8562$, $r_e = 1.3523$ Å for the $a'^3\Sigma^+$ state; $\omega_e = 1743.41$, $\omega_e x_e = 14.36$, $r_e = 1.20574$ Å for the $a^3\Pi_r$ state (Huber and Herzberg, 1976). For calculations of triplet FCF values we used the following spectroscopic constants: $\omega_e = 1171.94$, $\omega_e x_e = 10.635$, $\mu = 6.8562$, $r_e = 1.3696$ Å for the $d^3\Delta_i$ state; $\omega_e = 1743.41$, $\omega_e x_e = 14.36$, $r_e = 1.20574$ Å for the $a^3\Pi_r$ state (Huber and Herzberg, 1976).

6. Conclusions

We obtained medium-resolution spectroscopic observations of the bright comet C/2000 WM1 (LINEAR) at the Guillermo Haro Astrophysical Observatory in Mexico. We found emission lines of the following neutral radicals: C_2, NH_2, CN, C_3, CH, and possibly CO. We also observed the following ion species: H_2O^+ and presumably C_2^-. The possible presence of CO and C_2^- emissions lines in the spectra is discussed. Franck–Condon factors for C_2^- and CO (Asundi bands) were calculated and used for the analysis of the brightness distribution of the identified emission lines of these molecules. By analyzing the brightness distribution of C_2, C_3, CN emission lines along the spectrograph slit the escape velocity of these molecules from the nucleus and their lifetimes were determined.

Acknowledgement

Alexei Berezhnoi would like to acknowledge Conacyt (through project 32112-E) for provision of a postdoctoral fellowship at INAOE.

References

Barckholtz, C., Snow, T. P., and Bierbaum, V. M.: 2001, *ApJ* **547**, 171–174.
Brown, M. E., Bouchez, A. H., Spinrad, H., and Jons-Krull, C. M.: 1996, *AJ* **112**, 1197–1202.
Chaizy, P., Reme, H., Sauvaud, J. A., d' Uston, C., Lin, R. P., Larson, D. E., Mitchell, D. L., Anderson, K. A., Carlson, C. W., Korth, A., and Mendis, D. A.: 1991, *Nature* **349**, 393–396.
Churyumov, K. I., Chernyi, G. F., and Shapovalova, A. I.: 1993, *Astron. Lett.* **19**, 332–334.
Combi, M. R. and Fink, U.: 1997, *ApJ* **484**, 879–890.
Cosmovici, C. V., Barbieri, C., Bonoli, C., Bortoletto, F., and Hamzaoglu, E.: 1982, *Astron. Astrophys.* **114**, 373–387.
Czabo, Gy. M., Kiss, L. L., Sarneczky, K., and Sziladi, K.: 2002, *Astron. Astrophys.* **384**, 702–710.
Delsemme, A. H.: 1982, Comets, *IAUC* 61, 85 pp.
Jones, P. L., Mead, R. D., Kohler, B. E., Rosner, S. D., and Lineberger, W. C.: 1980, *J. Chem. Phys.* **73**, 4419–4432.
Halmann, M. and Laulicht, I.: 1965, *J. Chem. Phys.* **43**, 438–448.
Huber, K. P. and Herzberg, G.: 1976, 'Constants of Diatomic Molecules', *Molecular Spectra and Molecular Structure*, Van Nostrand Reinhold, New York, 716 pp.
Kuzmenko, N. E., Kuznetsova, L. A., and Kuzyakov, Yu. A.: 1984, *Frank–Condon Factors for Diatomic Molecules*, Moscow University Publishing, 342 pp. (in Russian).
Mead, R. D., Hefter, U., Schulz, P. A., and Lineberger, W. C.: 1985, *J. Chem. Phys.* **82**, 1723–1731.
Shulman, L. M.: 1970, 'Distribution of the Neutral Matter Density in the Cometary Atmosphere', *Astron. Astrophys.* **11**, 26–30 (in Russian).
Wekhof, A.: 1981, *Earth Moon Planets* **24**, 157–173.

OBSERVATIONS OF THE OH RADIO LINES OF COMETS AT NANÇAY – RECENT DEVELOPMENTS

JACQUES CROVISIER, PIERRE COLOM, ÉRIC GÉRARD and DOMINIQUE BOCKELÉE-MORVAN
Observatoire de Paris, F-92195 Meudon, France

(Received 18 February 2002; Accepted 19 May 2002)

Abstract. The 18 cm lines of OH have been observed in more than 70 comets with the Nançay radio telescope since 1974. The observations made from 1982 to 1999 are now organized in a data base. The Nançay radio telescope has been recently upgraded, and observations since 2000 are made with a sensitivity increased by about a factor of two. Twelve comets were observed from July 2000 to May 2002.

Keywords: Comets, OH radical, radio lines

The 18-cm lines of OH have been systematically observed in more than 70 comets with the Nançay radio telescope since 1974. The excitation mechanism of these lines – a weak maser process in the linear regime – is now well known and the object of reliable modelling. This allowed an evaluation of the cometary water production rates and their evolution with time, as well as a study of several physical processes such as the collisional quenching of the OH maser, the expansion of the cometary atmospheres, their anisotropy in relation with non-gravitational forces, and the Zeeman effect in relation with the cometary magnetic field.

These observations were the object of occasional publications. The observations made from 1982 to 1999 are now organized in a comprehensive data base (Crovisier et al., 2002). The detailed data, in the form of figures showing the individual and averaged spectra and tables giving observational characteristics and derived OH production rates, is available electronically from *Astronomy & Astrophysics* and from `http://wwwusr.obspm.fr/~crovisie/basecom/index.html`

The Nançay radio telescope was recently upgraded. The Hoghorn feeds at the focus were replaced by a double Gregorian wide-band system with corrugated horns. The sensitivity increased by about a factor of two, which gives us access to weaker objects, or allows us to follow them at larger heliocentric distances. Since 2000, 12 comets have been observed with the new system (Table I).

Some of these observations were coordinated with ground based millimetric and submillimetric observations (C/1999 S4, C/1999 T1, C/2001 A2, 19P, C/2000 WM_1, C/2002 C1) and with observations of the 557 GHz water line with the ODIN satellite (C/2001 A2, 19P, C/2000 WM_1, C/2002 C1). C/1999 S4 (LINEAR) was observed before and during the complete disruption of its nucleus (Bockelée-

Earth, Moon and Planets **90:** 369–370, 2002.

TABLE I

Comets observed at Nançay since 2000

Comet	Perihelion yymmdd	q [AU]	Date range of observations	N a	b
C/1999 S4 (LINEAR)	000726.17	0.765	000706–000803	20	D
C/1999 T1 (McNaught–Hartley)	001213.47	1.172	001115–010130	40	D
C/2000 W1 (Utsunomiya–Jones)	001226.56	0.321	001212–010107	11	D
73P/Schwassmann–Wachmann 3	010127.71	0.937	001201–010310	31	D
45P/Honda–Mrkos–Pajdušáková	010329.89	0.528	010103–010315	27	–
24P/Schaumasse	010502.66	1.205	010310–010630	47	D
C/2001 A2 (LINEAR)	010524.52	0.779	010402–010712	56	D
16P/Brooks 2	010719.82	1.835	010417–010510	18	–
19P/Borrelly	010914.73	1.358	010719–011129	54	D
C/2000 WM_1 (LINEAR)	020122.67	0.555	011004–020420	104	D
C/2002 C1 (Ikeya–Zhang)	020318.98	0.507	020227–	c	D
C/2002 F1 (Utsunomiya)	020422.90	0.438	020409–	c	D

[a]Number of observations (one per day).
[b]–: No detection; D: Clear detection.
[c]Observations still going on (May 2002).

Morvan et al., 2001). Observations of 19P/Borrelly were made in support of its flyby by the Deep Space 1 spacecraft (Bockelée-Morvan et al., 2002).

The Nançay Radio Observatory is the Unité scientifique de Nançay of the Observatoire de Paris, associated as Unité de service et de recherche (USR) No. B704 to the French Centre national de la recherche scientifique (CNRS). Its recent upgrade was financed jointly by the Conseil régional of the Région Centre in France, the CNRS and the Observatoire de Paris.

References

Bockelée-Morvan, D., Biver, N., Colom, P., Crovisier, J., Festou, M. C., Henry, F., Lecacheux, A., and Weaver, H. A.: 2002, these proceedings.

Bockelée-Morvan, D., Biver, N., Moreno, R., Colom, P., Crovisier, J., Gérard, E., Henry, F., Lis, D. C., Matthews, H., Weaver, H. A., Womack, M., and Festou, M. C.: 2001, *Science* **292**, 1339.

Crovisier, J., Colom, P., Gérard, E., Bockelée-Morvan, D., and Bourgois, G.: 2002, *Astron. Astrophys.* **393**, 1053.

SPIN TEMPERATURE OF AMMONIA DETERMINED FROM NH_2 IN COMET C/2001 A2 (LINEAR)

HIDEYO KAWAKITA
Gunma Astronomical Observatory, 6860-86 Nakayama, Takayama, Agatsuma, Gunma 377-0702, Japan

JUN-ICHI WATANABE and TETSUHARU FUSE
National Astronomical Observatory Japan, 2-21-1 Osawa, Mitaka, Tokyo 181-8588, Japan

REIKO FURUSHO
Japan Science and Technology Corporation, 5-3 Yonbancho, Chiyoda, Tokyo 102-0081, Japan

SHINSUKE ABE
Institute of Space and Astronautical Science, 3-1-1 Yoshinodai, Sagamihara, Kanagawa 229-8510, Japan

(Received 1 March 2002; Accepted 25 May 2002)

Abstract. The ortho-to-para ratio (OPR) of a cometary molecule is one of primordial character in comets. The OPR which is characterized by a spin temperature, is thought to reflect the formation conditions of the molecule. In this paper we show the high-dispersion spectrum of cometary NH_2 in Comet C/2001 A2 (LINEAR), from which the OPR of NH_2 is determined based on the fluorescence excitation model. Since the NH_2 is a photodissociation product of cometary ammonia, we applied the permutation group theory to the whole reaction system (i.e. the photodissociation reaction of ammonia to NH_2 and H) in order to derive the OPR of ammonia from that of NH_2.

The derived OPR of ammonia is 1.12 ± 0.03 in Comet C/2001 A2 (LINEAR). This value corresponds to a spin temperature of 30^{+3}_{-2} K. If this reflects the temperature where the comet formed in the protosolar nebula, our result indicates that this comet was formed in the region of the giant planets between Jupiter and Neptune.

Keywords: Ammonia, comet, comet formation, spectroscopy

1. Introduction

A molecule which has protons (or atoms with non-zero nuclear spins) at symmetrical positions is classified into different nuclear spin species. Usually these nuclear spin species are called "ortho" (all nuclear spins of protons are parallel) and "para" (nuclear spins of protons are anti-parallel) for the molecules with two or three identical protons like H_2O or NH_3. As reviewed by Mumma et al. (1993), the ortho-to-para ratio (OPR) of a cometary molecule which is characterized by a spin temperature is believed to be of primordial character for a comet, because the radiative and non-destructive collisional transitions are strictly forbidden between ortho and para species. Although the real meaning of the spin temperature has never been understood, it may represent the temperature of cometary molecule at the time

Earth, Moon and Planets **90:** 371–379, 2002.

of its chemical formation or it may reflect the re-equilibration of OPR according to the thermal history of molecules. Reliable OPRs of water have been derived for three comets: 1P/Halley, C/1995O1 (Hale–Bopp), and 103P/Hartley 2 (Irvine et al., 2000). These comets indicated spin temperature of about 30 K (25–35 K). Unfortunately, it is difficult to increase the number of samples for OPRs of water because it is required to observe comets from an airplane or an artificial satellite to avoid the interference by telluric water vapor. Moreover, there may be difficulties in modeling opacity effects in the cometary coma. OPRs have been tentatively measured in other molecules (H_2CO and CH_4; Biver et al., 1999; Womack et al., 1997; Weaver et al., 1997) but these results may be not as reliable as for water because of they require more sophisticated or extensive studies.

We reported the OPR of cometary ammonia in Comet C/1999S4 (LINEAR) in a previous paper (Kawakita et al., 2001). This result is derived from the OPR of NH_2 which is a photodissociation product of ammonia in the cometary comae.

In this article, we describe our results for the OPR of ammonia of another comet: C/2001 A2 (LINEAR) was observed on July 26, 2001 with the Subaru telescope and the high-dispersion spectrograph (HDS). We compare the spin temperatures of ammonia with those of water as well as with the temperature ranges estimated from other observations related to the solar nebula or pre-solar molecular cloud.

2. Observations of Comet C/2001 A2 (LINEAR)

In order to obtain high S/N emission lines of NH_2, we observed Comet C/2001 A2 (LINEAR) with the Subaru telescope and the High Dispersion Spectrograph (HDS) (Noguchi et al. 1998). Observational parameters for the comet are listed in Table I.

Comet C/2001 A2 (LINEAR) was discovered by the Lincoln Near Earth Asteroid Research (LINEAR) program (Stokes et al., 2000). The orbital parameters of the comet suggest that it may be belong to the group of Oort cloud comets. The object became bright near perihelion and it was thus easy to get high-S/N spectra for the comet. Comet C/2001 A2 (LINEAR) showed an outburst of about 4 magnitudes at 1.3 AU inbound. Our observation was performed when the comet was at 1.4 AU post-perihelion. Because this comet had a very high gas-to-dust ratio, high S/N spectra of gaseous emission lines could be obtained due to smaller contamination from the sunlight reflected by cometary grains.

The NOAO IRAF software package with the standard reduction procedures was used here. The intensity calibration of the spectrum was performed as follows. The cometary spectrum was normalized to the continuum (i.e., the sunlight reflected by the cometary dust grains). Then the normalized spectrum was divided by a normalized solar spectrum (Kurucz et al., 1984). The result was multiplied by the product of the solar spectrum and the reddening of the dust grains determined from a low dispersion spectrum. The reddening of cometary dust grains is typically about

TABLE I
Relevant observational parameters

	C/2001 A2 (LINEAR)
Date (UT)	2001 July 27.56
Heliocentric distance	1.402 AU
Heliocentric velocity	23.7 km s^{-1}
Geocentric distance	0.467 AU
Geocentric velocity	23.5 km s^{-1}
Total exposure time	3600 s
Slit size	7.0 arcsec × 1.0 arcsec[a]
Spectral resolution ($\lambda/\Delta\lambda$)	36000
Wavelength coverage	5100–7800 Å

[a] The slit was put on the central brightness peak in the coma of the comet.

10%/1000 Å (Jewitt and Meech, 1986). For C/2001 A2 (LINEAR), the reddening is determined from a low dispersion spectrum taken on 12 July 2001 (UT) by the 65cm telescope and the low dispersion spectrograph (GCS) at the Gunma Astronomical Observatory. A reddening of 4.3%/1000 Å was found for the comet. Note that our dust reddening measurement may be affected by an outburst of the comet around July 12. However, our conclusions remain unchanged by the possible error in the dust reddening as shown in the next section. The final cometary spectrum is relatively calibrated for its intensity (Figure 1).

3. Results

Many ortho- and para-lines are resolved in the observed spectrum of the comet. The OPR of NH_2 can be determined from these lines based on fluorescence model calculations (Kawakita et al., 2001). Emission lines with S/N ratio higher than ≈ 20 are used here. The relative intensity calibration of the spectrum may be incomplete because the calibration was performed based on the measured reddening of dust grains as observed at a different date from the HDS observation. Therefore, the OPR values were estimated for (0, 9, 0) band and (0, 7, 0) band individually (the reasons why we selected these bands are described in Kawakita et al., 2001). Due to the narrow wavelength range of each band, i.e., smaller than 100 Å, the influence of an error from the color of dust grains is thought to be negligible within a given band.

Using least χ^2 fitting we derived an OPR for NH_2 of 3.43 ± 0.09 in (0, 9, 0) band of C/2001 A2 (LINEAR). The error corresponds to ± 1-σ. The OPR derived from (0, 7, 0) band is consistent with above result for the comet. However, this

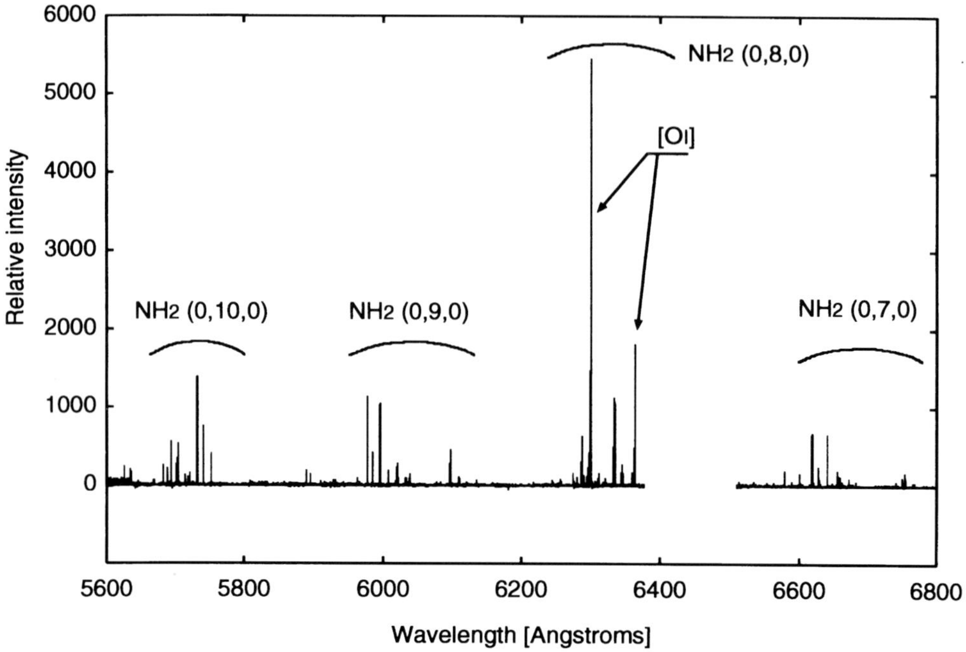

Figure 1. Observed spectra of NH_2 emission bands in comet C/2001 A2 (LINEAR) on 2001 July 27 (UT). The wavelength region includes the NH_2 (0, 10, 0), (0, 9, 0), (0, 8, 0), and (0, 7, 0) bands. The dust continuum has been subtracted. The Doppler shift due to the comet's velocity relative to the Earth has also been corrected. In the (0, 8, 0) band, there are two strong emission lines from forbidden transitions of atomic oxygen (at 6300 and 6364 Å).

value depends on 0_{00}–1_{10} line in (0, 7, 0) band. Namely, when 0_{00}–1_{10} line is eliminated from the least χ^2 fit, the OPR obtained is slightly different from the one with 0_{00}–1_{10} line. For C/2001 A2 (LINEAR), we obtained OPR values are 3.25 $\pm$ 0.08 and 3.08 $\pm$ 0.08 for the cases with and without this line, respectively. We suspect that the 0_{00}–1_{10} line is contaminated by yet another unidentified emission line. In fact, the profile of this line is not an exact Gaussian and and appears to be asymmetric with respect to the central wavelength of the line. The asymmetric line profile of 0_{00}–1_{10} is shown in Figure 2 together with the forbidden transition of atomic oxygen at 5577 Å for comparison. Since the line profile of atomic oxygen at 5577 Å is symmetric (close to a Gaussian function), we can exclude that the asymmetry in 0_{00}–1_{10} line is not caused by the instrumental line profile. Although 0_{00}–1_{10} line should be split into two components for a higher spectral resolution due to an unpaired electron (S = 1/2) of NH_2, the observed transittion profile seems to consist of three lines (these lines can not be clearly separated in our spectra). Hence, we conclude that the line is contaminated by other unidentified emission line.

Our final value of OPR is a weighted average of the OPRs determined from (0, 9, 0) and (0, 7, 0) bands, where for the latter the OPR is taken from emission

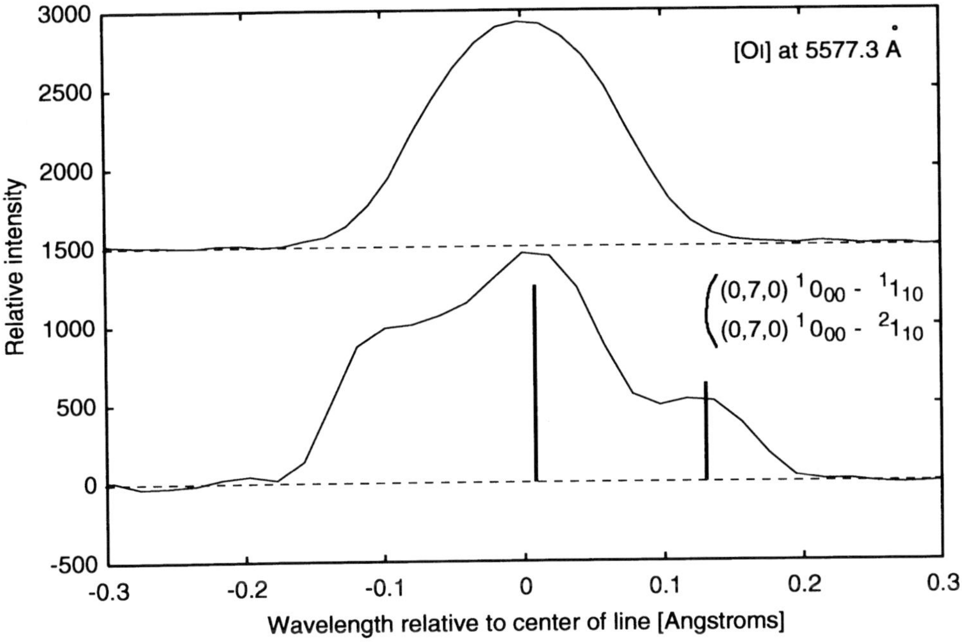

Figure 2. Line profiles of the forbidden transition of atomic oxygen (at 5577 Å) and NH_2 (0, 7, 0) 0_{00}–1_{10} para line in Comet C/2001 A2 (LINEAR). The line profile of atomic oxygen is symmetric relative to the line center. On the other hand, the 0_{00}–1_{10} emission seems to consist of three components. Two of them are identified as fine structure of the 0_{00}–1_{10} line (expected intensities are indicated in the figure). The origin of the third component is not identified.

without 0_{00}–1_{10} line. A reciprocal 1-σ error of each OPR is used as a weight factor for averaging. The simulated and observed spectra for (0, 9, 0) band are shown in Figure 3. The modeled spectrum reproduces well the observed spectrum. Measured ortho- and para-lines are marked in the figure. We applied the permutation group theory (Quack, 1977) to the full reaction system including the source molecule NH_3 and its products NH_2 and H. The theory predicts that ortho-NH_2 only is generated from ortho-NH_3, while ortho- and para-NH_2 are produced with a 1:1 ratio from para-NH_3. This prediction is supported by laboratory studies (Fuke et al., 1988). As a result, the OPR of ammonia are determined to be 1.12 ± 0.03 for C/2001 A2 (LINEAR).

The OPR of ammonia can be calculated for a given spin temperature based on the molecular constants of ammonia (Poynter and Kakar, 1975). It approaches the value 1.0 if ammonia ice formed in equilibrium conditions at high temperature ($\geq$ 40 K) while it becomes larger for low temperature conditions (e.g., more than 10 at 5 K). The OPR derived here indicates a spin temperature of 30^{+3}_{-2} K for C/2001 A2 (LINEAR).

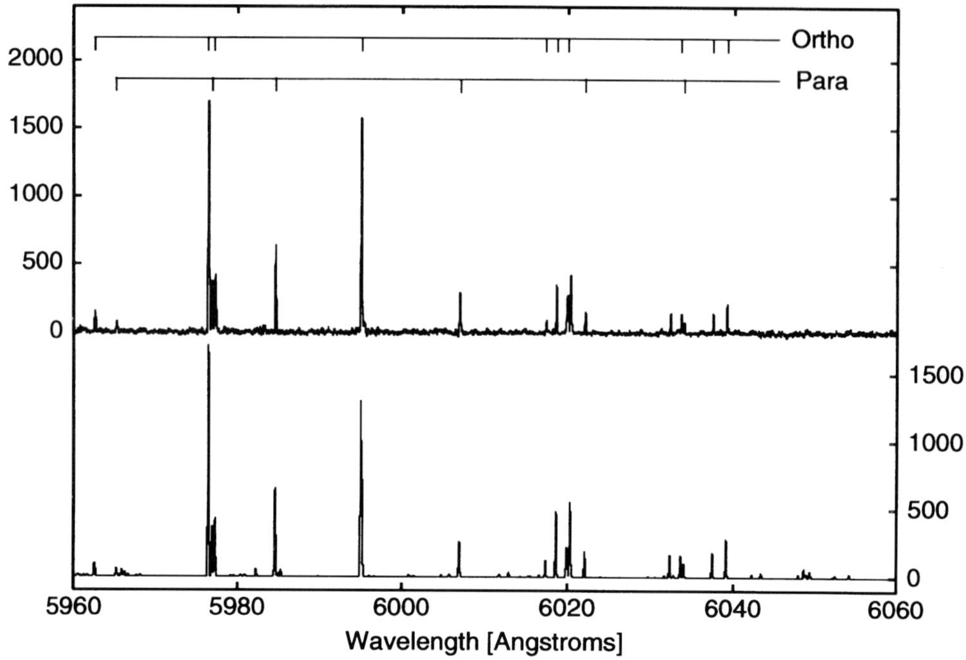

Figure 3. Observed (top) and calculated (bottom) spectra of NH_2 emission bands of (0, 9, 0) in Comet C/2001 A2 (LINEAR) on 2001 July 27 (UT). The ortho- and para-NH_2 lines used for the measurements are indicated in the figure.

4. Discussion

The spin temperature of ammonia in C/2001 A2 (LINEAR), about 30 K, is consistent with the ammonia spin temperature of C/1999S4 (LINEAR) (Kawakita et al. 2001), and consistent with the spin temperature of water observed in Oort cloud comets (comet C/1995O1 (Hale–Bopp) and comet 1P/Halley), $\approx$ 30 K.

Mumma et al. (1993) and Crovisier (1999) argued that the OPR could reflect the temperature at the moment of formation or condensation of the molecules. Our results exclude a high temperature regime (OPR = 1 for ammonia). This fact eliminates the formation of cometary ammonia through chemical reactions only in gas phase (e.g., reactions shown in Hiraoka et al., 1995). Because the chemical reactions in gas phase are exothermal, the OPR of ammonia formed in the gas phase should be unity according to the statistical weights of the nuclear spin in ortho and para species (Irvine, 1999). The OPR larger than the high temperature limit indicates the formation under low temperature conditions. Saito et al. (2000) suggested that a part of ammonia in cold dark clouds could originate from dust related reactions associated with energetic events. The formation of ammonia on the cold grain surface demonstrated by Hiraoka et al. (1995) seems to be important. It appears that ammonia was produced through the gas-grain chemistry in the pre-solar molecular cloud or in the solar nebula, in which ammonia formed on the icy mantle of grains (Hiraoka et al., 1995). In this case, the OPR of ammonia may reflect the temperature of icy grains where the molecules formed (Irvine, 1999). Alternatively, ammonia formed in the gas phase (with OPR = 1) and then frozen onto cold grains that contain paramagnetic species or magnetic nuclei could undergo ortho-para conversion in a relatively short time (cf. Mumma et al., 1993). The OPR of ammonia re-equilibrated according to the low temperature of icy grains in this case. Although this ortho-para conversion process is studied in laboratory, the possibility of ortho-para conversion in cometary nuclei is not yet clear because the actual conditions (especially, the temperature) of interior of cometary nuclei and of the cometary icy grain (its structure, composition, etc.) are not well known. In either case, OPR values for ammonia above 1 are indicative for grain processing.

The spin temperature of ammonia which is consistent with that of water, ≈ 30 K, indicates that the ammonia and water ices formed under the same temperature condition. This fact supports the scenario that ammonia and water ices formed or condensed on the icy grains at a certain temperature. This consistency is important for investigations on the molecular formation processes in the successive stages from the molecular cloud to protoplanetary disk. The next question is when and where the molecules formed during the evolution of the pre-solar cloud to the solar nebula. The re-equilibration of OPR in the interior of cometary nucleus during the long stay in Oort cloud or in the Kuiper Belt (about 4.5 Gyr), and the re-equilibration of OPR at the moment of sublimation from the cometary nucleus, are unlikely for the following reasons. Regarding to the re-equilibration of OPR in the interior of cometary nucleus, the spin temperatures range from 25 K to 35 K for water and ammonia in several comets whose orbital periods range from only 6 years to more than several thousands years. The temperature in the interior of cometary nucleus would be lower for the comet having longer period (namely, larger aphelion distance). It seems to be unlikely that all comets ever observed have the same nucleus temperature (this is also pointed out for cometary water by Irvine et al. 2000). For the re-equilibration of OPR at the moment of sublimation from the cometary surface, the sublimation temperatures expected for water and ammonia are much higher than 30 K. Therefore, it is difficult to attribute the re-equilibration of the OPR of cometary water or ammonia to the moment of sublimation.

The spin temperatures of cometary ammonia shown here and in our previous study (Kawakita et al., 2001) are consistent with temperatures determined for C/1995O1 (Hale–Bopp) from D/H ratios of water and hydrogen cyanide (Blake et al., 1999; Meier et al., 1998a, b). Bergin et al. (1999) suggested the possible formation of molecules in the pre-solar cloud and the solar nebula at ≈ 25–40 K. These results are consistent with the formation of cometary molecules at the temperature of ≈ 30 K. In addition, the depletion of neon in C/1995O1 (Hale–Bopp) may indicate the temperature range higher than 20–25 K (Krasnopolsky et al., 1997; Krasnopolsky and Mumma, 2001; Stern et al., 2000). Recent detection of argon in C/1995O1 (Hale–Bopp) (Stern et al., 2000) requires that the interior of the comet has never been exposed to temperatures of 35–40 K. In Comet C/2001 A2 (LINEAR), however, Feldmann et al. (2001) reported no detection of argon with an upper limit (for the argon abundance relative to oxigen) of more than 10 times lower than solar. This may indicate that this comet was exposed to a higher temperature in the past. Note that the original abundances of these species may changed due to effects from their orbital evolutions.

What could determine the temperature of $\approx$ 30 K? If the Sun formed as an isolated low mass star in the dense core of a dark cloud ($\approx$ 10 K), the obtained spin temperature seems to be too high. Therefore, the spin temperature of ammonia, about 30 K, is thought to represent the temperature at the place where the ammonia formed or condensed in the solar nebula. If it is assumed that the cometary molecules formed or re-processed in the solar nebula under certain temperature condition, the formation zone of the comets can be inferred from the observations. Although the temperature distribution depends on the physical model of the solar nebula (e.g., Hersant et al., 2001; Aikawa et al., 1999; and Willacy et al., 1998), the temperature of $\approx$ 30 K points towards the region of the giant planets in most cases. Comet C/2001 A2 (LINEAR) came from the Oort Cloud which is thought to originate from the cometesimals formed in the giant planets' region based on the statistical studies on orbital evolutions of cometesimals (e.g., Weissman, 1999). The formation zone inferred from spin temperatures is consistent with the area inferred from orbital evolutions. In our future work, we would like to increase the number of comets with OPR of ammonia measured and compare the results between Oort cloud comets and Jupiter family short period comets (originating most likely from the Kuiper Belt).

Acknowledgements

We would like to thank the anonymous referee for his comments to improve the manuscript. This report is based on data collected at the Subaru Telescope, which is operated by the National Astronomical Observatory of Japan. NSO/Kitt Peak FTS data used here were produced by NSF/NOAO.

References

Aikawa, Y., Umebayashi, T., Nakano, T., and Miyama, S. M.: 1999, *Astrophys. J.* **519**, 705–725.

Bergin, E. A., Neufeld, D. A., and Melnick, G. J.: 1999, *Astrophys. J.* **510**, L145–L148.

Biver, N., Bockeleé-Morvan, D., Crovisier, J., Davies, J. K., Matthews, H. E., Wink, J. E., Rauer, H., Colom, P., Dent, W. R. F., Despois, D., Moreno, R., Paubert, G., Jewitt, D., and Senay, M.: 1999, *Astron. J.* **118**, 1850–1872.

Blake, G. A., Qi, C., Hogerheijde, M. R., Gurwell, M. A., and Muhleman, D. O.: 1999, *Nature* **398**, 213–216.

Crovisier, J.: 1999, in J. M. Greenberg and A. Li (eds.), *Formulation and Evolution of Solids in Space*, Kluwer Academic Publishers, Dordrecht, pp. 389–426.

Feldmann, P. D., Weaver, H. A., and Burgh, E. B.: 2001, *Bull. Amer. Astron. Soc.* **33**, 4301.
Fuke, K., Yamada, H., Yoshida, Y., and Kaya, K.: 1988, *J. Chem. Phys.* **88**, 5238–5240.
Hersant, F., Gautier, D., and Huré, J.-M.: 2001, *Astrophys. J.* **554**, 391–407.
Hiraoka, K., Yamashita, A., Yachi, Y., Aruga, K., Sato, T., and Muto, H.: 1995, *Astrophys. J.* **443**, 363–370.
Irvine, W. M.: 1999, *Space Sci. Rev.* **90**, 203–218.
Irvine, W. M., Schloerb, F. P., Crovisier, J., Fegley,Jr., B., and Mumma, M. J.: 2000, in V. Mannings, A. P. Boss, and S. S. Russell (eds.), *Protostars and Planets IV*, Univ. Arizona Press, Tucson, pp. 1159–1200.
Jewitt, D. and Meech, K. J.: 1986, *Astrophys. J.* **310**, 937–952.
Kawakita, H., Watanabe, J., Ando, H., Aoki, W., Fuse, T., Honda, S., Izumiura, H., Kajino, T., Kambe, E., Kawanomoto, S., Noguchi, K., Okita, K., Sadakane, K., Sato, B., Takeda-Hidai, M., Takeda, Y., Usuda, T., Watanabe, E., and Yoshida, M.: 2001, *Science* **294**, 1089–1091.
Krasnopolsky, V. A. and Mumma, M. J.: 2001, *Astrophys. J.* **549**, 629–634.
Krasnopolsky, V. A., Mumma, M. J., Abbott, M., Flynn, B. C., Meech, K. J., Yeomans, D. K., Feldman, P. D., and Cosmovici, C. B.: 1997, *Science* **277**, 1488–1491.
Kurucz, R. L., Furenlid, I., Brault, J., and Testerman, L.: 1984, *National Solar Observatory Atlas No. 1*.
Meier, R., Owen, T. C., Jewitt, D. C., Matthews, H. E., Senay, M., Biver, N., Bockelée-Morvan, D., Crovisier, J., and Gautier, D.: 1998, *Science* **279**, 1707–1710.
Meier, R., Owen, T. C., Matthews, H. E., Jewitt, D. C., Bockelée-Morvan, D., Biver, N., Crovisier, J., and Gautier, D.: 1998, *Science* **279**, 842–844.
Mumma, M. J., Weaver, H. A., and Larson, H. P.: 1987, *Astron. Astrophys.* **187**, 419–424.
Mumma, M. J., Weissman, P. R., and Stern, S. A.: 1993, in E. H. Levy and J. I. Lunine (eds.), *Protostars and Planets III*, Univ. Arizona Press, Tucson, pp. 1177–1252.
Noguchi, K., Ando, H., Izumiura, H., Kawanomoto, S., Tanaka, W., and Aoki, W.: 1998, in Sandro D'Odorico, (ed.), *Proceedings of SPIE, Optical Astronomical Instrumentation*, 3355, pp. 354–362.
Quack, M.: 1977, *Mol. Phys.* **34**, 477–504.
Saito, S., Ozeki, H., Ohishi, M., and Yamamoto, S.: 2000, *Astrophys. J.* **535**, 227–230.
Stern, S. A., Slater, D. C., Festou, M. C., Parker, J. W. M., Gladstone, G. R., A'Hearn, M. F., and Wilkinson, E.: 2000, *Astrophys. J.* **544**, L169–L172.
Stokes, G. H., Evans, J. B., Viggh, H. E. M., Shelly, F. C., and Pearce, E. C.: 2000, *Icarus* **148**, 21–28.
Weaver, H. A., Brooke, T. Y., Disanti, M. A., Mumma, M. J., Tokunaga, A., Chin, G., A'Hearn, M. F., Owen, T. C., and Lisse, C. M.: 1997, *Bull. Amer. Astron. Soc.* **29**, 1041.
Weissman, P. R.: 1999, *Space Sci. Rev.* **90**, 301–311.
Willacy, K., Klahr, H. H., Millar, T. J., and Henning, Th.: 1998, *Astron. Astrophys.* **338**, 995–1005.
Womack, M., Faith, D., Festou, M. C., Slater, D., and Stern, S. A.: 1997, *IAUC* 6542.

EVIDENCE FOR CHEMICAL HETEROGENEITY IN THE NUCLEUS OF C/1995 O1 (HALE–BOPP)

SUSAN M. LEDERER
NASA Johnson Space Center, Mail Code: SR, Houston, TX 77058, USA
E-mail: susan.m.lederer1@jsc.nasa.gov

HUMBERTO CAMPINS
Lunar and Planetary Laboratory, University of Arizona, Tucson, AZ 85721, USA

(Received 1 March 2002; Accepted 3 June 2002)

Abstract. We present an analysis of OH, CN, and C_2 jets observed in the coma of Comet Hale–Bopp on UT April 22, 23, and 25, 1997. Monte Carlo models designed to simulate the gas jets were employed to analyze the nuclear active areas responsible for the observed coma gas jets. Our results indicate that four active areas are necessary to reproduce the CN and C_2 jets while five active areas are required to simulate the OH jets. The additional OH active area must produce significant levels of OH, but cannot emit measurable quantities of either carbon radical. This difference suggests that the nucleus of Comet Hale–Bopp is chemically heterogeneous.

Keywords: C/1995 O1 (Hale–Bopp), comets, gas jets

1. Introduction

Understanding the composition, structure, and homo- or heterogeneity of cometary nuclei is of utmost importance if we are to gain insights into these building blocks of planetary formation. The direct study of cometary nuclei, however, is difficult. When the comet is near the Sun, the nucleus is obscured by the coma. When little or no coma is present, the comet is far from the Sun and faint. Instead, many of the chemical and physical properties of the nucleus and its active areas are inferred by analyzing the coma gas and dust jets, which originate in the nucleus. The apparition of Comet Hale–Bopp in 1997 provided us with an excellent opportunity to study the morphology of different species in the coma, including tracking the evolution of gas and dust jets over many days, and sometimes months (Lederer et al., 1999).

Gas jets in Comet Hale–Bopp were observed in five species. Laffont et al. (1999) reported sunward dust and C_2 jets. As observed in Comet Halley (A'Hearn et al., 1986a, b), CN jets were also detected in Comet Hale–Bopp (Larson et al., 1999, Mueller et al., 1999). In these observations, Comet Hale–Bopp's dust jets coincided with the CN jets in the sunward direction, but not in the anti-sunward direction. Lederer et al. (1999) extended the list of observed gas jets in Comet Hale–Bopp to include not only C_2 and CN, but also OH, NH, and C_3 jets. Again, the gas jets were evident in the anti-sunward direction where no dust jets could be

Earth, Moon and Planets **90:** 381–389, 2002.

TABLE I
Comet Hale–Bopp observing details

Dates	April 22, 23, 25, and 26, 1997–	
r_H (AU)		0.987–1.016
Δ (AU)		1.62–1.69
Phase angle		36.0–33.3
PA_s		240°–247°
Species	Dust, OH, CN, C_2	

detected. Further analysis of these gas images demonstrated that although dust jets have some overlap with the gas jets in the sunward direction, the morphology of the gas jets was significantly different from that of the dust jets.

These morphological and spatial differences help us characterize the gas and dust jets separately, and yield clues about the origin of the gas jets. Previous observational evidence for chemical inhomogeneities in other cometary nuclei (e.g., Comets Encke and Halley) has been reported by several authors (A'Hearn et al., 1983; Mumma et al., 1990). Until now, we did not know whether Comet Hale–Bopp's gas jets originated from a homogeneous or heterogeneous nucleus. Through Monte Carlo models presented here, the observed morphology of gas jets is used to constrain the location and composition of the nuclear active areas in Comet Hale–Bopp. Our results indicate that the nucleus of this comet is chemically heterogenous.

2. Observations and Data Reduction

Data were obtained at Lowell Observatory with the 42″ (1.1 m) Hall Telescope and the 2048 × 2048 SITe charge-coupled device (CCD) camera (2 × 2 binned). The scale of the resultant images is 1.14″/pixel. The Comet Hale–Bopp filter set was employed for all observations. Each filter isolates either the light that fluoresces from a particular gas species (OH at 3090 Å, CN at 3870 Å, and C_2 at 5141 Å), or the light scattered by dust for the continuum filters centered at 3448 Å, 4450 Å, 5260 Å, and 7128 Å (see Farnham et al., 2000 for further details). Table I summarizes the observing details, including the dates on which the observations were made, the heliocentric (r_H) and geocentric (Δ) distances of the comet, the phase angle (Sun–comet–Earth), the position angle of the Sun (PA_s), and the species that were observed. The plate scale for these observations was roughly 1390 km/pixel.

Standard data reduction procedures were applied to obtain an absolute calibration of the observations. Images were corrected for atmospheric extinction and flux calibrated using the standard stars and calibration method outlined in Farnham et al. (2000). The underlying continuum due to the scattered light from dust in the comet's coma was computed and subtracted from each gas image; the amount of gas contamination in relevant filters was computed and eliminated. The resulting gas images contain signal created by fluorescence of the target gas species. A Minimum Azimuthal Subtraction (MAS) routine was applied to enhance the jets. Here, a template representative of the average radial brightness profile is subtracted from the data to yield an image representative of the non-isotropic component. Further details of the data reduction and enhancement routine can be found in Lederer (2000).

3. Modeling Results

Monte Carlo modeling can be used to investigate the chemical and physical properties responsible for the gas jets we observe. Our modeling effort was simplified because the Sun was located at a high cometocentric latitude in April, 1997, which dictates that any active area that resides entirely below roughly 20°S latitude is never exposed to the Sun. This limits the number of active areas that must be included in the model, and the uniqueness of the model is better constrained.

Here, we present the results of models designed to characterize the active areas on the cometary nucleus and simulate the behavior of gases as they evolve from the nuclear active areas into the observed coma gas jets. The modeling was performed using the 3-D time-dependent Monte Carlo cometary coma model described by Lederer (2000). The final model is a 2-D image of the coma projected onto the plane of the sky as dictated by the geometry describing the observations (Table I). Figure 1 demonstrates the fit of the models to the data. On the left are the images of the data taken on April 26, 1997 (approximately one month following perihelion). The images on the right are the best-fit Monte Carlo models. All three gas species are presented in this figure (OH top, CN middle, C_2 bottom).

This Monte Carlo model includes: (i) Collisions of the observed radicals (OH, CN, C_2) with other gases in the coma (primarily water and OH because these are the most abundant coma gases), (ii) gas acceleration in the coma, (iii) radiation pressure from the Sun, and (iv) ejection of the precursors (parents) of the observed gases from discrete active areas on the nucleus that result in coma jets. The observed daughter radicals are produced when a parent decays. The parent's behavior has been modeled through a distribution representative of either an exponential decay or a triangular distribution. The former distribution represents the photodissociation of a parent gas (e.g., H_2O or HCN) while the latter represents production from an extended source (e.g., sub-micron organic grains). For all gas species, observations are best described by models that include radicals produced partially by parent gases and partially by an extended source (see Lederer, 2000 for further discussion).

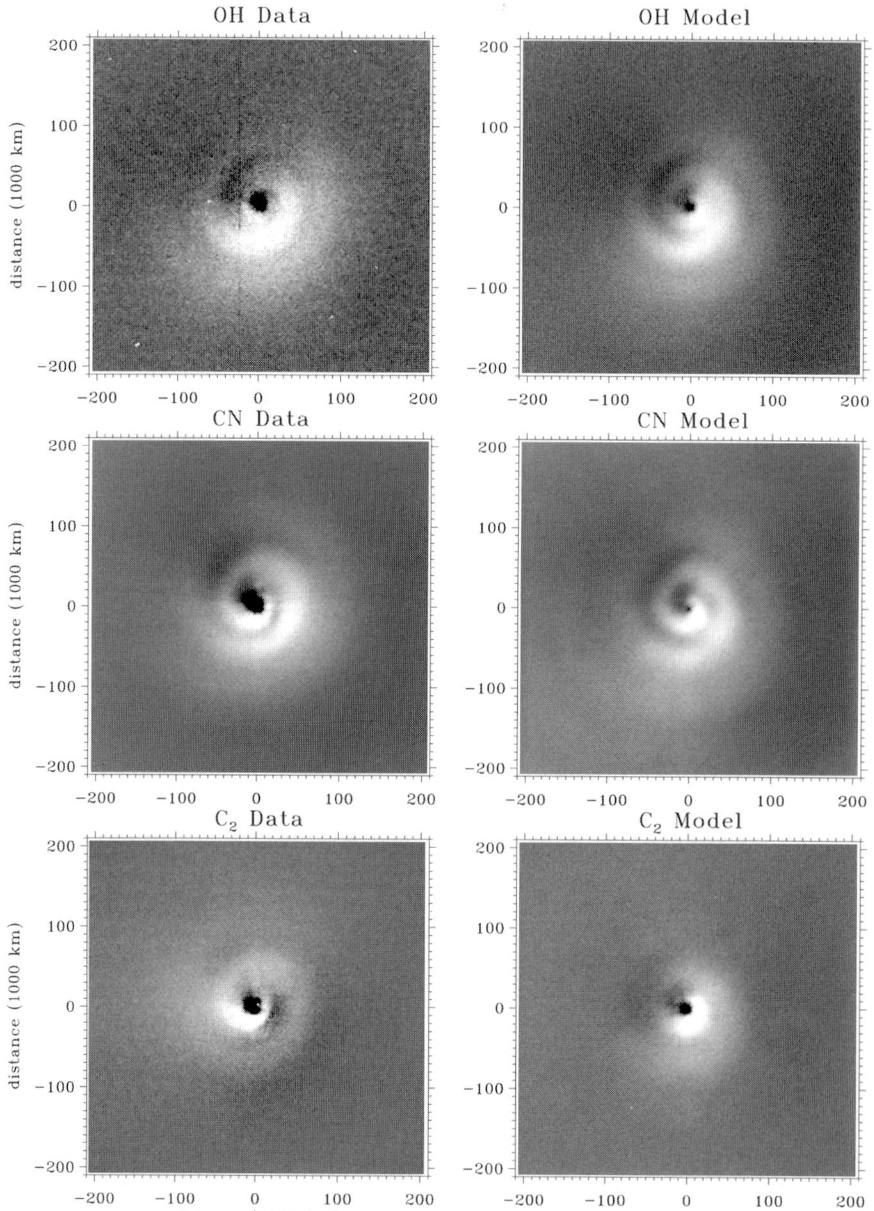

Figure 1. Data and models (April 26, 1997). The images on the left are OH, CN, and C_2 data. The images on the right are the Monte Carlo models that best reproduce the observations for each species. Greyscale translates to intensity. The jets originate from a triangular (grain) extended source. A diffuse (isotropic) gas source model must be combined with the jet model to reproduce the morphology. North is up and east is left. The position angle of the sun is 247°.

TABLE II
Parameters describing the model active areas

Species	lat (°) (± 5)	long (°) (± 10)	Open. Ang. (°) (± 10)	Strength (± 20%)	Q_g (%) (± 5%)
OH (AA 1)	60	175	20	1.0	9
OH (AA 2)	20	240	40	3.0	22
OH (AA 3)	15	120	60	1.2	8
OH (AA 4)	−22	225	60	5.3	17
OH (AA 5)	−12	120	120	10.0	44
CN (AA 1)	60	175	40	1.0	16
CN (AA 2)	13	240	40	4.0	43
CN (AA 3)	15	120	60	1.3	16
CN (AA 4)	−22	225	40	4.0	25
C_2 (AA 1)	60	175	50	1.0	16
C_2 (AA 2)	13	240	60	4.0	43
C_2 (AA 3)	15	120	90	1.3	16
C_2 (AA 4)	−22	225	90	4.0	25

Table II lists the free parameters in our model that describe the discrete nuclear active areas responsible for the coma jets, including: (a) The cometocentric latitude and longitude of the center of the active area (AA), (b) the opening angle (*Open. Ang.*) of the active area, (c) the relative strength of the active area (*Strength*), and (d) the contribution (%) that each individual active region makes to the total production rate of radicals by grains (Q_g). The relative strength denotes the number of particles of a given species that one active area would produce, compared with the other areas producing that species if all active areas were equally illuminated.

Our results indicate that three to four active areas, producing the same OH:CN:C_2 mixture, are necessary to reproduce the coma jets in all three gas species. Three of these active areas are located in the northern hemisphere ([60°N, 175°], Active Area 1; [20°N, 240°] for OH or [13°N, 240°] for CN and C_2, Active Area 2; and [15°N, 120°], Active Area 3). The third of these is necessary to reproduce the CN, but not the OH or C_2 (see Discussion). The fourth, Active Area 4, is located at [22°S, 225°]. A fifth region, Active Area 5([12°S, 120°]), must be included in the OH models to recreate the OH jet morphology, but *cannot* be included in the CN or C_2 models. This region is responsible for about half of the jet-source OH.

4. Discussion

The number of active areas required by our models depends on the species of interest. Our models best reproduce the coma morphology when three to four areas for CN and C_2, or four to five areas for OH are active on the comet's surface. The model parameters required to fit the CN and OH data are more restrictive than those needed to fit the C_2 data. Therefore, the models more rigorously constrain the locations of the active areas producing OH and CN jets than the C_2 jets. We base our conclusions primarily on our OH and CN analyses, and include the C_2 study as it does offer additional information.

In all species, care should be taken when directly comparing the observation's inner coma with the model's inner coma. The processing applied to (a) emphasize the jet morphology (applied to both the observation and the model) and (b) subtract the underlying continuum (applied only to the observation), can affect the morphology. The combination affects the intensities (brightness) within 10–20,000 km of the central condensation, so we did not place great emphasis on morphology fits within $\approx$ 20,000 km. The continuum subtraction can create residual effects within 40–70,000 km, especially in the SW (lower right) quadrant where the dust jets are the most visible. The underlying continuum contributes 10–20% of the signal in the OH data, 20–30% in CN, and 60–70% in C_2. The continuum subtraction, then, will most notably affect the C_2 data, as can be seen in the C_2 data image presented here. We believe that these effects are largely responsible for the intensity differences between the C_2 data and model (e.g., the innermost C_2 jet appears strongest in the south (data) versus the southeast (model), and a dark trough is visible in the southeast portion of the data, but not in the model).

In addition, our models only produce C_2 as a daughter product, not as a granddaughter or great-granddaughter. Studies by Lederer (2000) suggest that changing the parent and daughter lifetimes (as would be necessary to include these processes) would most strongly affect how the intensity of the jets changes with distance from the nucleus, but should not affect the locations of the coma jets. This may partially explain the discrepancies between the morphology of the C_2 data and model. A full analysis of how each input parameter affects the models can be found in Lederer (2000). Based on this analysis, we can say with confidence that although the C_2 fit is not as good, it does not alter our conclusions.

Vasundhara and Chakraborty (1999; hereafter VC99) report that three active areas are required to model the dust jets in their April 10 and May 2, 1997 images. However, incorporating the exact locations of the active areas employed by VC99 produced significant discrepancies between our model's coma gas jet locations and widths when compared with our observations. Hundreds of combinations of active area parameters were subsequently tested. Fortunately, the data are remarkably limiting and only the combination of jet parameters presented here results in a model that fits the observations to within the errors listed (Table II).

Although there are some differences in the exact locations of the active areas between the models in this study and those in VC99, there are also significant similarities. Both models require one active area at a high latitude (Active Area 1) and at least two active areas located near the comet's equator (Active Areas 2 and 4). The region located at the high northern latitude (60°N; Active Area 1) is likely physically much smaller than the near equatorial active areas. Unlike VC99, our gas models require an additional active area at a very low latitude (22°S; Active Area 4). The gas production from this region, which is always exposed to near-twilight conditions, may not be strong enough to lift relatively large (optically scattering) dust particles from the surface, but could lift sub-micron grains in the gas flow, which may be the extended source that produces the gas jets. This would explain why VC99, whose data detects only the larger, optically scattering dust particles, did not invoke a source at this latitude to reproduce their dust jets.

Comparison of the strengths and locations of Active Areas 1, 2, and 4 shows that they are emitting roughly the same relative mixture of OH:CN:C_2. An ambiguity in the total number of active areas arises here because Active Area 3 is emitting only a relatively minor amount of gas, but is required to simply increase the amount of CN in the sunward trough closest to the central condensation. In contrast, the OH and C_2 images display jet peaks that coincide with the location of this CN trough. Although Active Area 3 is not required to reproduce the morphology (shape or brightness levels) of the OH and C_2 jets, it is included in the OH and C_2 models to verify that the same mixture of OH:CN:C_2 *can* be emitted from all four active areas required to reproduce the CN morphology. Alternatively, if some of the CN is a granddaughter product as opposed to a daughter product (an effect not currently included in our models), the effect might also result in filling in this trough. Hence, we do not consider the presence of Active Area 3 to be as well constrained as the other active areas.

If only these active areas existed (1–4), then one could reasonably assume that Comet Hale–Bopp was homogeneous because the same OH:C_2:CN mixture can be emitted from each nuclear region. However, important evidence suggests otherwise: the cometocentric locations of two active areas producing the OH jets do not coincide with those that recreate the CN and C_2 jets. Although one varies only slightly in latitude (Active Area 2), the other requires an entire additional region (Active Area 5). In contrast, the dissimilarities in the CN and C_2 morphologies are created by the differences in the opening angles, daughter velocities, and daughter lifetimes of these two species, not by any variations in active area locations.

Our models indicate that the latitude of Active Area 2, which produces a relatively small portion of the gas jets, is somewhat higher for OH compared with the carbon species. It is unclear whether this difference is real because (i) it resides at the same longitude for all three species (240°), and (ii) it differs only slightly in latitude (20° for OH; 13° for CN and C_2). Attempts were made to resolve the difference in latitude by shifting either the OH-producing region to a more southerly location consistent with the carbon-species, or the reverse. Shifting either source by only $\sim 5^\circ$ caused a noticeable (detrimental) change in the morphology, suggesting that the difference in locations is real. Alternatively, an additional (unknown) effect not included in our model may be responsible for this difference.

The existence of Active Area 5, producing only OH, is of particular interest because it implies that the nucleus of Comet Hale–Bopp is not homogeneous. This active region produces roughly half of the OH that originates from the extended source. However, no appreciable amount of CN or C_2 can be emitted from this region without negatively altering the morphology of the carbon species jets. This additional active area creates the bulk of the morphology differences between OH and the carbon species (e.g., variations in coma jet width, anti-sunward OH peaks coinciding with troughs observed in the carbon species). These results suggest that the composition of the nucleus is heterogeneous.

5. Conclusions

Our models suggest that Comet Hale–Bopp's coma gas jets, observed in April, 1997, originated from a heterogeneous nucleus. Four to five active areas on the surface of Comet Hale–Bopp are necessary to recreate the gas jet morphology evident in our observations of OH, CN and C_2. Three of the active areas can be modeled as having been produced by the same mixture of these gas species; the mixture of an additional active area is indeterminate, but may be the same. Significantly, the fifth active area produces almost half of the jet-source OH, but cannot produce measurable amounts of CN or C_2. This active area is responsible for the variations observed between the morphology of the OH coma jets and that of the carbon species. The extreme enhancement of OH originating from this active area suggests that the coma is not produced by a completely homogeneous nucleus.

6. Acknowledgements

We wish to thank Dr. Eleanor Dixon and an anonymous referee for discussions and comments that have improved this manuscript. This work was supported in part by a NASA GSRP Fellowship, a National Research Council Fellowship, and grants from NASA and NSF.

References

A'Hearn, M. F., Hoban, S., Birch, P. V., Bowers, C., Martin, R., and Klinglesmith, D. A.: 1986a, *Nature* **324**, 649–651.

A'Hearn, M. F., Birch, P. V., and Klinglesmith, D. A.: 1986b, in *Proceedings of the 20th ESLAB Symposium on the Exploration of Halley's Comet*, Heidelberg, West Germany, **1**, pp. 483–486.

A'Hearn, M. F., Millis, R. L., and Thompson, D. T.: 1983, *Icarus* **55**, 250–258.

Laffont, C., Rousselot, P., Clairemidi, J., Moreels, G., and Boice, D. C.: 1999, *Earth Moon Planets* **78**, 211–217.

Larson, S. M., Hergenrother, C. W., and Brandt, J. C.: 1997, *Bull. Am. Astron. Soc.* **29**, 1036.

Lederer, S. M: 2000, 'The Chemical and Physical Properties of the OH, CN and C_2 jets in Comet Hale–Bopp (1995 O1)', Ph.D. Thesis, University of Florida.

Lederer, S. M., Campins, H., Osip, D. J., and Schleicher, D. G.: 1999, *Earth Moon Planets* **78**, 131–136.

Mueller, B. E. A., Samarasinha, N. H., and Belton, J. J. S.: 1999, *Earth Moon Planets* **77**, 181–188.

Mumma, M. J., Reuter, D., and Magee-Sauer, K.: 1990, *Bull. Am. Astron. Soc.* **22**, 1088.

Vasundhara, R. and Chakraborty, P.: 1999, *Icarus* **140**, 221–230.

OPTICAL SPECTROSCOPY AND NEAR-INFRARED OBSERVATIONS OF COMET C/2000 WM1 (LINEAR) IN DECEMBER 2001 FROM CHILE AND BRAZIL

ENOS PICAZZIO[1], AMAURY A. DE ALMEIDA[1], KLIM I. CHURYUMOV[2], SERGEI M. ANDRIEVSKI[1,3] and IGOR V. LUK'YANYK[2]

[1]*Department of Astronomy, Institute of Astronomy, Geophysics and Atmospheric Sciences, University of São Paulo, Main Campus, Rua do Matão 1226, CEP 05508-900, São Paulo, SP, Brazil;* [2]*Astronomical Observatory, Kyiv National Shevchenko University, Kyiv, Ukraine;* [3]*Department of Astronomy, Odessa Mechnikov National University, Odessa Shevchenko Park, 65014, Ukraine*

(Received 5 March 2002; Accepted 21 June 2002)

Abstract. We report on the preliminary analysis of the high-resolution spectrum of Comet C/2000 WM1 (LINEAR), obtained on Dec. 1, 2001 with the Fiber fed Extended Range Optical Spectrograph (FEROS) installed on the 1.52-m telescope of ESO (Chile). Many emission lines of the molecules C_2, C_3, CN, CH, CH^+, NH_2, CO, CO^+, H_2O^+ and, presumably, C_2^- were identified in the spectral range 400–900 nm.

Also, near-infrared photometry was performed on Dec. 2 and 3, with the infrared camera (CamIV) attached to the 0.60-m Boller and Chivens telescope of the Pico dos Dias Observatory (LNA/MCT), Brazil. We report the preliminary and comparative analysis of the I−J and J−H color indices.

Keywords: Comet C/2000 WM1 (LINEAR), comets, infrared photometry, spectroscopy

1. Introduction

High-resolution spectra of comets allow to search for new cometary emissions and the presence of cometary luminescence continuum of non-solar nature, to investigate the chemistry and subsequently the origin and other problems of cometary physics. Near-infrared photometry allows to obtain an important information about color characteristics of comets.

2. Identification of the Optical Spectra

Five high-resolution spectra of comet C/2000 WM1 (LINEAR) were obtained on Dec. 1 and 2, 2001 (UT) with the Fiber fed Extended Range Optical Spectrograph (FEROS) installed on the 1.52-m telescope of European Southern Observatory (ESO), La Silla, Chile. The heliocentric and geocentric distances of the comet were 1.230 AU and 0.318 AU, respectively. All the CCD frames were processed with MIDAS computer programs.

Earth, Moon and Planets **90:** 391–400, 2002.

For the detailed identification we selected the spectrum obtained on Dec. 1, 2001. The Doppler shift was taken into account. For identification of emission lines we used the wavelength tables from the papers of Cochran (2001), Brown et al. (1996), Cosmovici et al. (1982), and Churyumov and Chorny (1992). We considered identification of cometary emissions to be valid if $\Delta\lambda$ - the difference between observed λ_{obs} and laboratory λ_{cat}, taken from the catalogues – did not exceed ± 0.1 Å. The identification reliability was verified using well known prominent cometary emission lines such as the lines of the Swan $C_2(0-0)$ and others.

Many emission lines of the molecules C_2, C_3, CN, CH, CH^+, NH_2, CO, CO^+, H_2O^+, and presumably C_2^- were identified in the spectral range 400-900 nm. The total number of identified emission lines is 4566: 2734 of C_2, 1195 of NH_2, 289 of CN, 158 of C_3, 60 of CO, 51 of H_2O^+, 50 of CH, 16 of CO^+, 8 of CH^+, and 5 of C_2^-. It should be noted that there are also many unidentified emission lines in the comet's spectrum. We determined the wavelengths for 21100 of them. One of the main peculiarities of the investigated spectrum is the presence of emission lines of the neutral molecule CO in the visible region (triplet bands - $d^3\Delta_i - a^3\Pi_r$, Asundi bands – $a'^3\Sigma^+ - a^3\Pi_r$, and Herman bands – $e^3\Sigma^- - a^3\Pi_r$). This is a very rare case of detection of these emission lines in cometary spectra.

Because of the space limitations Tables I–IV show only the identification of emission lines of CO^+, CO (triplet), CH^+, and H_2O^+ bands. The following designations are used in Tables I–IV: λ_{obs} – observed wavelength, I_{rel} – relative intensity, λ_{cat} – wavelength from catalogue, Mol. – molecule; ET, VT and RT – electronic, vibrational and rotational transitions (or branch), respectively. A high-resolution catalogue of emission lines is presently being under preparation and will be published elsewhere.

The identified emission lines of the CO^+ belong to the well-known comet-tail bands $A^2\Pi - X^2\Sigma^+$, presented over the whole observed region.

Seven identified emission lines of CH^+ belong to the electronic transition $A^1\Pi - X^1\Sigma^+$ which contains P, Q, and R branches. We identified 32 lines of the CO triplet bands (electronic transition $d^3\Delta_i - a^3\Pi_r$) which wavelengths matched the precision of ± 0.14 Å relatively to the theoretical wavelengths.

We also identified 51 H_2O^+ emission lines belonging to the electronic transition $\tilde{A}^2A_1 - \tilde{X}^1B_1$.

The Figure 1 shows four sections of the investigated spectrum (406.2–406.7 nm, 515.0-516.5 nm, 541–542 nm, and 525.0–526.7 nm) where the emissions lines of the molecules C_2, NH_2, CO, and presumably C_2^- were detected.

3. Near-Infrared Observations

The near-IR observations were made during Dec. 01-03, 2001 at the Pico dos Dias Observatory (LNA/MCT), Minas Gerais (Long.: +45° 34′ 57″, Lat. :−22° 32′ 04″),

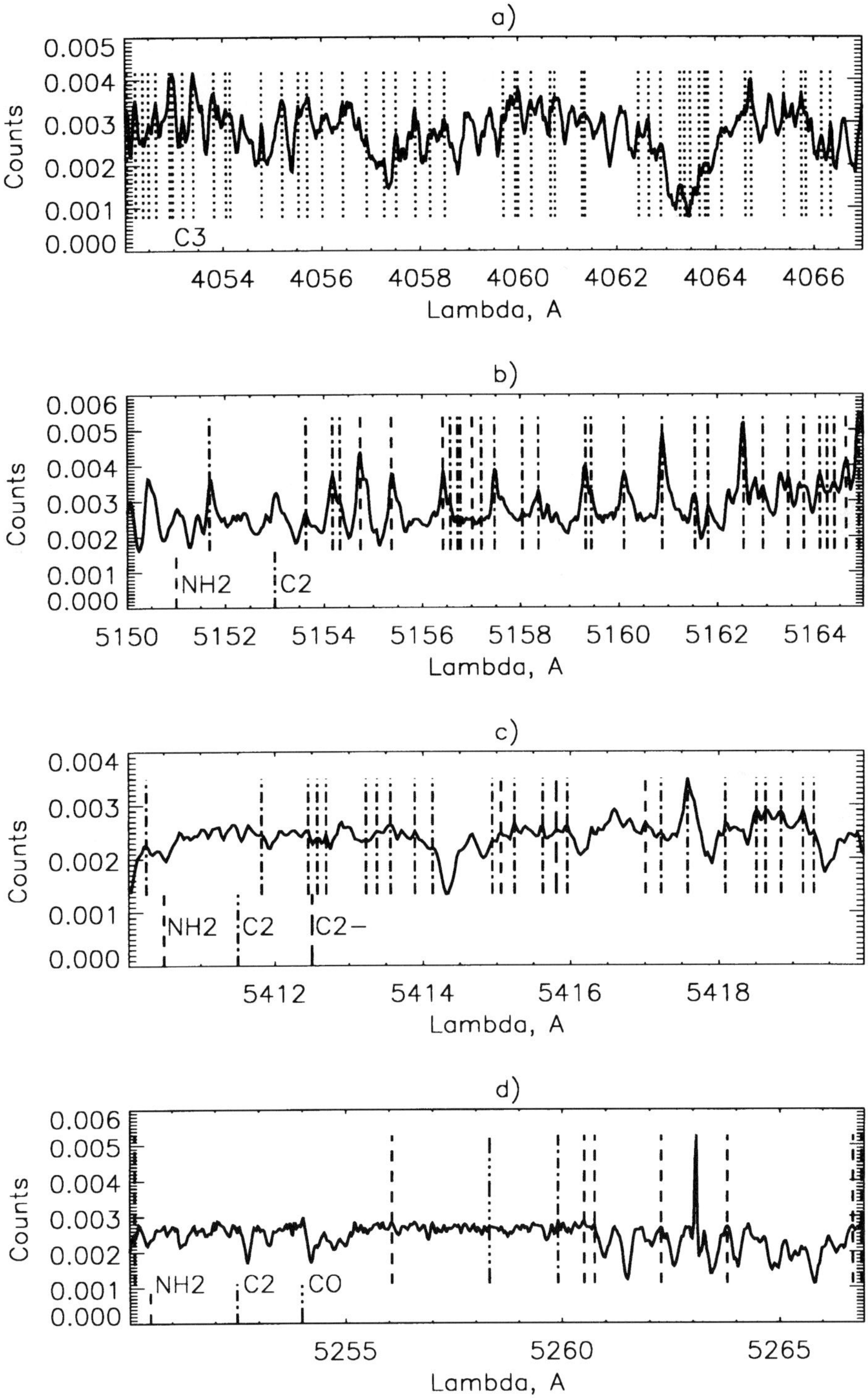

Figure 1. Four sections of the comet spectrum with line identifications, presumably including C_2^-.

TABLE I
Emission lines of CO^+

λ_{obs}	I_{rel}	λ_{cat}	Mol.	ET	VT	Branch
4252.135	0.00223	4252.097	CO^+	A–X	2–0	R2+Q21(4)
4252.465	0.00210	4252.469	CO^+	A–X	2–0	Q2+P21(0)
4252.585	0.00209	4252.558	CO^+	A–X	2–0	R2+Q21(5)
4253.065	0.00217	4253.041	CO^+	A–X	2–0	Q2+P21(1)
4253.875	0.00110	4253.903	CO^+	A–X	2–0	P2(1)R2+Q21(7)
4254.656	0.00218	4254.645	CO^+	A–X	2–0	Q2+P21(3)
4257.027	0.00269	4257.051	CO^+	A–X	2–0	R2+Q21(10)
4258.198	0.00150	4258.159	CO^+	A–X	2–0	P2(4)Q2+P21(6)
4258.378	0.00201	4258.402	CO^+	A–X	2–0	R2+Q21(11)
4259.609	0.00169	4259.621	CO^+	A–X	2–0	Q2+P21(7)
4261.200	0.00140	4261.235	CO^+	A–X	2–0	Q2+P21(8)
4261.560	0.00166	4261.542	CO^+	A–X	2–0	R2+Q21(13)
4261.770	0.00189	4261.818	CO^+	A–X	2–0	P2(6)
4263.300	0.00240	4263.341	CO^+	A–X	2–0	R2+Q21(14)
4263.811	0.00197	4263.843	CO^+	A–X	2–0	P2(7)

TABLE II
Emission lines of CH^+

λ_{obs}	I_{rel}	λ_{cat}	Mol.	ET	VT	Branch
4227.102	0.00133984	4227.054	CH^+	A–X	0–0	R(2)
4232.564	0.00214348	4232.535	CH^+	A–X	0–0	R(0)
4242.140	0.00140268	4242.106	CH^+	A–X	0–0	Q(3)
4245.741	0.00233737	4245.769	CH^+	A–X	0–0	Q(4)
4247.603	0.00193419	4247.568	CH^+	A–X	0–0	P(2)
4254.355	0.00116055	4254.380	CH^+	A–X	0–0	P(3)
4270.774	0.00141577	4270.774	CH^+	A–X	0–0	P(5)

Brazil. The infrared camera (CamIV) was attached to the 0.60-m (f/13.5) Boller & Chivens telescope. The comet and the standard star were observed with filters I (λ = 1.04 μm, FWHM = 0.13 μm), J (λ = 1.25 μm, FWHM = 0.285 μm) and H (λ = 1.65 μm, FWHM = 0.290 μm). The K filter was not available. Gl 105.5 (I = 8.137, J = 7.240 and H = 6.635) from CIT/CTIO list of Standard Stars was selected as calibration star for the observations. An aperture of 1.9″ was used for the photometry.

TABLE III
Emission lines of CO $d^3\Delta_i - a^3\Pi_r$ (triplet)

λ_{obs}	I_{rel}	λ_{cat}	Mol.	ET	VT
4437.111	0.00168247	4437.0	CO	$d^3\Delta_i - a^3\Pi_r$	
4445.395	0.00188244	4445.5	CO	$d^3\Delta_i - a^3\Pi_r$	
4462.861	0.00185889	4462.9	CO	$d^3\Delta_i - a^3\Pi_r$	(15, 3)
4478.768	0.00177397	4478.8	CO	$d^3\Delta_i - a^3\Pi_r$	(15, 3)
4488.401	0.00144916	4488.4	CO	$d^3\Delta_i - a^3\Pi_r$	(13, 2)
4494.403	0.00160623	4494.4	CO	$d^3\Delta_i - a^3\Pi_r$	(15, 3)
4505.478	0.00181624	4505.5	CO	$d^3\Delta_i - a^3\Pi_r$	(13, 2)
4520.783	0.00163164	4520.7	CO	$d^3\Delta_i - a^3\Pi_r$	(13, 2)
4834.959	0.00198630	4835.0	CO	$d^3\Delta_i - a^3\Pi_r$	(11, 2)
4897.470	0.00213416	4897.5	CO	$d^3\Delta_i - a^3\Pi_r$	(11, 2)
4917.086	0.00212238	4917.2	CO	$d^3\Delta_i - a^3\Pi_r$	(11, 2)
4958.894	0.00219081	4959.0	CO	$d^3\Delta_i - a^3\Pi_r$	(9, 1)
4979.046	0.00215302	4979.0	CO	$d^3\Delta_i - a^3\Pi_r$	(9, 1)
4997.099	0.00225675	4996.9	CO	$d^3\Delta_i - a^3\Pi_r$	(9, 1)
5033.202	0.00225106	5033.0	CO	$d^3\Delta_i - a^3\Pi_r$	(7, 0)
5052.901	0.00220360	5052.7	CO	$d^3\Delta_i - a^3\Pi_r$	(7, 0)
5071.100	0.00230556	5070.9	CO	$d^3\Delta_i - a^3\Pi_r$	(7, 0)
5215.911	0.00286605	5216.0	CO	$d^3\Delta_i - a^3\Pi_r$	(8, 1)
5238.539	0.00281854	5238.4	CO	$d^3\Delta_i - a^3\Pi_r$	(8, 1)
5258.319	0.00270296	5258.3	CO	$d^3\Delta_i - a^3\Pi_r$	(8, 1)
5307.979	0.00262822	5308.0	CO	$d^3\Delta_i - a^3\Pi_r$	(6, 0)
5330.430	0.00290220	5330.5	CO	$d^3\Delta_i - a^3\Pi_r$	(6, 0)
5351.112	0.00314623	5351.2	CO	$d^3\Delta_i - a^3\Pi_r$	(6, 0)
5508.196	0.00249691	5508.0	CO	$d^3\Delta_i - a^3\Pi_r$	(7, 1)
5533.496	0.00247337	5533.5	CO	$d^3\Delta_i - a^3\Pi_r$	(7, 1)
5554.360	0.00248857	5554.1	CO	$d^3\Delta_i - a^3\Pi_r$	(7, 1)
5624.011	0.00278107	5624.0	CO	$d^3\Delta_i - a^3\Pi_r$	(5, 0)
5647.690	0.00315385	5647.6	CO	$d^3\Delta_i - a^3\Pi_r$	(5, 0)
5870.440	0.00458341	5870.5	CO	$d^3\Delta_i - a^3\Pi_r$	(5, 0)
5981.891	0.00428262	5982.0	CO	$d^3\Delta_i - a^3\Pi_r$	(4, 0)
6010.536	0.00430975	6010.5	CO	$d^3\Delta_i - a^3\Pi_r$	(4, 0)
6036.768	0.00422900	6037.0	CO	$d^3\Delta_i - a^3\Pi_r$	(4, 0)

TABLE IV
Emission lines of H_2O^+

λ_{obs}	I_{rel}	λ_{cat}	Mol.	ET	VT	RT
5799.605	0.00472159	5799.559	H_2O^+	A–X	(0, 4, 0) – (0, 0, 0)	303 – 313
5804.927	0.00459278	5804.917	H_2O^+	A–X	(0, 4, 0) – (0, 0, 0)	000 – 110
5805.228	0.00424989	5805.220	H_2O^+	A–X	(0, 4, 0) – (0, 0, 0)	000 – 110
5837.851	0.00457254	5837.856	H_2O^+	A–X	(0, 4, 0) – (0, 0, 0)	303 – 413
5838.182	0.00481665	5838.174	H_2O^+	A–X	(0, 4, 0) – (0, 0, 0)	303 – 413
5846.690	0.00463459	5846.692	H_2O^+	A–X	(0, 3, 0) – (0, 0, 0)	221 – 211
5913.422	0.00468941	5913.376	H_2O^+	A–X	(0, 3, 0) – (0, 0, 0)	221 – 331
5924.962	0.00453192	5924.942	H_2O^+	A–X	(0, 3, 0) – (0, 0, 0)	321 – 431
5936.771	0.00470443	5936.791	H_2O^+	A–X	(0, 3, 0) – (0, 0, 0)	423 – 533
5938.724	0.00468401	5938.772	H_2O^+	A–X	(0, 3, 0) – (0, 0, 0)	423 – 533
5939.986	0.00460109	5939.973	H_2O^+	A–X	(0, 3, 0) – (0, 0, 0)	422 – 532
6135.559	0.00433093	6135.545	H_2O^+	A–X	(0, 3, 0) – (0, 0, 0)	312 – 202
6140.920	0.00422614	6140.917	H_2O^+	A–X	(0, 3, 0) – (0, 0, 0)	211 – 101
6146.850	0.00399751	6146.802	H_2O^+	A–X	(0, 3, 0) – (0, 0, 0)	110 – 000
6155.563	0.00378561	6155.556	H_2O^+	A–X	(0, 3, 0) – (0, 0, 0)	111 – 101
6156.041	0.00414650	6156.038	H_2O^+	A–X	(0, 3, 0) – (0, 0, 0)	111 – 101
6158.646	0.00400321	6158.637	H_2O^+	A–X	(0, 3, 0) – (0, 0, 0)	212 – 202
6162.567	0.00356231	6162.597	H_2O^+	A–X	(0, 3, 0) – (0, 0, 0)	313 – 303
6166.998	0.00400706	6166.961	H_2O^+	A–X	(0, 3, 0) – (0, 0, 0)	414 – 404
6179.417	0.00429608	6179.395	H_2O^+	A–X	(0, 3, 0) – (0, 0, 0)	211 – 303
6185.221	0.00390289	6185.198	H_2O^+	A–X	(0, 3, 0) – (0, 0, 0)	211 – 221
6186.597	0.00394041	6186.586	H_2O^+	A–X	(0, 3, 0) – (0, 0, 0)	312 – 322
6198.710	0.00405165	6198.747	H_2O^+	A–X	(0, 3, 0) – (0, 0, 0)	110 – 220
6199.428	0.00389251	6199.388	H_2O^+	A–X	(0, 3, 0) – (0, 0, 0)	111 – 221
6209.415	0.00410778	6209.417	H_2O^+	A–X	(0, 3, 0) – (0, 0, 0)	211 – 321
6210.850	0.00409977	6210.869	H_2O^+	A–X	(0, 3, 0) – (0, 0, 0)	212 – 322
6221.730	0.00408896	6221.774	H_2O^+	A–X	(0, 3, 0) – (0, 0, 0)	312 – 422
6527.873	0.00439825	6527.916	H_2O^+	A–X	(0, 3, 0) – (0, 0, 0)	202 – 110
6573.970	0.00438159	6573.977	H_2O^+	A–X	(0, 2, 0) – (0, 0, 0)	321 – 211
6576.734	0.00409410	6576.769	H_2O^+	A–X	(0, 3, 0) – (0, 0, 0)	202 – 312
6592.754	0.00411773	6592.792	H_2O^+	A–X	(0, 3, 0) – (0, 0, 0)	303 – 413
6593.371	0.00431298	6593.328	H_2O^+	A–X	(0, 2, 0) – (0, 0, 0)	321 – 313
6686.584	0.00467230	6686.557	H_2O^+	A–X	(0, 2, 0) – (0, 0, 0)	221 – 331
6699.962	0.00453253	6699.929	H_2O^+	A–X	(0, 2, 0) – (0, 0, 0)	321 – 431
6700.138	0.00441337	6700.178	H_2O^+	A–X	(0, 2, 0) – (0, 0, 0)	322 – 432
6958.207	0.00368915	6958.228	H_2O^+	A–X	(0, 2, 0) – (0, 0, 0)	312 – 202
6963.298	0.00323200	6963.331	H_2O^+	A–X	(0, 2, 0) – (0, 0, 0)	211 – 101
6965.452	0.00371237	6965.470	H_2O^+	A–X	(0, 2, 0) – (0, 0, 0)	211 – 101
6973.684	0.00381286	6973.720	H_2O^+	A–X	(0, 2, 0) – (0, 0, 0)	110 – 000
6987.704	0.00439073	6987.737	H_2O^+	A–X	(0, 2, 0) – (0, 0, 0)	212 – 202
7024.227	0.00358590	7024.196	H_2O^+	A–X	(0, 2, 0) – (0, 0, 0)	312 – 322
7038.210	0.00332267	7038.188	H_2O^+	A–X	(0, 2, 0) – (0, 0, 0)	111 – 221
7052.744	0.00361645	7052.723	H_2O^+	A–X	(0, 2, 0) – (0, 0, 0)	211 – 321
7053.180	0.00364694	7053.196	H_2O^+	A–X	(0, 2, 0) – (0, 0, 0)	212 – 322
7069.456	0.00344846	7069.496	H_2O^+	A–X	(0, 2, 0) – (0, 0, 0)	312 – 422
7070.444	0.00280392	7070.407	H_2O^+	A–X	(0, 2, 0) – (0, 0, 0)	312 – 422
7446.828	0.00266996	7446.836	H_2O^+	A–X	(0, 2, 0) – (0, 0, 0)	303 – 211
7467.723	0.00266819	7467.763	H_2O^+	A–X	(0, 2, 0) – (0, 0, 0)	101 – 111
7469.064	0.00258802	7469.071	H_2O^+	A–X	(0, 2, 0) – (0, 0, 0)	202 – 212
7492.090	0.00260724	7469.257	H_2O^+	A–X	(0, 2, 0) – (0, 0, 0)	202 – 212
7494.372	0.00263643	7494.344	H_2O^+	A–X	(0, 2, 0) – (0, 0, 0)	101 – 211

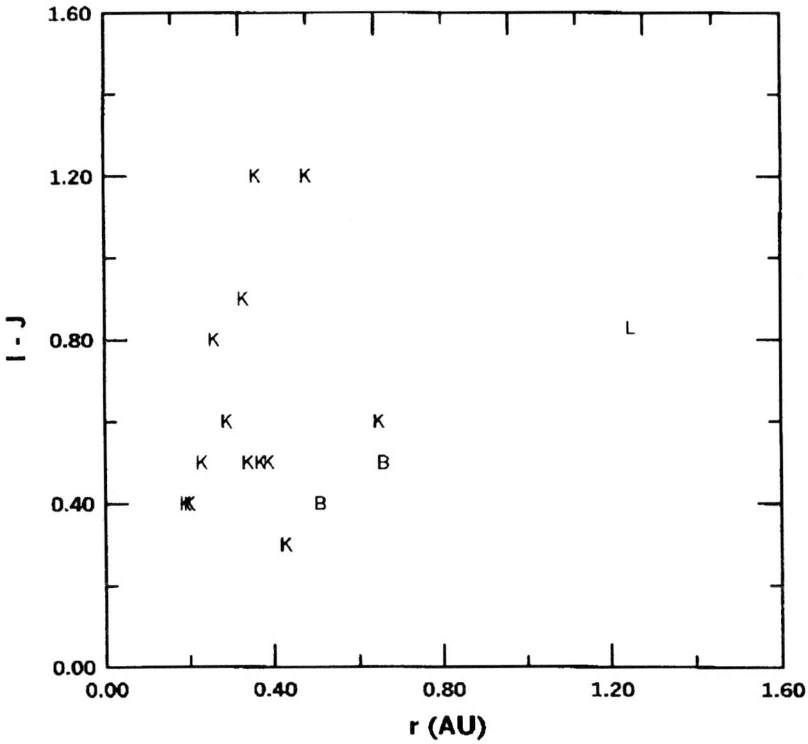

Figure 2. I–J color vs heliocentric distance. Symbols are B: C/Bradfield (1974 C1); K: C/Kohoutek (1973 E1), Ney, 1974, and L: C/2000 WM1 (LINEAR).

Due to the high velocity of the comet in the sky and due to restrictions in telescope guiding system, the exposures could never exceed 10 s. We secured several series of short exposures, but for this work we selected a sample with only a few images of 10 s exposure, taken on Dec. 01–02. The heliocentric and geocentric distances of the comet were 1.214 AU and 0.316 AU, respectively. The corresponding phase angle was 38.6 degrees. For the standard star Gl 105.5 the exposure time was 2 s.

4. Near-Infrared Results

Our preliminary results are summarized in Figures 2–5.

As mentioned above, it was not possible to observe Comet C/2000 WM1 (LINEAR) in K band. This fact made it difficult to compare our data with those for other comets for which H−K measurements are available. Although our data in the I band are rather scarce to draw any reliable conclusion, they suggest that Comet C/2000 WM1 (LINEAR) seems to be similar to C/Kohoutek (1973 E1) and redder

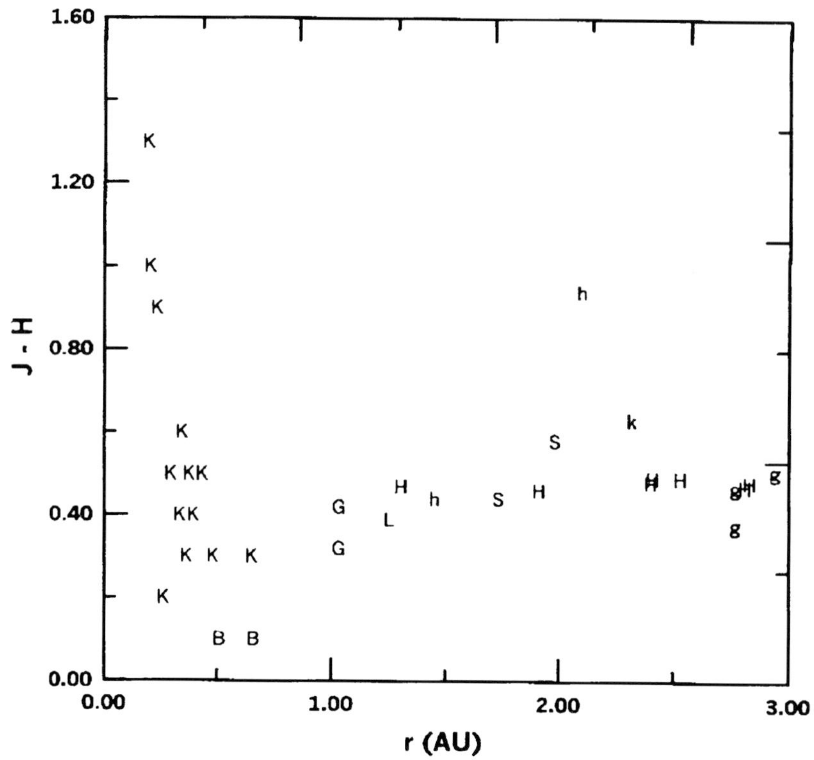

Figure 3. J−H color vs heliocentric distance. Symbols are B: C/Bradfield (1974 C1); G: 26P/Grigg–Skjellerup (1982 IV); g: 65P/Gunn (1982 X); H: 1P/Halley (1986 III) (Tokunaga et al., 1986); h: 1P/Halley (1986 III) (Matsuura et al., 1987); K: C/Kohoutek (1973 E1); k: 59P/Kearns–Kwee (1981 XX), L: C/2000 WM1 (LINEAR) and S: 38P/Stephan–Oterma (1980 X). Adapted from Hanner et al., 1984.

than C/Bradfield (1974 C1) (Figure 2). The I−J colors discrepancy is too high to be simply due to filter characteristics.

Although the J−H color of Comet C/2000 WM1 (LINEAR) is a little bluer, their values are similar to those published J−H data for other comets (Figure 3). Tokunaga et al. (1986) and Jewitt and Meech (1988) pointed out that in most cases the J−H color index of comets is almost constant and independent of the heliocentric distance. Moreover, if we consider the same heliocentric distance for the comets, the J−H values of C/2000 WM1 (LINEAR), 1P/Halley (1986 III), and 26P/Grigg-Skejellerup (1982 IV) appear to be very similar.

According to Figure 5 the J−H colors seem to be fairly uniform and independent of the scattering angle.

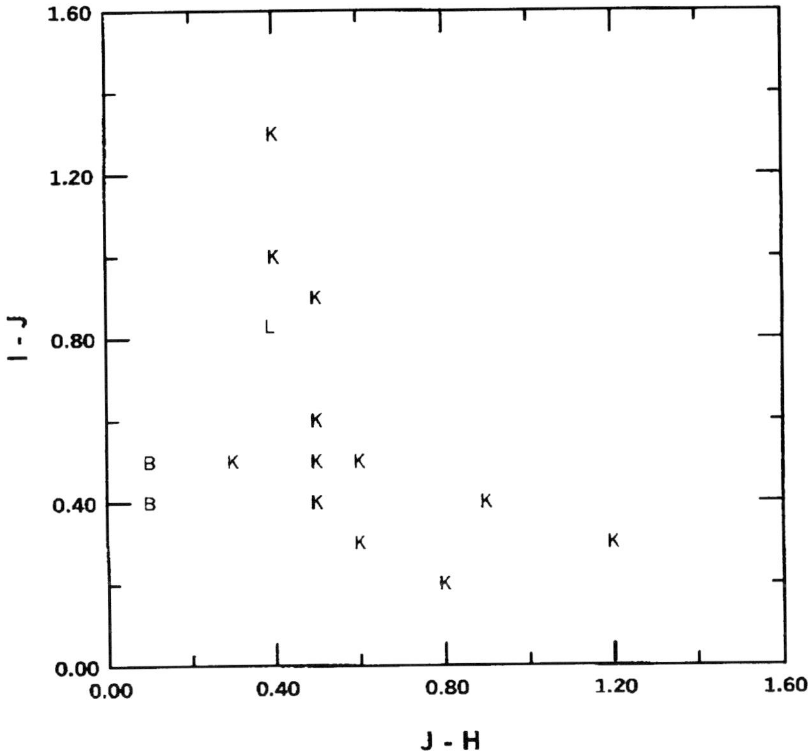

Figure 4. I−J vs J−H color diagram. Symbols are defined in previous figures.

5. Conclusions

In 2001, we performed high-resolution spectroscopic observations of the bright comet C/2000 WM1 (LINEAR) at ESO. We found emission lines of the following neutral radicals: 2734 of C_2, 1195 of NH_2, 289 of CN, 158 of C_3, 60 of CO, 50 of CH. We observed also lines of the following ionic species: 51 of H_2O^+, 16 of CO^+, 8 of CH^+, and presumably 5 of C_2^-.

In the near-infrared the I−J colors discrepancy is too high to be considered as originating from the filter characteristics. J−H colors of comets are almost constant and independent of the heliocentric distance and scattering angle.

Acknowledgements

We are grateful to Dr. Anita Cochran for giving us the opportunity to use her excellent Spectral Atlas of comet de Vico prior to publication. The FEROS observations at ESO were carried out under the Observatório Nacional ON/ESO and ON/IAG agreements, supported by FAPESP through the project no. 1998/10138-8.

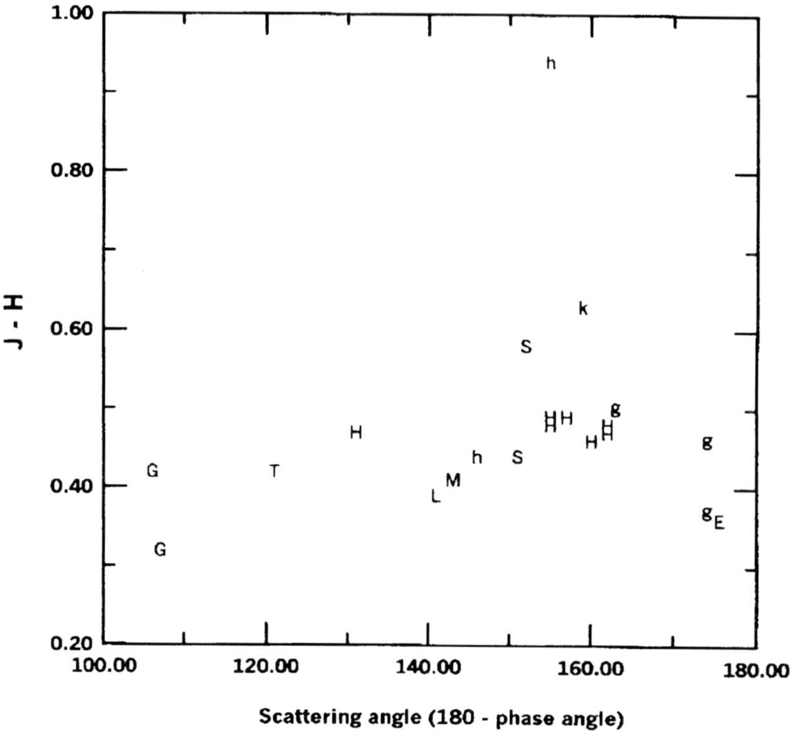

Figure 5. J–H color vs scattering angle. Symbols are defined in previous figures.

References

Brown, M. E., Bouchez, A. H., Spinrad, H., and Jons-Krull, C. M.: 1996, 'A High-Resolution Catalog of Cometary Emission Lines', *Astron. J.* **112**, 1197–1202.

Churyumov, K. I., and Chorny, G. F.: 1992, 'The First Identification of C_2^- Emission Bands in Comet Scorichenko–George (1989e1) Spectrum', in *Proceedings of the International Conference "Asteroids, Comets, Meteors 1991"*, pp. 117–120.

Cochran, A. and Cochran, W. D.: 2001, 'A High Spectral Resolution Atlas of Comet 122P/de Vico', *Icarus*, in press.

Cosmovici, C. D, Barbieri, C., Bonoli, C., Bortoletto, F., and Hamzaoglu, E.: 1982, 'On the Spectrum of Comet Bradfield 1980f', *Astron. Astrophys.* **114**, 373–387.

Hanner, M. S., Tokunaga, A. T., Veeder, G. J., and A'Hearn, M. F.: 1984, 'Infrared Photometry of the Dust in Comets', *Astron. J.* **89**, 162–169.

Jewitt, D. and Meech, K. J.: 1988, 'The Absence of a Color-Distance Trend in Comets', *Astron. J.* **96**, 1723–1730.

Matsuura, O. T., Picazzio, E., and Kawara, K.: 1987, 'Preperihelion Photometry of the Comet Halley in the Near Infrared', *PASP* **99**, 154–157.

Ney, E. P.: 1974, 'Multiband Photometry of Comets Kohoutek, Bennett, Bradfield, and Encke', *Icarus* **23**, 551–560.

Tokunaga, A. T., Golisch, W. F., Griep, D. M., Kaminski, C. D., and Hanner, M. S.: 1986, 'The NASA Infrared Telescope Facility Comet Halley Monitoring Program I. Preperihelion Results', *Astron. J.* **92**, 1183–1190.

PHOTOMETRY AND IMAGING OF COMET C/2000 WM1 (LINEAR)

DAVID G. SCHLEICHER and LAURA M. WOODNEY
Lowell Observatory, 1400 W. Mars Hill Rd., Flagstaff, AZ 86001, USA
PETER V. BIRCH
Perth Observatory, Walnut Rd., Bickley 6076, Western Australia

(Received 25 February 2002, Accepted 21 June 2002)

Keywords: C/2000 WM1, C/LINEAR (2001 WM1), comet, imaging, narrowband photometry

We present preliminary results from our on-going observational campaign of Comet LINEAR (2000 WM1). These ground-based observations include both narrowband photometry and CCD imaging. We began monthly photometric observations of LINEAR in August 2001 ($r_H = 2.7$ AU) using the Hall 1.1-m telescope at Lowell Observatory. These continued from Lowell until mid-December, when LINEAR became a southern hemisphere object and observations began from the 0.6-m Planetary Patrol telescope at Perth Observatory. Production rates for the measured gas species and for dust (see A'Hearn et al., 1995 for model parameters) are shown in Figure 1 for dates ranging from 23 August 2001 to 11 January 2002, only eleven days prior to perihelion (0.56 AU). Our most recent production rate for OH was obtained on 13 December (1.05 AU), with a Haser value of 3.8×10^{28} mol/s, corresponding to a vectorial equivalent water production rate of 5.2×10^{28} mol/s. (The near-UV OH and NH bands are not measured from Perth due to its relatively low elevation.) The r_H-dependence of water closely follows a standard ice vaporization model (cf. Cowan and A'Hearn, 1979), and the water production implies an effective active area of 10 km^2.This corresponds to a lower limit of the effective nucleus radius of about 0.9 km. Relative abundances of the minor species yield a classification of "typical" in the A'Hearn et al. (1995) database.

The apparently anomalous behavior of dust production with distance shown in Figure 1 is actually a consequence of phase angle effects. The phase angle varied between 21° and 26° from August to October, decreased to 15° during the November observations, and reached 72° in December. Therefore, enhanced backscattering at small phase angles contributes to the high values in November near 1 AU. However, the primary cause of the artificially high values in November is LINEAR's strong dust tail, shown in Figure 2. At small phase angles, the tail is nearly behind the coma and therefore partially within the photometer's entrance apertures. The measurements obtained at large phase angles in December are, therefore, more representative of LINEAR's actual dust production. This research is supported by NASA.

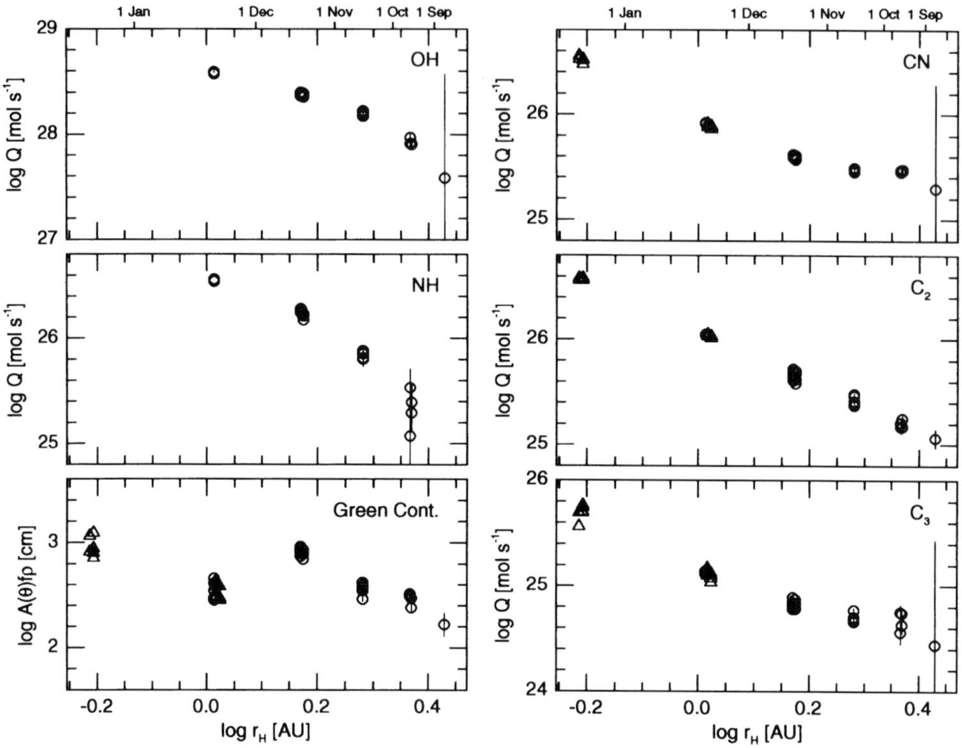

Figure 1. Log of gas production rates and log $Af\rho$ (a measure of dust production) at λ5260 plotted as a function of the log of heliocentric distance. Monthly observations from Lowell Observatory are shown as circles, while triangles represent observations obtained from Perth Observatory. The apparent decrease in dust production in December near 1 AU is a consequence of artificially high apparent dust production values in November and early, when the comet was at smaller phase angles.

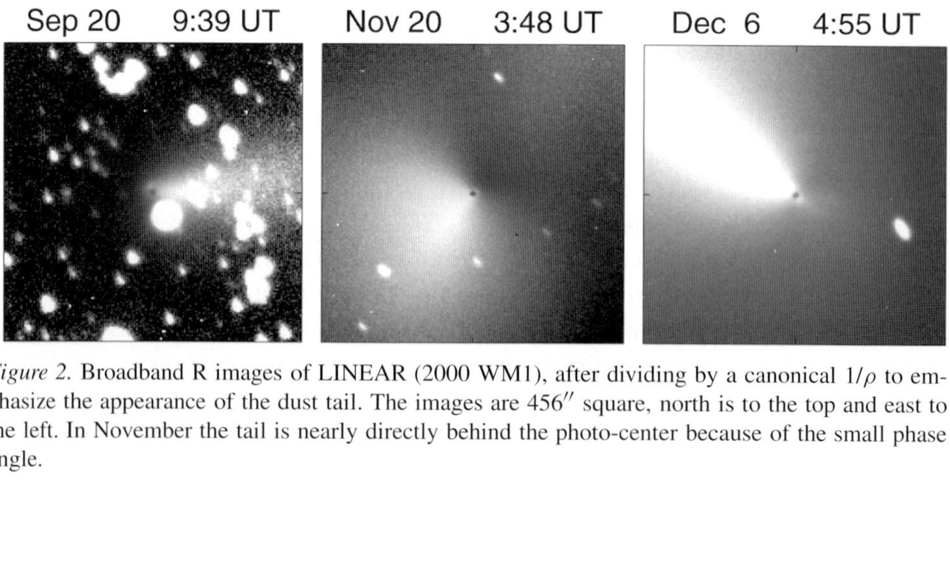

Figure 2. Broadband R images of LINEAR (2000 WM1), after dividing by a canonical $1/\rho$ to emphasize the appearance of the dust tail. The images are 456″ square, north is to the top and east to the left. In November the tail is nearly directly behind the photo-center because of the small phase angle.

References

A'Hearn, M. F., Millis, R. L., Schleicher, D. G., Osip, D. J., and Birch, P. V.: 1995, 'The Ensemble Properties of Comets: Results from Narrowband Photometry of 85 Comets, 1976–1992', *Icarus* **118**, 223–270.

Cowan, J. J. and A'Hearn, M. F.: 1979, 'Vaporization of Cometary Nuclei: Light Curves and Lifetimes', *Moon Planets* **21**, 155–171.

DISCONNECTION EVENTS IN COMETS

MARCOS RINCON VOELZKE
Department of Informatic, Cruzeiro do Sul University, Regente Feijo Avenue 1295, 03342-000 Sao Paulo, SP, Brazil
E-mail: mrvoelzke@zipmail.com.br

(Received 6 March 2002; Accepted 3 July 2002)

Abstract. Cometary and solar wind data are compared with the purpose of identifying the solar wind conditions which are associated with comet plasma tail disconnection events (DEs), i.e., when the plasma tail appears disconnected from the cometary head. The cometary data are from *The International Halley Watch Atlas of Large-Scale Phenomena* (Brandt et al., 1992a). A systematic visual analysis of the atlas images (Voelzke and Matsuura, 1998) revealed, among other morphological structures, 47 DEs along the plasma tail of comet P/Halley. This work compares the current competitive theories, based on the triggering mechanisms, in order to explain the cyclic phenomena of DEs, i.e., the ion production effects, the pressure effects and the magnetic reconnection effects are analysed. The distribution of the DEs in time or heliocentric distance presents a bimodal character possibly associated with the cometary passage through the magnetic sector boundaries in the interplanetary medium. The 47 DEs documented in 47 different images allowed the estimation of 19 onsets of DEs, i.e., the time when the comet supposedly crossed a frontier between magnetic sectors of the solar wind. The solar wind data are taken from *in situ* measurements of IMP-8 (King, 1982), which is used to construct the actual variation of solar wind speed, density and dynamic pressure during the analysed interval. These *in situ* measurements are referenced to the comet by standard co-rotation methods. The preliminary results of this research reveal that the DEs onsets of comet P/Halley are correlated with pressure effects only in 23% of the analysed cases, therefore these effects should not be the principal triggering mechanism of DEs.

Keywords: Comets, disconnection events

1. Introduction

This work is based on a systematic analysis of 531 images of comet P/Halley available in *The International Halley Watch Atlas of Large-Scale Phenomena* (Brandt et al., 1992a). They were selected on the basis of their many conspicuous morphological structures along the cometary tail and they were visually and systematically analysed for the purpose of identification and classification of optical structures (Voelzke and Matsuura, 1998).

Occasionally, the entire plasma tail or a part of it separates from the comet and drifts away (antisunward), followed by simultaneous renewal of the plasma tail. The cyclic nature of this phenomena, called a disconnection event (DE), was noted long ago (Barnard, 1899, 1920). Disconnection events (Niedner and Brandt, 1979; Jockers, 1985; Celnik et al., 1988; Delva et al., 1991; Brandt et al., 1992b;

Voelzke and Matsuura, 1998, 2000) are common. They occur over wide ranges in heliocentric distance, heliospheric latitude, and phases of the solar cycle. Although the driving mechanism remained unresolved, the variable solar wind was suspected to play a major role in the phenomena of comet plasma tail DEs. Many ideas were suggested to explain the cyclic phenomena of DEs (Brandt, 1990). The current competitive theories to explain the onset of DEs can be grouped into three classes based on the triggering mechanisms:

- Ion production effects. If the ion production rate were drastically reduced, the ionosphere of the comet could shrink temporarily and allow the magnetic field lines to slip inwards. The change of ionization rate in the comet could be strongly influenced by a change in the solar UV radiation (Wurm and Mammano, 1972).
- Pressure effects. A large dynamic pressure increase in the solar wind could compress the comet's ionosphere and change the formation of the magnetic field lines. If this is the relevant mechanism, one would expect DEs to be associated with high-speed streams or high-density regions (Ip and Mendis, 1978; Ip, 1980; Jockers, 1985; Wegmann, 1995, 1998, 2000).
- Magnetic reconnection. The sunward magnetic reconnection occurs as the comet crosses the interplanetary magnetic field (IMF) sector boundary (Niedner and Brandt, 1978; Brandt et al., 1992c, 1999; Yi et al., 1996, 1998; Brandt and Snow, 2000; Voelzke and Matsuura, 2000), while the tailside reconnection might be triggered by either an interplanetary corotating shock or a high-speed stream (Russel et al., 1986).

In this work DEs in comet P/Halley will be interpreted in terms of the variation of the dynamical pressure effects, in order to investigate whether these effects play an important role as triggering mechanism of DEs.

2. Data

A disconnection event was found in 47 out of the selected 531 images from *The International Halley Watch Atlas of Large-Scale Phenomena* (Brandt et al., 1992a), covering the period from 1985 November 13 to 1986 April 15. They were visually and systematically analysed to derive their onset times (Table I). The 47 DEs documented in the 47 different images allowed the estimation of onset times of 19 DEs as listed in Table I. The images show the evolution of different DEs as a function of time, i.e., they ilustrate a sequence of events involving the separation of the plasma tail and the re-growth of a new one. Reports by other authors on a DE occuring within $\pm$ 0.5 day were considered to be possibly related, and they are cited in the Table.

The solar wind data are *in situ* measurements from IMP-8 (MIT Space Plasma Group-IMP 8 Home Page), which are used to construct the variation of solar wind speed, density and dynamic pressure during the analysed interval. These values

TABLE I

Onset time of 19 DEs. Number density of proton (N_p), proton bulk velocity (V_p) and dynamic pressure (DP) as a function of onset time

Onset time (UT)	N_p (N cm^{-3})	V_p (km s^{-1})	DP (nPa)	Possibly related reports
1985 Nov 12.76	↑14.52	↓451.8	4.96	
1985 Dec 05.61	↓7.88	↑457.2	2.75	
1985 Dec 13.54	*	*	*	Niedner (1986); Yi et al. (1994b)
1985 Dec 14.16	*	*	*	
1986 Jan 01.90	↓4.34	↑689.0	3.47	
1986 Jan 07.52	*	*	*	
1986 Jan 09.63	*	*	*	Niedner (1986); Tomita et al. (1987); Niedner et al. (1991)
1986 Jan 10.16	↓0.65	↓441.0	0.21	Niedner (1986); Tomita et al. (1987); Niedner et al. (1991)
1986 Jan 10.76	↑11.40	↓421.0	3.38	
1986 Feb 22.17	↑3.81	↑626.0	2.50	Niedner (1986); Yi et al. (1994b)
1986 Mar 08.90	↑7.42	↓593.7	4.37	Niedner (1986); Wu and Qiu (1987); Niedner and Schwingenschuh (1987)
1986 Mar 14.05	↑7.58	↓312.0	1.23	
1986 Mar 15.87	↓13.00	↑369.0	2.96	Feldman et al. (1986); Celnik et al. (1988); Yi et al. (1994a)
1986 Mar 20.01	↑5.19	↓406.3	1.43	
1986 Apr 03.38	↑6.99	↓416.0	2.02	
1986 Apr 06.45	*	*	*	Niedner (1986); Celnik et al. (1988)
1986 Apr 07.70	*	*	*	
1986 Apr 11.75	↑10.58	↓341.4	2.06	
1986 Apr 15.34	↑28.20	↑337.0	5.36	

given in Table I are derived from IMP-8 measurements at the Earth which are corotated, by standard methods, to the comet's location.

3. Onset Analysis

The MIT Faraday cup experiment on IMP-8 (King, 1982) measures currents from solar wind ions, and from these measurements the velocity, density, and the temperature of the solar wind are derived.

The number density of protons (N_p in N cm^{-3}) for each DE onset time, listed in Table I, was compared with the respective proton bulk velocity (V_p in km s^{-1}), looking for a possible correlation between the DEs onset time and dynamic pressure effects.

The dynamic pressure (DP) was calculated from the proton density and velocity in order to depict its variation as a function of time, using the equation:

$$DP = 1.6726 \times 10^{-6} \times N_p \times V_p^2. \quad (1)$$

At the time of six onsets (1985 December 13.54; 14.16; 1986 January 07.52; 09.63; April 06.45 and 07.70) the IMP-8 does not have measurements. Ten of the other thirteen onsets (77%) show an anticorrelation between the velocity and the density, i.e., when one increases (↑), the other decreases (↓) (Table I), therefore in these cases the pressure effects do not play an important role in the formation of DEs because the solar wind dynamic pressure does not vary strongly.

Only three DE onsets (23%), namely on 1986 January 10.16; February 22.17 and April 15.34, show the same trend between density and velocity, i.e., both increase (two cases) or decrease (one case) at the same time.

Wegman (1995) showed that a strong interplanetary shock, with a Mach number in the frame of the ambient solar wind larger than two, can make a density hump in the plasma tail and concluded that about 25% of all tail disconnections must be caused by interplanetary shocks.

The solar wind dynamic pressure does not vary strongly because the solar wind density for the DEs analysed here, in most of the cases, increases when the solar wind velocity decreases and vice versa. Hence these DEs were found to be uncorrelated with high-speed streams and high-density regions.

The preliminary results of this work agree with the earlier works from Yi et al. (1994a, b) and Brandt et al. (1992b, c, 1999), which also analysed comet Halley's DEs in order to understand the mechanism by which these DEs occur. Yi et al. (1993) considered the possible association of changes in the Alfvén Mach number with DEs and concluded that such an association was unlikely. The coronal mass ejection data in *Solar Geophysical Data* (Wagner, 1984) are also not supportive of a correlation between the interplanetary shocks and DEs (Yi et al., 1994a; Brandt et al., 1999). Much effort has been spent to demonstrate that there is a close association between DEs and pressure effects (Ip and Mendis, 1978; Ip, 1980; Jockers,

1985; Wegmann, 1995, 1998, 2000). It is now widely accepted that such an association exists, but it is less obvious that these effects play a major role in plasma tail DEs. In fact, it is clear that not all observed DEs can be caused by schocks, since comets are not hit by schocks frequently enough (Wegmann, 2000).

Wegmann (2000) studied the effect of several solar wind disturbances on the plasma tail of a comet by means of MHD model calculations and showed that high speed streams, non compressive density enhancements and shocks can cause DEs, i.e., in this work it is argued against a single cause as a possible triggering mechanism for DEs formation, and apparently against Occam's razor as well. Yet, Wegmann (2000) proposes that multiple causes (mostly pressure effects) should explain the DEs.

According to Wegmann, although there is no consensus in the literature on the formal definition of a DE, the main feature is that *a major portion of the tail disconnects from the coma and falls away* (Farnham and Meech, 1994). Wegmann's simulations produce disturbances, wavy structures (undulations or a train of waves), density enhancements or knots (Matsuura and Voelzke, 1990; Voelzke, 1996; Voelzke et al., 1997), solitary waves or solitons usually denominated kinks (Tomita et al., 1987), *Swan-like* tails (Hyder et al., 1974; Niedner and Brandt, 1980; Jockers, 1985) and also DEs (Niedner and Brandt, 1979; Niedner, 1981, 1982; Jockers, 1985; Russel et al., 1986; Brossius et al., 1987; Lundstedt and Magnusson, 1987; Celnik et al., 1988; Delva et al., 1991; Brandt et al., 1992b; Yi et al., 1993, 1994a; Voelzke and Matsuura, 1998, 2000) therefore it is not obvious that his simulations have a one-to-one correspondence, i.e., that high speed streams, non compressive density enhancements and shocks should be the principal triggering mechanism of tail disconnections.

4. Conclusions

The solar wind plays a major role in plasma tail DEs, which are characterized by the plasma tail being uprooted from the comet head and convected downstream in the solar wind while being replaced by a new tail constructed from folding rays. The comparison of the solar wind conditions and the DEs onsets of P/Halley's comet shows that these DEs are apparently not primarily associated with pressure effects.

The analysis of the DE onsets in comet P/Halley reveals that they are not correlated with pressure effects in 77% of the cases, therefore these effects do not play an important role to explain the origin of P/Halley DEs.

This is a research in progress and this communication only reports the preliminary results obtained from the analysis of the DE onsets of comet P/Halley. In a forthcoming work this research will be expanded to other comets and the DEs present in them will be correlated with solar wind data not only from IMP-8 (King,

1982) but also with ICE, VEGA 1, PVO and Ulysses (Wenzel et al., 1992; Smith and Wenzel, 1993).

Of course better solar wind data, more comet images from the ground and *in situ* measurements from cometary flyby missions will greatly contribute to our better understanding of the triggering mechanism of cometary DEs.

Acknowledgements

The support obtained by M.R.V. from *Coordenação de Aperfeiçoamento de Pessoal de Nível Superior*, CAPES, under grant BEX2582/97-0 and from Cruzeiro do Sul University, UNICSUL, are gratefully acknowledged. The author is grateful to the original investigators and the NSSDC, who kindly provided the IMP-8 data, and also to the anonymous referee for helpful and valuable comments.

References

Barnard, E. E.: 1899, *MNRAS* **59**, 354 pp.
Barnard, E. E.: 1920, *ApJ* **51**, 102–106.
Brandt, J. C.: 1990, in J. W. Mason (ed.), *Comet Halley: Investigation, Results, Interpretation 1,* Ellis Horwood, New York, pp. 33–55.
Brandt, J. C. and Snow, M.: 2000, *Icarus* **148**, 52–64.
Brandt, J. C., Caputo, F. M., Hoeksema, J. T., Niedner, M. B. Jr., Yi, Y., and Snow, M.: 1999, *Icarus* **137**, 69–83.
Brandt, J. C., Niedner, Jr. M. B., and Rahe, J.: 1992a, *The International Halley Watch Atlas of Large-Scale Phenomena*, Printed by: Johnson Printing Co., Boulder, CO, University of Colorado-Boulder, 1 p.
Brandt, J. C., Randall, C. E., Yi, Y., and Snow, M.: 1992b, 'Asteroids, Comets, Meteors 1991', in *Lunar and Planetary Institute*, LPI, Houston, pp. 93–96.
Brandt, J. C., Caputo, F. M., and Yi, Y.: 1992c, *BAAS* **24**, 1270–1270.
Brosius, J. W., Holman, G. D., Niedner, Jr. M. B., Brandt, J. C., Slavin, J. A., Smith, E. J., Zwickl, R. D., and Bame, S. J.: 1987, *Astron. Astrophys.* **187**, 267–275.
Celnik, W. E., Koczet, P., Schlosser, W., Schulz, R., Svejda, P., and Weissbauer, K.: 1988, *A&AS* **72**, 89–127.
Delva, M., Schwingenschuh, K., Niedner, Jr. M. B., and Gringauz, K. I.: 1991, *Planet. Space Sci.* **39**, 697–708.
Farnham, T. L. and Meech, K. J.: 1994, *ApJS* **91**, 419–460.
Feldman, P. D., AHearn, M. F., Festou, M. C., McFadden, L. A., Weaver, H. A., and Woods, T. N.: 1986, *Nature* **324**, 433–436.
Hyder, C. L., Brandt, J. C., and Roosen, R. G.: 1974, *Icarus* **23**, 601–610.
Ip, W.-H.: 1980, *ApJ* **238**, 388–393.
Ip, W.-H. and Mendis, D. A.: 1978, *ApJ* **223**, 671–675.
Jockers, K.: 1985, *A&AS* **62**, 791–838.
King, J.: 1982, in C. T. Russel and D. J. Southwood (eds.), *Availablity of IMP-8 Data for IMS Period, IMS Source Book*, American Geohysical Union, Washington, DC, p. 10.
Lundstedt, H. and Magnusson, P.: 1987, *Astron. Astrophys.* **187**, 261–263.
Matsuura, O. T. and Voelzke, M. R.: 1990, *Rev. Mexicana Astron. Astrof.* **21**, 609–612.

MIT Space Plasma Group: MIT Space Plasma Group–IMP 8 Home Page. ftp://space.mit.edu/pub/plasma/imp/www/imp.html

Niedner, Jr. M. B.: 1981, *Astrophys. J. Suppl. Ser.* **46**, 141–157.

Niedner, Jr. M. B.: 1982, *Astrophys. J. Suppl. Ser.* **48**, 1–50.

Niedner, Jr. M. B.: 1986, *Adv. Space Res.* **6**, 315–327.

Niedner, Jr. M. B. and Brandt, J. C.: 1978, *ApJ* **223**, 655–670.

Niedner, Jr. M. B. and Brandt, J. C.: 1979, *ApJ* **234**, 723–732.

Niedner, Jr. M. B. and Brandt, J. C.: 1980, *Icarus* **42**, 257–270.

Niedner, Jr. M. B. and Schwingenschuh, K.: 1987, *A&A* **187**, 103–108.

Niedner, Jr. M. B., Brandt, J. C., and Yi, Y.: 1991, 'The 10 January 1986 Disconnection Event in Comet Halley', in *Cometary Plasma Processes*, American Geophysical Union, AGU, Washington, D.C., pp. 153–159.

Russel, C. T., Saunders, M. A., Phillips, J. L., and Fedder, J. A.: 1986, *J. Geophys. Res.* **91**, 1417–1423.

Smith, E. J. and Wenzel, K.-P.: 1993, *J. Geophys. Res.* **98**, 21,111–21,127.

Tomita, K., Saito, T., and Minami, S.: 1987, *A&A* **187**, 215–219.

Voelzke, M. R.: 1996, *Publ. Astron. Soc. Pacific* **108**, 1063–1063.

Voelzke, M. R. and Matsuura, O. T.: 1998, *Planet. Space Sci.* **46**, 835–841.

Voelzke, M. R. and Matsuura, O. T.: 2000, *A&AS.* **146**, 1–5.

Voelzke, M. R., Schlosser, W., and Schmidt-Kaler, Th.: 1997, *Astrophys. Space Sci.* **250**, 35–51.

Wagner, W. J.: 1984, *ARA&A* **22**, 267–289.

Wegmann, R.: 1995, *A&A* **294**, 601–614.

Wegmann, R.: 1998, *J. Geophys. Res.* **103**(A4), 6633–6635.

Wegmann, R.: 2000, *A&A* **358**, 759–775.

Wenzel, K.-P., Marsden, R. G., Page, D. E., and Smith, E. J.: 1992, *A&AS* **92**, 207–219.

Wu, M. C. and Qiu, P. Z.: 1987, *A&A* **187**, 264–266.

Wurm, K. and Mammano, A.: 1972, *Ap&SS* **18**, 273 pp.

Yi, Y., Brandt, J. C., Randall, C. E., and Snow, M.: 1993, *ApJ* **414**, 883–891.

Yi, Y., Brandt, J. C., Randall, C. E., and Snow, M.: 1994a, *AJ* **107**, 1591–1596.

Yi, Y., Caputo, M. F., and Brandt, J. C.: 1994b, *Planet. Space Sci.* **42**, 705–720.

Yi, Y., Walker, R. J., Ogino, T., and Brandt, J. C.: 1996, *J. Geophys. Res.* **101**(A12), 27,585–27,601.

Yi, Y., Walker, R. J., Ogino, T., and Brandt, J. C.: 1998, *J. Geophys. Res.* **103**(A4), 6637–6639.

POLARIMETRIC STUDY OF COMETS C/1995 O1 (HALE–BOPP) AND C/2000 WM1 (LINEAR)

UMESH C. JOSHI, KIRAN S. BALIYAN and SHASHIKIRAN GANESH
Physical Research Laboratory, Navrangpura, Ahmedabad 380 009, India
E-mail: joshi@prl.ernet.in

(Received 1 March 2002; Accepted 25 June 2002)

Abstract. We made polarimetric observations of comet Hale–Bopp covering a wide phase angle range, from 18.8 to 47°. At certain phase angles the heliocentric distance of the comet was less than 1 AU during its pre and post perihelion passages. One of the important findings, based on the data in the visual bands, is the higher polarization with stronger wavelength dependence compared to comet Halley, indicating the presence of much finer grains in comet Hale–Bopp. It may also be noted that comet Hale–Bopp has shown highest degree of polarization known so far for any comet and hence falls in the class of high polarization comets. Polarimetric observations were made of comet C/2000 WM1 (LINEAR) using narrow band (IHW) filters 4845 Å and 7000 Å and broad bands filters BVR during November 23–26, 2001 when the phase angle ranged from 15 to 22°. Some of the results based on these observations are presented and discussed.

Keywords: Comets, dust characteristics, linear polarization

1. Introduction

Comets stay away from the Sun for most of their lives and hence face minimal weathering. They, therefore, act as fossils in revealing the primitive composition of the solar system. The early phase of accumulation of grains and ice with other volatile elements produced first proto cometary nuclei, in which the solid grains remained embedded in ice. At times these cold bodies are deflected towards the Sun. When coming closer to the Sun, the temperature of the nucleus increases, and ices close to the surface are sublimated releasing the gas and dust into space, often violently. The grains which are ejected from the interior are pristine and those blown out from the crust may have been reworked.

The study of the dust grains in comets has been an active area of investigation for quite some time, but the exact nature and composition of cometary grains is still an enigma. The space missions to comet Halley made useful in-situ measurements and have added important information on the nature of grains in that comet (Levasseur-Regourd et al., 1986). However, on the basis of ground based observations it has been noticed that there is variation in the detailed behavior of the grains in different comets. Until more in-situ measurements are carried out at other comets, the information about dust grains in comets has to come mainly from

Earth, Moon and Planets **90:** 413–421, 2002.

the ground based observations in conjunction with theoretical models (Krishna Swamy, 1986).

The continuum emission received at the Earth is sunlight scattered by the dust grains in the coma and the tail. The emission from the molecules which fluoresce in sunlight is superposed on this continuum. Information regarding the physical and chemical conditions at the time of the formation of solar nebula is imprinted on the dust grains and the ices. In the present paper we are mainly confined to the study of grains which is vital to understand the formation and evolution of the solar system. The light scattered by the dust grains is polarized and the degree of polarization depends on the phase angle, wavelength of radiation, grain size distribution and its complex refractive index. However, there are other factors which modify the nature of polarization, e.g., shape of the grains and their alignment, segregation etc. As there are several unknown parameters, it is difficult to find a unique solution. In order to constrain the parameters characterizing the grains, it is very important to have polarization observations over a wide phase angle range at several continuum wavebands, free from contamination by molecular emission.

In the past, linear and circular polarization measurements have been made by many investigators for several comets. Extensive polarimetric observations were made of comet P/Halley (Bastien et al., 1986; Brooke et al., 1987; Dollfus and Suchail, 1987; Kikuchi et al., 1987; Lamy et al., 1987; Le Borgne et al., 1987; Metz and Haefner, 1987; Sen et al., 1988, 1991 etc). The photopolarimeter on board the GIOTTO spacecraft recorded in-situ polarization measurements in the coma at a phase angle of 73° (Levasseur-Regourd et al., 1986). Most of these studies were aimed at understanding the nature of the dust through the study of polarization of light. However, systematic polarization measurements of comets over a large range of heliocentric distances and/or scattering angles forming a homogeneous database are lacking. Most of the studies are based on the data obtained from different telescopes using different instruments leading to a possibility of systematic differences between various data sets. Several authors (Dollfus, 1989; Krishna Swamy and Shah, 1988) compiled the data for comet Halley and some other bright comets to construct a complete phase-angle vs polarization curve. Such data, being inhomogeneous are not very suitable to compare with theoretical models. However, due to the lack of complete knowledge of cometary grains, the predictions made by the current models also have limitations. The span of variation of the calculated polarization (adjusting various parameters) within the wavelength range 0.4–0.7 μm is large (Krishna Swamy and Shah, 1988). Therefore, the use of composite polarization phase curve appears to be the best choice available at present.

Hale–Bopp has the distinction of being the brightest comet of the past two decades. Perihelion was on 1 April 1997 at r = 0.9AU. Advances in detector technology, long lead time from the discovery to perihelion and convenient locations afforded an excellent opportunity to study this comet. At the Physical Research Laboratory, Ahmedabad, we planned detailed photometric and polarimetric observations, covering a wide range in phase angle in three continuum bands defined by IHW filters. In addition to this we also attempted imaging polarimetry to study the nature of polarization in the inner coma. In this communication we report the polarization behavior of comet Hale–Bopp and describe grain characteristics based on polarimetric data.

In the recent past another comet, C/2000 WM1 appeared in the sky and was bright enough to make good S/N polarization observations. In the month of November 2001 we carried out polarimetric observations when the phase angle was between 14–22°. At low phase angles polarization is expected to be negative and observations at such phase angles are rather meager, which makes the present observations very useful. Some of the results based on these observations are also reported here.

2. Observations

Observations of Comet C/1995 O1 (Hale–Bopp) were taken with the optical polarimeter (Deshpande et al., 1985; Joshi et al., 1987, 1997) mounted at the Cassegrain focus of the 1.2 m telescope at Mt. Abu Infrared Observatory operated by the Physical Research Laboratory, India. The polarimeter was equipped with the IAU/IHW filters. IHW filters were acquired about a decade ago for P/Halley campaign and were carefully preserved in a dry atmosphere. However, to be sure of their characteristics, we determined their transmission curves before the start of present observations and found that except for the CO^+ (4260 Å) filter, curves for all other filters compared very well with the transmission characteristics supplied by the manufacturers without any significant deterioration. The polarization measurements in the three continuum bands (3650/80, 4845/65, 6840/90 Å) are discussed here. Though some of these continuum filters are slightly contaminated by molecular emission (D. G. Schleicher, private communication), the degree of polarization and position angle are not significantly affected. Also the observations with these filters facilitate comparison with comets observed earlier with the same set of filters, hence their continued use is justified. The details of the observing run and data analysis are given by Ganesh et al. (1998). In addition to the aperture polarimetry, we also carried out imaging polarimetry of comet Hale–Bopp on May 9, 1997 using a CCD camera (pixel scale 0.56 arcsec/pixel) equipped with a red filter (8000–10000 Å) and a Polaroid sheet which could be rotated in steps of 45°. Observing procedure and other details are discussed elsewhere.

Photopolarimetric observations of Comet C/2000 WM1 were also made on the 1.2 m telescope of Mt. Abu Observatory during November 23–26, 2001. We attempted observations through IHW filters in continuum bands 3650/80 (UC), 4845/65 (BC) and 6840/90(RC). However, the comet was too faint to achieve good S/N ratio with these narrow band filters, even for the integration time as long as 10 min. Therefore, we resorted to observations with broad bands – BVR filters of Johnson and Morgan system. We made a few observations through IHW filter

4845/65 (BC) to facilitate comparison with the data through the broad bands. During the observing run, the phase angle ranged between 14–22° when the degree of polarization is, in general, small ($\leq 2\%$). In the early part of the night, sky was not only bright but was also highly polarized due to the presence of moon which hampered the polarimetric observations. We performed the polarization observations after the Moon had set and the sky was dark. This, however, reduced the total observing time (between moonset and comet setting) considerably. The sky remained very good throughout the observing run, and we succeeded in obtaining good data through the broad band BVR filters and limited observations through IHW filters. Observations were taken through an aperture of 26.5 arcsec.

In the present paper we report results based on the observations made through the narrowband filters (4845 Å (IHW) and 7000 Å) and the broad band filter R. We defer discussion of the observations made with the other broad band filters (B, V) to a separate paper.

3. Discussion

3.1. Comet C/1995 O1 (Hale–Bopp)

The observed polarization-phase curves in three continuum filters are plotted in Figure 1. For comparison we have also plotted the degree of polarization of comet P/Halley (Kikuchi et al., 1987), one of the most extensively observed comets, at various phase angles. Note that while the 3650 Å and 4845 Å filters were identical for the two comets, Kikuchi et al., 1987 used a relatively much broader filter in place of the 6840 Å. Though the use of broad band by them may have resulted in contamination by molecular emission, the comparison of the polarization values obtained by Kikuchi et al. (1987) are not significantly different from the values reported at 6840 IHW-band (Dollfus et al., 1988). Figure 1 clearly shows a higher degree of polarization and steeper wavelength dependence in Hale–Bopp as compared to Halley. In fact comet Hale–Bopp turned out to be a comet having the highest degree of polarization in the last decade. The observations of Hale–Bopp were mostly made with an aperture of 26.5 arcsec diameter. However, because the Earth-comet distance was different for different set of observations, the sampled area at the comet differs. This might cause some change in the degree of polarization. The inner coma region of comets, including Hale–Bopp, has been found to be quite heterogeneous in spatial distribution of polarization by various researchers (Eaton et al., 1991; Renard et al., 1996; Kolokolova et al., 2001; Jockers, 1997; Hadamcik et al., 1997).

In the case of Halley, polarization phase curves for the inner coma (radius $<$ 5000 km) and the outer region (radius $>$ 10000 km) appear different (Dollfus et al., 1988). This may be due to either a difference in the microtexture of the dust grains (Dollfus, 1989) or a difference in dust size distribution. But in the case of

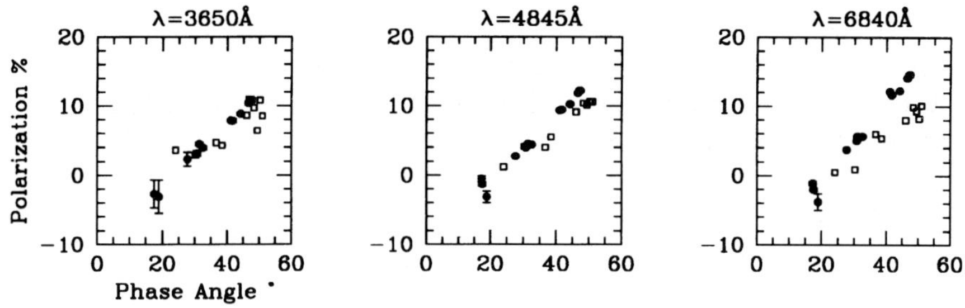

Figure 1. Phase dependence of polarisation of Comet Hale–Bopp (filled symbols) for three different continuum wavelengths. Open squares mark the corresponding data for Comet Halley (Kikuchi et al., 1987).

Halley, a fair degree of agreement was noticed among all the observers, as long as the polarization was estimated over the whole coma with a large aperture (which averages out the effect of the heterogeneity of the inner coma) centered on the comet's nucleus. In all our observations the linear scale sampled at comet Hale–Bopp is more than 13000 km radius, which is much larger than the inner coma region.

In order to check the polarization behavior with respect to the area sampled on the comet, we observed Hale–Bopp on March 1, 1997 through two different apertures, 26.5 and 52.4 arcsec, corresponding to 14318 and 28313 km radii respectively. We note that within the errors of observation there is no significant difference in measured polarization values for the two apertures. This means that the comparison of polarization observations on different dates is meaningful as long as the sampled linear size is in the range of 14000 to 28000 km radius. On the other hand, during the October–November 1996 observations, the sampled area is much larger (linear scale $\geq$ 55000 km radius), and therefore the tail region also contributed to the observed polarization. During these (October–November, 1996) observations, the phase angles were between 19 and 17° and the observed polarization was found to be negative. The comet being faint during this period, the errors in polarization measurements are large, especially in the U-band(3650 Å). In the other two continuum bands, S/N ratio is better than three and, therefore, the numbers are significant. The polarization values measured on November 1 and November 2, 1996 differ significantly, though the phase angles and comet distances from Sun and Earth are nearly the same. The difference in the polarization values could be due to different levels of activity.

Figure 1 indicates that there is a significant difference between the observed polarization values plotted for comet Hale–Bopp and comet Halley, especially at the larger phase angles. Levasseur-Regourd et al. (1996) have classified comets in two classes on the basis of available polarization data: low polarization and high polarization comets. Comparing the present observations with Figures 3a and 3b of their paper, it is evident that comet Hale–Bopp belongs to the high polarization class. In fact for phase angles above 30°, the observed polarization in Hale–Bopp is higher than for any other comet. In this sense comet Hale–Bopp appears to be different than other comets classified as high polarization class.

Vasundhara (2002) has attempted to characterize the dust grains in Hale–Bopp based on the polarization data, the shell structure seen in the comet during November 1996–May 1997 and the colour gradients. One of the conclusions drawn in that paper is that Mg-poor dust grains are required to explain, simultaneously, the shell structure and polarization phase curve. However, the mid-IR observations (Wooden et al., 1999; Hayward et al., 2000) refer to altogether contradictory values of Mg-rich silicates. The reason for the discrepancy may be the choice of the model by Vasundhara (2002) for describing the shell structure based on the dust particles trajectories. While fitting the calculated values of polarization to the observed polarization-phase curve, several parameters have been adjusted to achieve the required fit. The major discrepancy, however, appears to be due to the choice of model predicting the morphology of the coma. The model by Samarasinha (2000) is more successful in explaining the coma morphology due to an extended active region. A fresh look at the dust modelling is required to explain satisfactorily the coma morphology and the polarization-phase curve.

3.2. Comet C/2000 WM1 (LINEAR)

The polarization behavior of comet C/2000 WM1 in the phase angle range 14–22° is displayed in Figure 2, where we have plotted our observed polarization values as a function of phase angles for 7000 Å and 4845 Å narrowband and R band filters. It should be noted that the data are very rare in this important range. The polarization behavior for $\alpha < 22°$ provides clues to the refractory nature of the grains. Observations were largely made through the broad band filters to achieve good S/N ratio. However, some observations were made through narrowband filters: 4845 and 7000 Å. Here we discuss the results based on the observations made through the filters: 4845, 7000 and R-band. The degree of polarization in the negative branch of the P vs α curve is in general wavelength independent (Petrova et al., 2001b). One interesting thing to note is that the polarization at 7000 Å band, wherein H_2O emission lies, is not very different from that in the R-band and the continuum band at 4845Å. It appears that during the observing run in November 23–26, 2001, the comet was not very active though its heliocentric distance was ≈1.3 AU. Our photometric data (unpublished) which was obtained simultaneously with the polarimetric data also support this view. Therefore, the observed degree of polarization at 7000 Å can be treated close to the value of continuum polarization. There are no strong molecular emission bands in the spectral region covered by R-band. Some weak molecular bands may still contaminate the continuum in R-band resulting in lower polarization compared to 7000 Å band. The polarization angle on November 26 showed significant deviation from one filter to the other. The polarization near 22° phase angle is expected to be close to zero for which the polarization angle is not

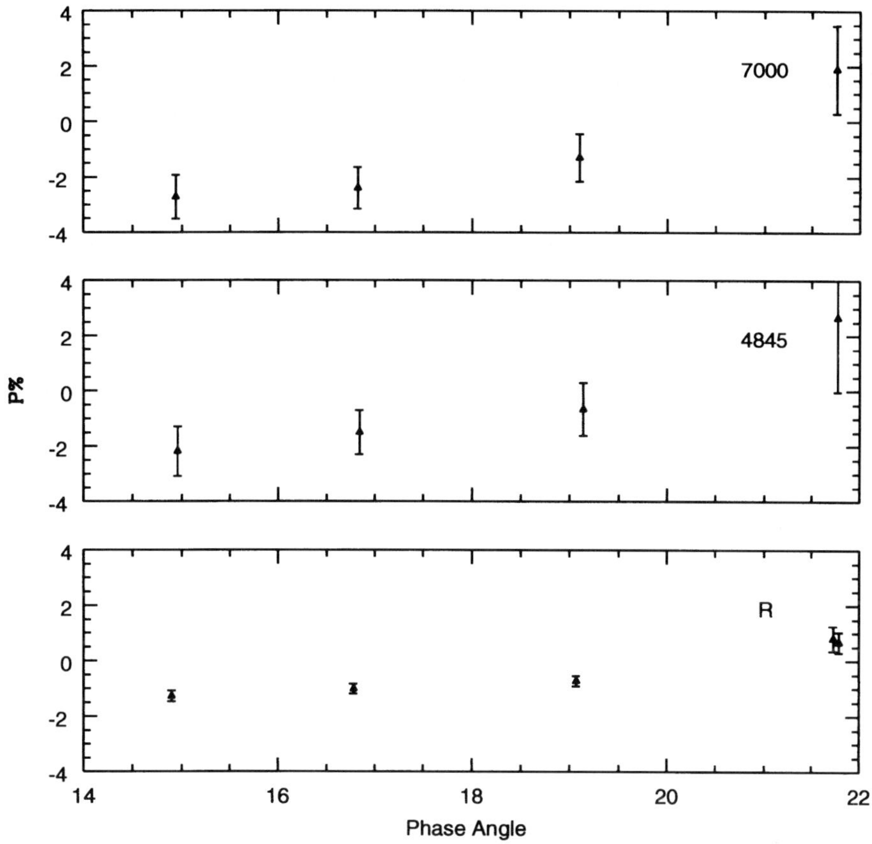

Figure 2. Phase dependence of the degree of polarization for comet C/2000 WM1.

well defined. A small contribution from molecular emission or from aligned grains may introduce large changes in the polarization angle under these conditions.

We also note from Figure 2 that the inversion angle($\alpha_{\rm min}$), where the polarization changes from negative to positive value, weakly depends on the wavelength; α_{min} being lower for shorter wavelengths. Petrova et al. (2001a, b) have shown that combined mutual influence of monomers (of sizes comparable to the wavelength of incident radiation) which are composed of aggregate cometary particles, leads to negative polarization at small phase angles. Comparison of the model calculations of Petrova et al. (2001b) with the present observations indicates that the aggregates of random structure composed of monomers of size parameter close to 1.5 and the complex refractive index ($1.65 + i0.05$) fits closely to the observations. The detailed discussions are given elsewhere.

4. Conclusions

In the present work we have discussed polarimetric observations on comet Hale–Bopp covering a wide phase angle range with the important finding that, compared to comet P/Halley, it shows higher degree of polarization with stronger wavelength dependence indicative of the presence of finer grains in Hale–Bopp. A careful study of the various models makes us believe that the inference drawn by Vasundhara (2002) on the grains to be Mg-poor could be due to an inadequate morphological model of the cometary coma. We also report on polarimetric observations of comet C/2000 WM1 when the phase angle was below 22°. The measurements are in agreement with a model involving dust aggregates of random structure and composed of monomers of size parameter close to 1.5 and complex refractive index $(1.65 + i0.05)$.

Acknowledgements

The work reported here is supported by the Department of Space, Government of India. The partial financial support from the European Space Agency made it possible for the first author to participate in the IAU Coll. 186 at Tenerife.

References

Bastien, P.; Menard, F., and Nadeau, R.: 1986, *Mon. Not. Roy. Astron. Soc.* **223**, 827–834.
Brooke, T. Y., Knacke, R. F., and Joyce, R. R.: 1987, *Astron. Astrophys.* **187**, 621–624.
Deshpande, M. R., Joshi, U. C., Kulshrestha, A. K. et al.: 1985, *Bull. Astron. Soc. India* **13**, 157–161.
Dollfus, A.: 1989, *Astron. Astrophys.* **213**, 469–478.
Dollfus, A. and Suchail, J. L.: 1987, *Astron. Astrophys.* **187**, 669–688.
Dollfus, A., Bastien, P., Le Borgne, J.-F. et al.: 1988, *Astron. Astrophys.* **206**, 348–356.
Eaton, N., Scarrott, S. M., and Wolstencroft, R. D.: 1991, *Mon. Not. Roy. Astron. Soc.* **250**, 654–656.
Ganesh, S., Joshi, U. C., Baliyan, K. S. et al.: 1998, *Astron. Astrophys. Suppl. Ser.* **129**, 489–493.
Hadamcik, E., Levasseur-Regourd A. C., and Renard, J. B.: 1997, *Earth Moon Planets* **78**, 365–371.
Hayward, T. L., Hanner, M. S., and Sekanina, Z.: 2000, *Astrophys. J.* **538**, 428–455.
Jockers, K.: 1997, *Earth Moon Planets* **79**, 221–245.
Joshi, U. C., Baliyan K. S., Ganesh, S. et al.: 1997, *Astron. Astrophys.* **319**, 694–698.
Joshi, U. C., Deshpande, M. R., Sen, A. K. et al.: 1987, *Astron. Astrophys.* **181**, 31–33.
Kikuchi, S., Mikami Y., Mukai, T. et al.: 1987, *Astron. Astrophys.* **187**, 689–692.
Kolokolova, L., Jockers, K., Gustafson, Bo, Å. S. et al.: 2001, *J. Geophys. Res.* **106**, 10113–10128.
Krishna Swamy, K. S.: 1986, 'Physics of Comets', *World Scientific*, Singapore.
Krishna Swamy, K. S. and Shah, G. A.: 1988, *Mon. Not. Roy. Astron. Soc.* **233**, 573–579.
Lamy, P. L., Grün, E., and Perrin, J. M.: 1987, *Astron. Astrophys.* **187**, 767–773.
Le Borgne, J. F., Leroy, J. L., and Arnaud, J.: 1987, *Astron. Astrophys.* **187**, 526–530.
Levasseur-Regourd, A. C., Bertaux, J. L., Dumont, R. et al.: 1986, *Nature*, **321**, 341–344.
Levasseur-Regourd, A. C., Hadamcik, E., and Renard, J. B.: 1996, *Astron. Astrophys.* **313**, 327–333.
Metz, K. and Haefner, R.: 1987, *Astron. Astrophys.* **187**, 539–542.
Petrova, E. V., Jockers, K., and Kiselev, N. N.: 2001a, *Sol. Sys. Res.* **35**, 57–69.

Petrova, E. V., Jockers, K., and Kiselev, N. N.: 2001b, *Sol. Sys. Res.* **35**, 390–399.
Renard, J. B., Hadamcik, E., and Levasseur-Regourd, A. C.: 1996, *Astron. Astrophys.* **316**, 263–269.
Samarasinha, N. H.: 2000, *Astrophys. J.* **529**, 107–110.
Sen, A. K.; Joshi, U. C.; Deshpande, M. R. et al.: 1988, *Astron. Astrophys.* **204**, 317–318.
Sen, A. K., Deshpande, M. R., Joshi, U. C. et al.: 1991, *Astron. Astrophys.* **242**, 496–502.
Vasundhara, R.: 2002, *Astron. Astrophys.* **382**, 342–358.
Wooden, D. H., Harker, D. E., Woodward, C. E. et al.: 1999, *Astrophys. J.* **517**, 1034–1058.

POLARIMETRIC AND PHOTOMETRIC OBSERVATIONS OF SPLIT COMET C/2001 A2 (LINEAR)

VERA ROSENBUSH[1], NIKOLAI KISELEV[2] and SERGEY VELICHKO[2]
[1]*Main Astronomical Observatory of the National Academy of Sciences, 27 Zabolotny Street, 03680 Kyiv, Ukraine;* [2]*Institute of Astronomy, Kharkiv National University, 35 Sumskaya Street, 61022 Kharkiv, Ukraine*

(Received 8 March 2002; Accepted 29 June 2002)

Abstract. We present the results of polarimetric and photometric observations of split Comet C/2001 A2 (LINEAR), which were obtained at the 70-cm telescope of the Astronomical Observatory of Kharkiv National University between 30 June and 31 July 2001. The brightest fragment of the comet, nucleus B, was observed. Eight narrowband cometary filters in the continuum and in emission bands and a wideband red filter have been used. The comet was observed through apertures of 88, 33, and 19 arcsec. Polarization phase curves were obtained for the continuum and for the first time, for $NH_2(0, 7, 0)$ emission. The degree of polarization of the light scattered by the dust decreases with the increase of aperture size. An important temporal variation of the polarization with a rotation of the polarization plane was observed at two phase angles (26.5° and 36.2°). Molecular column densities and production rates of CN, C_2, C_3, and NH_2 species are calculated in the framework of the Haser model. A comparative analysis of the temporal variations of the visual magnitudes, gas and dust production rates, dust color and polarization are presented.

Keywords: Comet C/2001 A2 (LINEAR), comets, dust, NH_2, polarization

1. Introduction

A partial fragmentation or complete disintegration of comets should be accompanied by ejection of the fresh internal material and therefore, provide a good opportunity to study the inner composition of nucleus and physical processes occurring in the atmosphere of splitting comets. Taking into account the unpredictable nature of such events, dust and gas monitoring observations of fragmenting comets are very sparse.

Comet A2 (LINEAR), hereafter called C/A2, displayed a series of sporadic outbursts due to disruption of its nucleus. According to Sekanina (2001), the initial splitting of the nucleus on fragments A and B occurred on March 29.9, 2001. The following separation of components C, D, E, and F from the nucleus B occurred respectively on May 11, June 3.5, June 9.5, and June 11.3. The favorable brightness predictions made this comet a suitable target for polarimetric and photometric observations even with small-sized telescopes.

Therefore, the purpose of our observations was to determine gas and dust production rates of the nucleus and to measure the linear polarization of light scattered by dust in order to find their possible variations caused by disruption of the cometary nucleus.

2. Observations

Comet C/A2 was observed after its perihelion passage that occurred on 24 May 2001, when its closest approach to Earth was at 0.24–0.51AU. Polarimetric observations were carried out for 17 nights from 30 June to 31 July 2001 with narrowband and wideband filters. The comet was observed by narrowband photometry during three nights. Measurements were obtained with the 70-cm telescope of the Astronomical Observatory of Kharkiv National University. A single-channel photoelectric photometer–polarimeter with a rapidly rotating polaroid (Kiselev and Velichko, 1998) was used. The HB and ESA set filters were used in order to isolate emission bands CN (λ387/6.2 nm), C_3 (λ406.2/6.2 nm), C_2 (λ514.1/11.8 nm), NH_2 (λ663/4 nm) and the ultraviolet λ344.8/8.4 nm (UC), blue λ445/5 nm (BC), green λ526/5.6 nm (GC), and red λ712.9/6.2 nm (RC) continuum ranges. The wideband filter λ722.8/114 nm (WRC) was also used for polarimetric observations. The 19, 33 and 88 arcsec diaphragms were employed. Several standard stars taken from (Farnham et al., 2000) were observed to convert the cometary counts to absolute fluxes.

3. Results

3.1. Photometry

The photometric reductions and computations of the Haser-model production rates Q for CN, C_3, and C_2 molecules were performed in a standard manner (Kiselev and Velichko, 1998). The absolute fluxes for standard stars in NH_2 filter were taken from (T. Bonev, personal communication). To determine the gas production rate for NH_2 we used a fluorescence efficiency value $g = 2.29 \times 10^{-15}$ (erg s^{-1} mol^{-1}) for the (0, 7, 0) band as calculated by (Kawakita et al., 2001). The lifetimes for NH_2 and its parent (1.0×10^5 s and 5.0×10^3 s, respectively) have been provided by (Korsun and Jockers, 2002).

The photometric results are listed in Table I with the log of the observations (time of observations, heliocentric r and geocentric Δ distance, phase angle α, projected diaphragm radius at the comet ρ).

The dust-equivalent A(α)fρ production rates and color of dust were calculated according to (Farnham et al., 2000) and are presented in Table II. The last column in this table gives the equivalent width W_{4845} of the C_2(0, 0) emission band characterizing the ratio of gas-to-dust in the cometary atmosphere. One can see that the gas-to-dust ratio in the large diaphragm was about 1000 Å for all dates of

TABLE I

Gas production rates for Comet C/2001 A2 (LINEAR)

Date, UT	r	Δ	α	ρ	log Q (mol s^{-1})			
July 2001	(AU)	(AU)	(deg)	(10^4 km)	CN	C_3	C_2	NH_2
14.960	1.231	0.320	42.0	1.020	25.60	25.06	25.94	24.65
14.994	1.232	0.320	41.9	0.388	25.63	25.08	25.94	24.73
17.888	1.271	0.349	37.6	1.114	25.53	24.83	25.74	24.46
31.990	1.463	0.529	26.1	1.688	25.52	24.89	25.74	24.51

TABLE II

Dust data for Comet C/2001 A2 (LINEAR)

Date, UT	log A(α)fρ (cm)				CE (%/1000 Å)			W_{4845}(Å)
July 2001	UC	BC	GC	RC	UC–BC	BC–GC	GC–RC	
14.960	2.67	2.59	2.55	2.73	−18.5	−9.6	21.5	1083
14.994	2.55	2.58	2.52	2.70	8.4	−16.8	21.1	562
17.888	2.54	2.51	2.46	2.65	−7.4	−14.4	23.1	898
31.990	–	2.52	2.46	2.65	–	−16.2	22.8	958

observations. Such value of the equivalent width is typical for gas-rich comets (Chernova et al., 1993). At the same time, there is the significant reduction of W_{4845} for small diaphragm. Thus, this comet seem to be gas-rich for large apertures and dust-rich for small apertures. It means that the concentration of the dust is higher in the circumnuclear region.

Dust in the comet was bluer relatively to the Sun in the wavelength range λ345–526 nm and redder in the wavelength range λ526–712.8 nm. There are two possible ways for an explanation of this observation: (1) Inherent properties of dust in this comet, and (2) the incomplete correction of contamination from gas emission, especially for the UC and BC filters.

As one can see from Table I, there is the decrease of the molecular and dust production rate from 14 to 17 July. The gas-to-dust ratio also decreased during this period. Note that this decrease falls on a descending branch of outburst, which occurred near 13 July. It is clearly seen in Figure 1, where the brightness evolution of Comet C/A2 during our observating campaign is shown.

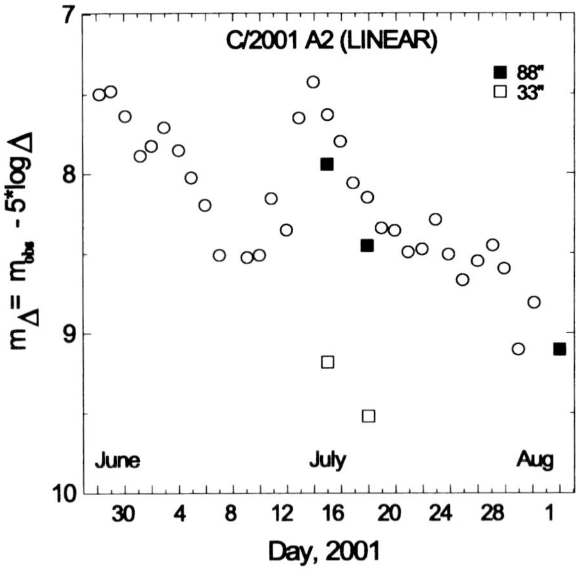

Figure 1. Brightness evolution of Comet C/A2 during our observations. The circles show averaged visual magnitudes (m_{obs}) taken from (ICQ, 2001) and reduced to the geocentric distance Λ=1 AU (m_Δ). The squares denote our relative photoelectric magnitudes in GC filter.

3.2. Polarimetry

The results of the polarimetric observations are given in Table III, which contains the degree of linear polarization P, the position angle Θ of the polarization vector, their statistical mean-square errors, the position angle of the scattering plane φ and the difference between the plane of polarization and the normal to the scattering plane Θ_r for different moments, phase angles, filters, and diaphragms.

The phase angle dependence of polarization for C/A2 is presented in Figure 2. One can see, that at large phase angles (between 31 June and 7 July), the polarization degree of the comet was lower than that usually observed for dusty comets. Besides, the degree of polarization measured with large aperture, which corresponds to the projected diaphragm radius $\approx$ 8000 km at the coma, is distinctly less than that with smaller apertures. In the near-nuclear region (ρ = 3250 km, on July 8.95 UT), the polarization of C/A2 was similar to that of dusty comets. Such effect was also observed for two other gas-rich comets: Comet D/1996 Q1 (Tabur) (Jockers, 1997), which was also a split comet, and Comet 2P/Encke (unpublished data by N. N. Kiselev et al.). A strong diaphragm dependence of polarization for gas-rich comets may be caused by incomplete correction of contaminations from gas emission bands or by change of intrinsic dust properties with distance from nucleus. The polarization degrees of dust and gas are practically equal at phase angles less than 40°. Therefore, in the first case, the difference between gas-rich

TABLE III
Polarization of Comet C/2001 A2 (LINEAR)

Date, UT 2001	Filter	ρ (km)	α (deg)	P (%)	σ_P (%)	Θ (deg)	σ_Θ (deg)	φ (d eg)	Θ_r (deg)
June									
30.978	WRC	7780	74.8	13.68	0.23	156.1	0.5	246.9	–0.8
July									
01.976	WRC	2928	72.1	16.26	0.24	158.4	0.4	247.1	1.3
02.967	WRC	1696	69.3	17.38	0.35	157.1	0.6	247.0	0.1
03.951	RC	7930	66.6	16.04	0.85	150.4	1.5	246.6	–6.2
03.987	NH_2	7935	66.5	9.09	0.40	155.0	1.2	246.6	–1.6
04.933	NH_2	8032	64.0	7.40	0.59	149.8	2.3	246.0	–6.2
04.975	RC	8036	63.9	9.12	0.70	162.2	2.2	245.9	6.2
06.946	RC	8307	58.7	14.02	0.50	153.5	1.0	244.0	–0.5
08.950	RC	3252	53.9	14.07	1.11	149.7	2.3	241.1	1.4
15.894	RC	3935	40.5	5.42	0.66	137.8	3.5	226.3	1.5
16.873	NH_2	10808	39.0	4.23	0.37	125.7	2.5	223.8	–8.1
16.961	RC	10838	38.9	4.68	0.39	131.6	2.4	223.6	–2.1
18.859	BC	11488	36.3	3.91	0.39	125.8	2.8	218.6	–2.8
18.954	RC	11521	36.2	5.61	0.44	123.6	2.2	218.3	–4.7
19.908	RC	2563	35.0	–4.86	1.22	49.3	7.2	215.7	103.6
20.873	RC	4584	34.0	4.06	1.02	156.9	7.2	213.1	33.8
20.975	RC	12638	32.8	2.14	0.55	116.4	7.4	210.0	–3.6
21.990	NH_2	12638	32.8	4.69	0.66	113.1	4.0	209.9	–6.8
26.925	RC	14616	28.8	2.67	0.64	79.2	6.9	196.0	–26.8
26.972	WRC	14648	28.7	2.51	0.25	114.6	2.8	195.9	8.7
28.887	WRC	15478	27.6	1.82	0.26	86.6	4.1	190.5	–13.9
28.892	WRC	5804	27.6	1.80	0.24	106.1	3.9	190.5	5.6
28.951	WRC	3348	27.6	2.18	0.25	120.4	3.3	190.3	2.1
29.939	NH_2	15956	27.0	4.11	1.15	95.4	8.0	187.6	–2.2
30.952	RC	16403	26.5	–2.36	1.03	0.5	12.5	184.8	85.7

and dust-rich comets as well as the diaphragm dependence of polarization should disappear with decreasing phase angle.

We have found that the spectral gradient dP/dλ of dust polarization for Comet C/A2 was 0.63 ± 0.22%/100 nm at $\alpha = 36.3°$. For comparison, the mean value of the spectral gradient for dusty comets is equal to 0.84 ± 0.03%/100 nm at phase angle 45° (Kiselev, 2002). Taking into account that the gradient dP(λ)/dα = 0.025 ± 0.001%/100 nm per 1° is constant within the phase angle range 30–65° (Kiselev,

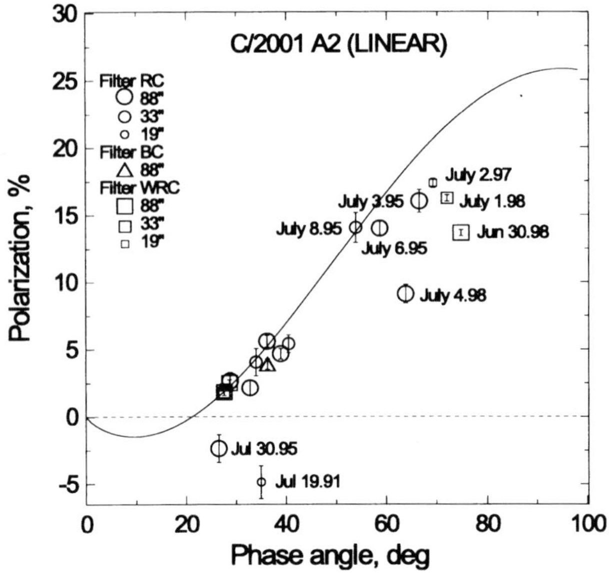

Figure 2. Polarization degree of Comet C/2001 A2 (LINEAR) as compared with the trigonometric fit for dusty comets in the red continuum filter (solid curve).

2002), we conclude that the wavelength dependence of polarization for C/A2 is close to that commonly observed for dusty comets.

Several interesting features were detected in polarization data of C/A2. On July 4.98 at phase angle 63.9°, a very small polarization of about 9% has been observed. Moreover, usually the plane of polarization is perpendicular to the scattering plane at phase angles larger than 22°, e.g., positive polarization is observed. It is surprising, that a change of the polarization plane relative to the scattering plane was detected at phase angles 35° and 26° on July 19.91 and 30.95. We could not explain these results by instrumental errors. The normal operation of the polarimeter was controlled by observations of standard stars with zero and large polarization during all observing nights. However, it is important to note, that these polarimetric features coincide in time with the outburst activity of the comet.

The most important molecule in the red spectral region is NH_2, which falls within the passbands of the used continuum filters. Theoretical as well as observational data on polarization of light emitted by these molecules are practically absent. Five measurements of polarization were obtained through the filter, which transmits radiation of molecules $NH_2(0, 7, 0)$ and continuum. The continuum flux F_{cont} in the NH_2(663/4 nm) filter was determined from the continuum fluxes in the GC and RC filters. It allowed us to define the relative contribution of the $NH_2(0,7,0)$ emission F'_{NH_2} and continuum F'_{cont} in the NH_2(663/4 nm) filter. Using equation $P_{(NH_2+cont)} = P_{NH_2} F'_{NH_2} + P_{cont} F'_{cont}$, a “clear” polarization P_{NH_2} of the $NH_2(0, 7, 0)$ radiation was calculated. In this equation $P_{(NH_2+cont)}$ is the observed polarization degree in the NH_2(663/4 nm) filter and P_{cont} is polarization degree of dust in the

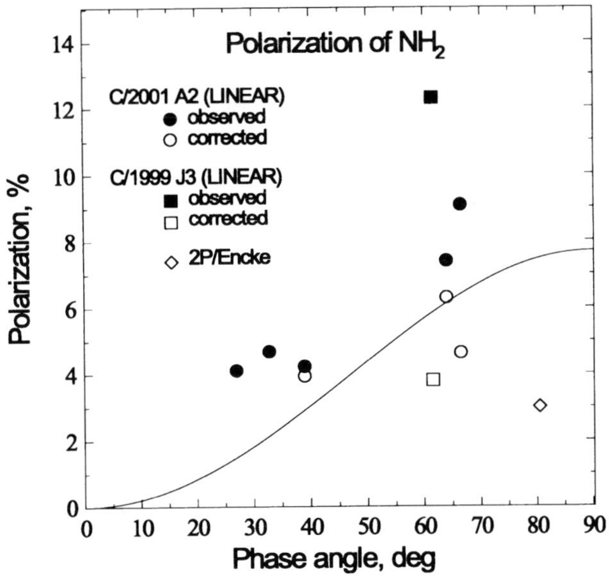

Figure 3. Polarization measurements of the cometary radiation of $NH_2(0, 7, 0)$ molecules in Comet C/2001 A2 (LINEAR). For comparison, the available data for comets C/1999 J3 (LINEAR) (Jockers, unpublished) and 2P/Encke (Kiselev et al., unpublished) are given. The solid curve is the phase angle dependence of polarization for the resonance fluorescence scattering (Le Borgne and Crovisier, 1987).

NH_2(663/4 nm) filter obtained by interpolation of the polarization degree in the GC and RC filters. The continuum contribution in the observed polarization of NH_2 reached 5% for C/A2. The observed and corrected degree of polarization for comets C/A2, C/1999 J3 (LINEAR) and 2P/Encke can be found in Figure 3. This figure shows that polarization of NH_2 is about the same as the polarization of the diatomic molecules C_2 and CN (Le Borgne and Crovisier, 1987).

4. Discussion

The study of Comet A2 (LINEAR) is important from two points. First, it is a split comet. And second, this comet is a gas-rich one. We have observed C/A2 (namely, nucleus B) in the active phase when its brightness varied rather considerably, with amplitude up to 1.5^m (Figure 1). Although we do not have an extensive set of photometric data for this comet, we can mark some its peculiarities.

To compare gas and dust production rates in split comet C/A2 with those in typical comets, we have determined the following ratios of production rates: $Q(C_3)/Q(CN)$, $Q(C_2)/Q(CN)$, $Q(NH_2)/Q(CN)$, and $A(\alpha)f\rho/Q(CN)$ (Table IV). The production rate ratios $Q(C_3)/Q(CN)$ and $Q(C_2)/Q(CN)$ in comet C/A2 are close to the upper limits for other comets (A'Hearn et al., 1995). The production rate

TABLE IV
Production rate rations of Comet C/2001 A2 (LINEAR)

Ratio	C/A2	Limits	Comets
Log [Q(C_3)/Q(CN)]	–0.61	$-1.74 \div -0.51$	Typical comets[a]
Log [Q(C_2)/Q(CN)]	0.27	$-0.20 \div 0.24$	Typical comets[a]
Log [Q(NH_2)/Q(CN)]	–0.98	0.26	C/1998 J3 (LINEAR)[b]
Log[A(α)fρ/Q(CN)]	–23.02	$-24.63 \div -21.87$	Typical comets[a]

[a]A'Hearn et al. (1995).
[b]Korsun and Jockers (2002).

ratio Q(NH_2)/Q(CN) was compared with that for the gas-rich Comet C/1999 J3 (LINEAR). This comet was a unique comet, which was observed in the same (0,7,0) band (Korsun and Jockers, 2002). It turned out that the relative content of NH_2 in comet C/A2 was about 18 times lower than that for J3 (LINEAR). Finally, we have estimated the averaged ratio A(α)fρ/Q(CN), which is within the limits for typical comets.

Variations among the continuum colors of comets do not appear to be correlated with heliocentric distance, phase angle, or dust production (Jewitt and Meech, 1986). Color differences apparently arise from intrinsic variations of dust grain properties in comets. Figure 4 shows that in the color diagram UC–BC versus BC–RC, comet C/A2 lies between very dusty comets C/1995 O1 (Hale–Bopp), C/1990 K1 (Levy), C/1996 B2 (Hyakutake) and very gas-rich comets 23P/Brorsen–Metcalf and C/1989 X1 (Austin). The light scattered by the dust in dusty comets is generally redder than the Sun. According to (A'Hearn et al., 1995), the light scattered by gas-rich comets seems to generally be bluer. These authors have noted that this effect may be artificial due to the gas contamination in the continuum filters. In particular, the ultraviolet continuum is contaminated by the short-wavelength wing of the C_3 emission. The C_2 emission can extend into the BC filter. New procedures developed by (Farnham et al., 2000) allow to remove contamination by different emissions. It is impotrant to note that comet C/A2 was more dusty in the small diaphragm (Table II). Simultaneously its color has changed: the UC–BC color became redder and the BC–GC color – bluer relatively to colors in the large diaphragm. It is possible that the correction for the gas emission contamination was incomplete. Moreover, the intrinsic properties of dust may be different throughout the coma. Therefore, the question of the blue color of gas-rich comets remains still open.

Since C/A2 was a gas-rich comet, it is interesting to compare its polarization with that for other gas-rich comets (e.g., C/1983 H1 (IRAS–Araki–Alcock), C/1989 X1 (Austin), 23P/Brorsen–Metcalf), and D/1996 Q1 (Tabur) (see Figure 5). For Comet Tabur, the observed polarization and the one corrected for the NH_2 contamination are given. The used data set for other comets are not free of emission by gas. The projected aperture for Comet IRAS–Araki–Alkock was 700–1900

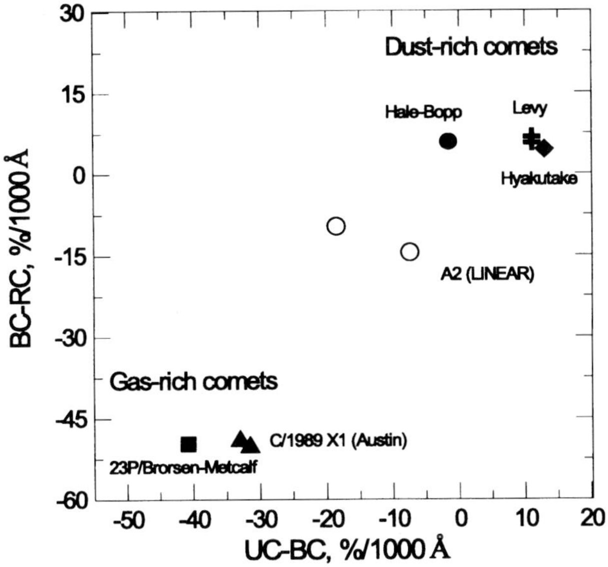

Figure 4. Color diagram UC–BC vs BC–RC for dusty and gas-rich comets. Colors of comets are taken from (Sen et al., 1991; Joshi et al., 1992) for Austin; (Kiselev and Velichko, 1998) for Hyakutake; (Kiselev and Velichko, 1997) for Hale–Bopp; (Kolokolova et al., 1997) for Levy; (Kiselev, 2002) for Brorsen–Metcalf.

km while for comet Brorsen–Metcalf – 6800–37,700 km, and for comet Austin – 75,700–4200 km. There is a clear difference between polarization of the near nucleus area and polarization obtained for large areas of the coma. Furthermore, the polarization degree measured in the wideband filtres is significantly less than that in the narrowband filters because of emission band contamination. In general, the phase angle dependence of polarization for C/A2 is very close to that for comets Tabur and Austin.

Gas-rich comets show an appreciable difference of the positive polarization branch from that for dusty comets. This difference is especially significant for large areas of the coma, few tens of thousand kilometers wide. In the near nucleus region (up to $\approx$1000–2000 km), the polarization degree of gas-rich comets is close to that for dusty comets. It is well seen from Figure 5 for comets Tabur, IRAS–Araki–Alcock as well as for C/A2. There are some possible explanations of these observational facts: (i) Gas contamination in the continuum filters; (ii) the differences in the dust properties in dusty and gas-rich comets; and (iii) different evolution of dust particles in dusty and gas-rich comets.

Outbursts and sudden increases of the dust outflow from the nucleus can cause a change of polarization parameters. Such variations of the polarization degree and plane were measured during the outburst activity in comets 29P/Schwassmann–Wachmann 1 (Kiselev and Chernova, 1979) and C/1990 K1 (Levy) (Rosenbush et al., 1994). Dollfus (1988) has also observed the variations of polarization degree

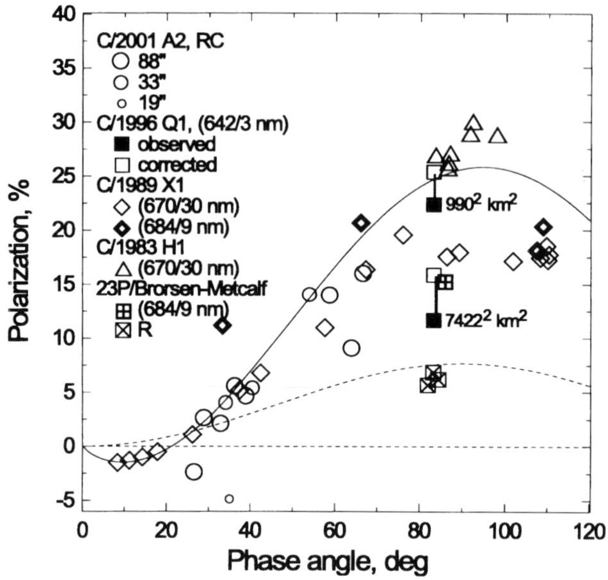

Figure 5. Polarization of Comet C/2001 A2 (LINEAR) as compared with polarization of gas-rich comets Austin (Sen et al., 1991; Joshi et al., 1992; Kikuchi, personal communication), Tabur (Jockers, 1997; Kiselev et al., 2001), IRAS–Araki–Alcock (Kikuchi, personal communication), and Brorsen–Metcalf (Chernova et al., 1993) in the red spectral range. Solid curve is the trigonometric fit to polarization data for dust-rich comets in the RC filter. The dash curve is the phase angle dependence of polarization for the resonance fluorescence scattering (Le Borgne and Crovisier, 1987).

for Comet Halley between the phase angles 50° and 65°. He has interpreted the temporary deviations of polarization in terms of an increased contribution of small dust particles in the cometary atmosphere. We have observed so far unexplained changes of polarization plane and of decrease of polarization degree in C/A2.

5. Conclusions

From our photometric and polarimetric observations of split comet C/2001 A2 (LINEAR), the production rates of CN, C_3, C_2, and NH_2, the quantity $A(\alpha)f\rho$, the dust color, gas-to-dust ratio, and polarization of dust and the NH_2 molecule emission band are determined. The results obtained allow us to make the following conclusions:

- The comet had high gas-to-dust ratio and therefore can be classified as a gas-rich comet;
- according to the relative abundances of C_3/CN and C_2/CN, comet C/A2 lies on upper limit of the range defined for a typical comets;
- the partial fragmentation of the comet has not resulted in global change of its phase dependence of polarization;

- there is short-term variability of the photometric data as well as polarimetric data on a time scale of day. Variations of the polarization degree and plane were detected;
- polarization of the $NH_2(0, 7, 0)$ emission was found, which is close to the theoretical phase-angle dependence of polarization for diatomic molecules.

Acknowledgements

We are very grateful to T. Bonev for helpful comments, M. A'Hearn, D. Schleicher, T. Farnham, and R. Schulz for the set of cometary filters, which were used in our observations of Comet A2 (LINEAR) as well as of other comets. We also thank an anonymous referee for careful review and valuable remarks. V.K.R. and N.N.K. thank IAU, SOC and LOC of IAU Colloquium No. 186. for the financial support. N.N.K. is supported by INTAS grant No. 652.

References

A'Hearn, M. F., Millis, R. L., Schleicher, D. G., Osip, D. J., and Birch, P. V.: 1995, *Icarus* **118**, 223–270.

Chernova, G. P, Kiselev, N. N., and Jockers, K.: 1993, *Icarus* **103**, 144–158.

Dollfus, A., Bastien, P., and Le-Borgne, J.-F.: 1988, *Astron. Astrophys.* **206**, 348–356.

Farnham, T. L., Schleicher, D. G., and A'Hearn, M.-F.: 2000, *Icarus* **147**, 180–204.

International Cometary Quarterly **23**, 182–195, 2001.

Jewitt, D. C. and Meech, K. J.: 1986, *Astrophys. J.* **310**, 937–952.

Jockers, K.: 1997, *Earth Moon Planets* **79**, 221–245.

Joshi, U. C., Sen, A. K., Deshpande, M. R., and Chauhan, J. L.: 1992, *J. Astrophys. Astron.* **13**, 267–277.

Kawakita, H., Watanabe, J., Kinoshita, D., Abe, S., Furusho, R. Izumiura, H., Yanagisawa, K., and Masuda, S.: 2001, *Publ. Astron. Soc. Japan* **53**, L5–L8.

Kiselev, N. N.: 2002, 'Observations and Interpretation of Light Scattered on Dust Grains of Comets, Asteroids, and Circumstellar Shells', Doctoral Thesis, Kharkiv National University.

Kiselev, N. N. and Chernova, G. P.: 1979, *Sov. Astron. Lett.* **5**, 294–299.

Kiselev, N. N. and Velichko, F. P.: 1997, *Earth Moon Planets* **78**, 347–352.

Kiselev, N. N. and Velichko, F. P.: 1998, *Icarus* **133**, 286–292.

Kiselev, N. N., Jockers, K., Rosenbush, V. K., and Korsun, P. P.: 2001, *Solar Syst. Res.* **35**, 480–495.

Kolokolova, L., Jockers, K., Chernova, G., and Kiselev N.: 1997, *Icarus* **126**, 351–361.

Korsun, P. P. and Jockers, K.: 2002, *Astron. Astrophys.* **381**, 703–708.

Le Borgne, J. F. and Crovisier, J.: 1987, in *Symposium on the Diversity and Similarity of Comets*, 6–9 April 1987, Brussels, Belgium, ESA SP-278, pp. 171–175.

Rosenbush, V. K., Rosenbush, A. E., and Dement'ev, M. S.: 1994, *Icarus* **108**, 81–91.

Sekanina, Z.: 2001, *IAUC* 7630, 7656.

Sen, A. K., Joshi, U. C., and Deshpande, M. R.: 1991, *Mon. Not. Roy. Astron. Soc.* **253**, 738–742.

THE NEAR-NUCLEAR COMA OF COMET HALLEY IN MARCH 1986

KAROLY SZEGÖ

KFKI Research Institute for Particles and Nuclear Physics, P.O. Box 49, H-1525 Budapest, Hungary

E-mail: szego@rmki.kfki.hu

JEAN-FRANÇOIS CRIFO

Service d'Aéronomie du CNRS, BP 3, F 91371 Verrières le Buisson Cedex, France

ALEXANDER V. RODIONOV

Central Research Institute on Machine Building (TsNIIMASCH), Pionyerskaya St., 4, Korolev, Moscow Region, 141070, Russia

MARCO FULLE

Osservatorio Astronomico di Trieste, Via Tiepolo, 11, I-34131 Trieste, Italy

(Received 25 February 2002; Accepted 3 July 2002)

Abstract. The cameras carried onboard the flyby missions to comet P/Halley in 1986 imaged the near nuclear jet activity from several spatial directions. The observed, very structured near nuclear dust jets were considered at that time as the result of dust emission from well localized active surface regions (without supporting 3-D model computations, however). Based on the first, recently developed 3-D gas dynamical model of P/Halley's activity, we have been shown that jet features can be reproduced assuming a homogeneous dusty ice nucleus surface. The dust in the collisional near nuclear coma is concentrated along the gas flow discontinuities resulting from the complicated surface orography, creating the visual impression of dust jets. We present here the results of these calculations for the near nucleus dust distributions, and we compare them with the direct observations made during the three Halley flybys (Vega 1, Vega 2, and Giotto).

Keywords: 3-D gas dynamics, dust jets, P/Halley

1. Introduction

Recently Rodionov et al. (2002) have developed the first 3-D gas dynamical model capable to calculate the near nuclear coma of comets for which the nucleus shape is known; an outline of this model is given in this volume (Crifo et al., 2002). Whereas the model can accommodate any degree of surface inhomogeneity (i.e., active and inactive areas), our starting point here is trying to reproduce observations within this model assuming that the surface is homogeneous; this minimizes the number of free parameters. When the nucleus is homogeneous, differences in surface flux result only from the differences in solar illumination due to shape and orography: this is sufficient to create quite an inhomogeneous gas and dust coma (Crifo and Rodionov, 1999). If the results were not satisfactory, then (and only then) it will be necessary to introduce surface inhomogeneities. An essential ingredient of the model is the nucleus shape. For the time being (and prior to the publication of a topographic map of the recently observed nucleus of comet P/Borrelly), the method can be applied only to P/Halley's nucleus. Its shape was derived from the 63 images taken during the VEGA 1 flyby on 6 March, 1986; the view directions of these images covered a $\simeq 160°$ angular range around the nucleus (Merenyi et al., 1990), a detailed comparison of the model and the images can be found in Szegö et al. (1995), a brief description of the flyby and image processing is also given in Crifo et al. (2002). To satisfy the requirements of the numerical code, the shape was smoothed by expanding it into a series of spherical harmonics, in the present calculation we used the first ten harmonics which means that we filtered out the topographic noise of spatial dimensions smaller than several 100 m.

In a nuclear centred ecliptic frame of reference (where the X-axis is parallel to the vernal equinox, the Z-axis points to ecliptic North, the longitude and latitude are denoted by l, b, respectively) the long axis of the nucleus – oriented from the centre towards the big end – pointed to $l = 79°$, $b = 15°$, and to $l = 310°$, $b = 9°$ during the VEGA 1 and VEGA 2 encounters, respectively (Szegö et al., 1995). During the GIOTTO flyby the angle between the Sun and long axis was $-32° \pm 10°$ (Keller et al., 1995). From this and other observations (Sadeev et al., 1995) concluded that the long axis orientation during the GIOTTO encounter was $l = 276°$, $b = -27°$. The error is about $\pm 5°$, higher for l than for b. The nucleus rolled about its long axis between the encounters, the roll angular values were $0°$, $30°$, $-10°$ at the time of the VEGA 1, 2, and GIOTTO encounters. For the computation of the inner coma (i.e., in the strict fluid regime) we define a nucleus-attached longitude-colatitude system (λ, β): the origin of the longitude, λ is at the big end, and the colatitude $\beta = 180°$ ($0 \leq \beta \leq 180°$) corresponds to the Northward pole of the nucleus. In this coordinate system the subsolar points during the encounters were ($\beta = 73°$, $\lambda = 344°$) for VEGA-1, ($\beta = 74°$, $\lambda = 112°$) for VEGA-2, and ($\beta = 75°$, $\lambda = 134°$) for GIOTTO. As can be seen, the Sun positions (subsolar points) between the VEGA 2 and GIOTTO encounters differed only by about 22°.

2. Results

The gas coma was computed using only one adjustable parameter, the icy area fraction f (the relative fraction of the exposed ice and dust in any element of the surface), which was set to 0.4 to meet the measured total gas production. Then, 0.9 μm radius spherical dust grains with specific mass 1 g cm^{-3} were released (with a constant dust-to-gas mass ratio) into the flow, to create a single-size dust coma.

The very irregular nucleus shape creates a quite structured coma. We can illustrate this by presenting the dust and gas parameters on three spheres centred on the nucleus, with radii 9, 18, and 36 km, respectively; because, as can be seen, the major coma structures actually develop within less than ten nucleus radii. We present the gas density results by contour plots, the gas flux results are exhibited by vector plots.

Figure 1 shows the gas density during the VEGA 1, 2, and GIOTTO encounters, respectively, on the three spheres. The GIOTTO and VEGA 2 densities are very similar, (though not identical) as can be expected from the fact that they have close subsolar points. Each sequence shows that the gas evolves very dynamically with increasing distance and the density distribution forms narrower structures. In all cases the gas coma is structured, and in fact it can be expected that the gas coma remains structured farther out from the nucleus as well. The gas distribution is of course not symmetric to the Sun-comet line.

The gas fluxes for the three encounters are shown on Figure 2, only within a region close to the (nucleus-attached) equatorial plane* for the sake of figure clarity. The flux patterns, as opposed to the density patterns, are broadening outwards: the strong variation seen at 9 km from the nucleus centre is smoothed outwards.

We present the dust results in a different way because we wish to compare them with the images obtained during the encounters. The plots show the nucleus surface in addition to the three previously used spheres, and we view the nucleus and these surfaces from the actual view direction of the cameras. Furthermore, we show that fraction of each sphere only, where the dust density exceeds 2/3 of the maximal value it assumes on that sphere. Finally, vectors are attached to the outermost sphere, proportional to the dust flux vectors at each of the plotted points.

Figure 3 compares two model results and the corresponding two images obtained during the VEGA 1 encounter; Figure 4 does the same for three VEGA 2 images. The image numbering convention, Sun direction, exposure times and spacecraft positions of the VEGA 1 and 2 images, as well as the comparison of the images with the nucleus shape model can be found in Szegö et al. (1995). Figure 5 shows the comparison of our model result with the synthetic image taken by the GIOTTO camera (Keller et al., 1995). It should be noted that whereas the model shows a 36 km wide region around the nucleus in 3-D form, the images cover a much broader volume around the nucleus, and are evidently column densities. Also, the optical signal is integrated over the full dust size spectrum. Yet, the images and the model results coincide surprisingly well, even though certain deviations are also evident. More dust jets are visible in the model than in the images, and especially for VEGA 1 the two computed lower jets are missing on the images. This, however, might be due to the significant camera intensity offset (cf. Szegö et al. (1995)). In all cases the dust distribution is smoother closer to the nucleus, however, jets are formed quickly as we move farther away.

* Note that the nucleus-attached coordinates bear no relation to the nucleus rotation, so the word "equator" has here a purely geometrical, and not a dynamical meaning.

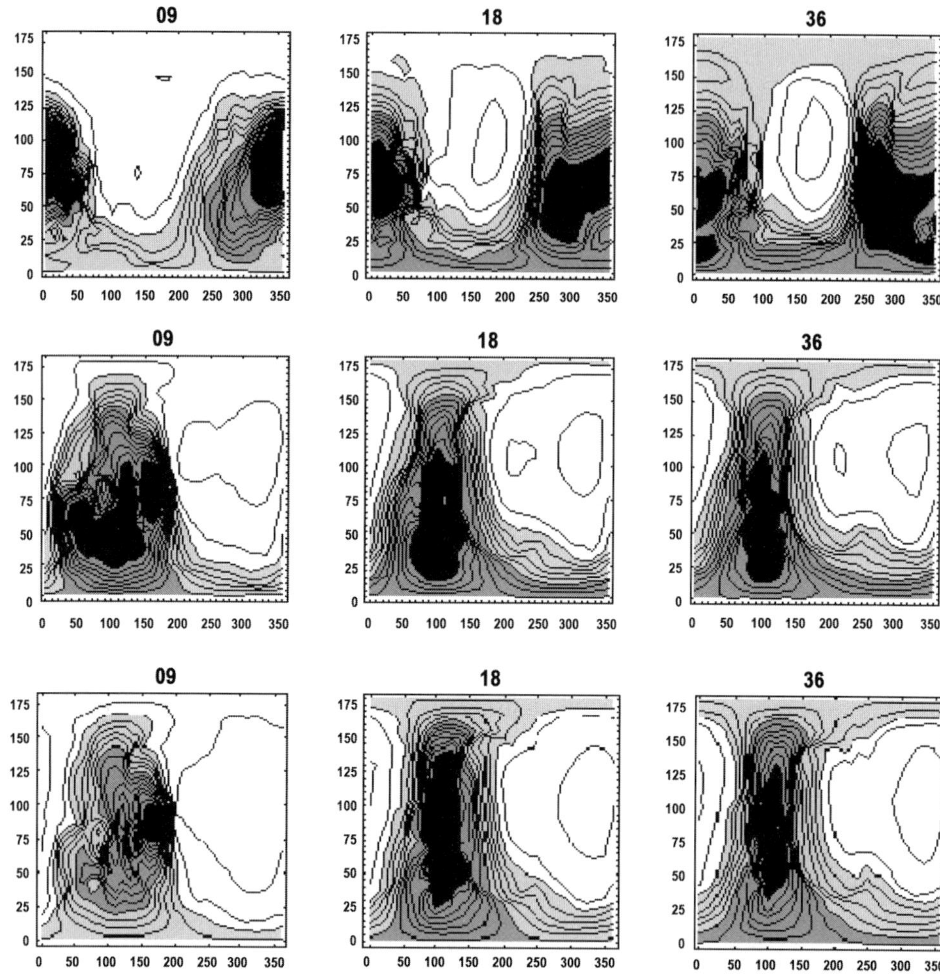

Figure 1. The gas density contour plots are shown during the VEGA 1 (top), VEGA 2 (middle), and GIOTTO (bottom) encounters, on three spheres centred on the nucleus, with radii 9 km (left), 18 km (center), and 36 km (right). The horizontal axis indicates longitude values, the vertical axis is colatitude. The greyscale for all plots is linear between the following minimum (light gray) and maximum (dark) values, in multiples of 10^{-6}g/m^3: for VEGA 1 at 9 km: (1.8, 130), at 18 km (0.24, 28), at 36 km (0.059, 6.8); for VEGA 2 at 9 km: (2, 150), at 18 km: (0.21, 16), at 36 km: (.036, 3.5); for GIOTTO at 9 km: (1.24, 136), at 18 km: (0.17, 21), at 36 km: (0.034, 5.2).

3. Discussion and Conclusions

The most important conclusion of this study is that there is no need to introduce active and inactive areas on the surface of P/Halley nucleus to reproduce the observed structured near-nucleus coma. It is sufficient to allow for the observed nucleus shape, and to take into account properly the collisional dusty gas dynamics in its vicinity. Whereas any non-spherical shape will always produce coma structures,

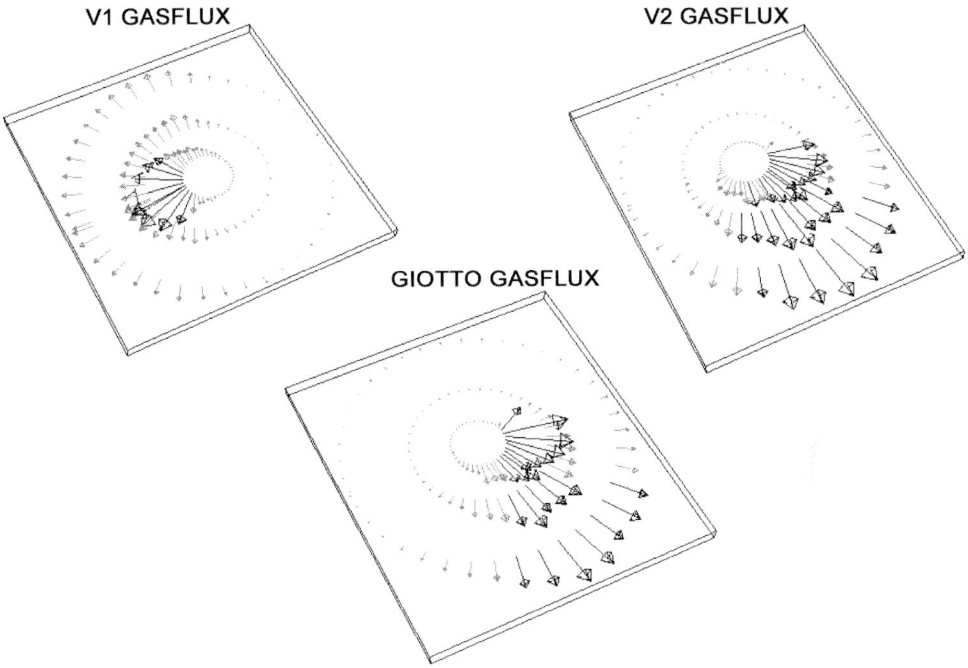

Figure 2. The gas flux vectors are shown around the nucleus during the VEGA 1, VEGA 2, and GIOTTO encounters, respectively, on three circles centred on the model nucleus of radii 9, 18, and 36 km. The circles lie in the equatorial plane of the model nucleus, and provide the feet of the vectors. The vector lengths, heads and gray scale densities are proportional to the gas flux. The square frame width indicates the magnitude of the off-plane Z-components.

there is simply no way known to us to interpret such structures when the nucleus shape is unknown. It seems to us that to fit large scale dust structures (spirals, ...) assuming trivial rectilinear emissions from a rotating sphere with discrete areas is a totally misleading exercise: not only because, as already shown long since by Kitamura (1990) the interaction between gas jets from postulated active areas results in dust density patterns totally different from the trivial rectilinear emission, but also because here we have shown that unless the nucleus is really a smooth sphere, the topography will still structure the coma (in addition to the active-inactive area network).

We believe that these results will open a new chapter in cometary studies: it hopefully will change the way how observations will be related to nuclear properties.

A further indication that the present jet reconstruction is correct, is provided by a special event during the VEGA 2 flyby on 9 March, 1986 when the spacecraft crossed a dust jet at about 7:20 UT. This jet was clearly detected by the SP-1 dust counter onboard (Vaisberg et al., 1986). The spacecraft in this moment was about 10,000 km from the nucleus. According to the dust counter data, dust of

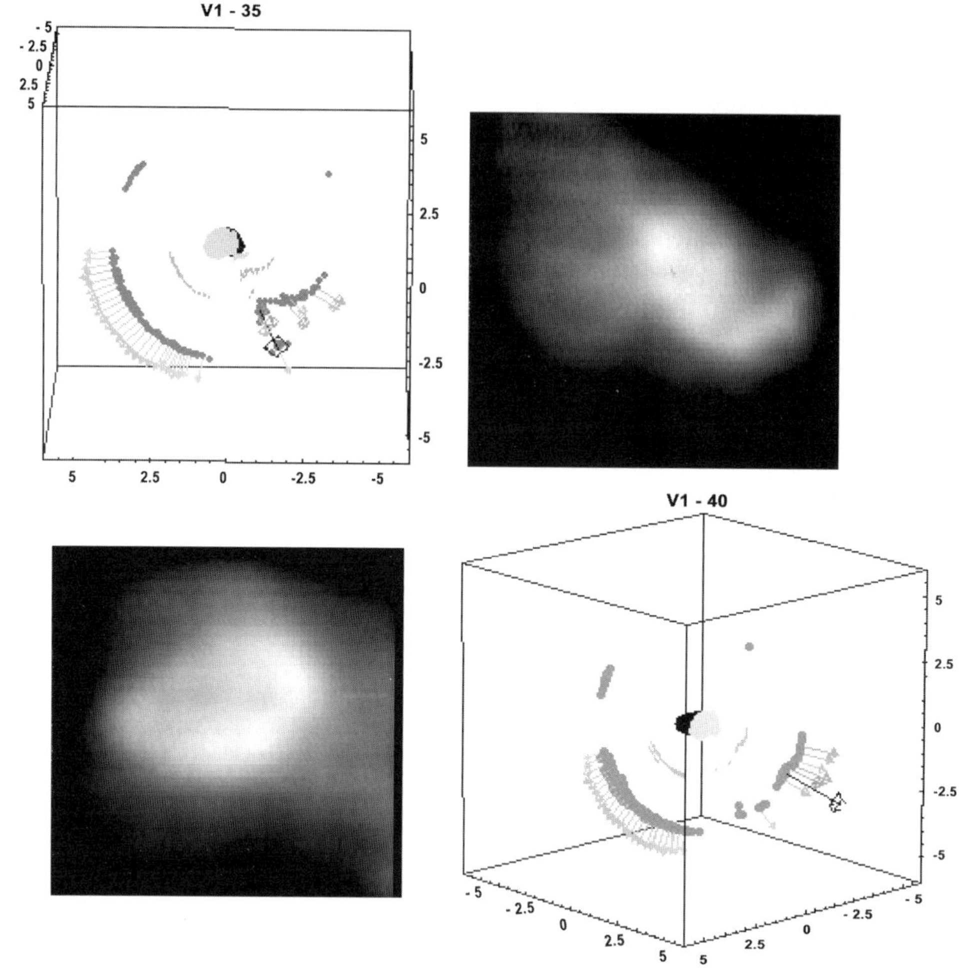

Figure 3. Two images taken by the VEGA 1 cameras and the corresponding model prediction of the jets. See the text for explanations.

different masses arrived at different times, in harmony with their expected different terminal velocity and with the prograde rotation of the nucleus. Within the same time interval image #94 was obtained; on that image the nucleus is dim and definitely obscured, this is what we expect if we had been inside the jet during the exposure. The raw version of this image contains several bright pixels, very likely light scattered from close dust particles. Our model shows that around that time the dust jets were oriented towards the spacecraft.

The three dimensional jet structure seen around Halley's nucleus on 9 March, 1986 was derived in Sagdeev et al. (1987) based on the 11 VEGA 2 images that covered a wide angular range around the nucleus, on those images the jets are very clearly identifiable. In that paper we have already concluded that the jets during the

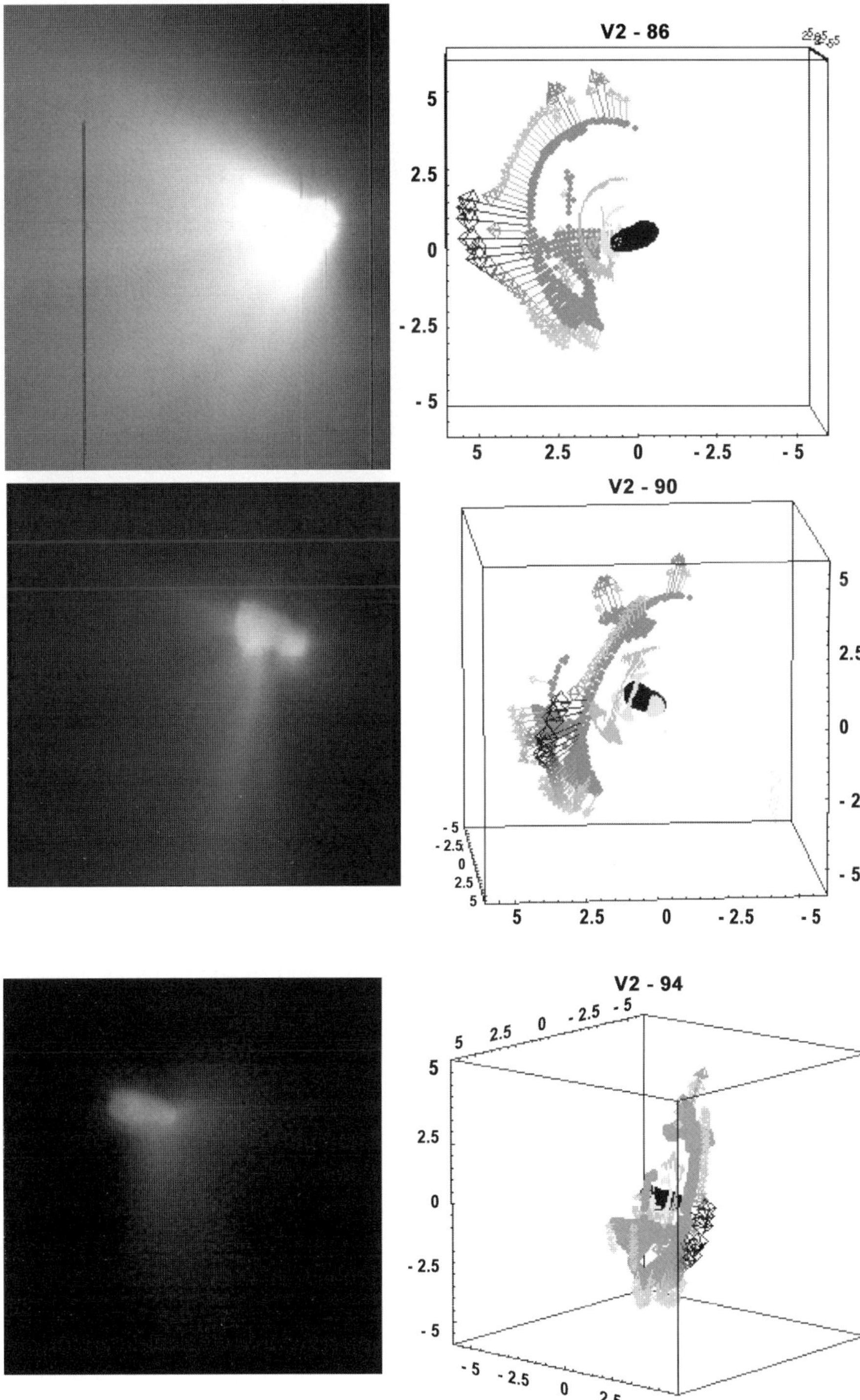

Figure 4. Three images taken by the VEGA 2 cameras and the corresponding model prediction of the jets. See the text for explanations.

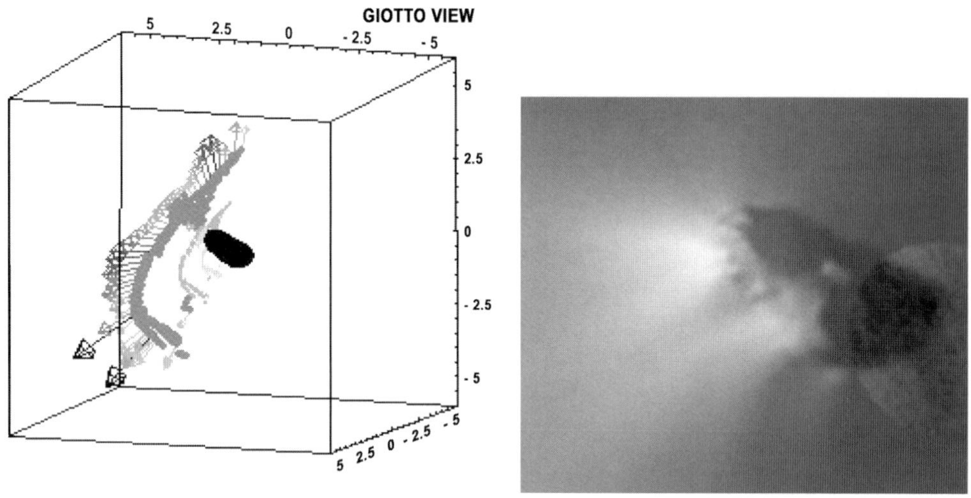

Figure 5. The GIOTTO composite image and the corresponding model prediction of the jets. See the text for explanations.

VEGA 2 formed a long, linear structure; this is very well reproduced by the model results presented here.

A limitation of our present calculations is the use of a single size dust fluid, which cannot reproduce the richness of the real dust distribution in a coma; it is not difficult to overcome this in the future. Another limitation (much more difficult to overcome) is that the coma we computed is time-stationary: We rotated the nucleus into its observed position taken during the encounters, and we calculated the near nuclear environment as if it were time-independent. That is the reason why we limited ourselves to the very close regions where the time dependence is not yet important. The real images, however, collected light from a much larger volume, a few tens of square km width in the image plane, and of the order of 10,000 km perpendicular to it. We plan to work out time-dependent solutions in the future.

We suspect that because the calculated gas coma is structured close to the nucleus, this might yield a structured coma farther at large distances as well. To confirm this conjecture numerically, we have to add gas-phase chemistry to the model, because the ground based observations do not identify water directly. This is also a task for the future.

We have shown that at the top of the collisional regime dust particles are accumulated into "jets"; these accumulations correspond to the feet of the dust jets seen from ground. Higher up dust particles already move freely with their terminal velocity on Keplerian orbits. The yet unresolved question is how their velocity perpendicular to the radial direction can be related to the rotation of the nucleus. As we have investigated static cases only, we cannot answer it yet. It is evident, however, that if such relationship holds, it is not simple, because inside the gas-dust interaction region perpendicular momentum is exchanged between dust and gas. Accordingly, to assume that the transverse dust velocity is equal to the initial angular velocity of the nucleus is on shaky grounds, therefore it is a non-trivial, open question how one can derive the nucleus rotation from dust (or gas) jet curvatures.

Acknowledgements

The work of K. Sz. was funded by the Hungarian OTKA grant T-32634; the contributions of J.-F. C. and A. V. R. were supported by French CNRS/INSU and CNES/CT contracts.

References

Crifo, J.-F. and Rodionov, A. V.: 1999, *Planet. Space Sci.* **47**, 797–826.

Crifo, J.-F., Fulle, M., Rodionov, A. V., Szegö, K., and Fulle, F.: 2002, 'Challenging a Paradigm: Do We Need Active and Inactive Areas to Account for Near-Nuclear Jet Activity?', these proceedings.

Keller, H. U., Curdt, W., Kramm, J. R., and Thomas, N.: 1995, *ESA SP-1127*, Vol. 1.

Kitamura, Y.: 1990, *Icarus* **86**, 455–475.

Merenyi, E., Foldy, L., Szegö, K., Toth, I., and Kondor, A.: 1990, *Icarus* **54**, 388–405.

Rodionov, A. V., Crifo, J.-F., Szegö, K., Lagerros, J., and Fulle, M.: 2002, *Planet. Space Sci.*, submitted.

Sagdeev R. Z., Smith, B., Szegö, K., Larson, S., Toth, I., Merenyi, E., Avanesov, G. A., Karasikov, V. A., Shamis, V. A., and Tarnapolski, V. I.: 1987, *Astron. Astrophys.* **187**, 293–298.

Sagdeev, R. Z., Szegö, K., Smith, B. A., Larson, S., Merényi, E., Kondor, A., and Toth, I.: 1989, *Astron. J.* **97**, 546–551.

Szegö, K., Sagdeev, R. Z., Whipple, F. L., Abergel, A., Bertaux, J. L., Merenyi, E., Szalai, S., and Varhalmi. L.: 1995, *ESA SP-1127*, Vol. 2.

Vaisberg, O. L., Smirnov, V. N., Gorn, L. S., Iovlev, M. V., Balichkin, M. A., Klimov, S. I., Savin, S. P., Shapiro, V. D., and Shevchenhko, V. I.: 1986, *Nature* **321**, 274–276.

THE EXPERIMENTS ONBOARD THE ROSETTA LANDER

JENS BIELE
Deutsches Zentrum für Luft- und Raumfahrt, Köln (E-mail: jens.biele@dlr.de)

(Received 5 March 2002; Accepted 12 June 2002)

Abstract. As a part of ESA's cornerstone mission "ROSETTA" to comet 46P/Wirtanen a 100 kg Lander will bring a scientific payload of almost 27 kg to the surface of the nucleus. After a first scientific sequence it will operate for a considerable fraction of the cometary orbit around the sun (between 3 AU and 2 AU). Ten experiments with a number of sub-experiments are foreseen; this paper presents the current status of the Lander development and reviews the scientific capabilities of each of the experiments at a time when the Flight Model (FM) of the Lander is already delivered.

Keywords: Comet nuclei, payload, space missions, Wirtanen

1. Introduction

The ROSETTA Lander mission will be the first to actually land on a comet's nucleus and to study it *in situ* for a period ofseveral months. The ROSETTA spacecraft (Lander+Orbiter) will be launched with an Ariane 5 from Kourou, French Guiana, in January 2003 and will approach comet 46/P Wirtanen in August 2011 (cf. Schwehm and Schulz, 1999). During its way to the comet the Lander will be mostly in hibernation mode. Occasional checkouts and magnetometer measurements (e.g., during asteroid fly-by) will be the only exceptions. After the rendezvous with the comet (at $\approx$ 4.5 AU) and about one year of observations with Orbiter instruments, a landing site will be chosen. The Lander will be activated and, after 7 days of delivery preparations, separated from the Orbiter with an altitude as low as possible, e.g., 1 km. Eject will be performed with lead screws with adjustable speed; under almost zero-g conditions a cold-gas push-down thruster and flywheel 1-axis stabilisation provide for a descent time of less than 3 hours and a hold-down thrust upon touch down on the comet. At touch-down with $\Delta v \approx 1$ m/s the Landing Gear will adjust itself to the terrain and dampen the impact momentum. At the same instant a harpoon connected to a tether is fired into the surface of the comet to anchor the Lander (see Ulamec, 1997a for a detailed discussion). On the comet the Landing Gear then provides for rotation, vertical adjustment and tilting of the Lander, thus the comet material can be probed on arbitrary positions on a working circle with a diameter of 0.8 m. The Lander subsystems are described in Ulamec et al. (1997b), Wittmann (1999).

The Rosetta Lander and its payload are designed to achieve the minimum mission goals within few days after landing, which will take place when the comet is at

Figure 1. The target: groundbased image of Comet Wirtanen (courtesy Rita Schulz).

a heliocentric distance, $r \approx 3$ AU, presumably before the onset of any major cometary activity. Subsequently, the long-term mission will study the evolution of the comet's activity while approaching the sun. Here the ROSETTA project combines two strategies for characterizing the properties of the cometary nucleus: on one hand, the comet's evolution along the orbit with decreasing r will be investigated with the Orbiter instruments by monitoring the physical and chemical properties of the nucleus and in situ analysis of the near-nucleus environment. On the other hand, the Lander will provide ground truth by analysing the nucleus material directly.

The target comet, 46P/Wirtanen (see Figure 1) was discovered in 1948; it is a short-period comet with an orbital period of 5.46 years, aphelion and perihelion distances of 5.46 and 1.06 AU, respectively. Its peak gas production rate at perihelion is of the order of 10^{28} mol/s; it is a relatively small comet with a radius of barely 600 m (assuming an albedo of 0.04). The comet's rotation period is not very well known (between 6 h and 10 d) while its rotation axis is presently unknown. See also Schulz and Schwehm, 1999. Tables I and II give an overview on the mission profile (cf. Schwehm and Schulz, 1999) and Orbiter/Lander characteristics.

Figure 2 shows the actual ROSETTA spacecraft (FM) with the Lander (on top) attached.

2. Lander Payload

There are 10 experiments on board of the Lander including the drill, sampling and distribution system SD^2, which serves other instruments; some of the experiments consist of several units. Table III gives an overview of the experiments, explaining the abbreviations and listing the providers. The scientific capabilities of each experiment are described in the following sections. – As this paper is meant to review and

TABLE I
Mission profile

Date	
13 Jan. 2003	Launch (Kourou)
05/2005	Mars flyby, gravity assist
10/2005	Earth flyby, gravity assist
07/2006	Otawara flyby
10/2007	Earth flyby, gravity assist
07/2008	Siwa flyby
11/2011	Rendezvous with the comet
10/2012	Lander delivery (r = 3 AU)
02/2013	Nominal end of longterm Lander mission (r = 2 AU)

TABLE II
Orbiter and Lander characteristics

Orbiter	Lander
Launch mass: 2900 kg (payload : 150 kg + 100 kg (Lander), 1600 kg propellants for a $\Delta V = 2.2$ km/s)	*Overall mass*: 100 kg (payload 26.7 kg). Practically a small autonomous spacecraft
Wing span: 32 m (solar array: 5.7 kW @ 1 A.U., 0.4 kW @ 5.25 A.U.). No RHUs.	*Energy/power* from primary batteries, solar array (about 10W at 3 a.u.) and secondary batteries.
On board memory: 42 Gb (solid state) – S&X-band TM/TC link	*On board memory*: 2*16 Mb
	S band TM/TC with the orbiter
Industrial team: DaimlerChrysler (prime) + MSS (platform and avionics) + Alenia (AIT)	*Designed and developed* by an European consortium under German (DLR) leadership.

summarize the Lander's scientific payload, it cannot replace the original research papers, which are referenced at the end of each section.

2.1. APX

APX is an Alpha-Xray-Spectrometer for elementary analysis (all major elements except H, He). It irradiates the comet's surface with an alpha source (30 mCi, 96-Cm) and measures the spectra of backscattered alpha particles (Rutherford

TABLE III
Experiment overview

Acronym	Full Name	Principal Investigator (PI)
APXS	Alpha Proton Xray Spectrometer	R. Rieder, MPI Chem. Mainz, Germany
ÇIVA	Comet Nucleus Infrared and Visible Analyser	J.P. Bibring, IAS, Orsay, France
– P/S	– panoramic incl. stereo	
– M(V/I)	– Microscope (V = visual, I = infrared spectra)	
COSAC	Cometary sampling and Composition Experiment	H. Rosenbauer, MPAe, Katlenburg-Lindau, Germany
– MS	– Mass Spectrometer	
– GC	– Gas chromatograph	
– PG	– Pressure gauge	
CONSERT	Comet Nucleus Sounding Exp. by Radiowave Transmission	W. Kofman, CEPHAG Grenoble, France
MUPUS	Multi-Purpose Sensors for Surface and Sub-Surface Science	T. Spohn, Univ. Münster, Germany
– PEN	– Penetrator	(SRC Warsaw)
– TM	– Thermal Mapper	(DLR Berlin)
– ANC	– Anchor sensors	(IWF Graz)
MODULUS/ Ptolemy	Method Of Determining and Understanding Light elements from Unequivocal Stable isotope compositions	I. Wright, Open University & Rutherford Appleton Laboratories, UK
– MS	– Mass Spectrometer	
– GC	– Gas chromatograph	
ROLIS	Rosetta Lander Imaging System	S. Mottola, DLR Berlin, Germany
ROMAP	Rosetta Lander Magnetometer and Spectrometer	U. Auster, MPE Munich and TU Braunschweig, Germany
SESAME	Surface Electrical, Seismic and Acoustic Monitoring Exp.	D. Möhlmann, DLR Cologne, Germany
– CASSE	– Comet Acoustic Surface Sounding Experiment	– DLR Cologne, Germany
– DIM	– Dust Impact Monitor	– I. Apathy, KFKI, Hungary
– PP	– Permittivity Probe	– W. Schmidt, FMI Finland
SD^2	Sample Drill and Distribution	Politecnico of Milan

Figure 2. The Lander mounted on top of the ROSETTA spacecraft while being installed in the thermal vacuum chamber at ESTEC, February, 2002.

scattering) and alpha-induced characteristic X-ray radiation. The design of the Rosetta Lander APX spectrometer is based on the experience gained with the APX spectrometers built for American and Russian Mars missions (Rieder et al., 1996).

The APXS provides information on the elemental composition of the material underneath the Lander which cannot be derived from the results of the more sophisticated evolved gas analysers (EGAs) because many elements may not be represented in the volatile fraction with their true abundance (the elemental and isotopic composition together should provide clues to the basic question of the comets origin). Only the composition of the uppermost atomic layers is determined by APX; the disadvantage is that these layers cannot be expected to show the composition of the bulk material. However, under sunlit conditions, the composition of the non-volatile fraction is probably well represented. Hence dynamic day/night changes in composition can be studied via the recondensation of volatiles at the surface (frost) which could happen in the "evening" and during "night". – The mechanism lowering the APXS to the optimum distance to ground (4 cm) is also used to get a ring in contact with the surface serving as the fourth electrode for the SESAME permittivity probe.

For more information on APX see (Rieder et al., 1996, 1997, 2000; Linkin et al., 1996; Economou et al., 1996, 1997).

2.2. ÇIVA

ÇIVA is an integrated set of imaging instruments, designed to characterise (1) the landing and sampling site, (2) the 360°panorama as seen from the Rosetta Lander, (3) samples collected and delivered by SD^2. It consists of a panoramic and stereo camera set (ÇIVA-P), and a microscope camera coupled to an IR spectrometer (ÇIVA-M(V/I)), sharing a common Command and Data Processing Unit (CDPU) with a 16 Mbytes mass memory (Beauvivre et al., 1999).

ÇIVA-P will characterise the landing site, from the landing legs to the local horizon. The camera system is composed of 5 identical cameras (5 mono cameras), mounted on the Lander sides plus a pair of the same micro-cameras (the stereo pair) at the balcony side of the Lander with their optical axes separated by 60°. Thus, 360°can be viewed at a time with the balcony side of the Lander being stereo-imaged. Lander rotation and repetitions of ÇIVA-P operations several times along the cometary activity will allow to characterise the 3-D surface topography, provide an albedo mapping of the landing site and detect manifestations of cometary activity and resulting surface changes at scales not achievable from the Orbiter.

The seven camera heads contain 6 elements objectives and 1024x1024 frame transfer CCD detectors and will be operated sequentially. The electronics driving each CCD has adequate intelligence to optimise the exposure time. All images will be compressed and stored before being transmitted, via the ROLIS DPU (see sec. 2.7), to the CDMS.

ÇIVA-M combines an ultra-compact and miniaturised visible light microscope (-M/V) and coupled IR spectrometer (-M/I), to characterise, by non-destructive analyses, the texture, albedo, mineralogical and in part also the molecular composition of the samples collected and distributed by SD^2. The resolution of ÇIVA-M/V images is 14 μm; the samples are illuminated sequentially with 3 LEDs of different colours. Following the microscope imaging, IR spectral images will be obtained, with a spatial resolution of 50 μm, with a 128 x 128 IR HgCdTe array operating at temperatures 120 to 170 K. The samples are illuminated by a monochromator operating between 1.0 and 4.0 μm at a resolution of 5 nm, using a rotating grating. The spectrometer should allow identifying the major organic chemicals in the volatiles and organic refractories. The entire process will take less than 5 minutes, to get all spectra with SNR $>$ 100, assuming IR albedos of 0.05. Then, the sample can be transferred to a subsequent experiment (COSAC or PTOLEMY). The process may be repeated for each sample obtained at different depths and/or location.

2.3. COSAC

The scientific goals of COSAC are: the identification of natural and pyrolytically generated volatile compounds and their parent molecules with emphasis on organic, potentially "pre-biotic" material. A special effort is devoted to chiral compounds with the objective to look for deviations from homochirality which is characteristic of life on Earth. The method: Cometary material from both the natural surface below the Lander and the ground of drill holes (collected immediately at and up to ~20 cm below the surface, provided by the SD^2 instrument (see sec. 2.10) is heated in steps up to 800°C. The evolving gases are analysed employing gas-chromatographic and mass-spectroscopic (GC/MS) methods.

The gas chromatograph (HETP$<$ 2mm) and the mass spectrometer will normally be used in a coupled mode. The separation and identification, respectively, of isobaric molecules is mainly accomplished by means of a high-resolution time-of-flight mass spectrometer ($M/\Delta M$ $>$ 3000 in a high-resolution mode, $\approx$ 300 in low-res mode for M up to about 7000, sensitivity better 100 ppm, almost noise-free detection) employing a special geometry of synchronous ion trajectories allowing for long paths despite a small instrument dimension (multiple reflections). Many chemical compounds and their chiral properties and isotopic ratios can be separated and detected in a quantitative way. COSAC also comprises a pressure gauge (one Penning, one Pirani sensor) to measure the cometary atmosphere's pressure in the range 10^{-7} to 1 mbar. For further reference, see Rosenbauer et al. (1999).

2.4. CONSERT

With the CONSERT experiment the internal structure of the nucleus can be probed. A transmitter on the Orbiter sends wave packets through the nucleus to the Lander, which returns them back in a transponder manner. The measured dielectric properties of the nucleus constrain the composition and can detect large-size structures and stratification. The scientific objectives of the CONSERT experiment on the ROSETTA mission are described in Barbin et al. (1999) and Kofman et al. (2002). The experiment uses a wide bandwidth signal, which will allow measurements of the signal propagating through the main and secondary paths. These measurements will also provide the distribution of the secondary path for a deeper description of the comet interior and better spatial coverage. In order to process the signal globally, i.e., to combine the measurements corresponding to the different position on the orbit, the signals on the Orbiter and on the Lander must be coherent with stable relative phase and have a good signal to noise ratio. These conditions had put constraints on system structure, clock stability, antennas and operations which were all met. For further reference, see Benna et al. (2001).

2.5. MUPUS

The Multi-Purpose Sensor Experiment MUPUS consists of three parts:

1. A penetrator, PEN, $\approx$ 40 cm long, that will be hammered into the ground about 1 m away from the Lander. During the penetration process the mechanical strength of the material will be measured by means of a depth sensor and a densitometer. The penetrator is equipped with a series of temperature sensors for determining the temperature as a function of depth and insulation. The temperature sensors can also be used as heaters which will allow to measure the thermal conductivity of the surrounding material (Seiferlin et al., 1996). PEN also carries the fifth electrode for the SESAME permittivity probe.
2. An accelerometer and a temperature sensor accommodated in the harpoon(s) will allow thermal and strength measurements to (probably) larger depths than the PEN. The accelerometer measures the impact profile when the harpoons are shot (at touchdown) and from which mechanical properties of the material penetrated can be reconstructed.
3. A four-channel infrared radiometer (MUPUS-TM) measures surface temperatures in the vicinity of the Lander. An accuracy in the order of 1..10 K is envisaged, while the resolution is better than 0.1 K. For further reference see Seiferlin et al. (1996, 2000).

2.6. MODULUS/PTOLEMY

This instrument consists of an ion trap mass spectrometer designed for isotope analysis and to determine qualitative analytical data; chemical separation of isobars (δ to $\pm 0.1\%$ for H, C, N, O) will be done by gas chromatography. The mass spectrometer will operate in the range 10–200 AMU with better than unit mass resolution (in isotope mode: 3 times better). Operation: Like for COSAC, the sample volatiles are analysed by heating the oven where they are quantified, purified and chemically reacted (if necessary) to produce a relatively simple gas mixture. Gases are then passed to the ion trap mass spectrometer, either directly or through one of three analytical channels comprising gas chromatography columns and additional chemical processing reactors. For experiments requiring gas chromatography a constant supply of the helium carrier gas is delivered by a regulated supply, which ensures maintenance of the necessary pressure and flow-rate. In either mode of operation (direct, or chromatography) the ion trap mass spectrometer is set to perform continuous sweeps over the mass range of interest. Since Ptolemy aims to obtain isotope ratio measurements of the highest possible precisions, the ion trap instrument will be calibrated in situ during the same period of time over which the cometary analyses are made. For this, equivalent analyses will be made of a reference gas taken from Earth to the comet and delivered to the instrument through the gas management system. In this way analogous isotope ratio data will be acquired from the reference gas. With knowledge of the actual isotopic ratio of the reference it will then be possible to correct the measured cometary data in order to obtain an absolute value for the ratio of interest. More details can be found in Wright and Pillinger (1998) and Wright et al. (2002).

2.7. ROLIS

ROLIS is a miniature CCD imager (1024 × 1024, focal length 12 mm) located on the balcony of the Rosetta Lander and oriented in a downward-looking direction. From this position ROLIS can observe a region of about 30x30 cm of the nucleus surface located below the Lander with a spatial sampling of 0.3 mm/pixel. In order to illuminate the field to be imaged, ROLIS incorporates four independent arrays of light emitting diodes (LEDs) irradiating through the visible and near IR, in spectral bands centred at about 470, 530, 640 and 870 nm with a FWHM of about 100 nm. ROLIS will also operate during the descent phase, acquiring images of the landing site and its vicinity shortly before touch-down. Due to its location on the so called "instrument common working circle", ROLIS can inspect the sampling sites of the "in situ" Lander analysers, before and after the drilling operation. In addition to gathering surface colours and morphology, imaging of the bore-hole sides can possibly reveal signs of stratification, or give clues about the mechanical strength of the surface layer. During the extended mission ROLIS will search for signs of evolution of the surface features as the comet approaches the Sun. Both ROLIS and ÇIVA are managed on board by a common "Imaging Main Electronics", ROLIS-IME; wavelet transformation data compression (>1:10) is used for both instruments. See also Mottola et al. (2002).

2.8. ROMAP

Romap comprises a magnetic field sensor (MAG) and a plasma monitor (SPM). It's main scientific goals are long-term magnetic field and plasma measurements of the surface of Comet 46P/Wirtanen in order to study cometary activity as a function of heliocentric distance, and measurements during the Lander's descent to investigate the structure of the comet's remnant magnetisation. MAG and SPM complement the plasma packages aboard the Rosetta Orbiter. Both instruments investigate the comet/solar wind interaction, cometary activity and the onset of diamagnetic cavity formation. Specifically, SPM measures the major solar wind parameters such as density, speed, temperature and flow direction, while MAG determines the magnetic field vector. The ROMAP fluxgate magnetometer, electrostatic analyser and Faraday cup measure the magnetic field from 0–64 Hz, ions of up to 8000 keV and electrons of up to 4200 keV. For MAG, dynamic feedback fields as well as offset fields of up to 2000 nT can be generated to compensate for Lander and/or Orbiter DC stray fields. The SPM sensors are active in the surface mode only. In this mode, MAG and SPM work sequentially. During Rosetta's cruise phase and during the Lander descent, ROMAP will measure in the slow (1 Hz) magnetometer mode. Particularly during periods when the spacecraft is active (e.g., before and during asteroid flybys), magnetometer data from ROMAP will help to distinguish between external fields and spacecraft disturbances. The 'fall off' profile during descent will provide reliable information about the comet's internal magnetic structure, if any. SPM will start its measurements some minutes after landing. After a short initial measurement interval of 20 min, measurement cycles lasting typically 4 h are expected for the long-term mission on the surface. The fast magnetometer mode (64 Hz sampling rate) will be used only for measurements in parallel with the Permittivity Probe (E-field) of the Lander's SESAME-PP instrument (see below). Further details can be found in (Auster, 2002).

2.9. SESAME

2.9.1. *SESAME-CASSE*

The CASSE instrument is developed to study by acoustic sounding the mechanical properties of the upper surface layers of the comet. It consists of piezoelectric actuators (transmitters) and sensitive accelerometers (sensors) in the "feet" of the Landing Gear. The near-surface structural properties and the layer structure of the soil will be determined by acoustic and seismic probing (active and passive). The measurements can be used to improve the modelling the surface layers and their influence on cometary outgassing.

2.9.2. *SESAME-PP*

PP sounds the permittivity properties of the surface electromagnetically. Five electrodes, three at the Landing Gear's "feet" and two at the MUPUS-PEN and APX deployables, are used to probe the soil with a quadrupole array technique at various frequencies, measuring the complex permittivity, i.e., electrical conductivity and dielectric polarizabilitiy down to a depth of a few meters. With these measurements the water content (distribution of polar molecules) and its variation can be studied. In passive mode, the experiment has the additional capability of a plasma wave investigation: it detects the electric fields of electrostatic and electromagnetic waves with frequencies up to 10 kHz, which are generated by the interaction of the solar wind with the charged dust and ionized outgassing products of the nucleus. By sampling the total intensity of plasma waves every few second, these measurements will consequently provide a continuous monitoring of the nucleus activity. For further reference, see Grard et al. (1996).

2.9.3. *SESAME -DIM*

DIM is a three-dimensional dust-impact monitor. While larger particles will be sensed by the CASSE receivers in the Lander's feet, DIM will investigate the flow of small particles hitting the Piezo-sensor plates of this instrument. For a more detailed discussion of the SESAME sub-experiments, see Möhlmann (2002) and Kochan et al. (2000).

2.10. SD^2

The Sampling, Drill & Distribution System is capable of drilling and sampling anywhere on a circle of about 0.4 m radius around the Lander's rotation axis up to a drill depth of 25 cm. Drilling in material with a compressible strength of up to 3 MPa (equivalent to solid ice with stony inclusions) has been demonstrated. A sample (6–34 mm^3) can be gauged with the Volume Checker and be transferred to a carousel containing 28 Medium- and High-Temperature ovens (the former equipped with sapphire windows to allow optical and IR spectroscopic observations by ÇIVA-M-V/I) serving the two EGAs by means of resealable Docking Stations.

3. Mission Operations and Timeline

After commissioning only one operation is currently foreseen before the comet mission, which is a ROMAP magnetometer measurement during the asteroid 140 Siwa flyby. At the comet, one distinguishes between the first scientific sequence and the long-term mission.

3.1. First scientific sequence

This high-priority sequence is foreseen to fulfil the minimum scientific requirements under worst-case circumstances (e.g., no energy from solar cells). All short term science investigations are completed once within the first 120 h after touchdown including the drilling and analysis of 4 samples (two surface and two sub-surface). The results of the surface sample analysis and the panoramic images are sent as soon as possible to ground in order to be analysed on Earth and permit an optimisation of the ongoing first sequence (optimum orientation of Lander to the sun, attitude for deployment of MUPUS-PEN and APX). A strawman mission timeline for the first ca. 51 h after separation has been worked out and proves to be compatible with energy, power and telemetry constraints. During the first 60 h at least one full orbit of Rosetta shall be dedicated to perform CONSERT operations.

3.2. Long-term Mission

The long-term mission is baselined from $r = 3$ AU to $r = 2$ AU (corresponding to a period of about 3 months); it is foreseen to map the evolution of the comet's activity with changing heliocentric distance. During this period the Lander will perform mostly serial operations, relying entirely on the solar generator. Long term, low-power experiments like SESAME, MUPUS and ROMAP can be turned on periodically, while others, like COSAC and PTOLEMY, will analyse samples whenever the energy budget will allow to do so. The exact operational sequence will be worked out interactively.

3.3. Extended Mission

At some stage 1 AU $< r <$ 2 AU the Lander will overheat or be covered with dust defining the point where any extended mission will definitely end.

3.4. The ROSETTA Lander Ground Segment

This ground segment consists of the Rosetta Lander Science and Navigation Centre, SONC, (CNES) and the Rosetta Lander Control Centre, LCC, (DLR Cologne), jointly commanding the Lander via the Rosetta Mission Operations Centre (ESOC) and the Perth 35m ground station.

4. Schedule, Status and Outlook

The Lander Flight Model (FM) has been delivered to ESA and is undergoing extensive system tests. A Ground Reference Model (GRM) will be build by summer 2002 and the Lander Software testbed development is ongoing. Only one descope action (MUPUS PEN densitometer) had to be performed; mass and telemetry budgets are under control, and the energy budget will be improved by implementing additional batteries. However, refurbishment actions, critical system tests and verifications are still ahead.

Acknowledgements

I would like to thank the PI teams without whose contributions this paper would not have been possible. Furthermore I extend my thanks to the reviewer, whose remarks helped to improve this paper significantly.

References

Auster, H. et al.: 2002, 'ROMAP: Rosetta Magnetometer and Plasma Monitor', in *ESA Special Publication on the ROSETTA Mission, ESA-SP1165*, ESA.

Barbin, Y., Kofman, W., Nielsen, E., Hagfors, T., Seu, R., Picardi, G., and Svedhem, H.: 1999, 'The Consert Instrument for the Rosetta Mission', *Adv. Space Res.* **24**, 1127–1138.

Beauvivre, S., Lamy, P., Nguyen-Trong, T., and Reynaud, J.-L.: 1999, 'The Panoramic Camera of the Rosetta Mission: Performances of Prototype 3D Microcameras', *Adv. Space Res.* **24**, 1115–1126.

Benna, M., Barriot, J., and Kofman, W.: 2001, 'Three Dimensional Simulation of Wave Propagation Into the Comet 46P/Wirtanen Nucleus (ROSETTA Space Mission – CONSERT Experiment)', in *American Geophysical Union, Fall Meeting 2001, abstract #P32A-0545*, pp. A0545.

Economou, T., Turkevich, A., Rieder, R., and Wanke, H.: 1996, 'Chemical Composition of Martian Surface and Rocks on Pathfinder Mission: The Alpha-Proton-X-Ray Spectrometer', in *Lunar and Planetary Institute Conference*, Vol. 27, pp. 333.

Economou, T., Turkevich, A., Rieder, R., and Wanke, H.: 1997, 'Report on the Status of the APXS on Mars Pathfinder Mission', in *Lunar and Planetary Institute Conference*, Vol. 28, pp. 319.

Grard, R., Laakso, H., Hamelin, M., Goldstein, B., Winterhalter, D., Kochan, H., and Ulamec, S.: 1996, 'Determination of the Conductivity and Permittivity of the Surface Material and Monitoring of the Outgassing Activity of the Cometary Nucleus', in *Lunar and Planetary Institute Conference*, Vol. 27, pp. 449.

Kochan, H., Feibig, W., Konopka, U., Kretschmer, M., Möhlmann, D., Seidensticker, K. J., Arnold, W., Gebhardt, W., and Licht, R.: 2000, 'CASSE – The ROSETTA Lander Comet Acoustic Surface Sounding Experiment – Status of Some Aspects, the Technical Realisation and Laboratory Simulations', *Planet. Space Sci.* **48**, 385–399.

Kofman, W., Barbin, Y., and Berthe, M. et al.: 2002, 'CONSERT: Comet Nucleus Sounding Experiment by Radiowave Transmission', in *ESA Special Publication on the ROSETTA Mission, ESA-SP1165*, ESA.

Linkin, V., Lipatov, A., Makarov, V., Khloustova, L., Ekonomov, A., Kremnev, R., Martinov, B., Harri, A.-M., Pirjola, R., Siili, T., Pellinen, R., Kumpulainen, K., Runavot, J., Vienne, D., Laplace, H., Mouli, G., Pommereau, J.-P., Malique, C., Carpentier, T., Rieder, R., Economou, T., Lognonne, P., Menvielle, M., Karczewski, J.-F., Herring, M., and Levachier, P.: 1996, 'The Mars 96 Small Station – Implementation and Scientific Objectives', *Bull. Am. Astron. Soc.* **28**, 1068.

Möhlmann, D. et al.: 2002, 'SESAME: Surface Electric Soundingand Acoustic Monitoring Experiment', in *ESA Special Publication on the ROSETTA Mission, ESA-SP1165*, ESA.

Mottola, S., Arnold, G., Grothues, H.-G., Hirsch, H., Jaumann, R., Michaelis, J., Neukum, G., Wäsch, R., and Bibring, J.-P.: 2002, 'ROLIS: Rosetta Lander Imaging System', in *ESA Special Publication on the ROSETTA Mission, ESA-SP1165*, ESA.

Rieder, R., J. Brückner, J., Klingelhöfer, G., Gellert, R., Dreibus, G., Lugmair, G., Wänke, H., and the Athena Science Team: 2000, 'The Athena Alpha Proton X-Ray Spectrometer (APXS)', in *Workshop on Concepts and Approaches for Mars Exploration, July 18–20, 2000, Houston, Texas*, Abstract no. 6102, pp. 6102.

Rieder, R., Wänke, H., and Economou, T.: 1996, 'An Alpha Proton X-Ray Spectrometer for Mars-96 and Mars Pathfinder', *Bull. Am. Astron. Soc.* **28**, 1062.

Rieder, R., Wänke, H., Economou, T., and Turkevich, A.: 1997, 'Determination of the Chemical Composition of Martian Soil and Rocks: The Alpha Proton X-Ray Spectrometer', *J. Geophys. Res.* **102**, 4027–4044.

Rosenbauer, H., Fuselier, S. A., Ghielmetti, A., Greenberg, J. M., Goesmann, F., Ulamec, S., Israel, G., Livi, S., MacDermott, J. A., Matsuo, T., Pillinger, C. T., Raulin, F., Roll, R., and Thiemann, W.: 1999, 'The COSAC Experiment on the Lander of the ROSETTA Mission', *Adv. Space Res.* **23**, 333–340.

Schulz, R. and Schwehm, G.: 1999, 'Coma Composition and Evolution of Rosetta Target Comet 46P/Wirtanen', *Space Sci. Rev.* **90**, 321–328.

Schwehm, G. and Schulz, R.: 1999, 'Rosetta Goes to Comet Wirtanen', *Space Sci. Rev.* **90**, 313–319.

Seiferlin, K., Hagermann, A., Banaszkiewicz, M., and Spohn, T.: 2000, in N. I. Kömle, G. Kargl, A. J. Ball and R. D. Lorenz (eds.), *Using Penetrators as Thermal Probes: The MUPUS Case*, Austrian Academy of Sciences Press, Vienna, pp. 161–184.

Seiferlin, K., Kömle, N. I., Kargl, G., and Spohn, T.: 1996, 'Line Heat-source Measurements of the Thermal Conductivity of Porous H_2O Ice, CO_2 Ice and Mineral Powders under Space Conditions', *Planet. Space Sci.* **44**, 691–704.

Ulamec, S., Feuerbacher, B., Wittmann, K., Rosenbauer, H., Bibring, J. P., Moura, D., Mugnuolo, R., and Haerendel, G.: 1997, 'Rosetta Lander – *In Situ* Investigation of a Comet's Nucleus', in *Lunar and Planetary Institute Conference*, Vol. 28, pp. 1461.

Ulamec, S. et al.: 1997, 'RoLand: A Long Term Lander for the Rosetta Mission', *Space Technol.* **17**, 59–64.

Wittmann, K. et al.: 1999, 'Rosetta Lander, *In Situ* Characterization of a Comet Nucleus', *Acta Astronaut.* **45**, 389–395.

Wright, I. P. and Pillinger, C. T.: 1998, 'Modulus – An Experiment to Measure Precise Stable Isotope Ratios on Cometary Materials', *Adv. Space Res.* **21**, 1537–1545.

Wright, I., Barber, S., Maynard, J., Morgan, G., Morse, A., Sheridan, S., Evans, S., Pillinger, C., Kent, B., Waltham, N., Whalley, M., Heys, S., Sawyer, E., and Turner, R.: 2002, 'Ptolemy – An Instrument to Measure Stable Isotopic Ratios of Key Volatiles in a Cometary Coma', in *ESA Special Publication on the ROSETTA Mission, ESA-SP1165*, ESA.

RADIO INVESTIGATIONS OF 19P/BORRELLY IN SUPPORT TO THE DEEP SPACE 1 FLYBY

DOMINIQUE BOCKELÉE-MORVAN
ARPEGES/LESIA, Observatoire de Paris, 5 place Jules Janssen, F-92195 Meudon, France

NICOLAS BIVER
ESA, ESTEC SCI-SB, Keplerlaan 1, NL-2200 AG Noordwijk ZH, The Netherlands

PIERRE COLOM and JACQUES CROVISIER
ARPEGES/LESIA, Observatoire de Paris, 5 place Jules Janssen, F-92195 Meudon, France

MICHEL C. FESTOU
Observatoire Midi-Pyrénées, 14 avenue Edouard Belin, F-31400 Toulouse, France

FLORENCE HENRY and ALAIN LECACHEUX
ARPEGES/LESIA, Observatoire de Paris, 5 place Jules Janssen, F-92195 Meudon, France

HAROLD A. WEAVER
Department of Physics and Astronomy, The Johns Hopkins University, Baltimore, Maryland 21218, USA

(Received 14 February 2002, Accepted 25 May 2002)

Abstract. NASA's Deep Space 1 mission flew by Comet 19P/Borrelly on September 22, 2001. We present observations of molecular species obtained with the 30-m telescope of the Institut de Radioastronomie Millimétrique (IRAM) and the Nançay radio telescope at and near the time of this flyby. OH, HCN, and CS production rates were measured, while upper limits were deduced for CO, H_2CO and H_2S.

Keywords: 19P/Borrelly, molecules, radio lines

Our observations at IRAM 30-m were obtained on September 19–21 and 24, 2001, when 19P/Borrelly was at $\sim$ 1.36 AU and $\sim$1.48 AU from the Sun and the Earth, respectively. Bad weather prevented us to collect data on September 22 and 23, at the time of the Deep Space 1 encounter with the comet. HCN was detected on all dates by its J(1-0) (88.6 GHz) and/or J(3-2) (265.9 GHz) lines. The HCN J(3-2) line is significantly blueshifted ($\Delta v = -0.18 \pm 0.04$ km s^{-1}), which is indicative of molecules mostly expanding in the earthward portion of the coma. The CS radical was marginally detected on September 21 and 24 by its J(3-2) (147 GHz) and J(5-4) (245 GHz) lines. The H_2S (168 GHz), H_2CO (225 GHz) and CO (230 GHz) lines were not detected but useful upper limits of these species production rates were obtained. As part of the Nançay comet monitoring programme (Crovisier et al., 2002, and this colloquium), observations of the OH 18-cm lines

TABLE I

Molecular production rates and abundances relative to water in 19P/Borrelly (Q(OH) $\sim$ 0.9 $Q(H_2O)$; upper limits are 3σ values)

Date 2001	Species	Line	Production rate [mol s^{-1}]	Abundance 2001[a]	Abundance 1994[b]
19/07–01/08	OH	18 cm	$(1.2 \pm 0.3) \times 10^{28}$		
02–15/08	OH	18 cm	$(1.6 \pm 0.4) \times 10^{28}$		
16–22/08	OH	18 cm	$(2.6 \pm 0.5) \times 10^{28}$		
22/08–03/09	OH	18 cm	$(2.5 \pm 0.5) \times 10^{28}$		
19–24/09	HCN	J(1-0)	$(2.13 \pm 0.40) \times 10^{25}$	0.082%	0.11%
	HCN	J(3-2)	$(2.11 \pm 0.13) \times 10^{25}$		
19–24/09	CS	J(3-2)	$(2.1 \pm 0.6) \times 10^{25}$ [c]	0.10%	
24.3/09	CS	J(5-4)	$(3.3 \pm 0.8) \times 10^{25}$ [c]		
20.3/09	H_2S	1_{10}-1_{01}	$<1.5 \times 10^{26}$	<0.55%	
21.3/09	CO	J(2-1)	$<4.9 \times 10^{27}$	<19%	
19–20/09	H_2CO	3_{12}-2_{11}	$<1.4 \times 10^{26}$ [c]	<0.5%	0.46%
	CH_3OH				1.5%

[a]Assuming a water production rate of 2.6×10^{28} mol s^{-1} (Schleicher, 2001).
[b]From Bockelée-Morvan et al. (1995).
[c]Assuming distributed sources with parent scale lengths of 500 km (CS) and 11000 km (H_2CO).

were successfully obtained from July 19 until September 3, 2001. This monitoring was resumed in November, but the comet was then no longer detectable.

A summary of inferred production rates and abundances relative to water is given in Table I. These rates were calculated as explained in, e.g., Biver et al. (1999) and Crovisier et al. (2002). Molecular abundances inferred at the previous passage in 1994 (Bockelée-Morvan et al., 1995) are also given. The CS abundance in Comet Borrelly fall in the middle of values measured in comets, while the HCN and CH_3OH abundances are rather low, compared to the mean cometary values (Biver et al., 2002). CO and H_2S are underabundant compared to values measured in comets Hale–Bopp and Hyakutake.

The detailed report and analysis of these observations are to be published in *Icarus*.

References

Biver, N., Bockelée-Morvan, D., Crovisier, J., Colom, P., Henry, F., Moreno, R., Paubert, G., Despois, D., and Lis, D. C.: 2002, 'Chemical Diversity among 24 Comets Observed at Radio Wavelengths', these proceedings.

Biver, N., Bockelée-Morvan, D., Crovisier, J., Davies, J. K., Matthews, H. E., Wink, J. E., Rauer, H., Colom, P., Dent, W. R. F., Despois, D., Moreno, R., Paubert, G., Jewitt, D., and Senay, M.: 1999, *Astron. J.* **118**, 1850–1872.
Bockelée-Morvan, D., Biver, N., Colom, P. Crovisier, J., Gérard, E., Jorda, L., Davies, J. K., Dent, W. R. F., Colas, F., Despois, D., Paubert, G., and Lamy, P.: 1995, *BAAS* **27**, 1144.
Crovisier, J., Colom, P., Gérard, E., Bockelée-Morvan, D., and Bourgois, G.: 2002, *Astron. Astrophys.* **393**, 1053–1064.
Schleicher, D.: 2001, *IAUC* 7722.

VISIBLE LIGHTCURVE OBSERVATIONS OF COMET 19P/BORRELLY *

BÉATRICE E. A. MUELLER and NALIN H. SAMARASINHA
*National Optical Astronomy Observatory**, 950 N. Cherry Ave., Tucson AZ 85719*
(E-mail: muller@noao.edu, nalin@noao.edu)

(Received 8 March 2002; Accepted 26 May 2002)

Abstract. The final Deep Space 1 (DS1) mission target, comet 19P/Borrelly, was observed from July 28–August 1, 2000 at the CTIO-1.5 m telescope in the R filter. The observed lightcurve has a large peak to peak variation between 0.84 mag and 1.0 mag. A period of 26.0 $\pm$ 1 hr (assuming a double-peaked lightcurve) was found using all five nights. This is in good agreement with the period of 25.02 $\pm$ 0.5 hr quoted by Lamy et al. (1998) using only 6 points of HST data and is also consistent with HST data taken around the DS1 encounter time by Weaver et al. (2002). Using the mean magnitude R = 20.8 mag and assuming a 4% albedo, we derive an effective nuclear radius of 2.6 km. The large lightcurve amplitude translates to a long to intermediate axial ratio $a/b \geq 2.2$, in excellent agreement with the HST result of $a/b \geq 2.4$ (Lamy et al., 1998) and with DS1 images (Soderblom et al., 2002).

Keywords: 19P/Borrelly, comets, photometry, rotation

1. Introduction

19P/Borrelly was observed as the final target by NASA's Deep Space 1 (DS1) mission on September 22, 2001. A highly elongated and active nucleus was revealed from the spacecraft data (Soderblom et al., 2002). HST data taken around the encounter time (Weaver et al., 2002) were compatible with a long rotational period of about 25 hr indicated by earlier HST data (Lamy et al., 1998). Borrelly has been observed since the beginning of last century (Sekanina, 1979, and references therein) but no rotational period is available from those apparitions.

We discuss our visible observations in Section 2, the lightcurve analysis in Section 3, and summarize our conclusions in Section 4.

2. Observations and Data Analysis

Borrelly was observed for five nights from July 28 until August 1, 2000 at the CTIO-1.5m telescope at $f/7.5$ with the Tek2K which gives a plate scale of $0\farcs4334$/pix. The heliocentric distance r, geocentric distance Δ, the phase angle α,

* based on observations taken at CTIO.

** The National Optical Astronomy Observatory is operated by the Association of Universities for Research in Astronomy, Inc. (AURA) under cooperative agreement with the National Science Foundation.

TABLE I
Geometrical and observational circumstances for 19P/Borrelly in 2000

Date	r [AU]	Δ [AU]	α [°]	Sky condition
July 28, 2000	3.81	2.902	7.9	Not photometric
July 29, 2000	3.80	2.902	8.1	Partly photometric
July 30, 2000	3.80	2.901	8.3	Partly photometric
July 31, 2000	3.79	2.902	8.4	Photometric
August 1, 2000	3.79	2.902	8.6	Partly photometric

and the sky conditions are given in Table I. The standard broadband R filter, which is very close to the Kron-Cousins R, was used for all observations. The integration time was limited as not to trail the comet more than the seeing disk. The images were reduced with the IRAF package. Each image was bias subtracted and flat fielded with combined twilight flats. For the photometry we used a 47 pix (3.″0) digital aperture for the object and a 15 pix (6.″5) aperture for comparison stars in the same frames and also for the Landolt stars (SA 112) used for absolute calibration. We only used one filter for the reduction to absolute photometric magnitudes. According to Suntzeff et al. (1999) the color term in R is only 0.012 (V-R) mag, which makes the error in the absolute calibration by not including the color term much less than our photometric error. An aperture correction was applied to the cometary magnitudes to account for the different sized apertures used for the comet and the comparison stars. The comet moved slowly enough to have common stars in frames of adjacent days. The object was measured with respect to at least five stars in the same frame. Because of overlapping stars between nights, the relative photometry of the comet could be tied to one reference night (night four in this case), reducing the magnitude error in the lightcurve between the nights. The absolute calibration was done for night four.

3. Lightcurve Analysis

The absolute magnitude R(1,1,0) (the observed R-magnitude corrected to heliocentric and geocentric distances of 1 AU and to 0 deg phase angle using an assumed phase correction of 0.035 mag/deg) and photometric error are given in Table II. The time in Table II refers to the time at the middle of the integration.

We used a Fourier transformation followed by a WindowCLEAN algorithm (Belton and Gandhi, 1988) to extract the periodicities. This technique has been successfully used to extract periods from comets and asteroids in non-principal axis spin states like 4179 Toutatis (Mueller et al., 2002), 29P/Schwassmann–Wachmann 1 (Meech et al., 1993), and 1P/Halley (Belton et al., 1991), and for

TABLE II

Photometry for 19P/Borrelly, July 28 - August 1, 2000.

JD-2450000	R(1,1,0)	Error	JD-2450000	R(1,1,0)	Error
1753.5363	14.74	0.11	1753.7558	15.38	0.15
1753.5400	14.66	0.10	1753.7586	15.36	0.15
1753.5731	15.10	0.14	1753.7615	15.40	0.16
1753.5809	14.78	0.10			
1753.5855	15.15	0.15	1754.5310	14.97	0.13
1753.5882	14.84	0.10	1754.5364	14.91	0.11
1753.5909	15.00	0.12	1754.5391	14.87	0.12
1753.5936	14.88	0.11	1754.5531	14.96	0.11
1753.5962	15.07	0.12	1754.5561	15.03	0.10
1753.6083	14.62	0.08	1754.5592	14.90	0.09
1753.6229	14.84	0.10	1754.5622	15.04	0.11
1753.6264	14.80	0.10	1754.5652	15.04	0.11
1753.6292	14.92	0.11	1754.6068	15.12	0.11
1753.6321	15.09	0.12	1754.6094	14.85	0.09
1753.6350	14.93	0.10	1754.6120	15.04	0.12
1753.6378	14.94	0.10	1754.6147	14.82	0.09
1753.6427	15.18	0.15	1754.6173	14.71	0.08
1753.6455	14.96	0.12	1754.6358	14.86	0.08
1753.6484	15.03	0.12	1754.6389	14.88	0.09
1753.6512	14.91	0.10	1754.6420	14.86	0.08
1753.6541	14.79	0.08	1754.7052	14.86	0.08
1753.6580	14.96	0.10	1754.7083	14.92	0.08
1753.6609	15.10	0.12	1754.7114	14.91	0.08
1753.6637	14.91	0.09	1754.7145	15.12	0.10
1753.7004	15.16	0.13	1754.7382	15.15	0.10
1753.7048	15.07	0.11	1754.7413	15.07	0.10
1753.7076	15.08	0.12	1754.7444	15.25	0.12
1753.7219	15.56	0.16	1754.7475	15.01	0.11
1753.7248	15.29	0.13	1754.7506	15.11	0.12
1753.7276	15.16	0.11	1754.7548	15.04	0.11
1753.7305	15.25	0.13	1754.7578	14.76	0.10
1753.7358	15.92	0.25	1754.7687	15.12	0.12
1753.7386	15.27	0.13	1754.7718	14.94	0.09
1753.7415	15.26	0.14	1754.7749	14.91	0.09
1753.7443	15.37	0.15	1754.7900	15.30	0.13
1753.7472	15.50	0.17	1754.7931	15.45	0.14
1753.7501	15.36	0.14	1754.7962	15.20	0.13
1753.7529	15.35	0.15	1754.7993	15.25	0.13

TABLE II
Continued.

JD-2450000	R(1,1,0)	Error	JD-2450000	R(1,1,0)	Error
1754.8317	15.01	0.11	1755.7923	15.12	0.12
1754.8348	15.44	0.18	1755.7954	15.16	0.13
1754.8379	15.23	0.15	1755.7986	14.92	0.11
1754.8410	15.49	0.18			
1754.8441	15.36	0.16	1756.5206	15.77	0.22
			1756.5243	16.20	0.31
1755.5400	14.76	0.07	1756.5279	15.99	0.25
1755.5436	14.96	0.08	1756.5315	15.91	0.23
1755.5473	15.34	0.12	1756.5352	15.43	0.15
1755.5509	15.19	0.10	1756.5399	15.59	0.17
1755.5559	14.92	0.08	1756.5435	15.50	0.15
1755.5599	15.23	0.11	1756.5471	15.38	0.14
1755.5635	15.35	0.12	1756.5508	15.62	0.17
1755.5671	15.60	0.15	1756.5544	15.73	0.21
1755.5707	15.44	0.12	1756.5587	15.43	0.15
1755.5871	15.10	0.09	1756.5623	15.70	0.18
1755.5914	15.24	0.10	1756.5659	15.74	0.18
1755.6106	15.01	0.08	1756.5695	15.49	0.15
1755.6143	15.17	0.10	1756.5732	15.92	0.22
1755.6179	15.07	0.09	1756.6082	15.55	0.16
1755.6215	15.28	0.10	1756.6134	15.61	0.21
1755.6252	15.18	0.09	1756.6207	15.03	0.11
1755.6502	14.80	0.07	1756.6233	15.17	0.12
1755.6534	14.70	0.06	1756.6259	15.17	0.12
1755.6565	14.62	0.06	1756.6285	15.24	0.13
1755.6894	14.86	0.08	1756.6317	15.13	0.11
1755.6926	14.86	0.08	1756.6346	15.53	0.17
1755.6957	15.05	0.09	1756.6376	15.32	0.14
1755.6989	15.06	0.09	1756.6487	15.28	0.14
1755.7021	15.05	0.10	1756.6518	15.47	0.17
1755.7311	14.76	0.08	1756.6960	14.99	0.11
1755.7340	14.86	0.09	1756.6987	14.85	0.10
1755.7369	14.79	0.08	1756.7012	15.28	0.16
1755.7397	14.78	0.08	1756.7038	15.09	0.14
1755.7426	14.78	0.08	1756.7064	15.36	0.17
1755.7459	14.86	0.08	1756.7100	15.23	0.14
1755.7487	14.80	0.09	1756.7558	14.92	0.09
1755.7891	14.96	0.11	1756.7589	14.84	0.09

TABLE II
Continued.

JD-2450000	R(1,1,0)	Error	JD-2450000	R(1,1,0)	Error
1756.7621	14.95	0.10	1757.5977	15.31	0.13
1756.7653	14.83	0.09	1757.6011	15.95	0.24
1756.7694	14.82	0.09	1757.6041	15.73	0.22
1756.7730	15.07	0.11	1757.6070	15.37	0.15
1756.7766	15.15	0.12	1757.6129	15.61	0.19
1756.7803	14.90	0.08	1757.6443	15.53	0.17
1756.8001	14.80	0.09	1757.6473	15.58	0.19
1756.8037	14.86	0.09	1757.6502	15.88	0.23
1756.8074	14.74	0.08	1757.6531	15.36	0.14
1756.8172	15.17	0.17	1757.6561	15.63	0.18
1756.8201	14.81	0.12	1757.6595	16.01	0.25
1756.8230	15.07	0.15	1757.6653	16.01	0.26
1756.8260	14.97	0.15	1757.6683	15.51	0.16
1756.8289	14.97	0.15	1757.6712	15.61	0.18
1756.8330	14.86	0.11	1757.6758	15.80	0.20
1756.8362	14.82	0.11	1757.6787	15.48	0.17
1756.8394	14.95	0.13	1757.6816	15.25	0.13
1756.8425	14.63	0.09	1757.6846	15.29	0.13
1756.8457	14.93	0.12	1757.6875	15.32	0.14
			1757.7111	15.46	0.14
1757.5293	15.26	0.13	1757.7142	15.34	0.12
1757.5325	15.56	0.18	1757.7174	15.27	0.12
1757.5356	15.31	0.15	1757.7206	15.42	0.14
1757.5388	15.52	0.17	1757.7237	15.10	0.10
1757.5420	15.05	0.12	1757.7496	15.06	0.09
1757.5454	15.37	0.16	1757.7530	14.80	0.07
1757.5486	15.54	0.18	1757.7564	14.96	0.08
1757.5518	15.64	0.19	1757.7598	14.81	0.07
1757.5549	15.22	0.14	1757.7814	14.93	0.10
1757.5581	15.47	0.17	1757.7848	14.87	0.10
1757.5619	15.70	0.21	1757.7882	15.02	0.10
1757.5651	15.30	0.15	1757.7916	14.86	0.09
1757.5682	15.57	0.18	1757.7950	15.05	0.11
1757.5714	15.29	0.15	1757.7999	15.07	0.13
1757.5746	15.44	0.16	1757.8033	14.92	0.11
1757.5860	15.52	0.18	1757.8067	15.01	0.12
1757.5889	15.40	0.14	1757.8135	14.91	0.17
1757.5919	15.57	0.18	1757.8172	15.36	0.25

the discovery of a period change in 10P/Tempel 2 (Mueller and Ferrin, 1996). It uses a Fourier transform to calculate the power spectrum (called dirty spectrum) and a window function. Then, the clean algorithm coupled with the window function (hence WindowCLEAN) is applied to the dirty spectrum to yield the clean spectrum.

Figure 1a shows the dirty and clean spectra, while Figure 1b displays the window and the residual spectra. The WindowCLEAN algorithm cleaned all the frequency peaks out from the dirty spectrum except for one at $f = 1.847$ day^{-1}. Assuming the lightcurve variation is due to the nuclear shape (i.e. assuming a double peaked lightcurve), the corresponding rotational period is $P = 26.0 \pm 1.0$ hr. We used the standard deviation of the Gaussian representation of the frequency peak as a measure of error which is an upper limit. We are confident that the peak in the clean spectrum represents the correct rotational period of Borrelly. There is a peak, however, at lower frequency in the dirty spectrum at 0.853 day^{-1} which was identified as $(f - 1)$ but might also be $f/2$ which would yield a rotational period of 52 hr. The data were phased with 52 hr and the resulting plot is not compatible with a double peaked lightcurve. There is no indication of non-principal axis rotation but it cannot be entirely excluded due to the noisiness of the data. A description of what spin states are compatible with Borrelly is given in Samarasinha and Mueller (these proceedings). Lamy et al. (1998) extracted a period of 25.02 ± 0.5 hr using six HST data points. Our analysis of their data yields a period of 26.5 hr which is consistent with the period derived from our visible lightcurve. A search for periodicities in the HST data taken around the DS1 encounter time (Weaver et al., 2002) produces a period of 25.9 hr consistent with the periods found by the earlier HST data and our visible lightcurve data.

In Figure 2, a phase plot with a period of 26.0 hr is shown. The comet was very faint during the observations in July 2000 and the data are therefore noisy. We avoided error bars for clarity. Despite the scatter in the data due to noise, it is clear that the data can be phased well with a period of 26 hr.

The mean observed magnitude in the lightcurve (cf. Figure 2) is R = 20.8 mag which gives an effective nuclear radius of 2.6 km for an assumed albedo of 4%. The peak to peak variation is between 0.84 mag and 1 mag, depending on where exactly the minima and maxima are placed in magnitude. These variations give an axial ratio of $a/b \geq 2.2$ and $a/b \geq 2.5$, respectively. This is in good agreement with the ratio $a/b \geq 2.4$ derived by Lamy et al. (1998) from HST observations and with images from the DS1 encounter (Soderblom et al., 2002). The amplitude in the HST data (Lamy et al., 1998) is comparable to the one in our lightcurve data meaning that in both cases we observe (almost) the same illuminated fraction of the nucleus.

All the images in one night were added to increase the S/N for Borrelly to search for any low level coma. No coma could be detected by comparing spatial profiles of the comet and stars of comparable brightness within the errors. This agrees with the fact that the derived nuclear radius is 2.6 km, i.e. that we observed

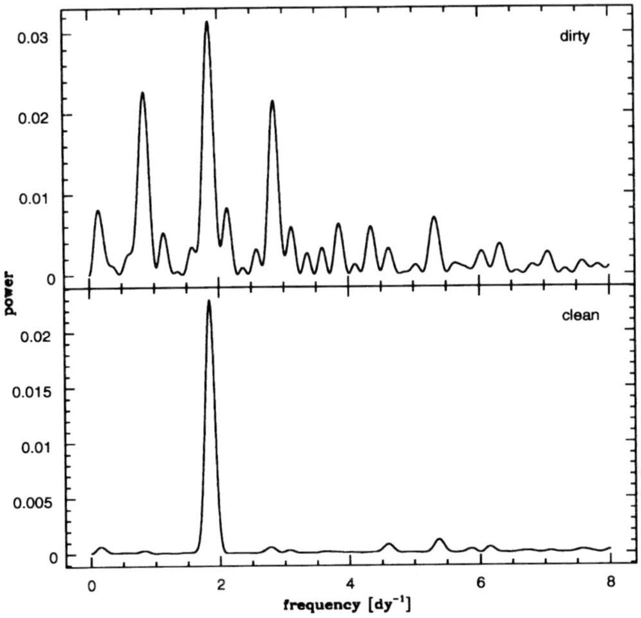

Figure 1a. Dirty (top) and clean (bottom) spectra of 19P/Borrelly.

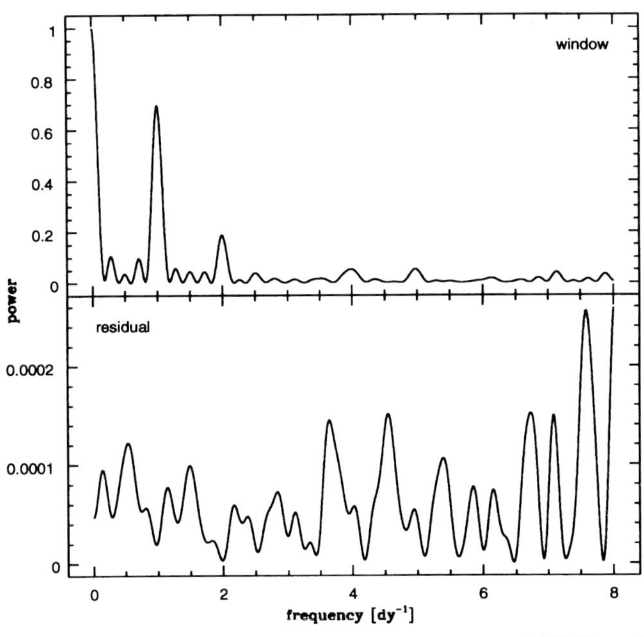

Figure 1b. Window (top) and residual (bottom) spectra of 19P/Borrelly.

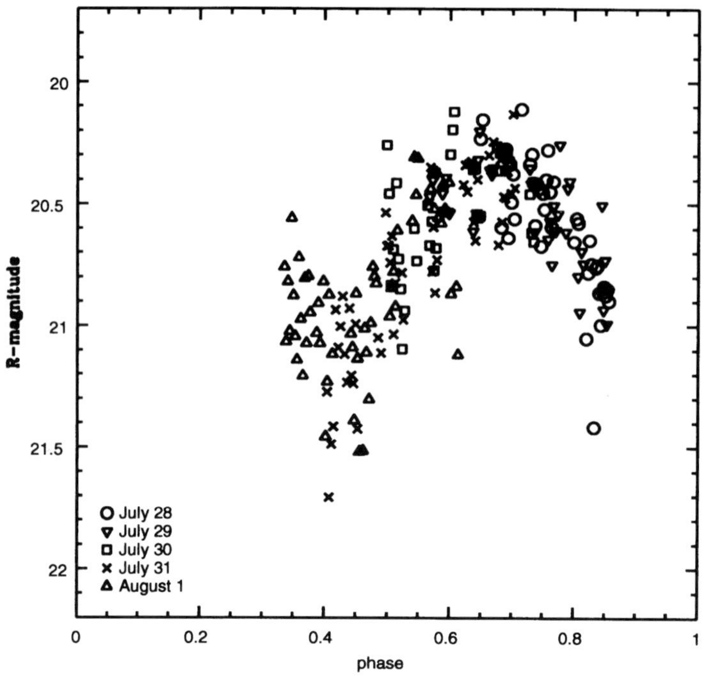

Figure 2. Phase plot with a period of 26.0 hr for the 2000 data of 19P/Borrelly.

an inactive nucleus. However, a contribution of an unresolved coma contributing less than 10% to the total flux would not be observable in our images and therefore cannot be excluded.

4. Discussion and Conclusion

A visible lightcurve of Borrelly was observed with a large peak to peak variation between 0.84 mag and 1.0 mag. Analysis of this lightcurve yields a rotational period of 26.0 ± 1 hr, an effective nuclear radius of 2.6 km (for an assumed albedo of 4%), and an axial ratio of $a/b \geq 2.2$. These numbers are consistent with values found in the literature.

No coma was detected. This is consistent with the derived nucleus size. It is also consistent with the fact that the comet was at heliocentric distance of 3.8 AU and that strong activity has not started yet. A faint or small coma (less than 10% of the total flux) cannot be excluded, however.

Borrelly is another comet with an elongated nucleus with a moderate size. Its rotational period is long but not atypical for Jupiter family comets.

Acknowledgements

We thank the Telescope Assistant at the CTIO-1.5m Roger Leiton and the CTIO support staff. This work was supported by a grant from the NASA Planetary Astronomy program.

References

Belton, M. J. S. and Gandhi, A.: 1988, 'Application of the CLEAN Algorithm to Cometary Light Curves', *Bull. Am. Astron. Soc.* **20**, 836.

Belton, M. J. S., Julian, W. H., Anderson, A. J., and Mueller, B. E. A.: 1991, 'The Spin State and Homogeneity of Comet Halley's Nucleus', *Icarus* **93**, 183–193.

Lamy, P. L., Toth, I., and Weaver, H. A.: 1998, 'Hubble Space Telescope Observations of the Nucleus and Inner Coma of Comet 19P/1904 Y2 (Borrelly)', *Astrophys. J.* **337**, 945–954.

Meech, K. J., Belton, M. J. S., Mueller, B. E. A., Dicksion, M. W., and Li, H. R.: 1993, 'Nucleus Properties of P/Schwassmann–Wachmann 1', *Astron. J.* **106**, 1222–1236.

Mueller, B. E. A. and Ferrin, I.: 1996, 'Change in the Rotational Period of Comet P/Tempel 2 between the 1988 and 1994 Apparitions', *Icarus* **123**, 463–477.

Mueller, B. E. A., Samarasinha, N. S., and Belton, M. J. S.: 2002, 'The Diagnosis of Complex Rotation in the Lightcurve of 4179 Toutatis and Potential Applications to Other Asteroids and Bare Cometary Nuclei', *Icarus* **158**, 305–311.

Samarasinha, N. H. and Mueller, B. E. A.: 2002, 'Spin Axis Direction of Comet 19P/Borrelly Based on Observations from 2000 and 2001', these proceedings.

Sekanina, Z.: 1979, 'Fan-Shaped Coma, Orientation of Rotation Axis, and Surface Structure of a Cometary Nucleus. I. Test of a Model on Four Comets', *Icarus* **37**, 420–442.

Soderblom, L. A., Becker, T. L., Bennett, G., Boice, D. C., Britt, D. T., Brown, R. H., Buratti, B. J., Isbell, C., Giese, B., Hare, T., Hicks, M. D., Howington-Fraus, E., Kirk, R. L., Lee, M., Nelson, R. M., Oberst, J., Owen, T. C., Rayman, M. D., Sandel, B. R., Stern, S. A., Thomas, N. and Yelle, R. V.: 2002, 'Observations of Comet 19P/Borrelly by the Miniature Integrated Camera and Spectrometer aboard Deep Space 1', *Science* **296**, 1087–1091.

Suntzeff, N. B., Phillips, M. M., Covarrubias, R., Navarrete, M., Pérez, J. J., Guerra, A., Acevedo, M. T., Doyle, L. R., Harrison, T., Kane, S., Long, K. S., Maza, J., Miller, S., Piatti, A. E., Clariá, J. J., Ahumada, A. V., Pritzel, B., and Winkler, P. F.: 1999, 'Optical Light Curve of the Type Ia Supernova 1998bu in M96 and the Supernova Calibration of the Hubble Constant', *Astron. J.* **117**, 1175–1184.

Weaver, H. A., Stern, S. A., and Parker, J. Wm.: 2002, 'HST/STIS Observations of 19P/Borrelly during the Deep Space 1 Encounter', *Astron. J.*, in revision.

SPIN AXIS DIRECTION OF COMET 19P/BORRELLY BASED ON OBSERVATIONS FROM 2000 AND 2001

NALIN H. SAMARASINHA and BÉATRICE E. A. MUELLER
National Optical Astronomy Observatory,★ *950 N. Cherry Ave., Tucson, AZ 85719, USA*
E-mail: nalin@noao.edu,muller@noao.edu

(Received 8 March 2002; Accepted 26 May 2002)

Abstract. We calculate the direction of the rotational angular momentum vector, **M**, of comet 19P/Borrelly based on rotational lightcurve data from 2000, groundbased imaging of the coma during the Deep Space 1 encounter, and the basic near-nucleus coma morphology as revealed by the Deep Space 1 spacecraft. For the most likely direction, we derive a family of solutions (with center at RA = 221°, Dec = −7°) if the direction of **M** is towards the sunward hemisphere during the Deep Space 1 encounter, whereas if the rotation is of opposite sense, the diametrically opposite family of solutions (with center at RA = 41°, Dec = 7°) would result. We argue that the coma morphology in September 2001 is consistent with the nucleus being a principal axis rotator or one observationally indistinguishable from it. Therefore, for all practical purposes, the direction of the rotational angular momentum vector coincides with the spin axis. We also discuss why the determination of the spin axis direction based on observations from the last apparition is in disagreement with the current result.

Keywords: Comet 19P/Borrelly, rotation

1. Introduction

Jupiter family comet 19P/Borrelly attracted considerable interest as it was the final flyby target of the Deep Space 1 (DS1) spacecraft, a technology validation mission under the NASA New Millennium program. The DS1 encounter with 19P/Borrelly occurred a few days after perihelion on 22 September 2001 (Soderblom et al., 2002). It is the second cometary nucleus to be imaged by a spacecraft after comet 1P/Halley.

We use groundbased observations together with the basic near-nucleus coma morphology of 19P/Borrelly as revealed by the DS1 spacecraft to determine the direction of the spin axis of the nucleus. The next section of this paper describes the groundbased observations we used, while Section 3 is focussed on deriving the direction of the spin axis. Section 4 addresses the likely spin state and a possible reason for the incompatible derivations of the rotation axis based on observations from prior apparitions. The last section summarizes the conclusions.

★ The National Optical Astronomy Observatory is operated by the Association of Universities for Research in Astronomy, Inc., under cooperative agreement with the National Science Foundation.

Earth, Moon and Planets **90:** 473–482, 2002.

2. Observations and Analysis

2.1. Nuclear Lightcurve Observations from 2000

Nuclear lightcurve observations of 19P/Borrelly were taken at the CTIO 1.5 m telescope in broadband R from July 28–August 1, 2000. The reader is referred to Mueller and Samarasinha (2002) for a detailed description of those data and the corresponding analysis. Based on those data, a period of 26.0 $\pm$ 1 h was derived assuming a double-peaked lightcurve. As described in Section 3, we use the brightness variation in that lightcurve (approximately 0.84–1.0 mag) to determine the spin axis of 19P/Borrelly. The lightcurve is consistent with a principal axis rotation around the short axis.

2.2. Groundbased R-band Coma Images from September 2001

We imaged the coma of comet 19P/Borrelly on September 21, 22, and 23 in 2001 bracketing the DS1 encounter of the comet. Observations were carried out at the KPNO 2.1 m telescope in broadband R. Table I lists the geometrical circumstances for these observations. Unfortunately as the comet was a morning object, each night's observations were limited to less than 2 h. Figure 1a represents an azimuthally renormalized image from September 22 (this enhancement technique destroys radial information while it enhances azimuthal variations). Figure 1b shows the same in (ρ, θ) representation, where ρ is the projected radial distance from the nucleus (optocenter) while θ is the azimuth. Remarkably, the strong sunward linear jet at position angle (PA) of 93° $\pm$ 1° shows no detectable change over the three nights. This jet is the brightest coma feature up to about 7 $\times$ 10^4 km from the nucleus after which the tail starts to dominate. The full-width at half-maximum for this jet (after background subtraction) is 35°. We identify this jet with the strong straight jet seen in the DS1 images, which is the same as the eastward jet at a PA of 93° in the near-simultaneous HST images (Weaver et al., 2002). Lack of curvature in this jet over 7 $\times$ 10^4 km, corresponding to dust released over a day or more for sub km s^{-1} grain outflow velocities (i.e., a time interval at least of the order of the 19P/Borrelly rotation period), points to a source region at or near the rotation pole of the nucleus for an assumed principal axis rotator. However, if a broad range of grain velocities are present, even for a jet source region clearly offset from the pole, the curvature of the jet can get smeared and appear straight except when close to the nucleus (cf. effect shown in figure 3 of Samarasinha et al., 1999). As the high-resolution HST images too show no detectable curvature (A. Stern, personal communication), we conclude that the jet is indeed emanating from the pole or from somewhere nearby. This is consistent with the conclusion of the DS1 observations (Soderblom et al., 2002).

TABLE I

Geometrical circumstances for 19P/Borrelly in September 2001

Date	UT range	r_h[a] (AU)	Δ[b] (AU)	α[c] (°)
September 21, 2001	10:48–12:15	1.360	1.479	41.1
September 22, 2001	10:54–12:06	1.361	1.475	41.2
September 23, 2001	10:48–11:09	1.363	1.471	41.3

[a] Heliocentric distance.
[b] Geocentric distance.
[c] Solar phase angle.

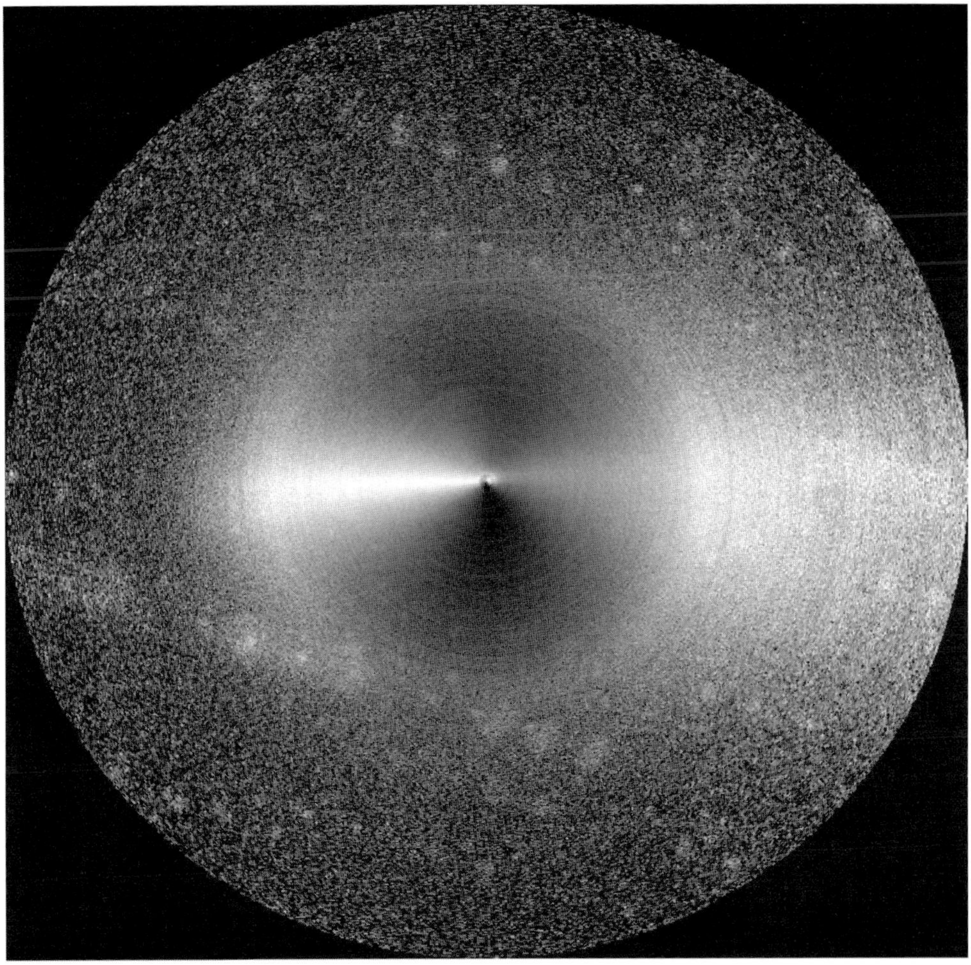

Figure 1a. Azimuthally renormalized 19P/Borrelly image from September 22 with nucleus at the center. Each side is 2.6×10^5 km. North is to the top and east is to the left. The projected solar direction is at PA of 101°. White denotes bright features. After about 7×10^4 km from the nucleus, the dust tail starts to dominate.

Figure 1b. (ρ, θ) representation of the image of Figure 1a. Leftmost azimuth is north and the PA increases to the right. Radial distance increases towards the top and the radial extent is 1.3×10^5 km. The jet (on the left) has a nearly constant width.

3. Constraining the Spin Axis

First, we use nuclear lightcurve observations from 2000 to place initial constraints on the spin axis. We assume (a) scattered light from the comet received at Earth is proportional to the cross sectional area of the nucleus exposed to sunlight (strictly speaking this holds true for Lambertian scattering occurring at zero solar phase angle), (b) the ratio between the long and intermediate axes of the nucleus, a/b, is approximately 2.5 (cf. results from the MICAS instrument aboard DS1 spacecraft, lightcurve from HST (Lamy et al., 1998)), and, (c) the nucleus is in principal axis spin state around the short axis (which we justify later). Then the approximate lightcurve variation of 0.84–1.0 mag corresponds to a spin axis inclined to the line of sight by approximately 60°–85° during the July 2000 lightcurve observations (two lightly shaded strips in Figure 2 correspond to directions in the sky compatible with this condition). We assumed that the phase angle effects on the lightcurve amplitude are negligible as the solar phase angle was near zero degrees (eight degrees).

The directions of the spin axis compatible with the PA of the polar jet ($93° \pm 1°$) seen in groundbased images of September 2001 are shown by the darkly shaded region in Figure 2. Therefore, the intersection of lightly shaded and darkly shaded regions provides the solutions for the directions of the spin axis (as well as that of the strong sunward jet) compatible with the before mentioned observations. It is worthwhile noting that there are two narrow bands of solutions with their centers at (RA = 221°, Dec = −7°) and (RA = 184°, Dec = 7°). Furthermore, as we do not have any evidence as to the sense of rotation, the diametrically opposite solutions with corresponding centers at (RA = 41°, Dec = 7°) and (RA = 4°, Dec = −7°) are also possible for the spin axis direction. We stress that the entire family of solutions are equally acceptable and mid points are provided only for convenience. Independent determinations by Soderblom et al. (2002), Schleicher et al. (2002),

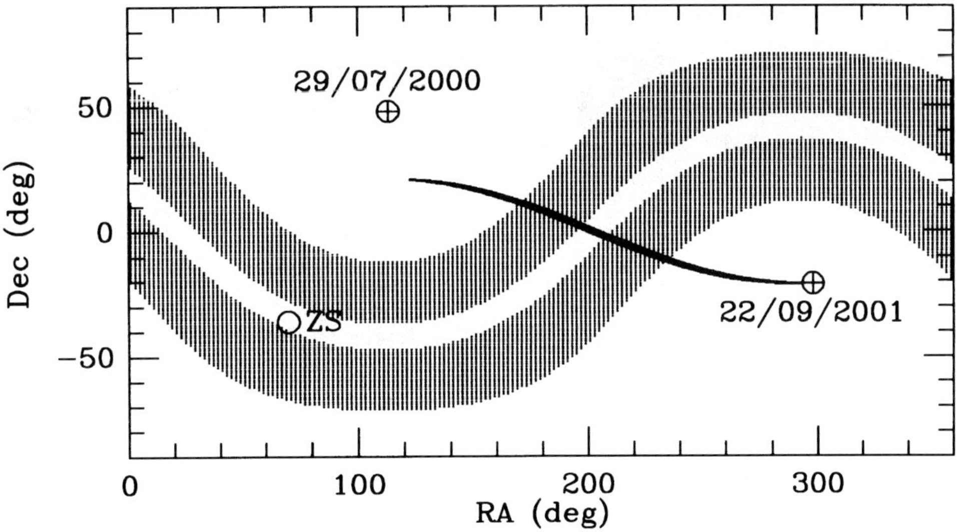

Figure 2. Spin axis direction of the nucleus of comet 19P/Borrelly. Earth symbol denotes the sub-Earth directions with respect to the comet for the dates listed. Intersection of lightly and darkly shaded regions (and the diametrically opposite directions) provides the solutions for the spin axis direction. The solutions are highly constrained in a direction which approximately aligns with Declinations but poorly constrained in the direction normal to it. ZS denotes the direction proposed by Sekanina (1979).

Farnham and Cochran (2002), Thomas et al. (2001), and Ho et al. (2002) based on different sets of observations point to directions of the spin axis consistent with the first value (i.e., RA = 221°, Dec = −7°) or that diametrically opposite of it. Interestingly, these spin axis directions are in the neighbourhood of asymptotic directions predicted for the long-term evolution of the rotational angular momentum vector due to torques induced by sublimation (e.g., Samarasinha, 1997). We urge observers to carefully perform enhancements of comet 19P/Borrelly images taken at different epochs to detect any coma feature with curvature which may be useful in eliminating the ambiguity due to the sense of rotation.

4. Discussion

4.1. Spin state of comet 19P/Borrelly

For calculating the spin axis direction, we assumed that the nucleus of 19P/Borrelly is in principal axis rotation around the short axis. Using the characteristics of the strong sunward jet of 19P/Borrelly, we show that indeed the nucleus is at this rotation state or at a slightly excited state. We argue that the latter is observationally indistinguishable from the former. Figure 3 depicts different spin states of a nucleus and their primary observationally identifiable characteristics. Depending

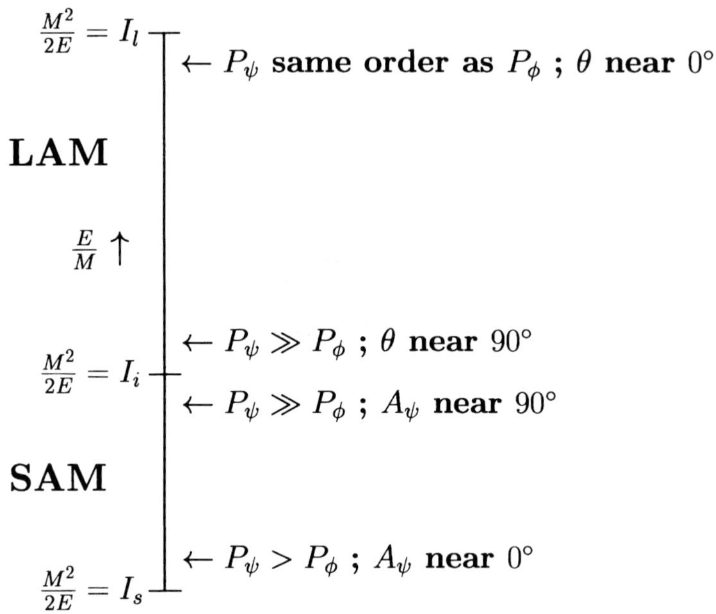

Figure 3. Basic characteristics of various spin states.

on whether the short axis or the long axis moves about the rotational angular momentum vector, **M**, rotation can be categorized into two modes. They are called Short Axis Mode (SAM) or Long Axis Mode (LAM), respectively (Julian, 1987; Samarasinha and A'Hearn, 1991). In Figure 3, M stands for the rotational angular momentum, and E for the rotational kinetic energy, while I_l, I_i, and I_s represent the moments of inertia around long, intermediate, and short axes, respectively. P_ψ denotes the rotation or oscillatory period (depending on LAM or SAM) around the long axis, whereas P_ϕ stands for the mean precession period of the long axis around **M**. θ is the angle between the long axis and **M**, while A_ψ is the amplitude of the oscillatory motion of the long axis for SAM. The angle θ is 0° for the principal axis rotation around the long axis while it is 90° for the principal axis rotation around the short axis. The following is an analysis of the spin states compatible with observations for 19P/Borrelly.

Principal axis rotation around the long axis ($\frac{M^2}{2E} = I_l$): As the DS1 images show that the sunward linear jet is emanating from a direction nearly normal to the long axis, the nucleus could certainly not be in this highest energy state. Otherwise, the sweeping motion of the jet would result in a curvature representative of the rotation rate.

Non-principal axis rotation near $\frac{M^2}{2E} = I_l$: Since θ is near 0°, the character of the coma morphology will be similar to the previous case. Therefore, we rule out these states.

Non-principal axis rotations near $\frac{M^2}{2E} = I_i$: For these states, $P_\psi \gg P_\phi$ and for coma morphology, effects due to precession of the long axis dominate. As the groundbased images do not indicate any signature of curvature nor changes in PA of the jet over a few days (i.e., there is no indication of the rotation/oscillatory motion of the long axis) we conclude that these states are unlikely.

Non-principal axis rotations near $\frac{M^2}{2E} = I_s$: Since A_ψ is near 0°, coma morphology due to a jet will be indistinguishable from that due to a nucleus undergoing principal axis rotation around the short axis.* In other words, a slightly excited non-principal axis rotation state cannot be excluded based on coma morphology. Furthermore, such a state cannot be identified from a groundbased lightcurve either.

Therefore, a nucleus undergoing principal axis rotation around the short axis or one slightly excited from this state together with a source region at or nearby to the pole can reproduce the strong sunward jet of comet 19P/Borrelly. Coma simulations indicate that the proposed spin axis direction together with a polar jet is consistent with the groundbased observations during the DS1 encounter.

4.2. Prior Determination of the Spin Axis Direction

There are derivations of the spin axis direction of 19P/Borrelly based on observations from prior apparitions. Fulle et al. (1997) follow the coma structure of comet 19P/Borrelly during its 1994 perihelion passage and in early 1995 to derive the direction of the spin axis. When the comet was near perihelion in late 1994, there is a clear sunward structure similar to the sunward jet of September 2001. Fulle et al. (1997) used the PA of this structure and its evolution over time in their derivation of the spin axis direction. They find that a single spin axis direction cannot explain the evolution of the PA, and therefore they adopt a precession model and derive a precession period of 2.5 yr for the spin axis.

As the orbital period of 19P/Borrelly is 6.87 yr (i.e., close to an integer number) between the 1994 and 2001 perihelion passages, the observing geometries from the Earth during the 1994 and 2001 apparitions are qualitatively similar. Also, the same source region causing the strong sunward jet of September 2001 should be exposed to constant sunlight around the 1994 perihelion passage (assuming there was no significant change in the direction of the angular momentum vector during the two apparitions – a reasonable assumption for a nucleus as large as 19P/Borrelly under the corresponding outgassing torques due to the polar jet). Table II lists the observing circumstances for the Fulle et al. (1997) data for our spin axis direction of (RA = 221°, Dec = −7°). The penultimate column of the table lists the PAs of the projected **M** whereas the PAs of the sunward structure as measured by Fulle et al. (1997) are given in the last column. The least agreement between these two columns can be found for the last few entries in Table II. Note that the sun–comet–

* There is a another rotation component to the spin state represented by a nodding motion of the long axis; however as the amplitude of that component is even smaller than A_ψ, it is not considered in the discussion.

angular momentum angle is increasing with time. Especially during February and March 1995, the angular momentum vector, and therefore the source region for the strong sunward jet, is near or below the terminator. If the spin axis direction is near (RA = 184°, Dec = 7°), the strong sunward jet would turn-off even earlier than February 1995. Therefore, we argue that the evolving structure of Fulle et al. (1997) does not represent the same jet during the entire duration of their observations. We think that the evolution of that structure was caused by contributions from other source regions on the nucleus in addition to the projection effects. Indeed, evidence for other possible jet activity on the nucleus is present in DS1 images (Soderblom et al., 2002).

For the 1994 HST images of Lamy et al. (1998), the PA of the spin axis is 96° for a spin axis direction of (RA = 221°, Dec = −7°). We measure a PA of 98° ± 2° for the sunward structure in 1994 HST image shown in figure 1 of Lamy et al. (1998). Therefore, this PA is in good agreement with the PA in the previous sentence. A similar agreement was also found independently by L. Jorda (personal communication). The 1994 HST images were taken only less than 2 days prior to the third entry in Table II. However, the PA measured by Fulle et al. (1997) for the sunward structure in the corresponding image for the third entry is 102°, leading us to suspect that the errors in the measured PAs quoted in Fulle et al. (1997) are more than 1° or 2°.

Sekanina (1979) derived a spin axis direction from the PAs of the sunward structures quoted in the literature. These data correspond to early twentieth century apparitions of the comet. The Sekanina solution is marked as ZS in Figure 2 for reference. However, this solution is not compatible with the PA of the strong sunward jet seen in September 2001. If the polar source region seen in DS1 images is directly facing the Sekanina spin axis direction, we find that the source region should be in the night side during the DS1 encounter. If we use the diametrically opposite direction, the PA of the polar jet as seen from the Earth during the DS1 encounter should be 46° – in direct conflict with the measured value of 93°.

5. Conclusions

The main conclusions of this study are:

(a) The rotation state of comet 19P/Borrelly is either in the principal axis rotation around the short axis or in a state slightly excited which is observationally indistinguishable from the former. For the principal axis spin around the short axis, the nucleus rotates around the short axis with a period of 26.0 h.

(b) The strong sunward jet seen in groundbased images taken around the DS1 encounter is emanating from a source region at or near the pole.

(c) The most likely spin axis direction for comet 19P/Borrelly is either near (RA = 221°, Dec = −7°) or near the diametrically opposite direction (RA =

TABLE II

Geometrical circumstances for Fulle et al. (1997) data

Year	Date (month)	Day	RA[a] (deg)	Dec[b] (deg)	r_h (AU)	Δ (AU)	PA_S[c] (deg)	ECS[d] (deg)	ECM[e] (deg)	SCM[f] (deg)	$PA_{\mathbf{M}}$[g] (deg)	PA_F[h] (deg)
94	10	21	288	−9	1.37	0.80	99	46	66	21	94	90
94	11	06	298	−18	1.37	0.70	102	44	76	32	93	95
94	11	30	313	−36	1.40	0.62	96	37	88	50	97	102
94	12	01	314	−37	1.41	0.62	95	37	88	51	97	102
94	12	03	315	−38	1.41	0.62	94	36	89	53	98	102
94	12	04	316	−39	1.42	0.62	94	36	89	53	98	105
94	12	07	317	−42	1.43	0.62	91	35	90	55	99	104
94	12	15	321	−48	1.45	0.63	84	33	92	61	102	108
94	12	24	325	−54	1.49	0.65	74	30	92	67	105	109
94	12	27	326	−56	1.51	0.67	70	30	92	69	106	110
95	1	03	328	−60	1.54	0.70	60	29	92	73	108	110
95	1	08	328	−63	1.57	0.73	52	28	91	76	108	110
95	1	09	328	−63	1.57	0.74	50	28	91	77	108	109
95	1	29	324	−68	1.69	0.89	16	27	88	88	105	111
95	2	01	323	−69	1.71	0.91	11	27	88	89	104	110
95	2	03	323	−69	1.72	0.93	8	27	88	90	103	110
95	2	06	322	−69	1.74	0.96	3	27	87	92	102	110
95	3	11	317	−64	1.96	1.33	322	27	86	106	98	120

[a]RA of the sub-Earth direction. [b]Dec of the sub-Earth direction. [c]PA of the projected solar direction. [d]Earth–comet–Sun angle. [e]Earth–comet–angular momentum angle. [f]Sun–comet–angular momentum angle. [g]PA of projected **M**. [h]PA of the evolving sunward structure from Fulle et al. (1997)

41°, Dec = 7°). One of these solutions can be ruled out if the sense of rotation is known.

(d) Prior determination of the spin axis direction based on the previous apparition, illustrates the difficulties and traps that one should be aware of when the observations used consist only of coma morphology.

Acknowledgements

We thank Dr. Tod R. Lauer for assistance during the September 2001 observing run and the KPNO support personnel for their prompt responses. This work was supported by a grant from the NASA Planetary Astronomy program.

References

Farnham, T. L. and Cochran, A. L.: 2002, 'McDonald Observatory Observations of Comet 19P/Borrelly: Placing the Deep Space 1 Observations into a Broader Context', *Icarus*, in press.

Fulle, M., Milani, A., and Pansecchi, L.: 1997, 'Tomography of a Sunward Structure in the Dust Tail of Comet 19P/Borrelly', *Astron. Astrophys.* **321**, 338–342.

Ho, T. M., A'Hearn, M. F., Boice, D. C., Bonev, T., Jockers, K., Meech, K. J., Soderblom, L. A., and Thomas, N.: 2002, 'Comparison of Earth-Based and Deep Space 1 Observations of Comet 19P/Borrelly', in *Abstracts for IAU Coll. 186: Cometary Science after Hale–Bopp*, p. 22.

Julian, W. H.: 1987, 'Free Precession of the Comet Halley Nucleus', *Nature* **326**, 57–58.

Lamy, P. L., Toth, I., and Weaver, H. A.: 1998, 'Hubble Space Telescope Observations of the Nucleus and Inner Coma of Comet 19P/1904 Y2 (Borrelly)', *Astrophys. J.* **337**, 945–954.

Mueller, B. E. A. and Samarasinha, N. H.: 2002, 'Visible Lightcurve Observations of Comet 19P/Borrelly', these proceedings.

Samarasinha, N. H.: 1997, 'Preferred Orientations for the Rotational Angular Momentum Vectors of Periodic Comets', *Bull. Am. Astron. Soc.* **29**, 743.

Samarasinha, N. H. and A'Hearn, M. F.: 1991, 'Observational and Dynamical Constraints on the Rotation of Comet P/Halley', *Icarus* **93**, 194–225.

Samarasinha, N. H., Mueller, B. E. A., and Belton M. J. S.: 1999, 'Coma Morphology and Constraints on the Rotation of Comet Hale–Bopp', *Earth Moon Planets* **77**, 189–198.

Schleicher, D. G., Woodney, L. M., and Millis, R. L.: 2002, 'Comet 19P/Borrelly at Multiple Apparitions: Seasonal Effects in Gas and Dust', *Icarus*, submitted.

Sekanina, Z.: 1979, 'Fan-Shaped Coma, Orientation of Rotation Axis, and Surface Structure of a Cometary Nucleus. I. Test of a Model on Four Comets', *Icarus* **37**, 420–442.

Soderblom, L. A., Becker, T. L., Bennett, G., Boice, D. C., Britt, D. T., Brown, R. H., Buratti, B. J., Isbell, C., Giese, B., Hare, T., Hicks, M. D., Howington-Fraus, E., Kirk, R. L., Lee, M., Nelson, R. M., Oberst, J., Owen, T. C., Rayman, M. D., Sandel, B. R., Stern, S. A., Thomas, N., and Yelle, R. V.: 2002, 'Observations of Comet 19P/Borrelly by the Miniature Integrated Camera and Spectrometer Aboard Deep Space 1', *Science* **296**, 1087–1091.

Thomas, N., A'Hearn, M. F., Boice, D. C., Britt, D. T., Meech, K. J., Sandel, B. R., Soderblom, L. A., and Yelle, R. V.: 2001, 'Jet Morphology in the Inner Coma of Comet 19P/Borrelly Observed by the Deep Space 1 MICAS Imaging System', *Bull. Am. Astron. Soc.* **33**, 1074.

Weaver, H. A., Stern, S. A., and Parker, J. Wm.: 2002, 'HST/STIS Observations of Comet 19P/Borrelly during the Deep Space 1 Encounter', *Astron. J.*, submitted.

DIRECTIONAL DISTRIBUTION OF OBSERVABLE COMETS INDUCED BY A STAR PASSAGE THROUGH THE OORT CLOUD

PIOTR A. DYBCZYŃSKI
Astronomical Observatory, A. Mickiewicz University, Słoneczna 36, 60-286 Poznań, Poland
E-mail: dybol@amu.edu.pl

(Received 28 February 2002; Accepted 3 July 2002)

Abstract. We simulated the passage of a star through the Oort cometary cloud and analyzed the resulting sample of observable long period comets, noting strong asymmetries in the directional distribution of the perihelion points of those comets. We discuss the results previously published by Weissman (1996) for the same case. An explanation is suggested why the isotropic output can be obtained only for a very peculiar case. The "cometary shower" density variation with time is also presented and the time-dependence of the directional distribution is discussed.

Keywords: Comets, Oort cloud, stellar passages

1. Introduction

The results of simulating the passage of a star through the Oort cloud by Dybczyński (2002) and Weissman (1996) show some significant discrepancies. The present paper explains the source of these discrepancies and describes in more detail the time-dependence of several characteristics of the "cometary shower" induced by a single stellar passage through the Oort (1950) cometary cloud.

In his paper (hereafter WP) Weissman (1996) simulated the spherically symmetric cloud of comets according to the distributions found by Duncan et al. (1987) using a sample of 10^7 comets. He discussed an example of a close stellar passage, the effects of which were calculated by means of the classical impulse approximation. To present the results, Weissman used several plots, showing the instantaneous location of comets (in rectangular coordinates) at the moment of the stellar passage, separately for the ejected subsample and for the subsample perturbed to the perihelion distances less than 10 AU, which was the observability threshold assumed in WP.

During our extended investigation of the effects of the stellar and galactic perturbations on the Oort cloud of comets we attempted to repeat the simulation presented in WP, reproduce results published there and then investigate in detail the obtained observable subsample of comets to verify Weissman's conclusion, i.e., that there are no anisotropies in its distributions.

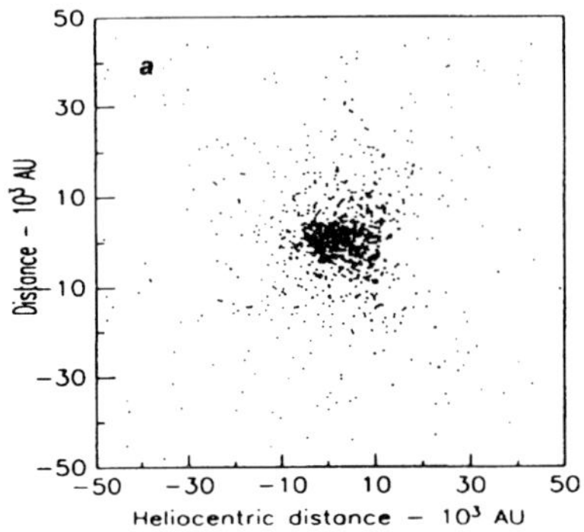

Figure 1. Original Weissman's plots, published in WP as Figures 2c and 2d (with kind permission from Kluwer Academic Publishers). Both axes describe the heliocentric distance of a comet in two different projections: (a) Onto the XY plane, (b) onto the YZ one (see text).

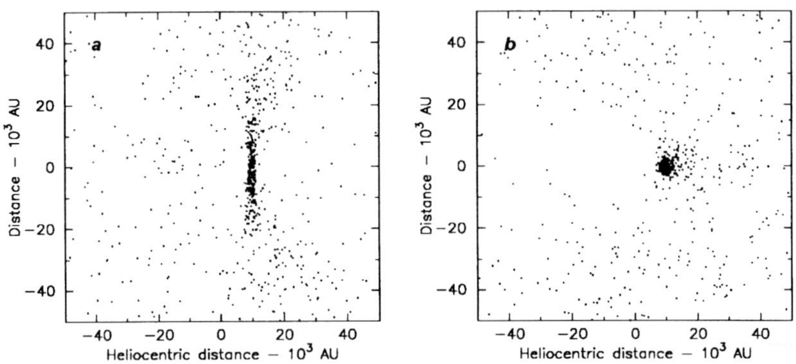

Figure 2. The positions of comets perturbed to $q < 10$ AU, obtained from our simulation for IB = 50 AU, with the same scale and projections as in Figure 1.

2. Observable Subsample

In WP the results of the close stellar passage are discussed in detail based on the example of one solar mass star passing through the simulated Oort cloud at 10 000 AU from the Sun with a velocity of 30 km/s. In his Figures 2a and 2b, Weissman showed the positions of comets ejected out of the cloud by the perturbation from the passing star in two different projections (on XY and YZ planes respectively). The coordinate system is oriented such that the stellar straight line path is parallel to the OX axis and the XY plane contains both the stellar path and the Sun. These plots are not very sensitive to the parameter values and can be easily reproduced.

A problem was encountered when we tried to repeat the results presented as Figures 2c and 2d in WP (reproduced here as Figure 1). *Using the same model and method as applied in WP we could not obtain plots even similar to those in WP.* We used the distribution of semimajor axis and the eccentricity from Duncan et al. (1987) and assumed spherical symmetry for the cloud and a uniform distribution of the mean anomaly. All these parameters are identical with those used in WP, except of probably negligible differences coming from the numerical reconstruction of the semimajor axis and eccentricity distributions on the basis of the plots presented in Duncan et al. (1987). We used the classical impulse approximation and an inner boundary (IB) of the simulated cloud equal to 50 AU whereas the value for the outer boundary (OB) was 2×10^5 AU, the same as in WP. The resulting distributions are presented in Figure 2.

Weissman stated in his paper, that comets observable as a result of the close stellar passage come to the solar vicinity from all directions. From Figure 2 it is clear that this is not the case. The anisotropy in the position distribution is clearly visible here and we will discuss the resulting anisotropies in the perihelion direction distribution for this subsample later. Comparing Figure 1 and Figure 2 we asked for the reason for such a drastic difference. The most peculiar feature in Figures 1a and 1b (and one very sensitive for IB changes) is the strong concentration of points, surrounding the position of the Sun. It is impossible to reproduce it using the IB value published in WP. It is almost impossible to obtain an observable cometary orbit, applying the stellar impulse to comets in the close solar vicinity but on Weissman's plots the majority of points are concentrated near the Sun. Varying the value of the inner boundary, we were able to reproduce the plots published in WP when IB was set to 10 AU.

We emphasize that all four parts of WP's Figure 2 can be reproduced perfectly with our code only when the inner boundary value is the same as the observability threshold, in this case 10 AU. Perhaps even more surprising is the sensitivity of the result to the precise value of the IB; even for IB = 10.5 AU the plots obtained are substantially different. The reason for this is that if IB is equal to the observability limit, arbitrarily small perturbations can make an orbit observable. The simulation described in WP seems to have been performed for an inner boundary value of 10 AU (instead of 50 AU as it is stated).

3. Time Dependence of the Directional Distribution of Perihelion Points

The asymmetries are clearly visible in the corrected plots (Figure 2). The use of the rectangular coordinates of comets at the moment of the stellar passage is however not very convenient and it is difficult to compare such statistics with the observed sample of long-period comets. In Figure 3 we therefore present the distribution of the perihelion directions (as points on the celestial sphere in an equi-areal projection) for the observable subsample of comets obtained in the same manner as for Figure 2, but enlarged in number to obtain a better readable plot. Stellar path parameters are shown in the upper left corner. The central part of this figure shows the cometary shower density histogram, which adds the time-dependence. This histogram shows the number of comets passing perihelion over each 20 000 year sub-interval. It clearly shows a significant cometary influx increase in the first

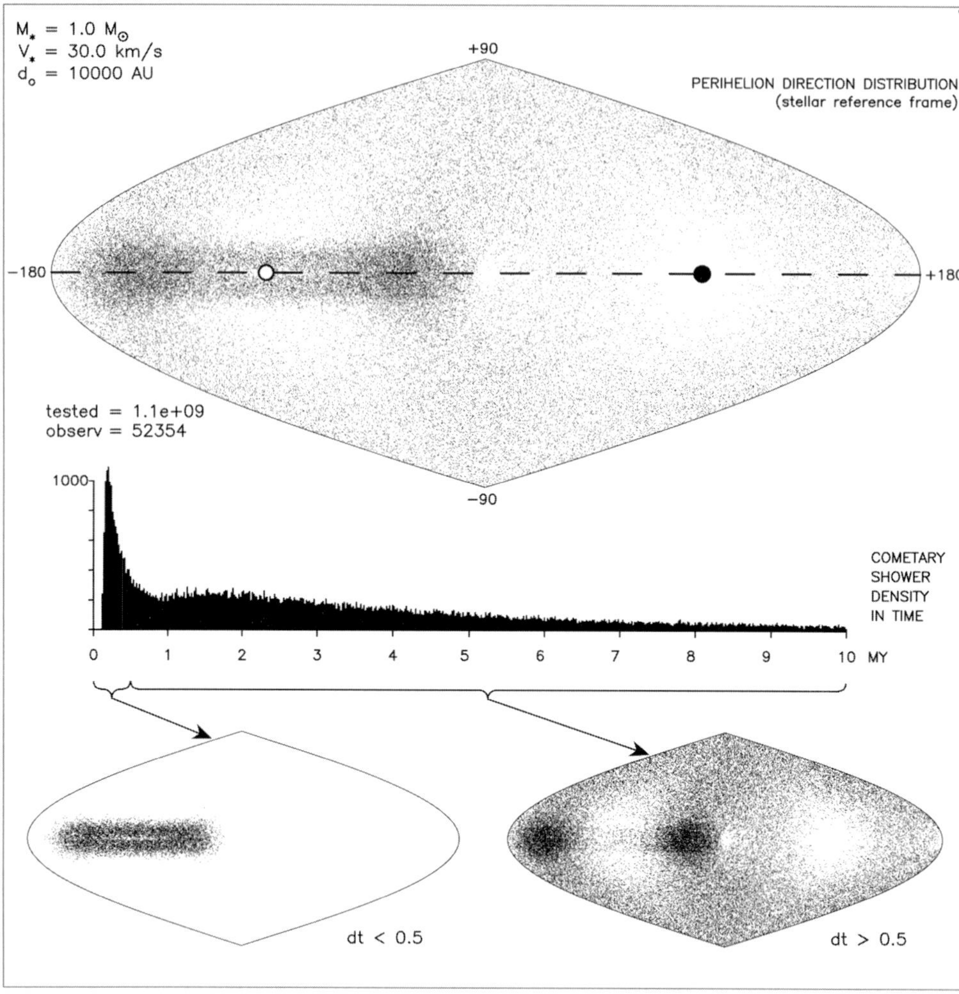

Figure 3. Perihelion points direction distribution.

500 000 years after the stellar passage. For a total number of comets of 1.1×10^9 in the simulated cloud the long-period observable comet influx reaches a number of 1000 observable comets (perihelion distance below 10 AU) per 20 000 years. When scaled to the widely accepted number of comets in the Oort cloud (about two orders of magnitude higher) it gives as much as five "shower" comets per year in the part of highest intensity of the "cometary shower". However, as the number of comets in the cloud is estimated on the basis of the current long-period cometary influx, the numbers presented above cannot be used to judge, whether we are currently experiencing an increased flux of comets owing to a recent stellar perturbation or a steady state one.

The answer may come from the lower part of Figure 3. It consists of two parts extracted from the main, upper plot: the left part presents the perihelion direction points only for those comets which arrive at perihelion no later than 500 000 years after the stellar passage. It should be stressed that the left plot consists of almost 20% of all simulated observable long-period comets obtained from the simulation. The rest of the simulated sample is presented in the lower-right part. From the lower-left part of Figure 3 one may conclude that the observed long-period comet population does not belong to the intense "cometary shower" part, because such a strong perihelion points direction concentration is not observed. However, reliable cometary observations are available only for the last two centuries which makes the "probing" subinterval very short in comparison to the duration of the shower.

From the above cometary influx estimations one should expect no more than about 50 "shower" comets in the observed population, taking into account that the observability threshold value used here (10 AU) is significantly larger than the practical instrumental limitations even a few years ago. More sophisticated studies of the observed long-period comet sample are probably needed to separate possible "shower" comets from the background flux.

The presented example of a stellar passage is by no means the best to produce perihelion direction asymmetries. On the contrary, more distant (and far more likely) stellar passages produce asymmetries that are much more prominent.

4. Conclusions

We presented the results of simulating the output of a close stellar passage through the Oort cometary cloud from the point of view of the directional distribution of the observable comets induced by this passage. We showed that except for a singular case the resulting distribution is significantly anisotropic. The concentrations of perihelion points align with the stellar heliocentric orbit and its anti-perihelion. This finding may be used to search for the recent stellar perturber fingerprints based on the observed long-period comet perihelion distribution irregularities. It should be stressed, however, that the directional characteristic of the "cometary shower" as well as its time dependence strongly depend on the geometry of the star path, mainly on the minimum distance between the star and the Sun. Before any attempts can be made to compare the simulation results with the observed long-period comet population, it is necessary to also include galactic perturbation into the model. A paper covering these aspects of the problem is in preparation.

A more detailed analysis of the presented example of the stellar passage as well as a discussion of the validity of the classical impulse approximation in this study (based on the comparison with the direct numerical integration) and discussion of the various end-state probabilities for the simulated sample of comets may be found in Dybczyński (2002).

Acknowledgements

The author wishes to thank Dr. Paul Weissman for valuable comments during the presentation of this paper at IAU Coll. No. 186. The author is also indebted to an anonymous referee, whose valuable suggestions allowed him to improve the quality and readability of this paper.

References

Duncan, M., Quinn, T., and Tremaine, S.: 1987, *AJ* **94**, 1330–1338.
Dybczyński, P. A.: 2002, *A&A* **383**, 1049–1053.
Oort, J. H.: 1950, *Bull. Astron. Inst. Neth.* **11**, 91–110.
Weissman, P. R.: 1996, *Earth Moon Planets* **72**, 25–30.

MODELING OF THE ORBITAL EVOLUTION OF 2060 CHIRON

NATALIYA S. KOVALENKO, YURY G. BABENKO and KLIM I. CHURYUMOV
Astronomical Observatory of Kyiv Shevchenko National University, Observatorna Str., 3, Kyiv, 04053, Ukraine
E-mail: nkovalenko@observ.univ.kiev.ua

(Received 1 March 2002; Accepted 26 June 2002)

Abstract. The origin of Centaurs is one of the most interesting problems of Solar system science, and it has not yet been solved. To shed light on this problem one can investigate Centaurs' past and future orbital evolution. In this paper we discuss the results of Chiron's orbital evolution modeling. It was the first discovered Centaur and is the brightest one. Numerical integration was produced for 1 Myr forward and backward from the present time. A program based on the Everhart single sequence method for integrating orbits was used.

Keywords: Centaurs, Chiron, numerical integrations, orbital evolution

1. Introduction

In recent years object 2060 Chiron has been a subject of permanent interest. It is the first discovered and the brightest representative of Centaurs – a group of objects that have perihelia and semi-major axis located between the orbits of Jupiter and Neptune. Chiron moves on its orbit with revolution period $P = 50.2$ years, semi-major axis $a = 13.6$ AU, eccentricity $e = 0.38$, and inclination $i = 6.94$. At aphelion Chiron is 18.8 AU away from the Sun, and in perihelion it approaches the Sun at 8.5 AU. 2060 Chiron was discovered by Kowal in 1977.

Cometary activity of Chiron was detected by Tholen in 1988 which appears to be atypical for objects at such large heliocentric distances. Sublimation of water ice as the most common constituent of comets can account for the activity up to 5–6 AU. At larger distances cometary nucleus activity is controlled by sublimation of volatile ices such as CH_4, CO, CO_2, N_2, and CN. However, the question arises whether the activity at large heliocentric distances is driven by real-time sublimation of surface volatiles, by sublimation delay (due to slowly propagating thermal waves inside the nucleus), or by non-thermal processes.

The detection of CN emission from Chiron was made by Bus et al. (1991) and that of CO by Womack and Stern (1995). Water ice in Chiron has also been detected (Luu et al., 2000). Nevertheless, the mechanism that drives Chiron's activity is still in debate.

Continued activity in Chiron implies that it has been in its present orbit for a very long time. The origin of Chiron and other Centaurs is one of the intriguing

Earth, Moon and Planets **90:** 489–494, 2002.

problems of Solar System physics, and it has not yet been solved. Transformation into short-period comets is even regarded as the future dynamical evolution for Centaurs. The Oort cloud and the Kuiper belt have been proposed as the sources of short-period comets. It is estimated that approximately 10^5 objects with diameters of 100–300 km exist in the Kuiper belt. The first Kuiper belt object was discovered by Jewitt and Luu in 1992, and many others have been discovered since.

The diameter of Chiron was estimated to be at least 140 km, so it is one order of magnitude larger than the nuclei of typical comets. The similarity in diameters of Chiron and discovered Kuiper belt objects makes the Kuiper belt a likely source of Chiron and other Centaurs. On this basis, the integration of Chiron's orbit evolution is of great interest as a way to constrain Chiron's origin in the Kuiper belt. In this paper we will discuss the results of the long-term orbital evolution of the Centaur object 2060 Chiron that possibly originated from the Kuiper belt.

2. Chiron's Dynamical Behavior

The investigation of Chiron's dynamical evolution has a long history. Numerical integrations were made for different time intervals both backward and forward (Oikawa and Everhart, 1979; Scholl, 1979; Kowal et al., 1979; Hahn and Bailey, 1990; Nakamura and Yoshikawa, 1993). It was shown that Chiron is presently under "control of Saturn's gravity", and it follows a chaotic (unstable) orbit pattern (Oikawa and Everhart, 1979).

The general conclusion is that the past and future orbit of Chiron remains unknown over a long time scale and numerical integrations are of statistical significance only. Chiron's dynamical life time in its present orbit is considered to be approximately 1–10 million years. Marsden (Kowal et al., 1979) remarked that predictibility of the orbit is lost if the integration is extended past any close encounter with Saturn. Slightly different starting values may yield a completely different orbit (Scholl, 1979).

Evolution of Chiron's orbit both forward and backward in time are of similar character, although some asymmetry in time can be observed. Two evolutionary patterns can possibly be introduced: A considerable increase of semi-major axis a without any change of perihelion distance q as the first pattern; and a decrease of q which can move Chiron into Jupiter's gravitational domain with a possible subsequent decrease of the semi-major axis and of the perihelion distance q (to 1–2 AU). Furthermore, a considerable increase of the semi-major axis is also possible.

Since such an evolution of Chiron's orbit, in accordance with the numerical integration, is possible in the past as well as in the future, possibilities in the past are of great interest to us. Chiron's cometary activity can be chosen as the criterion for solving this problem. Its activity at large heliocentric distances implies the existence of volatiles with low sublimation temperatures on Chiron's surface. In fact, if Chiron had been in its present orbit of small q for some longer period of time, volatiles that were introduced as a possible explanation of Chiron's activity, would have vaporized completely. Thus, we conclude that Chiron is not in its present orbit with small q for a long period of time.

It is also of interest to us to find out whether one can get an idea about past changes in Chiron's orbit, and to trace its origin by integrating its present orbit in the past. There are two options to be considered: the first one is when q has changed from more than 18 AU to its present value during one rotational period, and the second one is when q was changing gradually in the past so that the main gravitational influence on Chiron was transferred, say, from Neptune to Saturn.

Since in the course of numerical integration of Chiron's orbital evolution backward in time both the probability of the second evolutionary path and the occurrence of the first option are quite low, we conclude that numerical modeling of Chiron's orbit both backward and forward in time describes, in fact, only future orbital evolution and does not have much significance for the origin of the object.

3. Numerical Integration of Chiron's Orbital Evolution

The numerical integration of Chiron's orbital evolution was performed over 1 million years backward and forward from the present time. The implicit single sequence methods for integrating orbits proposed by Everhart were used. The step-size of integration is variable. The relative truncation error was set to 1×10^{-12}. Starting orbital elements were taken from Ephemerids of Minor Planets for 2000. The program is based on the seven-body problem solution (with planets from Jupiter through Pluto). It enables simulating orbital evolution, calculating changes in orbital elements with time, finding the minimum distance between Chiron and giant planets during each period of revolution, and finding the correlation between close approaches to planets and orbital evolution.

Thirteen numerical integrations with slightly different, initial orbital elements equal to those of Chiron (± 1 in the last decimal position) were carried out. As a result, plots of long-term evolution of Chiron's orbital elements and planetary close approaches were obtained.

Detailed investigation of the orbital history of Chiron has been made earlier by Hahn and Bailey (1990), and Nakamura and Yoshikawa (1993). They inferred Chiron's orbital evolution statistically in terms of probability. In Nakamura and Yoshikawa's calculations over 200 kyr toward the past Chiron's orbit was decreasing from 20–40 AU to its present value in about 75% of the cases. In particular there were 3 cases of direct capture via a single planetary encounter from a = 100–400 AU. During the future 200 kyr about 70% of orbits increased to 20–30 AU from the present value. Hahn and Baily pointed out the past/future asymmetry which could not be so clearly seen in Nakamura and Yoshikawa's calculations. Their evolution of orbit perihelion distance q seems to be more complicated than that of the semi-major axis a. About a half of orbits showed very chaotic vari-

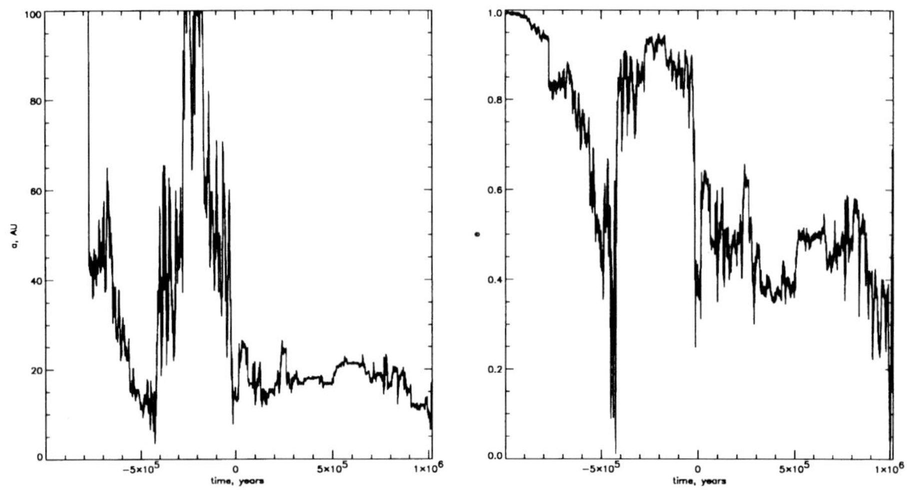

Figure 1. Time evolution of the semi-major axis and eccentricity of Chiron's orbit.

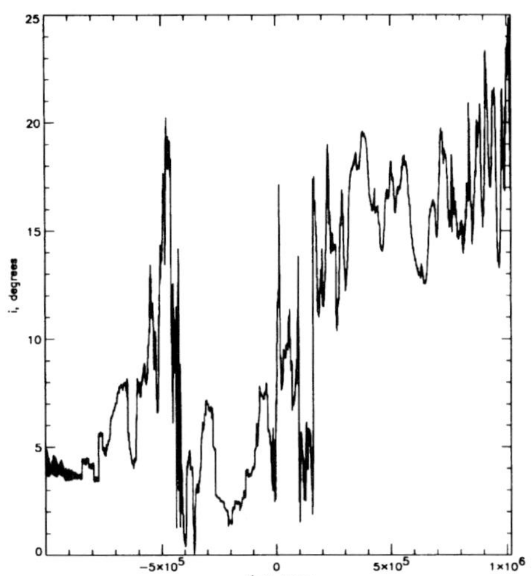

Figure 2. Time evolution of the inclination of Chiron's orbit.

ations between the present value and 1–2 AU, although such captures into orbits of $q = 1$–2 AU are temporary and rarely stable for more than 2–3 kyr.

Our results also show this past/future asymmetry. Figures 1 and 2 show one of the 13 orbit integrations in greater detail: The orbit with starting elements equal to those of Chiron. As we can see from Figure 1 the evolution of the semi-major axis in the past was more "dramatic" than it will be in the future. Two mean periods of capture into smaller orbits are present. Periodic modulations of about 30–40 AU are also found. In the future the behavior of a will be subject to smaller oscillations of about 5–10 AU and in general a will increase slightly compared to its present value. The analysis of all 13 orbit integrations shows that there are 6 cases of capture from orbits with a of more than 50–100 AU into orbits with a in the range of 5–15 AU. In the future there are 4 cases when a will increase to more than 100 AU, 2 cases with almost constant a close to the present value, and 7 integrations with a varying in the range of 5–30 AU. The evolution of the eccentricity in the past is also very chaotic while in the future it is not characterized by such dramatic variations and will remain within a moderate range of 0.3–0.6.

The maximum inclination (Figure 2) in the past slightly exceeds 20° and for about 400 kyr is around its present value. In the future there is a long-term tendency of increasing inclination. 400–750 kyr backward is the period of the most radical changes of a and e in the past. It coincides with the period of close approaches to Saturn and Jupiter as well as Uranus. The future orbit is determined by superposition of gravitational influences of Saturn (mostly) and Uranus.

It is also of great interest to us to trace the evolution of the perihelion distance. In our calculations 4 of 13 past orbit integrations did not show considerable changes in q, 2 cases showed decreasing q values compared the present one, and 7 orbits reached q values of less than 2 AU. Such captures are temporary that agree with Nakamura and Youshikawa's results. 7 of the future orbits show no chaotic variations of q, and q stays continuously in the range of 6–10 AU. In 5 orbits q decreases to 2–4 AU, and in 1 orbit it increases gradually to above 10 AU.

4. Discussion

The results of our orbit integration do not contradict the idea that Chiron may originate from the Kuiper belt. The orbit of Chiron is very chaotic, especially in the past. It evolves much more "quietly" in the future. Improvement of the accuracy of the orbital elements and numerical integration by decreasing the accumulation of the round-off error will allow us to make more confident conclusions about Chiron's orbital evolution only if small time steps are chosen for the integration of orbits. Thus we will be able to compare the new results with those obtained previously. To draw more confident conclusions, it will be useful to analyse the orbital evolution of other Centaurs with likely common origins, and to compare the results within this group of objects.

References

Bus, S. J., A'Hearn, M. F., Schleicher, D. G., and Bowell, E.: 1991, *Science* **251**, 774–777.

Everhart, E.: 1974, *Celest. Mech.* **10**, 35–55.

Hahn, G. and Bailey, M. E.: 1990, *Nature* **348**, 132–136.

Jewitt, D.: 2000, in A. Fitzsimmons, D. Jewitt, and R. M. West (eds.), *ESO Workshop on Minor Bodies in the Outer Solar System, 2–5 November 1998, Garching/Germany*, Springer-Verlag, Berlin, pp. 1–5.

Kowal, C. T., Liller, W., and Marsden, B. G.: 1979, in *Symposium on the Dynamics of the Solar System, 23–26 May 1978, Tokyo, Japan (A79-36276 15-89)*, D. Reidel Publishing Co., Dordrecht, pp. 245–250.

Luu, J., Jewitt, D., and Trujillo, Ch.: 2000, *Astrophys. J.* **531**, L151–L154.

Nakamura, Ts. and Yoshikawa, M.: 1993, *Celest. Mech. Dynam. Astron.* **57**, 113–121.

Oikawa, S. and Everhart, E.: 1979, *Astron. J.* **84**, 134–139.

Scholl, H.: 1979, *Icarus* **40**, 345–349.

Schor, V. A. (ed.): 1999, *Ephemerids of Minor Planets for 2000*, Institute of Applied Astronomy of RAS, S.-Petersburg, 1232.

Sheppard, Sc., Jewitt, D., Trujillo, Ch., Brown, M., and Ashley, M.: 2000, 'A Wide-Field CCD Survey for Centaurs and Kuiper Belt Objects', *Astron. J.* **120**, 2687–2694.

Womack, M. and Stern, S. A.: 1995, *IAUC* 6193.

THE SURFACE OF COMETARY NULCEI RELATED MINOR ICY BODIES

JAVIER LICANDRO and JUAN C. GUERRA
Centro Galileo Galilei & Telescopio Nazionale Galileo, P.O. Box 565, 38700 S/C de La Palma, Tenerife, Spain

HUMBERTO CAMPINS
University of Arizona, Tucson AZ, USA

MARIO DI MARTINO
Osservatorio Astronomico di Torino. I-10025 Pino Torinese, Italy

LUISA M. LARA
Instituto de Astrofísica de Andalucía, CSIC, Camino Bajo de Huétor 24, 18008 Granada, Spain

RICARDO GIL–HUTTON
Observatorio Astronómico Félix Aguilar, Av. Benavidez 8175 (oeste), 5407 Marquesado, San Juan, Argentina

GIAN P. TOZZI
Ossservatorio di Arcetri, Largo E. Fermi 5, I-50125 Firenze, Italy

(Received 8 March 2002; Accepted 25 May 2002)

Since the beginning of 2001 we are conducting a program to study the surface of Jupiter family cometary nuclei and other closely related populations of minor solar system objects like Trans-Neptunian objects (TNOs), Centaurs, and Trojan asteroids. They are probably the most pristine remaining planetesimals formed in the early stages of the protosolar nebula in the giant planets and transneptunian region. The study of their composition and surface evolution is crucial for understanding some cosmogonical aspects. The actual information on their surfaces is scarce but essential for the understanding not only of the formation processes, but also of the links between these populations. Visible, and, in particular, near-infrared spectroscopy provide important data for this kind of studies (Campins and Fernandez, 2002).

We obtained visible and infrared spectra of one comet (28P/Neujmin 1) and ten Trojan asteroids (588 Achilles, 911 Agamemnon, 2260 Neoptolemus, 1143 Odysseus, 1404 Ajax, 1437 Diomedes, 1868 Thersites, 3540 Protesilaos, 3548 Eurybates, and 4086 Podalirius), and infrared spectra of four TNOs (20000 Varuna, 1999 TC36, 2000 EB173, and 2001 KX76) and of one Centaur (1999 UG5).

Near-IR spectra were obtained with the infrared camera-spectrograph NICS installed on the 3.56 m Telescopio Nazionale Galileo (TNG) using its unique high throughput, low resolution spectroscopic mode with an Amici prism disperser (see Baffa et al., 2001). This device allowed us to obtain the first near-infrared spectra

Earth, Moon and Planets **90:** 495–496, 2002.

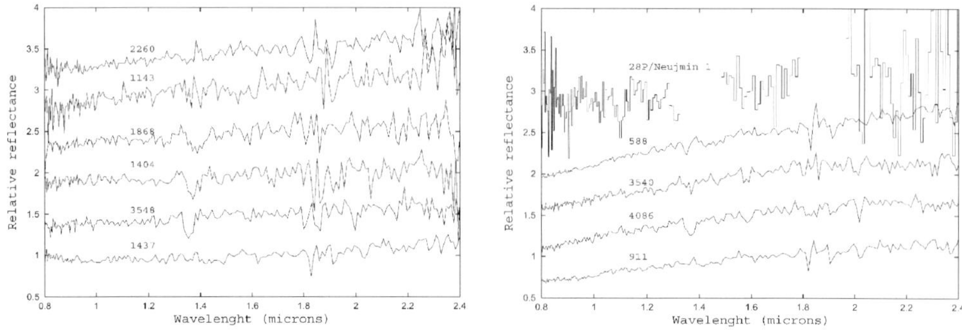

Figure 1. Near infrared spectra of 28P/Neujmin 1 and ten Trojan asteroids obtained in the course of this program. Using a 1″ slit, the resolving power is R $\simeq$ 50 in the whole spectral range. The spectra were obtained and reduced as described in Licandro et al. (2001). Relative reflectance was obtained by dividing the spectra of the objects by the spectra of solar analogues stars observed the same night at similar conditions, and finally normalized to unity at 1.6 μm. The regions around 1.4 and 1.9 μm, not plotted for Comet 28P, corresponds to strong telluric absorptian bands, resulting in much lower S/N as can be seen in the spectra of the Trojan asteroids.

of faint objects like TNOs ever obtained with this 4m class telescope (Licandro et al., 2001). Visible spectra were taken also with the TNG, using the visible camera-spectrograph DOLORES.

In Figure 1 we present, as an example of our results, the near-infrared spectra of Comet 28P/Neujmin 1 and the Trojan asteroids. 28P/Neujmin 1 has possibly the lowest activity of any Jupiter-family comet. At the time of our observations the heliocentric distance was r = 5.4 AU. No signs of coma activity were detected in the visible and infrared images obtained, hence, we believe our spectra are of its bare nucleus. The 28P/Neujmin 1 spectrum is slightly redder than the Sun, with no water ice signatures in the infrared. Trojan asteroid spectra do not show clear compositional features, and present colors from neutral to very red, typical of C, P, and D asteroids. The spectra shown here, together with the spectra obtained in the visible region, will be used to study the composition of the surfaces of these objects.

References

Baffa, C., Comoretto, G., Gennari, S., Lisi, F., Oliva, E., Biliotti, V., Checcucci, A., Gavrioussev, V., Giani, E., Ghinassi, F., Hunt, L. K., Maiolino, R., Mannucci, F., Marcucci, G., Sozzi, M., Stefanini, P., and Testi, L.: 2001, *Astron. Astrophys.* **378**, 722–728.

Campins, H. and Fernández, Y.: 2002, 'Observational Constraints on Surface Characteristics of Comet Nuclei', these proceedings.

Licandro, J., Oliva, E., and Di Martino, M.: 2001, *Astron. Astrophys.* **373**, L29–L32.

ON THE ROLE OF DUST MASS LOSS IN THE EVOLUTION OF COMETS AND DUSTY DISK SYSTEMS

CAREY LISSE
University of Maryland, Department of Astronomy, College Park, MD 20742, USA

(Received 1 March 2002; Accepted 21 June 2002)

Abstract. We present the results of an ongoing optical/thermal infrared photometric imaging survey of the dust mass loss and nuclear size of comets. We find an evolution with time of the kind of dust emitted from a comet's surface. Our results indicate that the mass loss rate from the short period comets alone is enough to supply the interplanetary dust (IPD) complex against losses. We conclude that the nature of the IPD cloud has changed over time, with small particles dominating early, and larger particles dominating in the present era. The rate of destruction of short period (SP) comets due to sublimation mass loss is found to be low compared to the rate of surface mantle formation, but fast compared to the rate of their dynamical removal from the inner solar system.

Keywords: Asteroids, comets, dust, evolution, interplanetary dust, nuclei

1. Introduction

Comets are thought to be the best-preserved relics of the first 100 Myr of the proto-solar nebula, when interstellar medium (ISM) dust particles of size $\sim$0.1 μm merged and accreted to form planetisimals of size $\sim$1 km. Unlike the details of the era of molecular cloud condensation and infall, and the era of the removal of the comets from the giant planet region to the Oort Cloud, the details of the mechanism of accretion are quite unclear. Without a detailed understanding of the mechanism driving this process, it is impossible to understand and predict how planetary cores are built and how they evolve, and what the planetisimal population in exo-solar systems of varying age might be (e.g., the recent reports of comets in the young stellar system β Pictoris by Lagrange et al. (1999), Roberge et al. (2000), and Lecavelier des Etangs et al. (2001); and in the late-type extreme carbon star system IRC+10216 by Melnick et al. 2001).

In this work, we report on an ongoing study of the emitted dust and exposed nuclei of a survey of the brightest near-Earth comets over the last 12 years. Combined optical and thermal infrared observations are analyzed using dynamical, spectral, and coma models (Lisse et al., 1998, 1999) to update the dust emission rate estimates of Kresak and Kresakova (1987), the nucleus size estimates of Jewitt (1991), and the evolutionary studies of Fernandez (1988). We present here our estimates for the dust mass loss rate, nucleus size, and lifetime versus sublimation for the survey comets. Studying the statistical trends in these observations, we find that there is

TABLE I

Dust properties for the comets in the IR survey

Comet	r_h AU	Q_{dust} kg/s	Dust/ gas	$T_c/$ T_{bb}	Dust albedo	Silicate strength
C/OLR 1989 (New)[a]	0.95	450	0.32	1.06	–	<10
C/Austin 1990 (New)[a]	0.94	2,400	0.79	1.08	0.06	~10
C/1999 LINEAR S4 (New)	1.07	6,300	5	0.86	0.15	N/A
C/IAA 1983 (LP) [a]	1.01	>200				
C/Levy 1990 (LP)	1.14	6,100	1.0	1.24	0.12	>30
C/Hyakutake(LP)	1.03	13,000	1.7	1.20	0.13	>50
C/Hale–Bopp(LP)	2.54	140,000	7.0	1.30	0.39	>200
C/2001 A2 (LP?)	1.30	~900	~0.85	1.13	–	
73P/SW-3 (SP)[a]	0.97	~120	>0.12	1.0	–	<20
2P/Encke (SP)[a]	1.17	120	2.3	1.03	0.05	<10
9P/Tempel 1 (SP)[a]	1.49	<390	[b]	1.06	0.08	<10
22P/Kopff (SP)[a]	1.91	1000	[b]	1.12	<0.08	<20
81P/Wild 2 (SP)[a]	1.77	<700	[b]	<1.08	<0.06	<20
55P/Tempel–Tuttle (H)[a]	1.15	210	2.9	1.03	0.05	<20
126P/IRAS (H)	1.71	300	3.3	1.25	0.16	>35
1P/Halley (H)	1.42	15,000	1.9	1.41	0.18?	~50

[a] Large particle dominated coma.
[b] Q_{gas} not well determined for this comet.

more than enough dust emitted from short period (SP) comets into bound solar system orbits to create and support the current interplanetary dust cloud (IPD), that this dust changes as the comet ages, and that the lifetime versus sublimation for the short period comets is much longer than their dynamical lifetime.

2. Observations

Over the last 12 years, we have obtained optical and infrared imaging photometry of the dust coma for a number of comets (Table I). From the photometry, we obtain both directly observable properties, such as the ratio of T_{color}/T_{local}, dust albedo, and silicate emission feature amplitude vs the underlying continuum; and model dependent properties such as the dust particle size distribution, dust composition, and dust mass emission rate.

By modeling the observed optical-infrared coma spectral energy distributions on the 1′ scale (SEDs), the observed coma and tail morphology (for comets with observed extent > 30′), and the long term light curves (for comets with well calibrated photometric observations over many months of time) we derive the particle size distribution (PSD), composition, and amount of dust Q_{dust} emitted from the nucleus into the coma and tail (Lisse et al., 1998). Cognizant of the pitfalls of modeling complexly shaped dust particles formed from agglomerations of small ($\sim$0.1 μm) interstellar dust particles, we have chosen the most phenomenological and least model dependent method possible of analyzing the dust photometry. We take a four step approach to reduce the complexities and narrow the phase space of our solutions:

1. We analyze the qualitative features (amplitude, ratio Afp/efp of reflected and thermal flux from the dust, superheat, silicate feature strength, long wavelength falloff) of the spectral energy distribution (SED) directly, in order to draw simple conclusions without applying any model. The observables $T_{\text{color}}/T_{\text{local}}$ (the ratio of the greybody temperature best-fitting the observed dust spectrophotometry to the equilibrium temperature for a greybody at the comet's heliocentric distance), dust albedo, and silicate emission feature amplitude vs the underlying continuum are minimal for dust distributions dominated by large (radius $>$ 20 μm) particles, irregardless of particle composition; these quantities are largest for dust dominated by small, highly absorbing carbonaceous and silicate particles (cf. Lisse et al., 1998; Harker et al., 1999, 2002).
2. We fit the SED using a Mie model modified for calculation of emission from non-spherical, porous, fractal particles. The efficacy of this model for cometary and interstellar dust composed of mixed silicates and carbonaceous material of fractal dimension $\sim$3 has been shown by Bazell and Dwek (1990). In essence, the 5–100 μm wavelength scales of IR emission photons sample the majority of the structure of a multi-unit fractal dust particle. Modeling the scattering of such at wavelengths $\sim$0.1–1.0 μm is much more problematic, as these wavelengths probe the interstellar grain fractal subunit dimension of the dust, and are much more sensitive to the exact geometry of the particles (Hanner and Giese, 1981). Model compositions and dust particle size distributions are derived from literature estimates of the in situ Halley flybys and measurements of interplanetary and lunar dust.
3. We analyze the qualitative features of the dust coma morphology directly: direction of emission (anti-solar, solar, parallel to the orbital velocity direction) and shape (linear or curved, strong or weak/no jets) to determine if the dust particle size distribution is small or large particle dominated.
4. Using a Monte Carlo/Finson Probstein model, we fit the shape of the observed coma using knowledge of the comet's orbit and emission rate history. Model dust particle size distributions and velocity fields are derived from literature estimates of the in situ Halley flybys and measurements of interplanetary, lunar, and cometary trail dust.

A detailed analysis of the resulting dust properties was presented at IAU Colloquium 181 (Lisse et al., 2002). To summarize, by plotting the derived properties by dynamical class (Figure 1), we found definite trends in the aggregate dataset of dust particle size distribution, total emission rate, and emission rate vs time as a function of cometary dynamical class. The short period comets slowly emit large, dark dust particles described by a shallow power law PSD (Lisse et al., 1998; Reach et al., 2000) while the long period comets emit most of their dust surface area (but not mass) rapidly in small, high albedo particles best described by the PSDs found for comet 1P/Halley by Lamy et al. (1987) and McDonnell et al. (1991). The observed differences are most probably due to the effects of cometary evolution on the structure of the cometary surface and the depletion of cometary volatiles, and are consistent with a larger sample of 46 infrared comet observations found in the literature, and with the grouping of "high" and "low" polarization comets by Levasseur-Regourd et al. (1996). The most likely operant physical mechanism is the creation of highly active surface areas as sub-surface regions of the comet are heated during repeated perihelion passage, leading to a change in emission character from the new comets to the long period comets (LP). The highly active areas emit material in a more violent fashion than usual, lead to increased dust particle fragmentation. We thus expect a correlation between active regions and jets and small particle dust emission. On the other hand, the known SP comets have, in the majority, endured numerous close perihelion passages, and are thus mostly devolatilized, leading to decreased activity, reduced active region emission, and large particle dominated comae.

New in this work is a different sort of analysis on the same dataset. Here we include our best estimates of the same comets' nuclear physical properties at the time of observation of the dust coma, and the long term rate of dust emission over a whole orbit (Table II). Using simultaneous photometric imaging of the nucleus and coma of a comet in the visible and infrared on the $1'$ scale, we have been able to model, fit, and remove the coma contribution, leaving a measure of the total flux from the nuclear surface, which is then used to determine the effective radiative surface (and thus the diameter) of the nucleus (e.g., Lisse et al., 1999; Fernandez et al., 2000). The rotation rates of the comets were either derived from photometric lightcurves obtained during our observations or from the literature where available, or, in the case of C/Hale–Bopp 1995 O1, from periodic morphological variations in dust coma structures. The mass lost by the comet per orbit for a dust production rate varying with heliocentric distance $Q_{\text{dust}} \sim r_h^{-\alpha}$ and emission cutoff at the ice line ($r_h = 5$ AU) is calculated using the measured Q_{dust} values, the known orbit of the comet, and the relation

$$M_{orbit} = \int m_{\text{comet}} Q_{\text{dust}}(1\ \text{AU})/(L_{\text{comet}} {}^{*} r^{\alpha-2})\ \text{dOrbit}.$$

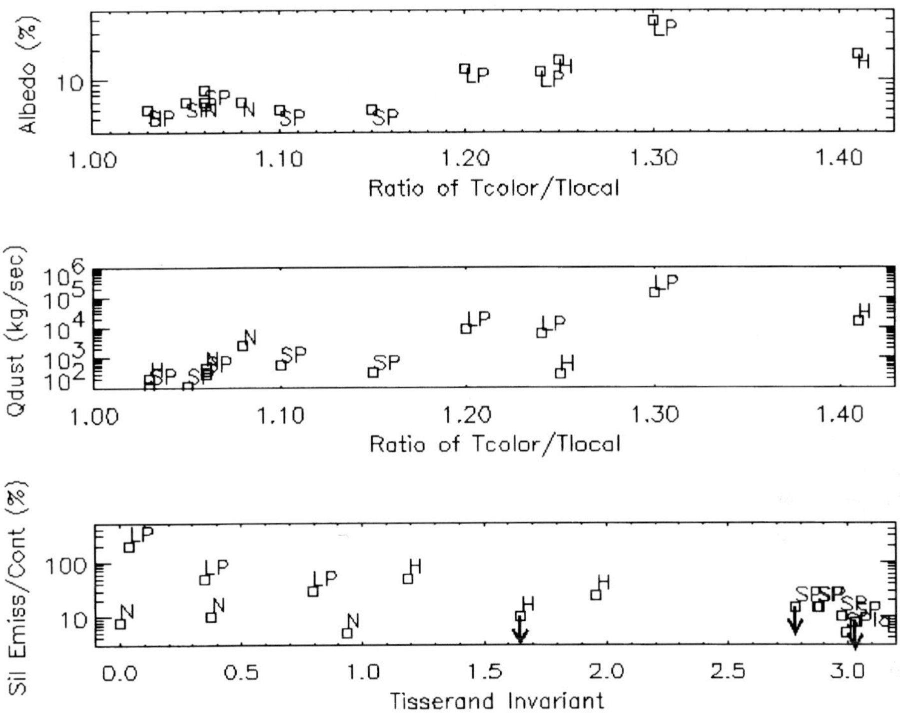

Figure 1. Observational trends for emitted cometary dust by dust albedo, silicate feature amplitude/continuum ratio, dust color temperature, Q_{dust}, and Tisserand invariant of the parent comet. There are clear trends and distinct groups by dynamical class – the LP comets produce the most dust with the hottest color temperature, highest albedo, and largest silicate feature. The New and SP comets produce little dust and what they do produce is cold, low albedo, and without much silicate feature. The Halley comets are between the two extremes.

3. Discussion and Conclusions

3.1. General Trends Found in the Survey

The majority of comets in our survey have radii of 1–5 km in size, and rotation rates slower than 5 rev day^{-1}. Taken together, the survey cometary nuclei tend to be smaller, darker, and rotating slower than the ensemble average of asteroids presented in the surveys of Donnison and Wiper (1999) and Pravec and Harris (2000). Our population agrees closely with the average size and rotaion rate of the "anomalously slow rotators" suggested by Pravec and Harris to be spun-down extinct cometary nuclei. The small size of the cometary nuclei as opposed to the asteroids may be due to the primitive state of the comet's accretion; collisional evolution of the planetisimal population; or rotational constraints on the size and rate of stably rotating fragile cometary material. We consistently find dust emission rates as large or larger than those published by Kresak and Kresakova (1987) using

TABLE II

Nucleus properties for the comets in the IR survey

Comet	R_{nuc} (km)	τ_{rot} (h)	Active area (%)	Q_{dust} (kg/s)	M_{orbit} (kg/orbit)
C/OLR (New)[a]	–	–	–	450	1.4×10^9
C/Austin (New)[a]	–	–	–	2400	1.1×10^{10}
C/1999 S4 LINEAR (New)	<4	>5.1	10–100	6300	$\sim 1 \times 10^{11}$
C/Levy (LP)	2	17–34	3	6100	1.8×10^{10}
C/Hyakutake (LP)	2.4	6.3	~100	1.3×10^4	5.2×10^{10}
C/Hale–Bopp (LP)	50	11.3		1.4×10^5	2.7×10^{12}
73P/SW-3 (SP)[a]	2.7	–		120	4.2×10^{10}
2P/Encke (SP)[a]	2.4	15.1	0.2	120	1.3×10^9
9P/Tempel 1 (SP)[a]	3.0	42	4.7	<390	6.8×10^8
P/Kopff (SP)[a]	1.5–2.5	–	16–44	1000	1.0×10^{10}
55P/Tempel–Tuttle (H)[a]	1.8	–		210	2.0×10^9
126P/IRAS (H)	2.86	–	>2	300	1.0×10^{12}
1P/Halley (H)	5	53,175	13	1.5×10^4	1.1×10^{11}

[a]Active area estimates are from A'Hearn et al. (1995); Lisse et al. (1999, 2002).

optical data alone, as expected for observations at longer wavelengths sensitive to larger particles (cf. Jewitt and Matthew, 1997a, b).

3.2. Comets as the Source of the IPD Cloud

The interplanetary dust (IPD) cloud extends throughout the inner ($r < 5$ AU) solar system. Composed of large ($a > 10\mu$m) particles of average density $\sim 10^{-5}$ cm^{-3} (Hauser et al., 1984; Zombeck, 1991), with high frequency structures ("bands") associated with a number of the known asteroidal families (Grogan et al., 1997; Sykes and Greenberg, 1986), its structure is still uncertain, although recent complete maps by the COBE/DIRBE instrument have greatly refined our understanding of the structure (Reach et al., 1997). The IPD cloud is not stable to collisions between dust particles on timescales of $\tau \sim 10^5$ years, or to inward spiraling of dust particles in the Sun's radiation field due to Poynting–Robertson drag on timescales of the order $10^5 * a(\mu m)$ years (Burns et al., 1979). Estimates by Delsemme (1975), Whipple (1975), and Gruen et al. (1985) require a fresh input of material to the IPD on the order of 10^4 kg/s (or 3×10^{13} kg/century) in order to provide a steady-state equilibrium. The solar system IPD cloud is a mature example of the dust shell expected to be found around any stellar system with a planetisimal population.

Previous estimates of mass loss by comets (Delsemme, 1975; Roser, 1975; Whipple, 1975; Gruen, 1985; Kresak and Kresakova, 1987), found that too little dust is lost by comets to support the IPD complex; current thought is that the bulk of the input material is asteroidal. However, these mass loss estimates were made using only visible observations of comets. For example, our IR survey estimated mass loss rate for comets C/OLR and C/Austin is approximately 1 order of magnitude greater than that estimated from visible observations. We note that for P/Encke, we have increased the estimated dust emission versus optical estimates by more than a factor of 50; we estimate that P/Encke alone contributes $> 2\%$ of the observed IPD dust. Since there are more than 150 known periodic comets, assuming this infrared mass enhancement is typical of all gassy comets and that most short period comets emit mainly large particles, it is certainly plausible that cometary dust is a major supplier of the IPD cloud.

More rigorously, separating the comets by the time required to complete an orbit (and thus roughly by the "boundedness" of the emitted dust), SP comets are much deeper in the Sun's gravitational potential well than the Oort cloud comets, so that $> 90\%$ of the emitted SP dust is placed into bound solar system orbits with PSD described by dn/dlog m $\sim m^{-0.70}$ (Lisse et al., 1998; Table I). $\sim$ 10% of an Oort cloud comet's dust is bound in a PSD that is either similar to Halley's (for the LP comets) or to the SP comets (for the New comets (Table I)). With these constraints, we find that for 150 SP comets, the total contribution is $> 5 \times 10^3$ kg s^{-1}, enough to support > 5090 of the IPD. The LP and New comets' total ranges from $> 1 \times 10^1$ to $> 1 \times 10^6$ kg s^{-1}, and the Oort cloud role in the IPD is highly uncertain. On the other hand, we know that there are structures in the IPD directly linked to main belt asteroids, so we must conclude that inputs from both kinds of minor bodies are important for supplying the IPD. The total loss rate of material into the IPD, over the age of the solar system, is approximately 4.5×10^9 yr * 3.2×10^7 s/yr * 1×10^4 kg/s = 1.4×10^{21} kg = 2.4×10^{-4} M_{earth}.

3.3. THE EVOLUTION OF THE IPD COMPLEX; THE NATURE OF DUST CLOUDS IN EXO-SYSTEMS

Assuming that the cometary contribution to the IPD is dominant, and considering the evolutionary trends found in the survey comets, we conclude that the IPD dust has evolved as the comets have evolved; the dominant input to the solar system IPD today is from SP comets slowly emitting large particles. This was not always the case, however; in the early solar system, most comets resembled the more pristine and unweathered LP comet surfaces. We thus expect many small IPD particles in the young solar system, with a gradual loss of small particle input, first from the rapidly aging SP comets, then from the more slowly aging LP comets, as weathering increases and mantles build up on the comets. This assertion is testable by obtaining spectra of the IPD clouds of systems of varying ages, and searching for effects due to particle size, such as the strength of the silicate emission feature at 11 μm. Consistent with our argument, using ISO SWS spectra of Beta Pictoris at 13 Myr, Heinrichsen et al. (1999) found a strong silicate feature, while ISO

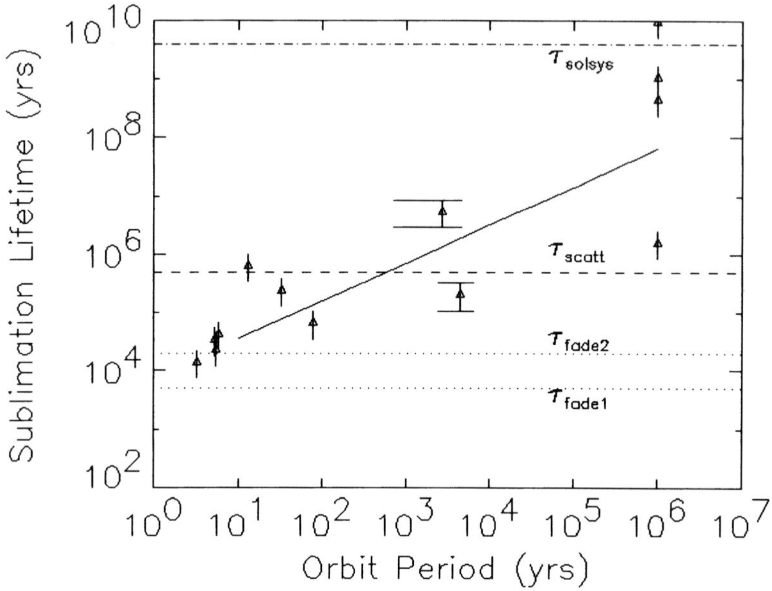

Figure 2. Time until turnoff of activity. The total lifetime versus sublimation for our survey comets, as derived using knowledge of the comet's orbit, initial mass, and Q_{dust} at time $t = 0$. The derived lifetimes correspond to 10^3 to 10^5 orbits. The SP survey comets are in the lower left of the figure. The mean dynamical lifetime estimated by Levison and Duncan (1994) for the SP comets is denoted by the dashed line, and their range of derived mantle formation times (or "fading times") are denoted by the dotted lines.

SWS measurements of our own zodiacal cloud at 4.5 Gyr show little to no silicate emission (Reach et al., 1996).

3.4. Timescales for the Evolution and Extinction of Short Period Comets

The present population of some 150 short period (SP) comets, linked as daughters to the Kuiper–Edgeworth (K–E) belt of planetisimals outside Neptune's orbit (Stewart 1990; Quinn and Tremaine, 1990), have also been shown to be dynamically unstable in their present day configuration. Ejection from the inner solar system for these bodies occurs on timescales of 10^5 years (Levison and Duncan, 1994). Therefore the current population of SP comets is a relatively recent sampling of the K–E belt; a constant input of comets from the K–E is required to supply the SP comets in a steady state.

The required rate of supply depends on the limiting rate of removal of SP comets from observabilty from Earth – i.e., removal by reducing their number, size, and/or increasing their effective distance from Earth. If the major mechanism for SP cometary "destruction" is dynamical removal from the inner solar system with mean lifetime $\sim 10^5$ yr as suggested by Duncan and Levison (1994), then we have the answer for what the input rate must be: $\sim$ 150 SP comets/10^5 yr = 1.5×10^{-3} new SP comets year^{-1} .

This argument only holds true if the average SP comet lifetime in the inner solar system is $>$ 10^5 yr. With the nuclear size and mass loss rates per orbit presented in this work, we are in a position to estimate the lifetime of cometary nuclei versus total evaporation due to sublimation mass loss. The results for our survey comets are shown in Figure 2. Our work suggests that $\tau_{\rm evap}$ due to cometary mass emission is significantly longer, $>$ 10^4 yr, than the Levinson–Duncan maximum "fading time", the time for mantle formation/devolatilization of the nuclear surface, but shorter than the $\sim$ 10^5 yr dynamical time for scattering. Thus most SP comets survive to be ejected: they turn off faster than they evaporate, and survive until they are dynamically scattered out of the inner solar system. This implies a population of extinct comets in the inner solar system, consistent with the detection of the anamolous slow rotators of Pravec and Harris (2000) and the discovery by Fernandez et al. (2001) of $>$ 10 extinct comet candidates in a survey of NEA's. On the other hand, the rate of mass loss due to sublimation is not negligible; the observed population of SP comets cannot be considered pristine, even beneath their mantles – a good fraction of their total radius, 1–10%, has been eroded away in one fading time.

References

Bazell, D. and Dwek, E.: 1990, *Astrophys. J.* **360**, 142–150.

Bregman, J. D., Campins, H., Witteborn, F. C. et al.: 1987, *Astron. Astrophys.* **187**, 616–624.

Chen, J. and Jewitt, D. J.: 1994, *Icarus* **108**, 265–271.

Delsemme, A. H.: 1976, 'Interplanetary Dust and Zodiacal Light, Lecture Notes in Physics 48', in *Proceedings of IAU Colloq.* 31, 10–13 June 1975, Heidelberg, Germany, Springer-Verlag, Berlin, pp. 314–318.

Donnison, J. R. and Wiper, M. P.: 1999, *MNRAS* **302**, 75–80.

Fernandez, J. A.: 1988, *Earth Moon Planets* **41**, 155–161.

Fernandez, Y. R., Lisse, C. M., Kaufl, H. U. et al.: 2000, *Icarus* **147**, 145–160.

Fernandez, Y. R., Jewitt, D. C., and Sheppard, S. S.: 2001, *Astrophys. J.* **553**, L197–L200.

Grogan, K., Dermott, S. F., Jayaraman, S., and Xu, Y. L.: 1997, *Planet. Space Sci.* **45**, 1657–1665.

Gruen, E., Zook, H. A., Fechtig, H. et al.: 1985, *Icarus* **62**, 244–272.

Hanner, M. S., Giese, R. H., Weiss, K., and Zerull, R.: 1981, *Astron. Astrophys.* **104**, 42–46.

Hanner, M. S., Lynch, D. K., and Russell, R. W.: 1994, *Astrophys. J.* **425**, 274–285.

Harker, D. E., Wooden, D. H., Woodward, C. E., and Lisse, C. M.: 2002, *Astron. J.*, in press.

Harker, D. E., Woodward, C. E., Wooden, D. H. et al.: 1999, *Astron. J.* **118**, 1423–1429.

Heinrichsen, I., Walker, H. J., Klaas, U. et al.: 1999, *MNRAS* **304**, 589–594.

Jewitt, D. C.: 1991, 'Comets in the Post-Halley Era', in *The 1989 Bamberg Meeting Proceedings*, Kluwer Academic Publishers, Dordrecht, pp. 19–65.

Jewitt, D. C. and Matthews, H. E.: 1997a, *Astron. J.* **117**, 1056–1062.

Jewitt, D. C. and Matthews, H. E.: 1997b, *Astron. J.* **113**, 1145–1151.

Kresak, L. and Kresakova, M.: 1987, in *Symposium on Diversity and Similarity of Comets*, 6–9 April 1987, ESA SP-278, pp. 739–744.

Lagrange, A.-M. et al.: 1998, *Astron. Astrophys.* **330**, 1091–1108.

Lamy, P. L., Toth, I., Jorda, L. et al.: 1998, *Astron. Astrophys.* **335**, L25–L29.
Lamy, P. L., Toth, I., A'Hearn, M. F., and Weaver, H. A.: 1999, *Icarus* **140**, 424–438.
Levasseur-Regourd, A. C., Hadamcik, E., and Renard, J. B.: 1996, *Astron. Astrophys.* **313**, 327–333.
Lecavelier des Etangs, A. et al.: 2001, *Nature* **412**, 706–708.
Levison, H. F. and Duncan, M. J.: 1994, in *Origins of Solar Systems Workshop: The Origin, Evolution, and Detectability of Short Period Comets* (see N94-11628 01-90), pp. 1–36.
Lisse, C. M., A'Hearn, M. F., Hauser, M. G. et al.: 1998, *Astrophys. J.* **496**, 971–991.
Lisse, C. M., Fernandez, Y. R., A'Hearn, M. F. et al.: 1999, *Earth Moon Planets* **78**, 251–257.
Lisse, C. M., Fernandez, Y. R., A'Hearn, M. F., and Peschke, S. B.: 2002, 'Dust in the Solar System and Other Planetary Systems', in *Proceedings of IAU 181*, April 2000, COSPAR, Kent, UK.
Lisse, C. M., Fernandez, Y. R., Kundu, A. et al.: 1999, *Icarus* **140**, 189–204.
McDonnell, J. A. M., Lamy, P. L., and Pankiewicz, G. S.: 1991, 'Comets in the Post-Halley Era', in *The 1989 Bamberg Meeting Proceedings*, Kluwer Academic Publishers, Dordrecht, pp. 1043–1073.
Melnick, G. J. et al.: 2001, *Nature* **412**, 160–163.
Pravec, P. and Harris, Alan, W.: 2000, *Icarus* **148**, 12–20.
Quinn, T., Tremaine, S., and Duncan, M.: 1990, *Astrophys. J.* **355**, 667–679.
Reach, W. T. et al.: 1996, *Astron. Astrophys.* **315**, L381–L384.
Reach, W. T., Franz, B. A., and Weiland, J. L.: 1997, *Icarus* **127**, 461–484.
Reach, W. T., Sykes, M. V., Lien, D., and Davies, J. K.: 2000, *Icarus* **148**, 80–94.
Roberge, A., Feldman, P. D., Lagrange, A. M. et al.: 2000, *Astrophys. J.* **538**, 904–910.
Stewart, G. R.: 1990, *Nature* **343**, 17–18.
Sykes, M. V. and Greenberg, R.: 1986, *Icarus* **65**, 51–69.
Walker, R. G., Aumann, H. H., Davies, J. K. et al.: 1984, *Astrophys. J.* **278**, L11–L14.
Whipple, F. L.: 1976, 'Interplanetary Dust and Zodiacal Light', Lecture Notes in Physics 48, in *Proceedings of IAU Colloq.* 31, 10–13 June 1975, Heidelberg, Germany, Springer-Verlag, Berlin, pp. 403–415.

CONFIRMATION OF COMETARY FEATURES OF NEWLY DISCOVERED BODIES

MILOŠ TICHÝ, JANA TICHÁ and MICHAL KOČER
Kleť Observatory, Zátkovo nábřeží 4, 370 01 České Budějovice, Czech Republic

(Received 20 February 2002; Accepted 25 May 2002)

Abstract. The majority of new ground-based discoveries of comets comes from large surveys devoted, predominantly, to Near Earth Asteroids. The first step in distinguishing these newly discovered members of the population of cometary bodies consists in confirmatory astrometric observations along with detection of their cometary features. Although both amateur and professional stations take part in such confirmatory observations, only some of them do a preliminary analysis of the cometary activity of a particular newly discovered body. A timely recognition of cometary features of a particular body having an unusual orbit can help in planning further observing campaigns. A main goals of the Kleť Observatory NEO astrometric follow-up programme consists of an analysis of possible cometary activity of newly discovered unusual bodies. Here, we describe several examples (comets C/1999 S4 (LINEAR) and so on). We also mention an extension of our programme to fainter objects after the completion of a new 1-m telescope at the Kleť Observatory.

Keywords: Astrometry, comet

1. Introduction

The number of newly discovered small bodies of the solar system has increased very rapidly in recent years, mainly due to the large surveys programmes LINEAR, Spacewatch, LONEOS, NEAT and CSS, devoted primarily to the search for Near Earth Asteroids. An efficient system enables us to select and present newly-discovered fast-moving or other unusual objects in need of astrometric confirmation due to the NEO Confirmation Page (the NEOCP) maintained by the Minor Planet Center (Marsden, 1998). It is possible that some of the objects listed on the NEOCP are comets, either recognized by discoverers or not yet recognized. Therefore it is necessary not only to find these new discoveries and provide astrometric measurements, but also to test them for possible cometary features and confirm their cometary nature, if any. For planning further physical observations it is necessary to detect cometary activity as soon as possible.

Earth, Moon and Planets **90:** 507–513, 2002.

2. Kleť Observatory Instrumentation

2.1. 0.57-METER F/5.2 REFLECTOR

The 0.57-m f/5.2 reflector is equipped with a CCD camera, Santa Barbara Instrumental Group ST-8 using Kodak KAF1600 detector of 1530 × 1020 elements; 9 × 9 microns. The field of view is 16.0 × 10.7 arcmin. Limiting magnitudes are $m_V = 19.5^m$ for 120-s exposure under standard sky conditions, and $m_V = 20.2^m$ for 120-s exposure under very good conditions with no moonlight.

2.2. KLENOT TELESCOPE

We are working on an extension of our follow-up program (Tichá, 2000) for fainter unusual objects using a larger telescope, which will be more suitable also for the analysis of possible cometary features of newly discovered bodies.

The KLENOT telescope is the 1.06-m f/2.7 reflector equipped with a CCD camera Photometrics Series 300 using SITe 003B CCD with 1024 × 1024 pixels, pixel size of 24 microns and with liquid nitrogen cooling. The field of view is $0.5° \times 0.5°$. This telescope is currently (Spring 2002) in a test operation.

3. Technology

The first step in distinguishing newly discovered members of the population of cometary bodies consists in confirmatory astrometric observations along with detection of their cometary features, i.e., if their images are diffuse (indication of a coma) or display tails.

The second step is the determination of reliable orbital elements of a newly discovered comet.

3.1. SOFTWARE

For the detection of cometary features we use the following software tools.

The **Ephem** software allows an observer to get ephemerides for an object including motion and magnitudes.

The **USNOview** software shows stars and catalog objects (minor planets/comets with the line showing their daily motion on the sky) on selected regions in the sky using USNO-A2.0 or GSC catalogs including magnitudes.

Astrometry is a program for the reduction of CCD images and the identification of stars using USNO-A2.0 or GSC catalogs. This program detects diffuse objects using Gaussian fits.

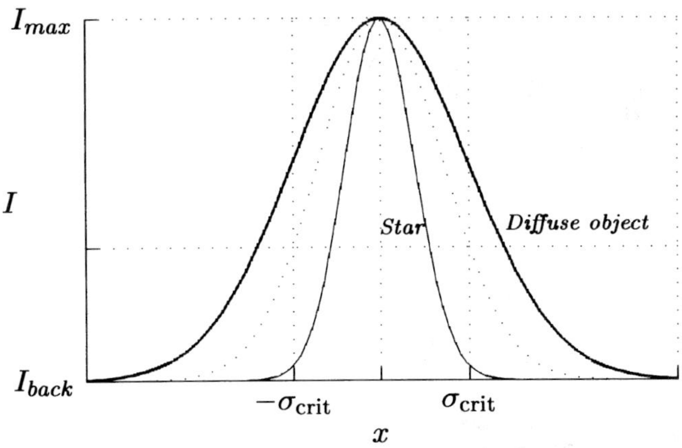

Figure 1. Profile of the star-like and diffuse object.

3.2. Detection

The important thing is the *detection* of cometary characteristics of the object image. For numerical approximation of the profile of objects on images we adopted the model of two-dimensional Gaussian equation in a simplified form:

$$I(x, y) = B + P \exp\left(-\frac{1}{2}\left[\left(\frac{x}{\sigma}\right)^2 + \left(\frac{y}{\sigma}\right)^2\right]\right),$$

where B is the background brightness and P is the peak value of the analyzed object.

There exists a critical value σ_{crit} of the width of a Gaussian such that an object having a profile with $\sigma > \sigma_{\text{crit}}$ can be considered as being diffuse (see Figure 1). For the analysis of most of our images we used the experimental value $\sigma_{\text{crit}} = 0.6$. This value was detected by an analysis of images of the same magnitude stars and comets for standard weather condition at Kleť. The value of σ_{crit} is site-dependent. It depends mainly on the stability and the transparency of the local atmosphere during observation.

4. Results

During the years 1999–2001 we found or confirmed cometary features of 54 objects and discovered one new fainter comet – P/2000 U6. The majority of confirmed cometary bodies were discovered by the large surveys LINEAR, NEAT, Spacewatch or LONEOS. Altogether 24 were discovered as apparently asteroidal objects with unusual motion, they were posted on the NEO Confirmation Page due to their unusual motion and have been found to be cometary. A further 30 objects were

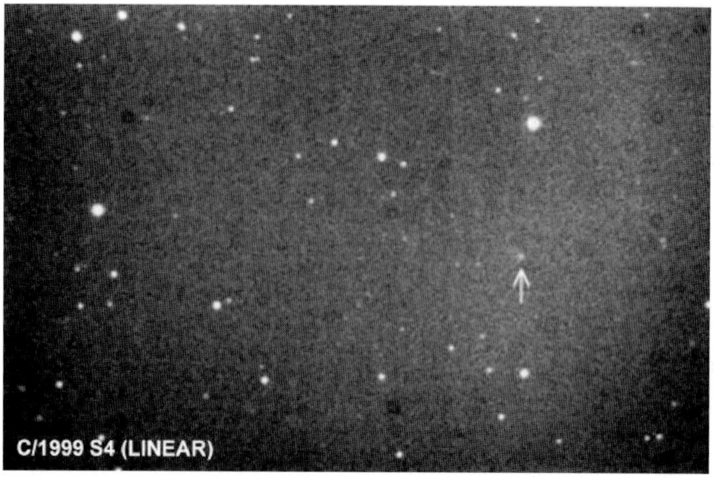

Figure 2. Comet C/1999S4 (LINEAR) – 30 December 1999, unfiltered 60 sec exposure, 16 × 10 arcmin, north on the top.

detected as cometary by their discoverers and we have confirmed their cometary features. Some of the most interesting results are presented in the following sections.

4.1. C/1999 S4 (LINEAR)

This object was posted on the NEO Confirmation Page as an unusual-moving object discovered by LINEAR on the 27 September 1999 and reported as asteroidal. It was confirmed by D. Durig (Sewanee) and J. Tichá and M. Tichý (Kleť) as a new comet (see Figure 2). The discovery magnitude was $m_1 = 16.2^{mag}$ (Green, 1999), in July 2000 the magnitude was $m_1 = 6^{mag}$. During July 2000 this comet showed a dramatic increase in activity and finally the comet was found split in many pieces.

4.2. C/1999 XS87 (LINEAR)

An object that was assumed to be asteroidal was found by LINEAR on the 7 and 8 December 1999 and it was later linked to observations by LINEAR on the 6 and 7 January 2000 by G. V. Williams, Minor Planet Center. Following a request from the Minor Planet Center, after seeing that the orbit appeared cometlike, M. Tichý and Z. Moravec obtained observations at Kleť on the 11 and 12 January that showed this object to be diffuse with a coma diameter of 15″ (Green, 2000a). The magnitude was $m_1 = 16.5^{mag}$.

4.3. C/2000 SV74 (LINEAR)

An apparently asteroidal object reported by LINEAR on two nights and published under the designation 2000 SV74, has been found to be cometary (diffuse with 16″

Figure 3. C/2000 SV74 (LINEAR) – 19 October 2000, unfiltered 60 s exposure, 16 × 10 arcmin, north on the top.

coma and 20″ tail at p.a. 150°) by M. Tichý on CCD images taken in October 2000. 19.8 UT with the 0.57-m f/5.2 reflector at Kleť (see Figure 3) after posting on the NEOCP (Green, 2000b). The magnitude was $m_1 = 16.6^{mag}$.

4.4. P/2000 U6 (Tichý)

An object found by Miloš Tichý on images taken at Kleť with J. Tichá and M. Kočer on 23 October 2000, originally reported as asteroidal, was subsequently noted to be diffuse on Kleť images taken during October 28.9–29.2 UT; Tichý also noted a 10″ coma in October 29.8 images (Figure 4). Images obtained on 1 November by S. Sanchez and M. Blasco at Mallorca and by D. T. Durig at Sewanee also showed diffuse appearance of the object (Green, 2000c). Although nine observing stations obtained early astrometry of this comet, only three of them (including the discoverer) mentioned cometary nature of this object. The magnitude was $m_1 = 17.8^{mag}$.

4.5. C/2001 G1 (LONEOS)

An apparently asteroidal object discovered on CCD images taken with the LONEOS 0.59-m Schmidt telescope and posted on the NEOCP has been found by other astrometric observers to be cometary on their CCD images. The object was first reported as being diffuse by J. Tichá, M. Tichý, and P. Jelínek at Kleť (April 2000 1.9 and 2.9 UT; 9″ coma in April 2.9) and by C. E. Lopez and M. R. Cesco at El Leoncito (April 2.2), and as having a 10″ coma by M. Busch and S. Kluegl at Heppenheim, Germany (April 1.9) and by J. Broughton, Reedy Creek, Queensland (April 2.5) (Green, 2001a). The magnitude was $m_1 = 17.1^{mag}$.

Figure 4. Comet P/2000 U6 (Tichý) – 2 November 2000, unfiltered 90 sec exposure, 8 × 5 arcmin, north on the top.

4.6. C/2001 RX14 (LINEAR)

The linkage by the Minor Planet Center of subsequent observations of an apparent main-belt minor planet observed by LINEAR on 10 and 11 September 2001 showed that this object, designated 2001 RX14, had a nearly parabolic orbit. When it was announced at the NEOCP, many further observations were received from 25 stations using from 0.2-m to 1-m class telescopes. Only M. Tichý (Kleť, 0.57-m f/5.2 reflector + CCD) indicated on October 18.7 UT that the comet appeared to be diffused with a coma of diameter 13″ and red mag 16.7. R-band images taken by C. Hergenrother and J. Barnes (Lunar and Planetary Laboratory) with the 1.54-m Catalina reflector on October 24.32 show a highly condensed coma of diameter 6″ and no hint of a tail (Green, 2001b).

4.7. P/2001 X2 (Scotti)

J. V. Scotti, Lunar and Planetary Laboratory, reported his discovery of a comet on CCD images taken with the 0.9-m Spacewatch telescope on 14 December 2001, noting a coma diameter of 5″ and a tail 0′.40 long in p.a. 283° (and $m_2 = 22.7$). When announced atthe NEOCP, no information on its cometary nature was given. Confirming observations were made at Kleť by J. Tichá, M. Tichý, and P. Jelínek, who reported a 8″ coma and a faint westward pointing tail (Green, 2001c). The magnitude was $m_1 = 18.4^{mag}$.

5. Discussion

The method used for the detection of cometary characteristics of the object on the image is based on a simplified form of the two-dimensional Gaussian function. This approach was successfully used for coma detection, but it was not satisfactory for the detection of objects with a tail and/or faint coma. For example, we were not able to apply this method successfully to verify the cometary nature comet 133P/Elst–Pizzarro (absence of a coma and the presence of the needle-like tail).

We are considering to extend our simple technique for the detection of cometary features involving a more complex numerical description.

Acknowledgements

This work has been sponsored by grants of the Agency of the Czech Republic Reg. No. 205/98/0266 and 2005/02/P114 and the 2000 NEO Shoemaker grant of the Planetary Society.

References

Green, D. W. E.: 1999, *IAUC* 7267.
Green, D. W. E.: 2000a, *IAUC* 7344.
Green, D. W. E.: 2000b, *IAUC* 7510.
Green, D. W. E.: 2000c, *IAUC* 7515.
Green, D. W. E.: 2001a, *IAUC* 7606.
Green, D. W. E.: 2001b, *IAUC* 7739.
Green, D. W. E.: 2001c, *IAUC* 7775.
Marsden, B. G. and Williams, G. V.: 1998, *Planet. Space Sci.* **46**, 299–302.
Tichá, J., Tichý, M., and Moravec, Z.: 2000, *Planet. Space Sci.* **48**, 787–792.

THE ROUND TABLE DISCUSSION: HOW SHOULD WE TAKE BEST ADVANTAGE OF THE NEW GENERATION OF LARGE TELESCOPES FOR COMETARY RESEARCH?

MARK R. KIDGER
Instituto de Astrofí sica de Canarias, E-38205 La Laguna, Tenerife, Spain

(Accepted 25 June 2002)

Abstract. The highlights of the round table discussion held to discuss the most important pending problems in cometary astrophysics are reported. A number of experts presented short summaries of what they regarded as the most pressing problems to be solved. These were taken up by a panel of five experts who discussed and analysed the best way to take the fullest advantage of the new telescopes and instrumentation that are becoming available and the requirements for collaboration with space missions.

Keywords: Comets, infrared, observations, visible

1. Introduction

The start of the 21st Century is an exciting moment for astrophysics in general and for cometary physics in particular. Over the next few years a series of large (8–10 m class) telescopes will become fully operational, with state of the art instrumentation in the visible and infrared. The Keck telescopes, the Very Large Telescope (VLT) and the Subaru telescope are already available, the two Gemini telescopes start to become operational, and the Large Binocular Telescope (LBT) and the Spanish Gran Telescopio CANARIAS (GTC) will be added soon. At the same time there will be a series of exciting cometry missions that will include both encounters, rendezvous with landers and sample return. These will visit a variety of targets including evolved and split comets, with the possibility of a serendipitous encounter of a high-activity object (although an encounter with C/2001 Q4 (NEAT), a comet that is expected to reach bright naked eye visibility in Spring 2004, has been ruled out).

It is thus a good time to look at the current state of cometary physics and examine the best way forward. The new generation of large telescopes offers a series of opportunities to cometary astronomers to advance enormously on the work that has been done previously. Given the competitive nature of time allocation on these telescopes it is important to establish what are the best and most important projects to be carried out.

Earth, Moon and Planets **90:** 515–522, 2002.

2. Being Competitive

The panel expressed some concern that cometary physics is not generally a high priority for most Time Allocation Committees (TACs). There is thus a possibility that although there are many fundamental advances to be made, cometary astrophysicists may be left behind unless they can make their case as strongly as the established fields of cosmology and related subjects that are often seen to be the natural beneficiaries of large telescopes. It should thus be a priority for cometary astrophysicists to search for high-priority and highly competitive projects.

A counter-argument was advanced that points out that cometary studies have never been a priority of the current generation of 3–5 m telescopes. While this suggests that history will repeat itself and access to the very largest telescopes will be difficult and very limited for the cometary community, it is likely that a large amount of time in medium-sized telescopes will be opened up as pressure shifts to the new, large telescopes. The cometary community should thus prepare itself to obtain extensive access on the medium-sized telescopes that have previously been difficult to access. At the same time, the very best projects can be prepared for the largest telescopes with the aim of competing successfully for time.

3. Why Do We Want to Observe with Large Telescopes?

3.1. The Visible Range

Why do we want to observe comets in the visible? The answer is that the visible is a good range for observing the gas-phase activity in comets as there are many important atomic and molecular transitions that may be observed. If we observe lines and bands with a high fluorescence efficiency we can detect very weak gas activity in distant comets and in objects with generally very low activity even when close to the Sun.

The two key activities to be carried out are thus:

- To follow the gas activity to large r.
- To observe the activity of Jupiter family comets all around their orbit.

The aim of this is to apply constraints to the sublimation process given that currently there are many models for explaining activity at large r, but none is completely convincing as yet.

Over the last few years though various automated programmes such as LINEAR, LONEOS and NEAT have also detected a large number of very low activity objects that would previously have not been detectable. These objects represent the tail of the cometary "luminosity function". These objects are very faint and often detected as asteroidal objects, before being seen to show a faint coma. In some cases the level of activity is so low that an object may be suspected to be a comet on the basis of its orbit, but months (or in extreme cases, years) may pass before cometary activity is detected.

These objects appear to be transition objects between comets and asteroids. Studying them provides information on the gas activity in faint objects and on the comet-asteroid transition.

3.1.1. *What is required?*

It was decided that the highest priority in the visible range is for long-slit spectroscopy with $R > 4000$. At present faint objects can only be observed at very low resolution due to the lack of available photons. There is also a need for very large surface area detectors to detect weak, and very extended activity.

The most critical problems that can be solved with such observations are:

- To measure isotopic ratios. At present very few ratios have been observed. There is strong evidence that the D:H ratio in comets is different to that in the Earth. Isotope ratios have strong implications for cosmogenetic models and for the formation of the solar system.
- To determine the spin temperature of the various atomic and molecular components that are detected. This allows the excitation conditions to be studied.
- To examine the composition of the coma. At present a significant number of constituents of the gas phase are known, but many of the components observed the millimetric and sub-millimetric cannot be detected in the visible. With the large increase in photon-gathering power that the large telescopes will provide many more gas-phase components will be detectable in the visible and their scale lengths and distribution can be mapped. This will allow parent molecules to be studied in detail.

3.2. The Infrared

It was pointed out that, at the time of the discussion, there are only three mid-IR instruments available on 8–10 m class telescopes anywhere in the world. This means that infrared and particularly mid-IR studies of comets are very limited, particularly compared to the visible range. The mid-IR will thus allow opportunities for the most spectacular advances in cometary studies.

Here there is a very wide range of projects where spectacular advances may be expected principally because the mid-infrared allows the nucleus to be studied more directly that in any other band using the nucleus's own thermal emission rather than reflected sunlight. The following are perhaps the most important:

- Physical evolution of cometary nuclei.
  The most critical observations that will be carried out are those related to albedos, near-IR colours and surface spectroscopy for taxonomy and classification.
- Thermal behaviour of cometary nuclei.
  Critical thermal observations include the study of the thermal inertia of the nucleus (and hence its surface properties) and the thermal phase effect.

- Dust phase.
  Many exciting and fundamental observations of dust can be carried out. In particular, characterisation of the dust properties as a function of the dynamical class of the comet and minerology of dust grains from spectroscopy. It has also been pointed out in the presentations that the physical properties of the grains (i.e., the presence of annealing) can be studied from mid-IR spectroscopy.
- Molecules.
  One of the greatest advances in cometary physics may be expected from the study of molecules, particularly in the mid-IR where many have observable transitions. To date very few species have been detected spectroscopically in the IR. A factor of 10 increase in sensitivity though should increase the number of detectable molecules in a factor of 6–8. As mid-IR observing is background limited though the sensitivity should scale as the 4th power of the telescope diameter rather than as the 2nd power (i.e., surface area). This implies that a telescope such as the GTC should give a gain of some two orders of magnitude in sensitivity over the NASA IRTF with the corresponding large increase in the number of detectable species. It was also felt that the large increase in sensitivity would allow deuterated species to be detected and thus to probe isotopic abundances.

4. Some Suggestions for Particular Projects on Large Telescopes

Given the example of a large telescope such as the LBT, some thought was given to the sort of projects that could profitably be carried out that would take the fullest advantage of the telescope's capabilities. The following are simply some suggestions.

It was felt that the greatest gain could be made in near-nucleus studies that require very high spatial resolution. In particular three problems were proposed:

- The study of new parent molecules in the inner coma, in particular the analogue to S_2.
- The study of inner coma chemistry.
- The study of short-lived solid grains, but icy and organic.

In general it was felt that it was of interest to look at ionisation mechanisms in more detail and in particular at fast processes very close to the nucleus that are extremely difficult to study with smaller telescopes and less sensitive instruments.

Considering the instrumentation that will be available on the 10.4-m Gran Telescopio CANARIAS (GTC) when it opens for scientific use in 2004, one obvious project for OSIRIS, the GTC's Day 1 imaging spectrograph would be to take very deep pencil-beam surveys to reach $R \sim 30$. The aim of these surveys would be to search both for TNOs down to the size of 1P/Halley in the inner Kuiper Belt and for faint asteroids. This would greatly improve knowledge of the size distribution of TNOs in the inner Kuiper Belt and would detect many scattered disk objects at much larger distances than have currently been observable.

5. The "Wish Lists" of the Panelists

The panel members, Dominique Bockelee-Morvan, Jack Brandt, Anita Cochran, Walter Harris, Walter Huebner, Horst-Uwe Keller and Hans Rickman, were selected with care to provide a wide range of interests and specialities. As these were reflected in the comments made, these are reproduced below more or less as they appeared in the discussion. This discussion was chaired and directed by Horst-Uwe Keller.

Three key projects were suggested that are of considerable interest. One of these could be considered of general interest to the scientific (and non-scientific community) and not necessarily purely a project for large telescopes, the other two reflect the panelist's interest in spectroscopy.

- The search for long period comets that may be a hazard (PHAs).
  There is a great deal of attention in the astronomical community to the danger of Near Earth Objects (NEOs) and Potentially Hazardous Asteroids (PHAs), but it is not often recognised that these are in many senses less dangerous than comets because the encounter velocities are lower (typically 10–15 km/s for an asteroid, but typically parabolic velocity, 41 km/s, for a comet), thus for the same mass a comet will have a typical kinetic energy, and thus destructive power, an order of magnitude greater than an asteroid.
  The larger asteroids in the PHA category are now all known and their orbits under control (although there are many smaller PHAs that are still to be found). Long period comets may though appear from any direction at any time and one could give rise to a disastrous impact with only a few months advance warning. Using larger telescopes to search for comets would allow potential impacts to be detected years in advance and a more effective warning to be given.
- The search for ice and dust in NEOs.
  It is generally accepted that at least in some cases NEOs could be the end states of comets when the dust mantle has become so thick that all activity is suffocated. The fact that some NEOs have orbits that are similar to short period comets is often regarded as indicative that some NEOs may have evolved from Jupiter family comets and thus represent a cometary end state.
  To investigate the link between NEOs and comets it is necessary to find evidence of low level residual activity in NEOs: either dust comae or ice. To date searches for a coma in different NEOs such as 1566 Icarus have been unsuccessful, but they do not rule out the possibility of low level sporadic activity as observed in 4015/Wilson–Harrington. As an alternative to observing a dust coma, the detection of surface ice would provide an alternative link between NEOs and comets of low activity.

– The study of the D/H ratio in comets.
  There is considerable interest in isotope ratios in comets in general, particularly for the cosmogenetic information that they can provide. One of the few isotope ratios to be measured to date is the D/H ratio. This has so far been measured for just three comets. The results show that the D/H ratio measured in these objects is different to the D/H ratio in the Earth's oceanic waters. This would suggest that, contrary to many expectations, the source of the water in the Earth's oceans is not cometary impact. This result needs to be confirmed though for more comets.
  It is also of considerable interest to measure the D/H ratio for Jupiter Family (JF) comets. Were the D/H ratio to be different in evolved JF comets to the ratio in long period comets then it is still possible that these could be the source of oceanic water, even if the D/H ratio does not match for new comets.

To this list we should probably add the study of nitrogen chemistry from spectroscopy as it is a solar-system model determining parameter. Using larger apertures and more sensitive instruments it will be possible to study additional nitrogen compounds and radicals and search for keys to the formation of the comets and from them the solar system itself.

Two principal lines of action suggest themselves. In line with what has been commented above, the shift of ground-breaking science to the 8 and 10-m telescopes provides a major opportunity for the cometary community. Up to now, most of the 4–6 m telescopes have been closed to comet observers, save in very specific cases of projects of particular interest (e.g., C/1995 O1 (Hale–Bopp)). Although the cometary community should attempt to gain time in the 8–10 m telescopes, its needs may be best served by shifting effort to the 4–6 m class telescopes and taking advantage of the greater time availability that the large telescopes will allow.

The second course of action is to use these "medium-sized" telescopes to improve sample of comets. At present, with most data taken in 2-m class telescopes the number of comets that can be observed in detail has been rather limited. In particular, the fainter JF objects cannot be studied in as much detail (i.e., with as high resolution, both spatial and spectral) as we would like. Using a 4-m telescope instead of a 2-m many studies of fainter JF comets can be carried out that to date have been impossible.

We really should be inverting the question. We often say just how important comets are for the study of the solar system because they are probes of how the solar system formed. On this basis observations of comets are as critical as any cosmological observations. Why do we not generally receive time on 10-m telescopes to observe comets? If we are right our observations should be competitive with any other programme.

What we do not do is think of how we can best use comets to get information about the solar system and its formation. This is the reason why usually a comets proposal will lose out to a cosmology proposal. We should be asking ourselves what we can really learn about the early solar system from studying comets and think carefully about exactly how we go about getting that information at the telescope.

The cometary community should remember that despite the big advances made in many aspects of cometary physics and reflected at this conference, there is a field that has been somewhat neglected and in which our state of understanding is somewhat limited still. This is of course the field of wide field data. Wide field data were an important part of the International Halley Watch effort and are critical to our understanding of plasma physics.

There is a tendency to aim for high spatial resolution and many of the smaller wide-field telescopes that could study the plasma tail of brighter comets are being closed. Of course the high-resolution data are very important and allow us to study what is happening close to the nucleus in great detail, but they should not be regarded as the be all and end all of cometary studies. The wide field data though are absolutely critical to understanding the interaction between the solar wind and the cometary plasma tail and as such we need to dedicate some time and effort to closing this gap in our observational capabilities. In part this can be covered by amateur astronomers who did a wonderful job during the IHW and continue to do it now, but some thought needs to be given to how we solve this particular problem.

6. Discussion Summary by Uwe Keller

Trying to select the most important questions to be answered is an extremely difficult problem and reflects the particular interests and biases of the summariser in the same way that each person's individual ideas will reflect their own particular interests and biases. However, given this caveat, there are a series of problems that appear to be of general interest to the community.

One particular suggestion is given below, and a suggestion for a possible way forward is made to round things off.

Many particular ideas have been suggested during this discussion and during the conference in general, but we can say that these all really boil down to a single problem: what is the structure and composition of the nucleus? It is the nucleus that is driving all the activity in the comet from coma chemistry to the formation of the tail and thus it must be the key to understanding comets in general.

What we really want to try to answer is a series of questions. In particular:

- How does cometary really activity work?
  and
- Why do periodic comets come back again and again showing the same area of the nucleus active?

The former cover the question of the structure and composition of the nucleus which dominates the emission from the nucleus: Chemistry, dust emission, etc. The latter determines how the dust and ice from within the nucleus are released. All of the questions that have been asked are really a microcosm of these fundamental questions.

Many people seem to think that the key to understanding the activity of comets lies in the very distant comets. These allow us to measure the size of the nucleus and its albedo and also to study the turn-on of cometary activity. By looking at activity at very large heliocentric distances pre- and post-perihelion where solar heating is at its weakest we can really examine the mechanisms of activity. At present this very distant activity is simply not understood and we have increasing evidence that many comets are active at far greater distances than was previously imagined, for example C/1995 O1 (Hale–Bopp) at 17 AU pre-perihelion and 1P/Halley with its surprising outburst at 17AU in 1995.

To wrap up this discussion, it seems timely to make a suggestion about how to proceed.

Over the next 5 years or so, space probes will visit 4–5 comets of very different types, including objects of moderate and very low activity and a comparatively recently fragmented object. Perhaps the best way forward at this time is to fix our aim on a series of concentrated and concerted programmes of observation to study these comets in detail and thus marry ground-based and spaceborn observations.

This has been suggested on various occasions and it is of course not as easy to get people excited by observing Comet Tempel 1 as with observing Halley, or Showmaker–Levy 9, but a concerted effort analogous to the International Halley watch would pay dividends. Such a programme may convince the TACs to dedicate the amounts of telescope time to space probe support observations that are required to get the data that is needed to answer these questions. The upcoming series of space missions may provide us with the impulse and the motivation to succeed.

LIST OF PARTICIPANTS IAU COLLOQUIUM 186: COMETARY SCIENCE AFTER HALE–BOPP, JANUARY 21–25, 2002

No.	Name	Institute	City	Country
1	A'Hearn, Michael F.	University of Maryland	College Park	USA
2	Artigue, Fernanda	University of Montevideo	Montevideo	Uruguay
3	Belton, Michael	BSEI	Tucson	USA
4	Bertaux, Jean-Loup	CNRS	Verrières Le Buisson	France
5	Biele, Jens	DLR	Koeln	Germany
6	Biver, Nicolas	ESTEC/Esa	Noordwijk	Netherlands
7	Bockelee-Morvan, Dominique	Observatoire De Paris-Meudon	Meudon	France
8	Boehnhardt, Hermann	European Southern Observatory	Santiago	Chile
9	Boice, Daniel	Southwest Research Institute	San Antonio	USA
10	Bonev, Tanyu	Bulgarian Academy of Sciences	Sofia	Bulgaria
11	Borysenko, Sergiy	Main Astronomical Observatory of NASU	Kiyv	Ukraine
12	Brandt, John C.	University of New Mexico	Albuquerque	USA
13	Campins, Humberto	Univ. Arizona & Research Corporation	Tucson	USA
14	Capria, Maria Teresa	Istituto Di Astrofisica Spaziale – CNR	Rome	Italy
15	Ceolin, Thierry	CS Systemes D'Iinformation	Toulouse	France
16	Charnley, Steven	NASA – Ames Research Center	Moffett Field	USA
17	Churyumov, Klim	Astronomical Observatyory	Kyiv	Ukraine
18	Clarr, Benton	Lockheed Martin Astronautics	Littleton	USA
19	Cochran, Anita	University of Texas	Austin	USA
20	Combi, Michael R.	University of Michigan	Ann Arbor	USA
21	Crovisier, Jacques	Observatoire De Paris-Meudon	Meudon	France
22	Davidsson, Björn	University of Uppsala	Uppsala	Sweden
23	Duncan, Martin J.	Queen's University	Kingston	Canada
24	Dybczynski, Piotr A.	Astronomical Observatory, Mickiewicz Univ.	Poznan	Poland
25	Dziak, Beata	Institute of Geophysics, Wwrsaw Univ.	Warszawa	Poland
26	Ehrenfreund, Pascale	Leiden Observatory	Leiden	Netherlands
27	Farnham, Tony	University of Texas	Austin	USA
28	Fernández, Julio Angel	Universidad De Montevideo	Montevideo	Uruguay
29	Fernandez, Yan R.	University of Hawaii	Honolulu	USA
30	Festou, Michel	Observatoire Midi-Pyrénées	Toulouse	France
31	Foing, Bernard H.	ESA	Noordwijk	Netherlands
32	Goldstein, Ray,pmd	Southwest Research Institute	San Antonio	USA
33	Gutiérrez, Pedro José	Instituto De Astrofisica De Andalucia, CSIC	Granada	Spain
34	Hadamcik, Edith	Service D'Aéronomie (CNRS)	Verrières Le Buisson	France
35	Harker, David E.	NASA – Ames Research Center	Moffett Field	USA
36	Harris, Walter	University of Wisconsin-Madison	Madison	USA
37	Helbert, Joern	DLR	Berlin	Germany
38	Henry, Florence	Observatoire De Paris-Meudon	Meudon	France
39	Ho, Tra-Mi	Max-Planck-Institut Für Aeronomie	Katlenburg-Lindau	Germany
40	Huebner, Walter F.	Southwest Research Institute	San Antonio	USA
41	Jackson, William M.	University of California	Davis	USA
42	Jehin, Emmanuel	European Southern Observatory	Santiago	Chile
43	Jorda, Laurent	Laboratoire D'Astrophysique De Marseille	Marseille	France
44	Joshi, Umesh C.	Physical Research Laboratory	Ahmedabad	India
45	Kawakita, Hideyo	Gunma Astronomical Observatory	Nakayama	Japan
46	Keller, Horst Uwe	Max-Planck-Institut Für Aeronomie	Katlenburg-Lindau	Germany
47	Kidger, Mark R.	Instituto De Astrofisica De Canarias	La Laguna	Spain

Earth, Moon and Planets **90:** 523–524, 2002.

48	Kissel, Jochen	MPI Für Extraterrestrische Physik	Garching	Germany
49	Koemle, Norbert	Austrian Academy of Sciences	Graz	Austria
50	Kolokolova, Ludmilla	University of Florida	Gainesville	USA
51	Kuehrt, Ekkehard	DLR	Berlin	Germany
52	Lazzaro, Daniela	Observatorio Nacional	Rio De Janeiro	Brazil
53	Lederer, Susan	NASA Johnson Space Center	Houston	USA
54	Leliwa-Kopystynski, Jacek	Institute of Geophysics, Warsaw Univ.	Warszawa	Poland
55	Licandro, Javier	Centro Galileo Galilei	La Palma	Spain
56	Lisse, Carey M.	University of Maryland	College Park	USA
57	Llodrá, Jaume	Observatorio Astronomico De Mallorca	Baleares	Spain
58	Lowry, Stephen	Jet Propulsion Laboratory	Pasadena	USA
59	Mäkinen, Teemu	Finnish Meteorological Institute	Vuorikatu	Finland
60	Morgenthaler, Jeffrey	University of Wisconsin-Madison	Madison	USA
61	Motschmann, Uwe	University of Braunschweig	Braunschweig	Germany
62	Motta, Verónica	Instituto De Astrofísica De Canarias	La Laguna	Spain
63	Mueller, Michael	The Open University	Milton Keynes	UK
64	Ortíz, José L.	Instituto De Astrofisica De Andalucia	Granada	Spain
65	Peschke, Sibylle B.	ISO Data Center – ESA-Vilspa	Madrid	Spain
66	Prialnik, Dina	Tel Aviv University	Tel Aviv	Israel
67	Rauer, Heike	DLR	Berlin	Germany
68	Rickman, Hans	Uppsala Astronomical Observatory	Uppsala	Sweden
69	Rodriguez, Juan	Observatorio Astronomico De Mallorca	Baleares	Spain
70	Rosenbush, Vera	Main Astronomical Observatory	Kyiv	Ukraine
71	Samarasinha, Nalin	National Optical Astronomy Observatory	Tucson	USA
72	Sánchez, Sslvador	Observatorio Astronomico De Mallorca	Baleares	Spain
73	Schleicher, David	Lowell Observatory	Flagstaff	USA
74	Schulz, Rita	ESA/RSSD	Noordwijk	Netherlands
75	Shulman, Leonid	NASU Main Astronomical Obs.	Kyiv	Ukraine
76	Szego, Karoly	KFKI Research Institute	Budapest	Hungary
77	Szutowicz, Slawomira	Space Research Centre-Polish Academy Sci	Warsawa	Poland
78	Tancredi, Gonzalo	Universidad De Montevideo	Montevideo	Uruguay
79	Ticha, Jana	Klet Observatory Ceske	Budejovice	Czech Republi
80	Tichy, Milos	Klet Observatory Ceske	Budejovice	Czech Republi
81	Toth, Imre	Konkoly Observatory	Budapest	Hungary
82	Tozzi, Gian Paolo	Osservatorio Astrofisico Di Arcetri	Firenze	Italy
83	Voelzke, Marcos Rincon	Universidade Cruzeiro Do Sul	Sao Paulo	Brazil
84	Watanabe, Jun-Ichi	National Astronomical Observatory	Tokyo	Japan
85	Weissman, Paul	Jet Propulsion Laboratory	Pasadena	USA
86	Wooden, Diane	NASA – Ames Research Center	Moffett Field	USA
87	Woodney, Laura	Lowell Observatory	Flagstaff	USA